Hyperbolic Functional Differential Inequalities and Applications

Mathematics and Its Applications

Volume 486

Hyperbolic Functional Differential Inequalities and Applications

by

Zdzislaw Kamont
Institute of Mathematics,
University of Gdańsk,
Gdańsk, Poland

SPRINGER-SCIENCE+BUSINESS MEDIA, B.V.

A C.I.P. Catalogue record for this book is available from the Library of Congress.

ISBN 978-94-010-5957-2 ISBN 978-94-011-4635-7 (eBook)
DOI 10.1007/978-94-011-4635-7

Printed on acid-free paper

Contents

Preface

This book is intended as a self-contained exposition of hyperbolic functional differential inequalities and their applications. Its aim is to give a systematic and unified presentation of recent developments of the following problems:

(i) functional differential inequalities generated by initial and mixed problems,

(ii) existence theory of local and global solutions,

(iii) functional integral equations generated by hyperbolic equations,

(iv) numerical method of lines for hyperbolic problems,

(v) difference methods for initial and initial-boundary value problems.

Beside classical solutions, the following classes of weak solutions are treated: Carathéodory solutions for quasilinear equations, entropy solutions and viscosity solutions for nonlinear problems and solutions in the Friedrichs sense for almost linear equations. The theory of difference and differential difference equations generated by original problems is discussed and its applications to the constructions of numerical methods for functional differential problems are presented.

The monograph is intended for different groups of scientists. Pure mathematicians and graduate students will find an advanced theory of functional differential problems. Applied mathematicians and research engineers will find numerical algorithms for many hyperbolic problems.

The classical theory of partial differential inequalities has been described extensively in the monographs [138, 140, 195, 225]. As is well known, they found applications in differential problems. The basic examples of such questions are: estimates of solutions of partial equations, estimates of the domain of the existence of solutions, criteria of uniqueness and estimates of the error of approximate solutions. Moreover, discrete versions of differential inequalities are frequently used to prove the convergence of approximation methods. The numerical method of lines and difference methods represents classical examples.

The present book offers a systematic treatment of functional differential problems.

Chapters I-III deal with initial problems for functional differential equations of the Hamilton-Jacobi type on the Haar pyramid. We begin by discussing functional differential inequalities. This theory is applied for estimations of the difference between solutions of two problems and to the formulation of the criteria of uniqueness of solutions of initial problems and results on continuous dependence of solutions on given functions. A monotone iterative method based on a linearization of the right-hand side of an equation is also considered. Next we prove existence theorems for classical solutions of nonlinear problems and Carathéodory solutions of quasilinear systems. Results are based on the method of bicharacteristics and on theorems on integral functional inequalities.

In recent years a number of papers concerned with difference and functional difference inequalities were published. Results on inequalities are based on the monotonicity of corresponding operators. These inequalities were usually applied to establish error evaluations or convergence results for suitable difference methods for corresponding functional differential problems. The considerations as a rule involved a lot of calculations to reach the result so that the main property of the corresponding operators was not easy to be seen. In Chapter III we show that the results on functional difference inequalities are consequences of a simple result concerned with recurrent inequalities for monotone operators.

The second part of Chapter III deals with difference methods for initial problems. Approximate methods for nonlinear first-order partial differential or functional differential equations were considered by many authors and under various assumptions. The main problem in these investigations is to find a suitable difference or functional difference equation which satisfies the consistency condition with respect to the original problem and which is stable. The method of difference inequalities or simple theorems on recurrence inequalities for functions of one variable is used in the investigation of stability. It has often been assumed that given functions have partial derivatives with respect to all arguments with the exception of (x, y). Our assumptions will be less restrictive. We will assume nonlinear estimates for given functions with respect to the functional variable.

In Chapter III we propose a general method for the investigation of the stability of difference schemes generated by initial value problems for hyperbolic functional differential equations. We prove a theorem on the error estimates of approximate solutions for functional difference equations of the Volterra type with unknown function of several variables. The error of an approximate solution is estimated by a solution of an initial problem for functional difference equation. We will apply this general and simple idea to the investigation of the stability of difference equations generated by various problems.

It is essential to our considerations that the right-hand sides of equations satisfy the nonlinear estimates of the Perron type with respect to the unknown functions. We will consider also nonlinear functional difference equations as comparison problems for original difference methods. Note that our assumptions in theorems on the stability of functional difference equations are identical with assumptions which guarantee uniqueness of solutions of initial problems.

Chapter IV deals with existence results for nonlinear equations

$$D_x z(x,y) = f(\,x, y, z_{(x,y)}, D_y z(x,y)\,) \tag{0.1}$$

in unbounded domains, where $D_y z = (\,D_{y_1} z, \ldots, D_{y_n} z\,)$. The symbol $z_{(x,y)}$ represents the functional variable and it denotes the restriction of the function z to the set $[x - r_0, x] \times [y - r, y + r]$ where $r_0 \in R_+ = [0, +\infty)$ and $r = (r_1, \ldots, r_n) \in R^n_+$. In Chapter IV we consider the Cauchy problem for (0.1) which is global with respect to y. We give uniqueness results for functional hyperbolic equations. First, we consider initial problems and generalized solutions in the Cinquini Cibrario sense. To be precise, we prove the uniqueness of the solution for Cauchy problem in the class of bounded and uniformly continuous functions, under the assumption of a Perron type comparison function for the right-hand side of the equation.

In the uniqueness theory for first-order partial differential equations, ordinary differential equations are considered as comparison problems of Perron or Kamke type. We wish to emphasize that functional differential comparison problems are the main tool in our research. An adequate example is given in Chapter 1 for nonlinear functional differential equations. It can be adopted for initial problems in unbounded domains and for mixed problems.

Functional differential inequalities generated by mixed problems are examined in Chapter V. Uniqueness of solutions and continuous dependence on given functions are consequences of comparison theorems. We consider also monotone iterative methods which are based on a linearization of the original equation with respect to the functional variable. Discrete versions of functional differential problems are considered as well.

In Chapter VI we present a theory of the numerical method of lines for functional differential problems. Furthermore, we consider the Cauchy problem in an unbounded domain, and mixed problems. The main theorems concerning the numerical method of lines will be based on comparison theorems where a function satisfying some differential difference inequalities in an unbounded domain is estimated by a solution of an adequate ordinary functional differential problem.

A classical solution to the hyperbolic functional differential problem denotes a function which is continuous on its domain (open connected subset of the Euclidean space) and which has its partial derivatives at every point of the domain and the functional differential equation is satisfied everywhere. Nonlinear equations with first-order partial derivatives have the following property: any classical solution to initial value problem exists locally with respect to x. This leads in a natural way to weak solutions. The following classes of weak solutions of hyperbolic functional differential problems are considered in the monograph: Carathéodory solutions, solutions in the Cinquini Cibrario sense, viscosity solutions and solutions in the Friedrichs sense. Functional differential inequalities and existence results are given in Chapter VII.

In Chapter VIII we give theorems concerning the existence and uniqueness of continuous solutions of functional integral equations in several variables

$$z(x) = F(\,x, \int_{H(x)} f(x, s, z(\alpha(s)\,)ds, z(\,\beta(x)\,)\,) \tag{0.2}$$

where $x = (x_1, \ldots, x_n)$. Equation (0.2) is considered in a Banach space. The Chapter is divided into three parts. In the first part we investigate equation (0.2) by means of the comparison method. If we assume that F and f satisfy the Lipschitz condition with respect to all variables except x or (x, s) respectively, then we prove under certain additional assumptions concerning the functions β, α and the Lipschitz coefficients, that there exists exactly one (in a certain class of functions) continuous solution of (0.2). This solution is the limit of a sequence of successive approximations. This result is obtained by the comparison method.

We shall consider a method of successive approximations, which has been used in various versions and under various assumptions in many papers, with respect to classical and generalized problems concerning partial or ordinary differential equations, integral differential equations, functional differential and functional integral equations and very general functional equations in some abstract spaces (see for instance: [80, 113, 114, 131, 176]). A general formulation of this method was given by T. Ważewski [232]. The second part of Chapter 8 concerns equation (0.2) considered in a finite dimensional Banach space. A theorem on the existence of at least one solution of (0.2) is proved. The last part of the Chapter deals with approximate methods for functional integral equations generated by the Darboux problem in two independent variables. We prove that there are natural difference schemes (implicit or explicit) corresponding to original problems, which are convergent.

Hyperbolic functional differential equations have applications in different branches of knowledge. We give a few examples. The simplest one is the following, which appears in the theory of the distribution of wealth. Let $z(t,x)$ denote the number of households at time t whose estates are $\leq x$, where $x > 0$. Suppose β is the growth rate at which assets of a household increase and γ is the mortality rate. The estate of each vanishing household will be equally distributed among n inheritors forming new households. Then z satisfies the differential equation with a deviated argument ([83])

$$D_t z(t,x) + \beta\, x\, D_x z(t,x) + \gamma\, z(t,x) = \gamma\, n\, z(t,nx).$$

The mathematical models of structured population dynamics give rise to differential functional equations. In these models one identifies one or more characteristics of individuals which are important in the development of the population as a whole. Such properties as age, size, nutritional state may be used as structure variables. Consider a population of individuals structured by two internal variables (x,y). Let $z(t,x,y)$ be the density function (which may be vector valued). Then

$$\int_0^{+\infty}\int_0^{+\infty} z(t,x,y)\,dx\,dy$$

gives the total population at time t. For each $t \geq 0$ the density $z(t,\cdot,\cdot)$ lies in a space X of R^n-valued functions defined on $(0,+\infty)\times(0,+\infty)$. The function z satisfies the so-called balance law of the population, which accounts for such processes as mortality, migration, or movement into population subclasses. If the density is sufficiently regular, then the balance law may be written in the following form ([190, 229])

$$D_t z(t,x,y) + D_x z(t,x,y) + D_y z(t,x,y) = F(\, z(t,\cdot,\cdot)\,)(x,y),$$

where F is a nonlinear operator and $F : X \to X$.

According to the nonlinear theory of age-structured populations, the evolution of a harvested population is described by the differential integral equation ([92])

$$D_t z(t,x) + D_x z(t,x) + f(\,x, \int_0^{+\infty} z(t,s)\,ds\,)z(t,x) + F(t,x)z(t,x) = 0,$$

where $z(t,x)$ is the number of individuals of age x at the time t, per unit age, $F(t,x)$ is the effort with which individuals of age x are harvested at time t and the term $f(\,t, \int_0^{+\infty} z(t,s)\,ds\,)$ is the rate at which individuals of age x die when the total population is $\int_0^{+\infty} z(t,s)\,ds$.

Quasilinear differential integral systems have been proposed in [19] as simple mathematical models for the nonlinear phenomenon of harmonic generation of laser radiation through piezo-electric crystals for non-dispersive materials and of Maxwell-Hopkinson type. Various nonlinear systems perturbed by a dissipative integral term of the Volterra type arise from laser problems in nonlinear optics [16 – 18], see also [164]. Nonlinear equations may be used to describe the growth of a population of cells which constantly differentiate (change their properties) in time [78]. A system of nonlinear differential integral equations which model an age-dependent epidemic of a disease with vertical transmission is investigated in [84]. Vertical transmission means the direct transfer of infection from a parent organism to its offspring. The paper [50] discusses with a model of proliferating cell population. Almost linear differential integral equation is considered in the paper. The model features the correlation of mother-daughter mitotic times. In [90] a differential integral evolution equation was considered. The study was motivated by a rheological model describing the dynamic of muscle contraction introduced by physiologists. In [79] the author considers hyperbolic conservation laws with fading memory i.e. quasilinear systems with integral terms of the Volterra type. A quasilinear wave equation with memory is considered in [209]. The differential integral equation described in the paper has it roots in a model for small amplitude vibrations of a string in which the dependence of the tension on the deformation cannot be ignored. Hyperbolic differential integral equations are considered in the nonlinear theory describing the motion of viscoelastic media [156]. First-order differential integral telegraphic equations were examined in [167]. The paper [194] deals with differential integral equation which describes the process of screening of granular bodies. Smoluchovsky's model with condensation in three dimensional space is considered in [48]. The problem leads to a quasilinear differential integral equation of the Volterra type. Our results are also motivated by applications of hyperbolic functional differential equations considered in [41, 82, 89, 203].

It should be noted that all problems considered in the book have the following property: the unknown function is the functional argument in equations. The partial derivatives appear in a classical sense. At the moment there are very few important results for functional differential equations with a functional argument at derivatives. The classical methods, such as the theory of bicharacteristics, differential inequalities, successive approximations, fail to work in this case. Existence results for equations with delays at derivatives can be found in [2 – 4] and [151, 226, 233]. They are based on the Nagumo lemma [169] or on power series expansions. These methods were used for proving the class of Cauchy-Kovalevskaya type theorems.

Many results included in this book have been discussed at seminars in the Institute of Mathematics of the Gdańsk University.

Chapter 1

Initial Problems on the Haar Pyramid

1.1 Introduction

The classical theory of differential inequalities generated by Hamilton-Jacobi equations or systems was developed in monographs [138, 140, 195] widely. In particular, uniqueness results for initial problems on the Haar pyramid with nonlinear a priori estimates, were obtained as consequences of suitable comparison theorems for differential inequalities. The authors deal with solutions which admit first order partial derivatives and are totally differentiable on a subset of the boundary of the Haar pyramid. For the Cauchy problem on an unbounded domain the assumption of total differentiability can be omitted provided the comparison function is linear ([21]). In this case the right-hand side of the differential equation satisfies the Lipschitz condition with respect to the unknown function and the uniqueness result is based on the fact that a first-order linear differential inequality can be solved. An interesting result for the global uniqueness of the Cauchy problem, when the right-hand side of the equation satisfies the Hölder condition can be found in [23]. Mixed inequalities between solutions of first order partial differential equations were considered in [173 – 175].

As it is well-known, solutions in Carathéodory sense of first order partial differential equations are not unique ([96, 136, 184]). Two natural subclasses are considered: the generalized entropy solutions and the solutions in the Cinquini Cibrario sense. The relative uniqueness results can be found in [55] and [136]. Differential inequalities for semilinear hyperbolic systems with two independent variables and applications to uniqueness theorems in the class of Friedrichs generalized solutions are presented in [142].

There is a wide literature on first-order functional differential problems, we wish to mention here some references on uniqueness criteria. Classical solutions on Haar pyramid can be found in [103], see also [27, 39, 40, 99]. Kamke functions in uniqueness theorems for functional differential equations are considered in

[1]. Functional differential inequalities and uniqueness results for initial-boundary value problems are presented in [28]. Infinite systems and comparison theorems for functional differential problems are investigated in [202]. Unbounded solutions of initial problems are discussed in [119, 147]. Uniqueness criteria for generalized entropy solutions of nonlinear functional differential problems are considered in [116]. Systems of functional differential inequalities in the Friedrichs sense can be found in [42].

Note that two models of the functional dependence in partial equations are used in literature. The first group of results is connected with initial problems for equations

$$D_x z(x,y) = G(x,y,z,D_y z(x,y)) \tag{1.1}$$

where the variable z represents the functional argument. This model is suitable for initial problems considered in the Haar pyramid. A new model of a functional dependence in partial equations is proposed in [98, 109, 110]. Partial equations have the form (0.1) where $z_{(x,y)}$ is a functional variable. This simple model is well known for ordinary functional differential equations ([81, 93, 94, 132]). It is also very general since equations (or systems) with retarded argument ([25, 26] and [124 – 127]), integral differential equations ([19, 97, 102, 168, 218]) and differential functional systems with operators of Volterra type ([27, 100, 123, 185] and [210 – 217]) can be obtained from (0.1) by specializing given functions. More detailed comparisons between different models of functional dependence are presented in [110]. The above discussion can be also extended on parabolic functional differential problems. Differential inequalities generated by parabolic nonlinear functional differential problems were first treated in [182] and [196 – 199]. Existence results can be found in [34, 36, 38, 171, 172, 183, 208, 230]. The monograph [230] contains a large bibliography.

1.2 Functional differential inequalities

Two different types of results on functional differential inequalities are taken into considerations in literature. The first type allows to estimate a function of several variables by means of an other function of several variables, while the second one, the so called comparison theorems, give estimates for functions of several variables by means of functions of one variable. In this Section we deal with the first kind results, while a comparison theorem will be proved in the next Sections. Classical results can be found in [138, 140, 195, 225].

For $y = (y_1, \ldots, y_n) \in R^n$ and $\eta = (\eta_1, \ldots, \eta_k) \in R^k$ we put

$$\|y\| = \sum_{i=1}^{n} |y_i| \quad \text{and} \quad \|\eta\| = \max\{\, |\eta_i| : \ 1 \le i \le k \,\}.$$

Unless otherwise noted, we use in the book the above norms and they are denoted by the same symbol $\|\cdot\|$. We will use vectorial inequalities with the understanding that the same inequalities hold between their corresponding components. For any metric spaces X and Y we denote by $C(X,Y)$ the class of all continuous functions from X into Y.

Let $a > 0$, $r_0 \in R_+$, $R_+ = [0, +\infty)$, be fixed and suppose that functions $\alpha, \beta : [0, a) \to R^n$, $\alpha = (\alpha_1, \ldots, \alpha_n)$, $\beta = (\beta_1, \ldots, \beta_n)$, and $\tilde{\alpha}, \tilde{\beta} : [-r_0, 0] \to R^n$, $\tilde{\alpha} = (\tilde{\alpha}_1, \ldots, \tilde{\alpha}_n)$, $\tilde{\beta} = (\tilde{\beta}_1, \ldots, \tilde{\beta}_n)$ satisfy the conditions:

(i) α and β are of class C^1 on $[0, a)$ and $\alpha(x) < \beta(x)$ for $x \in [0, a)$,

(ii) $\tilde{\alpha}, \tilde{\beta} \in C([-r_0, 0], R^n)$ and $\tilde{\alpha}(x) \leq \tilde{\beta}(x)$ for $x \in [-r_0, 0]$,

(iii) $\tilde{\beta}(0) = \beta(0) = b$ where $b = (b_1, \ldots, b_n)$, $b_i > 0$ for $1 \leq i \leq n$, and $\tilde{\alpha}(0) = \alpha(0) = -b$.

Let E be a generalized Haar pyramid

$$E = \{ (x, y) \in R^{1+n} : x \in (0, a),\ y = (y_1, \ldots, y_n),\ y \in [\alpha(x), \beta(x)] \} \tag{1.2}$$

and

$$E_0 = \{ (x, y) \in R^{1+n} : x \in [-r_0, 0],\ y \in [\tilde{\alpha}(x), \tilde{\beta}(x)] \}, \tag{1.3}$$

$$\partial_0 E = \partial E \cap ((0, a) \times R^n), \tag{1.4}$$

$$S_x = [\tilde{\alpha}(x), \tilde{\beta}(x)] \text{ for } x \in [-r_0, 0], \quad S_x = [\alpha(x), \beta(x)] \text{ for } x \in [0, a), \tag{1.5}$$

where the symbol ∂E denotes the boundary of E. Let $\Omega = E \times C(E_0 \cup E, R) \times R^n$ and suppose that $f : \Omega \to R$ and $\varphi : E_0 \to R$ are given functions. In this Section we discuss a number of questions referring to the Cauchy problem

$$D_x z(x, y) = f(x, y, z, D_y z(x, y)), \tag{1.6}$$

$$z(x, y) = \varphi(x, y) \text{ on } E_0, \tag{1.7}$$

where $D_y z = (D_{y_1} z, \ldots, D_{y_n} z)$. A function $z : E_0 \cup E \to R$ will be called the function of class D if z is continuous on $E_0 \cup E$, possesses the total differential on $\partial_0 E$ and has first partial derivatives in an interior of E. We will consider solutions of (1.6), (1.7) or solutions of functional differential inequalities generated by (1.6), (1.7) which are of class D on $E_0 \cup E$. Let

$$E_x = (E_0 \cup E) \cap ([-r_0, x] \times R^n), \quad -r_0 \leq x < a.$$

For every $x \in [-r_0, a)$ and $z \in C(E_0 \cup E, R)$ we write $\|z\|_x = \max\{ |z(t, s)| : (t, s) \in E_x \}$. If $w \in C([-r_0, a), R)$ then we write also $\|w\|_x = \max\{ |w(t)| : -r_0 \leq t \leq x \}$.

We will say that the function f satisfies the Volterra condition if for each $(x, y) \in E$ there is a set $E[x, y]$ such that

(i) $E[x, y] \subset E_x$,

(ii) if $z, \bar{z} \in C(E_0 \cup E, R)$ and $z(t, s) = \bar{z}(t, s)$ for $(t, s) \in E[x, y]$ then $f(x, y, z, q) = f(x, y, \bar{z}, q)$.

Note that the Volterra condition means that the value of f at the point (x, y, z, q) depends on (x, y, q) and on the restriction of z to the set $E[x, y]$ only.

Examples. Consider the equation with a deviated argument

$$D_x z(x, y) = F(x, y, z(x, y), z(\psi_0(x, y), \psi(x, y)), D_y z(x, y))$$

where $F : E \times R^2 \times R^n \to R$, $\psi_0 : E \to R$, $\psi : E \to R^n$ and $(\psi_0(x, y), \psi(x, y)) \in E_0 \cup E$ for $(x, y) \in E$. We assume that $-r_0 \leq \psi_0(x, y) \leq x$ on E. Then

$$f(x, y, z, q) = F(x, y, z(x, y), z(\psi_0(x, y), \psi(x, y)), q)$$

and $E[x,y] = \{ (x,y),\ (\psi_0(x,y), \psi(x,y)) \}$. For the above F consider the differential integral equation

$$D_x z(x,y) = F(\, x, y, z(x,y), \int_{E_x} z(t,s)\, dt\, ds, D_y z(x,y)\,).$$

Then $E[x,y] = E_x$.

For each $(x,y) \in E$ there exist sets (possibly empty) of integers $I_0[x,y]$, $I_+[x,y]$, $I_-[x,y]$ such that $I_+[x,y] \cap I_-[x,y] = \emptyset$, $I_0[x,y] \cup I_+[x,y] \cup I_-[x,y] = \{1, \dots, n\}$, and

$$y_i = \alpha_i(x) \ \text{ for } i \in I_-[x,y], \quad y_i = \beta_i(x) \ \text{ for } i \in I_+[x,y],$$

$$\alpha_i(x) < y_i < \beta_i(x) \ \text{ for } \ i \in I_0[x,y].$$

Let us denote by F the Niemycki operator corresponding to (1.6) i. e.

$$F[\,z\,](x,y) = f(\,x, y, z, D_y z(x,y)\,).$$

We start with a theorem on strong inequalities.

Theorem 1.1 *Suppose that*

1) the function $f : \Omega \to R$ *of the variables* (x,y,z,q), $q = (q_1, \dots, q_n)$, *satisfies the Volterra condition,*

2) the following monotonicity condition holds: if $(x,y) \in E$, $z, \bar z \in C(E_0 \cup E, R)$, $z(t,s) \le \bar z(t,s)$ *on* E_x *and* $z(x,y) = \bar z(x,y)$ *then* $f(x,y,z,q) \le f(x,y,\bar z,q)$ *for* $q \in R^n$,

3) for $(x,y,z,q) \in \Omega$, $\bar q \in R^n$ *we have*

$$f(x,y,z,q) - f(x,y,z,\bar q) + \sum_{i \in I_-[x,y]} \alpha_i'(x)(q_i - \bar q_i) + \sum_{i \in I_+[x,y]} \beta_i'(x)(q_i - \bar q_i) \le 0 \quad (1.8)$$

where $\bar q = (\bar q_1, \dots, \bar q_n)$ *and* $q_i \le \bar q_i$ *for* $i \in I_-[x,y]$, $q_i \ge \bar q_i$ *for* $i \in I_+[x,y]$, $q_i = \bar q_i$ *for* $i \in I_0[x,y]$,

4) the functions u, $v : E_0 \cup E \to R$ *are of class* D *and*

$$u(x,y) \le v(x,y) \ \textit{ for } \ (x,y) \in E_0, \quad u(0,y) < v(0,y) \ \textit{ for } \ y \in [\,-b, b\,], \qquad (1.9)$$

5) the functional differential inequality

$$D_x u(x,y) - F[\,u\,](x,y) < D_x v(x,y) - F[\,v\,](x,y) \qquad (1.10)$$

is satisfied on E.

Under these assumptions

$$u(x,y) < v(x,y) \ \text{ for } \ (x,y) \in E. \qquad (1.11)$$

PROOF. Suppose by contradiction, that assertion (1.11) fails to be true. Then the set

$$J_+ = \{ x \in [0,a) : \ u(x,y) \geq v(x,y) \text{ for some } y \in S_x \}$$

is not empty. If we put $\bar{x} = \min J_+$, from (1.9) it is clear that $\bar{x} > 0$ and there exists $\bar{y} = (\bar{y}_1, \ldots, \bar{y}_n)$ such that $(\bar{x}, \bar{y}) \in E$ and

$$u(x,y) < v(x,y) \text{ for } (x,y) \in E \cap ((0,\bar{x}) \times R^n) \text{ and } u(\bar{x},\bar{y}) = v(\bar{x},\bar{y}).$$

Then $D_{y_i}(u-v)(\bar{x},\bar{y}) \geq 0$ for $i \in I_+[\bar{x},\bar{y}]$, $D_{y_i}(u-v)(\bar{x},\bar{y}) \leq 0$ for $i \in I_-[\bar{x},\bar{y}]$ and $D_{y_i}(u-v)(\bar{x},\bar{y}) = 0$ for $i \in I_0[\bar{x},\bar{y}]$. Now, for $0 \leq x \leq \bar{x}$ put $\eta(x) = (\eta_1(x), \ldots, \eta_n(x))$ where

$$\eta_i(x) = \alpha_i(x) \text{ for } i \in I_-[\bar{x},\bar{y}], \quad \eta_i(x) = \beta_i(x) \text{ for } i \in I_+[\bar{x},\bar{y}], \tag{1.12}$$

$$\eta_i(x) = \bar{y}_i \text{ for } i \in I_0[\bar{x},\bar{y}], \tag{1.13}$$

and consider the composite function $\gamma(x) = (u-v)(x,\eta(x))$, $x \in [0,\bar{x}]$. It attains maximum at $\bar{x}$. Since u and v are of class D on $E_0 \cup E$ and $\gamma'(\bar{x}) \geq 0$ then we have

$$D_x(u-v)(\bar{x},\bar{y}) + \sum_{i \in J_-[\bar{x},\bar{y}]} \alpha_i'(\bar{x}) D_{y_i}(u-v)(\bar{x},\bar{y}) \tag{1.14}$$

$$+ \sum_{i \in J_+[\bar{x},\bar{y}]} \beta_i'(\bar{x}) D_{y_i}(u-v)(\bar{x},\bar{y}) \geq 0.$$

From assumptions 1), 2) and (1.8) - (1.10) we deduce that

$$D_x(u-v)(\bar{x},\bar{y}) < f(\bar{x},\bar{y},v,D_y u(\bar{x},\bar{y})) - f(\bar{x},\bar{y},v,D_y v(\bar{x},\bar{y}))$$

$$\leq - \sum_{i \in I_-[\bar{x},\bar{y}]} \alpha_i'(\bar{x})\, D_{y_i}(u-v)(\bar{x},\bar{y}) - \sum_{i \in I_+[\bar{x},\bar{y}]} \beta_i'(\bar{x})\, D_{y_i}(u-v)(\bar{x},\bar{y}),$$

which contradicts (1.14). Hence J_+ is empty and the statement (1.11) follows.

Remark 1.2 *The assumption 5) of Theorem 1.1 can be replaced by the following condition:*

5') denoted

$$\Sigma = \{ (x,y) \in E : \ u(t,s) < v(t,s) \text{ on } E \cap (0,x) \times R^n \text{ and } u(x,y) = v(x,y) \},$$

we have

$$D_x u(x,y) - F[u](x,y) < D_x v(x,y) - F[v](x,y), \ \ (x,y) \in \Sigma.$$

In the proof of Theorem 1.1 we apply functional differential inequality (1.10) at the point $(\bar{x},\bar{y})$ such that $(\bar{x},\bar{y}) \in \Sigma$.

Remark 1.3 *In Theorem 1.1 we can assume instead of condition* 5) *that*

$$D_x u(x,y) \leq F[u](x,y) \text{ and } D_x v(x,y) \geq F[v](x,y),$$

where $(x,y) \in E$ and for each (x,y) equality holds in at most one place.

Suppose now that the sets E_0 and E are given by

$$E_0 = [-r_0, 0] \times [-b, b] \tag{1.15}$$

and

$$E = \{\, (x, y) : \; x \in (0, a), \quad -b + Mx \le y \le b - Mx \,\} \tag{1.16}$$

where $b = (b_1, \ldots, b_n) \in R^n$, $b_i > 0$ for $1 \le i \le k$, and $M = (M_1, \ldots, M_n) \in R^n_+$. We assume that $b - Ma > 0$. Suppose that $\tilde{f} : E \times R \times C(E_0 \cup E, R) \times R^n) \to R$ and $\varphi : E_0 \to R$ are given functions. Now we give a theorem on functional differential inequalities generated be the Cauchy problem

$$D_x z(x, y) = \tilde{f}(\, x, y, z(x, y), z, \; D_y z(x, y)\,) \tag{1.17}$$

$$z(x, y) = \varphi(x, y) \;\text{ on }\; E_0. \tag{1.18}$$

Write $\tilde{F}[\, z \,](x, y) = \tilde{f}(\, x, y, z(x, y), z, \; D_y z(x, y)\,)$.

Theorem 1.4 *Suppose that*

1) the function $\tilde{f}$ satisfies the Volterra condition and it is nondecreasing with respect to the functional variable,

2) the Lipschitz condition

$$\Big| \tilde{f}(x, y, p, z, q) - \tilde{f}(x, y, p, z, \bar{q}) \Big| \le \sum_{i=1}^{n} M_i \, |q_i - \bar{q}_i| \tag{1.19}$$

is satisfied on $E \times R \times C(E_0 \cup E, R) \times R^n$,

3) the functions $u, v : E_0 \cup E \to R$ are of class D and $u(x, y) \le v(x, y)$ on E_0 and $u(0, y) < v(0, y)$ for $y \in [-b, b]$,

4) the functional differential inequality

$$D_x u(x, y) - \tilde{F}[\, u \,](x, y) < D_x v(x, y) - \tilde{F}[\, v \,](x, y)$$

is satisfied on E.

Then $u(x, y) < v(x, y)$ on E.

PROOF. The function $f(x, y, z, q) = \tilde{f}(x, y, z(x, y), z, q)$ and u, v satisfy all the assumptions of Theorem 1.1 and the assertion follows.

1.3 Weak functional differential inequalities

We denote by $\mathbf{F}(X, Y)$ the class of all functions defined on X and taking values in Y, where X and Y are arbitrary sets. We define a comparison problem corresponding to (1.6), (1.7). We will need the following operator $V : C(E_0 \cup E, R) \to \mathbf{F}(\,[-r_0, \, a), R_+)$ defined by: if $z \in C(E_0 \cup E, R)$ then

$$(Vz)(x) = \max\{|z(x, y)| : \; y \in S_x\}, \quad -r_0 \le x < a. \tag{1.20}$$

The following property of V is important in our considerations ([195], Theorem 33.1).

Lemma 1.5 *If $z \in C(E_0 \cup E, R)$ then the function $Vz : [-r_0, a) \to R_+$ given by (1.20) is continuous.*

Let $\sigma : [0,a) \times C([-r_0,a), R_+) \to R_+$ be a given function.

Assumption H $[\sigma]$. Suppose that

1) σ satisfies the Volterra condition and the following monotonicity condition is satisfied: if $x \in [0,a)$, $w, \bar{w} \in C([-r_0,a), R_+)$, $w(t) \le \bar{w}(t)$ for $t \in [-r_0, x]$ and $w(x) = \bar{w}(x)$ then $\sigma(x,w) \le \sigma(x,\bar{w})$,

2) σ is continuous and for $x \in [0,a)$ we have $\sigma(x,\theta) = 0$ where $\theta : [-r_0,a) \to R_+$ is defined by $\theta(x) = 0$ for $x \in [-r_0,a)$,

3) the function $\omega(x) = 0$, $x \in [-r_0,a)$, is the unique solution of the problem

$$\omega'(x) = \sigma(x,\omega), \quad \omega(x) = 0 \ \text{ for } x \in [-r_0, 0]. \tag{1.21}$$

Theorem 1.6 *Suppose that*

1) Assumption H $[\sigma]$ is satisfied,

2) for every $z, \bar{z} \in C(E_0 \cup E, R)$ if $\bar{z}(t,s) \ge z(t,s)$ on E_x then

$$f(x,y,\bar{z},q) - f(x,y,z,q) \le \sigma(x, V(\bar{z} - z)), \tag{1.22}$$

where $(x,y,q) \in E \times R^n$,

3) the functions $u, v : E_0 \cup E \to R$ are of class D and

$$u(x,y) \le v(x,y) \ \text{ on } E_0, \tag{1.23}$$

4) the functional differential inequality

$$D_x u(x,y) - F[u](x,y) \le D_x v(x,y) - F[v](x,y) \tag{1.24}$$

is satisfied on E.

Under these assumptions

$$u(x,y) \le v(x,y) \tag{1.25}$$

where $(x,y) \in E$.

PROOF. Let $0 < \tilde{a} < a$ be fixed. We prove (1.25) for $0 < x < \tilde{a}$, $(x,y) \in E$. For every $\varepsilon > 0$ we denote by $\omega(\cdot, \varepsilon)$ the right-hand maximum solution of the problem

$$\omega'(x) = \sigma(x,\omega) + \varepsilon, \quad \omega(x) = \varepsilon \ \text{ for } x \in [-r_0, 0]. \tag{1.26}$$

Then (see [93]) there exists $\varepsilon_0 > 0$ such that for every $0 < \varepsilon < \varepsilon_0$ the solution $\omega(\cdot, \varepsilon)$ is defined on $[-r_0, \tilde{a})$ and $\lim_{\varepsilon \to 0} \omega(x,\varepsilon) = 0$ uniformly with respect to $x \in [-r_0, \tilde{a})$. Let u_ε be the function defined by $u_\varepsilon(x,y) = u(x,y) - \omega(x,\varepsilon)$, $(x,y) \in E_{\tilde{a}}$, $0 < \varepsilon < \varepsilon_0$. It follows from (1.23) that $u_\varepsilon(x,y) < v(x,y)$ on E_0. Let $(x,y) \in E$, $0 < x < \tilde{a}$, be fixed. We will prove that

$$D_x u_\varepsilon(x,y) - F[u_\varepsilon](x,y) < D_x v(x,y) - F[v](x,y). \tag{1.27}$$

It follows from (1.22), (1.24) that

$$D_x u_\varepsilon(x,y) - F[u_\varepsilon](x,y) = D_x u(x,y) - F[u](x,y) - \sigma(x, \omega(\cdot, \varepsilon)) - \varepsilon$$

$$+\left[f(x,y,u,D_y u(x,y)) - f(x,y,u_\varepsilon,D_y u(x,y))\right]$$
$$\leq D_x v(x,y) - F[v](x,y) - \sigma(x,\omega(\cdot,\varepsilon)) - \varepsilon + \sigma(x,V(u-u_\varepsilon))$$
$$= D_x v(x,y) - F[v](x,y) - \varepsilon,$$

which completes the proof of inequality (1.27). It follows from Theorem 1.1 that

$$u(x,y) - \omega(x,\varepsilon) < v(x,y) \quad \text{on} \quad E \cap ([0,\tilde{a}) \times R^n).$$

From the above inequality we obtain in the limit, letting ε tend to 0, inequality (1.25) on $E \cap ([0,\tilde{a}) \times R^n)$. By the arbitrariness of $0 < \tilde{a} < a$, the assertion follows.

Remark 1.7 *The assumption 4) of Theorem 1.6 can be replaced by the following condition:*

4') denoted

$$\tilde{\Sigma} = \{(x,y) \in E : \ u(x,y) > v(x,y)\}, \tag{1.28}$$

for $(x,y) \in \tilde{\Sigma}$ we have

$$D_x u(x,y) - F[u](x,y) \leq D_x v(x,y) - F[v](x,y).$$

Then we use Theorem 1.1 with assumption 5') instead of 5), see Remark 1.2.

As an application of Theorem 1.6, the following uniqueness result can be derived.

Theorem 1.8 *Suppose that conditions* 1), 2) *of Theorem 1.6 hold.*

Then the Cauchy problem (1.6), (1.7) admits at most one solution of class D on $E_0 \cup E$.

Remark 1.9 *Suppose that $\tilde{\sigma} : [0,a) \times R_+ \to R_+$ is a given function. If we define σ by $\sigma(x,w) = \tilde{\sigma}(x,\|w\|_x)$, then the comparison problem (1.21) is equivalent to*

$$\omega'(x) = \tilde{\sigma}(x,\omega(x)), \quad \omega(0) = 0, \tag{1.29}$$

and assumption (1.22) has the form

$$f(x,y,\bar{z},q) - f(x,y,z,q) \leq \tilde{\sigma}(x,\|\bar{z}-z\|_x),$$

where $(x,y,z,q) \in \Omega$, $\bar{z} \in C(E_0 \cup E, R)$ and $\bar{z}(t,s) \geq z(t,s)$ on E_x. If $\omega(x) = 0$, $x \in [-r_0, a)$ is the unique solution of problem (1.29) and $\tilde{\sigma}$ is nondecreasing with respect to the second variable then the Cauchy problem (1.6), (1.7) admits at most one solution of class D on $E_0 \cup E$.

Remark 1.10 *A particular case of the comparison problem (1.21) is the following one. Let $\tilde{\sigma} : [0,a) \times R_+ \to R_+$ and $\alpha : [0,a) \to R$ be functions such that*

1) $\tilde{\sigma} \in C([0,a) \times R_+, R_+)$, $\tilde{\sigma}(x,0)$ for $x \in [0,a)$, and $\tilde{\sigma}$ is nondecreasing with respect to the second variable,

2) $\mu \in C([0,a), R)$ and $-r_0 \leq \mu(x) \leq x$ for $x \in [0,a)$,

3) for every $z, \bar{z} \in C(E_0 \cup E, R)$ *if* $\bar{z}(t,s) \geq z(t,s)$ *for* $(t,s) \in E_x$ *then*

$$f(x,y,\bar{z},q) - f(x,y,z,q) \leq \tilde{\sigma}(\,x, \|\bar{z} - z\|_{\mu(x)}\,),$$

where $(x,y,q,) \in E \times R^n$.

Then problem (1.21) is equivalent to the following one

$$\omega'(x) = \tilde{\sigma}(\,x, \omega(\mu(x))\,), \quad \omega(x) = 0 \text{ for } x \in [-r_0, 0].$$

Let E_0 and E be the sets given by (1.15) and (1.16) and suppose that the function $\tilde{f} : E \times R \times C(E_0 \cup E, R) \times R^n \to R$ is given. We prove a theorem on weak functional differential inequalities generated by equation (1.17).

Theorem 1.11 *Suppose that*

1) the function $\tilde{f}$ *satisfies the Volterra condition and it is nondecreasing with respect to the functional variable,*

2) the Lipschitz condition (1.19) is satisfied on $E \times R \times C(E_0 \cup E, R) \times R^n$,

3) there is a function $\tilde{\sigma} : [0,a) \times R_+ \times R_+ \to R_+$ *such that*

(i) $\tilde{\sigma}$ *is continuous and* $\tilde{\sigma}(\,x,0,0\,) = 0$ *for* $x \in [0,a)$,

(ii) $\tilde{\sigma}$ *is nondecreasing with respect to the last variable and the function* $\bar{\omega}(x) = 0$, $x \in [0,a)$, *is the unique solution of the problem*

$$\omega'(x) = \tilde{\sigma}(\,x, \omega(x), \omega(x)\,), \quad \omega(0) = 0,$$

(iii) for every $\bar{p} \geq p$ *and* $\bar{z} \geq z$ *where* $p, \bar{p} \in R$, $z, \bar{z} \in C(E_0 \cup E, R)$, *we have*

$$\tilde{f}(\,x,y,\bar{p},\bar{z},q\,) - \tilde{f}(\,x,y,p,z,q\,) \leq \tilde{\sigma}(\,x, \bar{p} - p, \|\bar{z} - z\|_x\,),$$

4) the function u, $v : E_0 \cup E \to R$ *are of class* D *and* $u(x,y) \leq v(x,y)$ *on* E_0,

5) the functional differential inequality

$$D_x u(x,y) - \tilde{F}[\,u\,](x,y) \leq D_x v(x,y) - \tilde{F}[\,v\,](x,y)$$

is satisfied on E.

Under these assumptions we have $u(x,y) \leq v(x,y)$ *on* E.

This theorem follows from Theorem 1.6 for $f(x,y,z,q) = \tilde{f}(x,y,z(x,y),z,q)$ and $\sigma(\,x,\zeta\,) = \tilde{\sigma}(\,x,\zeta(x),\|\zeta\|_x\,)$.

Assumption H $[\,\sigma_0\,]$. Suppose that

1) the function $\sigma_0 : [0,a) \times R_- \to R_+$, $R_- = (-\infty, 0]$, is continuous and $\sigma_0(\,x,0\,) = 0$ for $x \in [0,a)$,

2) the left-hand minimum solution of the problem

$$\omega'(x) = \sigma_0(\,x,\omega(x)\,), \quad \lim_{x \to a_-} \omega(x) = 0,$$

is $\omega(x) = 0$, $x \in [0,a)$.

Theorem 1.12 *Suppose that Assumption H $[\sigma_0]$ is satisfied and*

1) conditions 1) $-$ 3) *of Theorem 1.1 hold,*

2) the function $f : \Omega \to R$ satisfies the condition: if $(x,y) \in E$, $q \in R^n$, $z, \bar{z} \in C(E_0 \cup E, R)$, $z(t,s) \leq \bar{z}(t,s)$ on E_x and $z(x,y) < \bar{z}(x,y)$ then

$$f(x,y,z,q) - f(x,y,\bar{z},q) \leq \sigma_0(\,x, z(x,y) - \bar{z}(x,y)\,),$$

3) the functions $u, v : E_0 \cup E \to R$ are of class D and

$$D_x u(x,y) - F[\,u\,](x,y) \leq D_x v(x,y) - F[\,v\,](x,y) \quad on \quad E,$$

4) for $(x,y) \in E_0$ we have $u(x,y) \leq v(x,y)$ and $u(0,y) < v(0,y)$ for $y \in [-b,b]$.
Under these assumptions we have the strong inequality

$$u(x,y) < v(x,y) \tag{1.30}$$

for $(x,y) \in E$.

PROOF. At first we prove (1.30) for $(x,y) \in E$ and $0 < x < a - \varepsilon$ where $0 < \varepsilon < a$. Let

$$0 < p_0 < \min\{\, v(0,y) - u(0,y) : \; y \in [-b,b]\,\}.$$

For $\delta > 0$ denote by $\omega(\,\cdot\,,\delta)$ the right-hand minimum solution of the equation

$$\omega'(x) = -\sigma_0(\,x, -\omega(x)\,) - \delta$$

through $(0,p_0)$. If $p_0 > 0$ is fixed then to every $\varepsilon > 0$ there corresponds $\delta_0(\varepsilon) > 0$ such that for $0 < \delta < \delta_0(\varepsilon)$ the solution $\omega(\,\cdot\,,\delta)$ exists and it is positive on the interval $[0, a-\varepsilon)$. Let $\delta > 0$ be a such small constant that the function $\omega(\,\cdot\,,\delta)$ satisfies the above conditions. Let us denote by $\tilde{z} : E_0 \to R$ a continuous function such that $u(x,y) \leq \tilde{z}(x,y) \leq v(x,y)$ on E_0 and $\tilde{z}(0,y) = u(0,y) + p_0$ for $y \in [-b,b]$. Suppose that $\tilde{u} : E_0 \cup E \to R$ is defined by $\tilde{u}(x,y) = \tilde{z}(x,y)$ on E_0, $\tilde{u}(x,y) = u(x,y) + \omega(x,\delta)$ for $0 \leq x < a - \varepsilon$, $(x,y) \in E$. We will prove that

$$\tilde{u}(x,y) < v(x,y) \;\text{ for } 0 \leq x < a - \varepsilon,\; (x,y) \in E. \tag{1.31}$$

By Assumption H $[\,\sigma_0\,]$ and conditions 2), 3) we get for $0 < x < a - \varepsilon$, $(x,y) \in E$

$$\begin{aligned} D_x\tilde{u}(x,y) - F[\,\tilde{u}\,](x,y) &\leq D_x v(x,y) - F[\,v\,](x,y) + \omega'(x,\delta) \\ &+ f(x,y,u,D_y u(x,y)) - f(x,y,\tilde{u},D_y u(x,y)) \\ &\leq D_x v(x,y) - F[\,v\,](x,y) - \delta \end{aligned}$$

and consequently

$$D_x\tilde{u}(x,y) - F[\,\tilde{u}\,](x,y) < D_x v(x,y) - F[\,v\,](x,y), \;\; 0 < x < a - \varepsilon, \; (x,y) \in E.$$

By the definition of $\tilde{u}$ we have $\tilde{u}(x,y) \leq v(x,y)$ on E_0, $\tilde{u}(0,y) < v(0,y)$ for $y \in [-b,b]$, and hence we get from Theorem 1.1 that $u(x,y) < \tilde{u}(x,y) < v(x,y)$ for $0 \leq x < a - \varepsilon$, $(x,y) \in E$. Since $0 < \varepsilon < a$ is arbitrary, inequality (1.30) holds true on E. This completes the proof of Theorem 1.12.

Remark 1.13 *Suppose that the function $\tilde{f} : E \times R \times C(E_0 \cup E, R) \times R^n \to R$ satisfies the conditions*

1) $\tilde{f}$ is nondecreasing with respect to the functional variable,

2) the estimate

$$\tilde{f}(x, y, p, z, q) - \tilde{f}(x, y, \bar{p}, z, q) \leq \sigma_0(x, p - \bar{p})$$

holds, where $(x, y, z, q) \in \Omega$, $p, \bar{p} \in R$, $p < \bar{p}$ and σ_0 satisfies Assumption $H\,[\sigma_0\,]$.

Then the function $f(x, y, z, q) = \tilde{f}(x, y, z(x, y), z, q)$ satisfies condition 2) of Theorem 1.12.

Theorem 1.14 *Suppose that*

1) *assumptions* 1) – 3) *of Theorem 1.6 are satisfied,*

2) $\tilde{\Sigma}$ *is the set given by (1.28) and for $(x, y) \in \tilde{\Sigma}$ we have*

$$D_x u(x, y) - F[\,u\,](x, y) < D_x v(x, y) - F[\,v\,](x, y).$$

Under these assumptions we have $u(x, y) < v(x, y)$ on *E.*

This theorem can be proved by applying the weak functional differential inequalities Theorem 1.6 and then repeating the argument used in the proof of the theorem on strong inequalities (Theorem 1.1).

1.4 Comparison theorems for classical solutions

Let E, E_0, $\partial_0 E$, S_x be the sets defined by (1.2) - (1.5). Suppose that $g : E \times C([-r_0, a), R_+) \times R^n_+ \to R_+$ is a given function. In this Section we consider functional differential inequalities of the form

$$|D_x z(x, y)| \leq g\,(\,x, y, V z, [|D_y z(x, y)|]\,) \tag{1.32}$$

where

$$[|D_y z(x, y)|] = (\,|D_{y_1} z(x, y)|, \ldots, |D_{y_n} z(x, y)|\,)$$

and V is defined by (1.20).

Assumption H $[\,g, \sigma\,]$. Suppose that the function g satisfies the Volterra condition and there exists a function $\sigma : [0, a) \times C([-r_0, a), R_+) \to R_+$ such that

(i) σ is continuous and satisfies the Volterra condition,

(ii) the following monotonicity condition holds:
if $x \in [0, a)$, $w, \bar{w} \in C([-r_0, a), R_+)$, $w(t) \leq \bar{w}(t)$ for $t \in [-r_0, x]$ and $w(x) = \bar{w}(x)$ then $\sigma(\,x, w\,) \leq \sigma(\,x, \bar{w}\,)$,

(iii) for each $(x, y, w, q) \in E \times C([-r_0, a), R_+) \times R^n_+$ we have

$$g(x, y, w, q) - \sum_{i \in I_-[x,y]} \alpha_i'(x) q_i + \sum_{i \in I_+[x,y]} \beta_i'(x) q_i \leq \sigma(\,x, w\,), \tag{1.33}$$

where $q = (q_1, \ldots, q_n)$ and $q_i = 0$ for $i \in I_0[x, y]$.

Remark 1.15 *If $\tilde{\sigma} : [0, a) \times R_+ \times C([-r_0, a), R_+) \to R_+$ is nondecreasing with respect to the functional variable then $\sigma(x, w) = \tilde{\sigma}(x, w(x), w)$ satisfies the monotonicity condition (ii) in Assumption $H\,[g, \sigma\,]$.*

Let us state now a comparison lemma which will be used in the Section. We omit the proof (see [140, 195] for analogous results for equations without the functional dependence) and we just point out that it allows to estimate a function satisfying a functional differential inequality by means of the extremal solution of the corresponding functional differential problem.

Lemma 1.16 *Suppose that the function $\sigma : [0, a) \times C([-r_0, a), R_+) \to R$ satisfies the conditions (i), (ii) of Assumption $H\,[g, \sigma\,]$ and*

1) the functions $\tilde{\omega} \in C([-r_0, a), R_+)$ and $\lambda \in C([-r_0, 0], R_+)$ are such that $\tilde{\omega}(x) \le \lambda(x)$ for $x \in [-r_0, 0]$,

2) denoted by $\omega(\,\cdot\,, \lambda)$ the right-hand maximum solution of the problem

$$\omega'(x) = \sigma(x, \omega), \quad \omega(x) = \lambda(x) \ \ for \ \ x \in [-r_0, 0] \tag{1.34}$$

and

$$J^+ = \{x \in (0, a) : \tilde{\omega}(x) > \omega(x, \lambda)\},$$

suppose that

$$D_-\tilde{\omega}(x) \le \sigma(x, \tilde{\omega}) \ \ for \ \ x \in J^+,$$

where D_- is the left-hand lower Dini derivative.

Then we have $\tilde{\omega}(x) \le \omega(x, \lambda)$ for $x \in [-r_0, a)$.

Remark 1.17 *Suppose that E is the Haar pyramid given by (1.16) and*

$$g(x, y, w, q) = \sigma(x, w) + \sum_{i=1}^{n} M_i\, q_i, \tag{1.35}$$

where $(x, y, w, q) \in E \times C([-r_0, a), R_+) \times R_+^n$. Then condition (1.33) is satisfied.

The main theorem in this Section is the following.

Theorem 1.18 *Suppose that Assumption $H\,[g, \sigma\,]$ is satisfied and*

1) the function $u : E_0 \cup E \to R$ is of class D and there is $\lambda \in C([-r_0, 0], R_+)$ such that

$$|u(x, y)| \le \lambda(x) \ \ on \ \ E_0, \tag{1.36}$$

and the maximum solution $\omega(\,\cdot\,, \lambda)$ of problem (1.34) is defined on $[-r_0, a)$,

2) the function u satisfies the functional differential inequality (1.32).

Then

$$|\,u(x, y)\,| \le \omega(x, \lambda) \ \ on \ \ E. \tag{1.37}$$

PROOF. Let us define $\zeta(x) = \max\{\, |u(x,y)| : \ y \in S_x \,\}$, $-r_0 \le x < a$. Then $\zeta \in C([-r_0,a),R_+)$ and estimation (1.37) is equivalent to

$$\zeta(x) \le \omega(x,\lambda), \quad x \in [-r_0,a). \tag{1.38}$$

We will prove relation (1.38) by using Lemma 1.16. It follows from (1.36) that the initial inequality $\zeta(x) \le \lambda(x)$ for $x \in [-r_0,0]$ is satisfied. Put $J^+ = \{x \in (0,a) : \zeta(x) > \omega(x,\lambda)\}$. We prove that

$$D_-\zeta(x) \le \sigma(\,x,\zeta\,) \ \text{ for } x \in J^+. \tag{1.39}$$

Let $\bar{x} \in J^+$ be fixed. Then $\zeta(\bar{x}) > 0$ and there is $\bar{y} \in S_{\bar{x}}$ such that (i) $\zeta(\bar{x}) = u(\bar{x},\bar{y})$ or (ii) $\zeta(\bar{x}) = -u(\bar{x},\bar{y})$. Let us consider the case (i). We have

$$D_{y_i}u(\bar{x},\bar{y}) \ge 0 \ \text{ for } i \in I_+[\bar{x},\bar{y}], \quad D_{y_i}u(\bar{x},\bar{y}) \le 0 \ \text{ for } i \in I_-[\bar{x},\bar{y}] \tag{1.40}$$

and

$$D_{y_i}u(\bar{x},\bar{y}) = 0 \ \text{ for } i \in I_0[\bar{x},\bar{y}]. \tag{1.41}$$

Let $\eta = (\eta_1,\ldots,\eta_n) : [0,\bar{x}] \to R^n$ be given by (1.12), (1.13) and $\gamma(x) = u(x,\eta(x))$ for $x \in [0,\bar{x}]$. We have $\gamma(x) \le \zeta(x)$ for $x \in [0,\bar{x}]$ and $\gamma(\bar{x}) = \zeta(\bar{x})$. It follows that $D_-\zeta(\bar{x}) \le \gamma'(\bar{x})$. The above inequality and (1.32), (1.33), (1.40), (1.41) imply

$$D_-\zeta(\bar{x}) \le D_x u(\bar{x},\bar{y}) + \sum_{i=1}^{n} D_{y_i}u(\bar{x},\bar{y})\,\eta_i'(\bar{x}) \le g(\,\bar{x},\bar{y},Vu,[|D_y u(\bar{x},\bar{y})|]\,)$$

$$+ \sum_{i\in I_-[\bar{x},\bar{y}]} \alpha_i'(\bar{x})D_{y_i}u(\bar{x},\bar{y}) + \sum_{i\in I_+[\bar{x},\bar{y}]} \beta_i'(\bar{x})D_{y_i}u(\bar{x},\bar{y}) \le \sigma(\,\bar{x},\zeta\,)$$

which proves (1.39). If the case (ii) is satisfied then (1.39) can be proved in an analogous way. Thus, an application of Lemma 1.16 completes the proof of Theorem 1.18.

In the case when E_0 and E are given by (1.15) and (1.16) we have the following result.

Theorem 1.19 *Suppose that*

1) the function $\sigma : [0,a) \times C([-r_0,a),R_+) \to R_+$ satisfies conditions (i), (ii) of Assumption H $[g,\sigma]$,

2) the function $u : E_0 \cup E \to R$ is of class D and

$$|D_x u(x,y)| \le \sigma(\,x,Vu\,) + \sum_{i=1}^{n} M_i\,|D_{y_i}u(x,y)|, \ (x,y) \in E,$$

3) there is a function $\lambda \in C([-r_0,0],R_+)$ such that $|u(x,y)| \le \lambda(x)$ for $(x,y) \in E_0$ and the maximum solution $\omega(\,\cdot\,,\lambda)$ of problem (1.34) is defined on $[-r_0,a)$.

Under these assumptions we have $|u(x,y)| \le \omega(x,\lambda)$ for $(x,y) \in E$.

Theorem 1.20 *Suppose that the function* $\sigma_0 : [0,a) \times R_+^2 \to R_+$ *is given and*

1) σ_0 *is continuous and it is nondecreasing with respect to the last variable,*

2) *the function* $u : E_0 \cup E \to R$ *is of class* D *where* E_0 *and* E *are given by (1.15) and (1.16) respectively,*

3) *for* $(x,y) \in E$ *we have*

$$|D_x u(x,y)| \le \sigma_0(\, x, |u(x,y)|, \|u\|_x \,) + \sum_{i=1}^{n} M_i |D_{y_i} u(x,y)|,$$

4) *there is* $\xi \in R_+$ *such that* $|u(x,y)| \le \xi$ *for* $(x,y) \in E_0$ *and the maximum solution* $\omega(\,\cdot\,,\xi)$ *of the problem*

$$\omega'(x) = \sigma_0(\, x, \omega(x), \omega(x)\,), \qquad \omega(0) = \xi,$$

is defined on $[0,a)$.

Then $|u(x,y)| \le \omega(x,\xi)$, $(x,y) \in E$.

The above inequality follows from Theorem 1.19 for

$$\sigma(\, x, w\,) = \sigma_0(\, x, w(x), \|w\|_x\,).$$

1.5 Applications of comparison theorems

Now we give an estimation of solutions of the Cauchy problem (1.6), (1.7) by means of solutions of problem (1.34).

Theorem 1.21 *Suppose that the function* $f : \Omega \to R$ *satisfies the Volterra condition and*

1) there are functions

$$g : E \times C([-r_0, a), R_+) \times R_+^n \to R_+ \ \text{ and } \ \sigma : [0,a) \times C([-r_0,a), R_+) \to R_+$$

such that Assumption $H\,[g,\sigma\,]$ *is satisfied and*

$$|\, f(x,y,z,q)\,| \le g(\, x, y, Vz, [|q|]) \ \text{ on } \ \Omega,$$

where $[|q|] = (\,|q_1|, \ldots, |q_n|\,)$,

2) the function $u : E_0 \cup E \to R$ *is a solution of (1.6), (1.7) which is of class* D,

3) there is a function $\lambda \in C([-r_0, 0], R_+)$ *such that* $|u(x,y)| \le \lambda(x)$ *for* $(x,y) \in E_0$ *and the maximum solution* $\omega(\,\cdot\,,\lambda)$ *of problem (1.34) is defined on* $[-r_0, a)$.

Under these assumptions we have $|u(x,y)| \le \omega(x\,,\lambda)$ *on* E.

Proof. It follows that the function u satisfies all the assumptions of Theorem 1.18, and hence the assertion follows.

Remark 1.22 *Suppose that the sets E_0 and E are given by (1.15), (1.16) and*
1) $f : \Omega \to R$ satisfies the Volterra condition and there is a function

$$\sigma : [0, a) \times R_+ \times C([-r_0, a), R_+) \to R_+$$

such that
(i) the conditions (i), (ii) of Assumption $H\,[g, \sigma\,]$ hold,
(ii) the estimate

$$|\, f(x, y, z, q)\,| \le \sigma(\,x, Vz\,) + \sum_{i=1}^{n} M_i |q_i|$$

is satisfied on Ω,
2) the function $u : E_0 \cup E \to R$ is a solution of (1.6), (1.7), u is of class D and there is a function $\lambda \in C([-r_0, 0], R_+)$ such that $|u(x, y)| \le \lambda(x)$ on E_0 and the maximum solution $\omega(\,\cdot\,, \lambda$ of problem (1.34) is defined on $[-r_0, a)$.
Then $|u(x, y)| \le \omega(x, \lambda)$ on E.

Let us consider two problems, (1.6), (1.7) and the problem

$$D_x z(x, y) = \tilde{f}(\,x, y, z, D_y z(x, y)\,), \tag{1.42}$$

$$z(x, y) = \tilde{\varphi}(x, y) \;\text{ for }\; (x, y) \in E_0, \tag{1.43}$$

where $\tilde{f} : \Omega \to R$ and $\tilde{\varphi} : E_0 \to R$ are given functions. We prove a theorem on the estimate of the difference between solutions of (1.6), (1.7) and (1.42), (1.43).

Theorem 1.23 *Suppose that Assumption $H\,[g, \sigma\,]$ is satisfied and*
1) the functions f, $\tilde{f} : \Omega \to R$ satisfy the Volterra condition and

$$|\, f(x, y, z, q) - \tilde{f}(x, y, \bar{z}, \bar{q})\,| \le g(\,x, y, V(z - \bar{z}), [|q - \bar{q}|]\,) \;\; on \;\; \Omega, \tag{1.44}$$

2) the functions u, $\tilde{u} : E_0 \cup E \to R$ are solutions of problems (1.6), (1.7) and (1.42), (1.43) respectively,
3) u and $\tilde{u}$ are of class D and there is a function $\lambda \in C([-r_0, a), R_+)$ such that $|u(x, y) - \tilde{u}(x, y)| \le \lambda(x)$ on E_0 and the maximum solution $\omega(\,\cdot\,, \lambda)$ of problem (1.34) is defined on $[-r_0, a)$.
Then $|u(x, y) - \tilde{u}(x, y)| \le \omega(x, \lambda)$ on E.

PROOF. The function $u - \tilde{u}$ satisfies all the assumptions of Theorem 1.18 and the assertion follows.

The following uniqueness result is a consequence of Theorem 1.23.

Theorem 1.24 *Suppose that Assumption $H\,[g, \sigma\,]$ is satisfied and*
1) the function $f : \Omega \to R$ satisfies the Volterra condition and

$$|\, f(x, y, z, q) - f(x, y, \bar{z}, \bar{q})\,| \le g(\,x, y, V(z - \bar{z}), [|q - \bar{q}|]\,) \;\; on \;\; \Omega, \tag{1.45}$$

2) the function $\bar{\omega}(x) = 0$, $x \in [-r_0, a)$, is the maximum solution of (1.34) with $\lambda = 0$.
Under these assumptions the Cauchy problem (1.6), (1.7) admits at most one solution $u : E_0 \cup E \to R$ of class D.

Remark 1.25 *If E_0 and E are the sets defined by (1.15) and (1.16) respectively then we can assume in Theorem 1.24 that*

$$|\, f(x,y,z,q) - f(x,y,\bar{z},\bar{q})\,| \le \sigma(\,x, V(z-\bar{z})\,) + \sum_{i=1}^{n} M_i\,|q_i - \bar{q}_i| \;\; \text{on } \Omega \qquad (1.46)$$

where the function σ satisfies conditions (i), (ii) of Assumption $H\,[g,\sigma\,]$ and the maximum solution $\bar{\omega}$ of problem (1.34) with $\lambda = 0$ is $\bar{\omega}(x) = 0$ for $x \in [-r_0, a)$.

Remark 1.26 *Suppose that $\sigma_0 : [0,a) \times R_+ \to R_+$ is a given function. If assumption (1.46) has the form*

$$|\, f(x,y,z,q) - f(x,y,\bar{z},\bar{q})\,| \le \sigma_0(\,x, \|z-\bar{z}\|_x\,) + \sum_{i=1}^{n} M_i\,|q_i - \bar{q}_i| \;\; \text{on } \Omega,$$

then we have the following comparison problem

$$\omega'(x) = \sigma_0(\,x, \omega(x)\,), \qquad \omega(0) = 0.$$

Now we prove a theorem on the continuous dependence of solutions on initial data and on the right-hand sides of equations.

Theorem 1.27 *Suppose that Assumption $H\,[g,\sigma\,]$ is satisfied and*

1) the functions $f,\, \tilde{f} : \Omega \to R$ satisfy the Volterra condition and the estimate (1.45) holds,

2) the functions $u,\, \tilde{u} : E_0 \cup E \to R$ are solutions of (1.6), (1.7) and (1.42), (1.43) respectively, u and $\tilde{u}$ are of class D,

3) the function $\bar{\omega}(x) = 0$, $x \in [-r_0, a)$, is the maximum solution of (1.34) with $\lambda = 0$.

Let $0 < \tilde{a} < a$ be fixed. Under these assumptions, for every $\varepsilon > 0$ there exists $\delta > 0$ such that if

$$|\, f(x,y,z,q) - \tilde{f}(x,y,z,q)\,| < \delta \;\; \text{on } \Omega \qquad (1.47)$$

and

$$|\varphi(x,y) - \tilde{\varphi}(x,y)| < \delta \;\; \text{on } E_0 \qquad (1.48)$$

then

$$|u(x,y) - \tilde{u}(x,y)| < \varepsilon \;\; \text{for } 0 \le x < \tilde{a},\; (x,y) \in E. \qquad (1.49)$$

PROOF. For $\varepsilon > 0$ we can choose $\delta > 0$ so that the right-hand maximum solution $\omega(\,\cdot\,, \lambda, \delta)$ of the problem

$$\omega'(x) = \sigma(\,x, \omega\,) + \delta, \qquad \omega(x) = \lambda(x) \;\; \text{for } x \in [-r_0, 0],$$

where $\lambda \in C([-r_0,0], R_+)$ and $\lambda(x) \le \delta$ on $[-r_0, 0]$, is defined on $[-r_0, \tilde{a})$ and

$$\omega(x, \lambda, \delta) < \varepsilon \;\; \text{for } x \in [-r_0, \tilde{a}). \qquad (1.50)$$

Suppose that (1.47) and (1.48) hold true with the above chosen δ. Then we define

$$\lambda(x) = \max\{\, |\varphi(x,y) - \tilde{\varphi}(x,y)| : \;\; y \in S_x\,\}, \;\; x \in [-r_0, 0].$$

From (1.44), (1.47) it follows that

$$|f(x,y,z,q) - \tilde{f}(x,y,\bar{z},\bar{q})| \le g(x,y,V(z-\bar{z}),[|q-\bar{q}|]) + \delta \text{ on } \Omega.$$

Hence, by Theorem 1.18 the inequality $|u(x,y) - \tilde{u}(x,y)| \le \omega(x,\lambda,\delta)$, $0 \le x < \tilde{a}$, $(x,y) \in E$ holds. From condition (1.50) follows assertion (1.49).

Note that the connection with a functional differential comparison problem is essential for the uniqueness criterion stated in Theorem 1.24. The following example points out this property. Suppose that $0 < a \le 1$ and that $\bar{f} : \Omega \to R$ satisfies the estimation

$$|\bar{f}(x,y,z,q) - \bar{f}(x,y,\bar{z},\bar{q})| \le \sqrt{\|z-\bar{z}\|_{x^2}} + \sum_{i=1}^{n} M_i\,|q_i - \bar{q}_i|, \tag{1.51}$$

on Ω. Consider the problem

$$\omega'(x) = \sqrt{\omega(x^2)}, \quad x \in [0,a], \quad \omega(0) = 0. \tag{1.52}$$

Let $\tilde{\omega}$ be a solution of (1.52), then

$$\tilde{\omega}'(x) \le \sqrt{\tilde{\omega}(x)} \text{ for } x \in [0,a),$$

and consequently $\tilde{\omega}(x) \le x^2$ on $[0,a)$. Since

$$\tilde{\omega}(x) \le \int_0^x \sqrt{\tilde{\omega}(t^2)}\,dt \le x\,\sqrt{\tilde{\omega}(x^2)}, \quad x \in [0,a),$$

then $\tilde{\omega}(x) \le x^3$ on $[0,a)$. In analogous way we obtain the estimate

$$\tilde{\omega}(x) \le x^k, \quad x \in [0,a), \; k \ge 2.$$

Thus we have $\tilde{\omega}(x) = 0$ for $x \in [0,a)$ and (1.51) implies the uniqueness of the solution for a suitable initial problem. Note that this conclusion is not possible if we take x instead of x^2 in the right-hand side of (1.52).

1.6 Kamke functions

Let E be the Haar pyramid defined by (1.16). Write $E_0 = \{0\} \times [-b, b]$ and suppose that

$$f : E \times C(\bar{E}, R) \times R^n \to R \text{ and } \varphi : E_0 \to R$$

are given functions. Consider the initial value problem

$$D_x z(x,y) = f(x,y,z,D_y z(x,y)), \quad (x,y) \in E, \tag{1.53}$$

$$z(0,y) = \varphi(y) \text{ for } (0,y) \in E_0. \tag{1.54}$$

Let $\partial_0 E = \partial E \cap ((0,a) \times R^n)$ and

$$E_x = \{(t,s) \in \bar{E} : t \le x\}, \quad S_x = [-b + Mx, b - Mx], \quad 0 \le x < a.$$

We assume that f fulfils the Volterra condition.

A function $u : \bar{E} \to R$ is a solution of class D of problem (1.53), (1.54) if

(i) $u \in C(\bar{E}, R)$ and there exist first order partial derivatives of u on E,

(ii) there exists the total derivative of u on $\partial_0 E$,

(iii) u satisfies (1.53), (1.54).

A function $u : \bar{E} \to R$ is a solution of class D_0 of problem (1.53), (1.54) if the conditions (i) - (iii) are fulfilled and $D_x u$ is a continuous function on E_0.

We start with theorems on ordinary functional differential inequalities.

Assumption H $[\varphi, \sigma_0]$. Suppose that the functions φ and σ_0 satisfy the conditions:

1) $\varphi \in C([0, a_0], R_+)$, $\sigma_0 : (0, a_0] \times C([0, a_0], R_+) \to R_+$ where $a_0 > 0$ and $\sigma_0(t, \theta) = 0$ for $t \in (0, a_0]$, $\theta(t) = 0$ for $t \in [0, a_0]$,

2) the function σ_0 fulfils the Volterra condition,

3) $\varphi(0) = 0$, $\varphi(a_0) > 0$ and

$$D_+\varphi(t) \leq \sigma_0(t, \varphi) \text{ for } t \in (0, a_0), \tag{1.55}$$

where D_+ is the right-hand lower Dini derivative,

4) $\sigma_0 \in C((0, a_0] \times W, R_+)$ where

$$W = \{\, \omega \in C([0, a_0], R_+) : \ \omega(t) \leq L \text{ for } t \in [0, a_0] \,\}$$

and $L = 1 + \max\{\, \varphi(s) : \ s \in [0, a_0] \,\}$

5) the function $\sigma_0(t, \cdot)$ is nondecreasing on W for $t \in (0, a_0]$ and there is a function $\tilde{M} : (0, a_0] \to R_+$ such that

$$\sigma_0(s, \omega) \leq \tilde{M}(t) \text{ for } s \in [t, a_0], \ t \in (0, a_0], \ \omega \in W,$$

where $\tilde{M}$ is a locally bounded function on $(0, a_0]$.

Remark 1.28 *We can replace the derivative D_+ in (1.55) by any derivative D^+, D_-, D^- then we obtain the equivalent condition. Really, if we assume condition (1.55) then for the Picard transformation*

$$(P\varphi)(t) = \varphi(t) - \int_{t_0}^{t} \sigma_0(s, \varphi)\, ds, \ \ t \in (0, a_0],$$

with fixed $t_0 \in (0, a_0]$, we obtain

$$D_+(P\varphi)(t) = D_+\varphi(t) - \sigma_0(t, \varphi) \leq 0, \ \ t \in (0, a_0),$$

so, $P\varphi$ being continuous on $(0, a_0]$ is nonincreasing on $(0, a_0]$. Therefore

$$0 \geq D(P\varphi)(t) = D\varphi(t) - \sigma_0(t, \varphi) \text{ for } t \in (0, a_0),$$

where D means any Dini derivative. This was to be shown.

Theorem 1.29 *Suppose that Assumption H $[\varphi, \sigma_0]$ is satisfied. Then for every $b_0 \in (0, \varphi(a_0))$ there exists $\bar{\omega} \in C([0, a_0], R_+)$ such that*

$$\bar{\omega}(0) = 0, \ \bar{\omega}(a_0) = b_0, \ \bar{\omega}' \textit{ exists on } (0, a_0], \ \bar{\omega}' \in C((0, a_0], R),$$

and $\bar{\omega}'(t) = \sigma_0(t, \bar{\omega})$ for $t \in (0, a_0]$.

The proof of the theorem is given in [1]. This is the main theorem in the paper. In applications of Theorem 1.29 for investigations of uniqueness of solutions for functional differential equations presented in this Section, it is important to know that condition $D^+\varphi(0) = 0$ implies $D^+\bar{\omega}(0) = 0$. Unfortunately, the author cannot prove this fact in a general case. However, if the function σ_0 is of special form, then the mentioned fact holds true, what is shown in the next theorem.

Theorem 1.30 *Suppose that the assumptions of Theorem 1.29 are fulfilled and there exists a constant $K \in (0, 1]$ such that*

$$\sigma_0(t, w) = g(t, w|_{[Kt,t]}, \int_0^t h(t, s)w(s)\, ds), \quad t \in (0, a_0], \quad w \in C([0, a_0], R_+),$$

where the function $g : (0, a_0] \times C([0, a_0], R_+) \times R_+ \to R_+$ is nondecreasing with respect to the second and third variables, $h : [0, a_0] \times [0, a_0] \to R_+$, $h(t, \cdot)$ is an integrable function on $(0, t)$ for $t \in (0, a_0]$ and $w|_A$ means the restriction of the function w to the set A. Assume that

$$L_0 = \sup\left\{ \left[\int_0^\tau h(\tau, t)dt\right]^{-1} \int_0^{K\tau} h(\tau, t)dt \; : \; \tau \in (0, a_0] \right\} < 1.$$

Then the condition $D^+\varphi(0) = 0$ implies $D^+\bar{\omega}(0) = 0$, where $\bar{\omega}$ is the function mentioned in the assertion of Theorem 1.29.

PROOF. Suppose that the assertion of this theorem is not true. Then $D^+\varphi(0) = 0$ and there exists a number $\bar{a}$ such that

$$0 < \bar{a} < \limsup_{t \to 0+} \frac{\bar{\omega}(t)}{t}. \tag{1.56}$$

Our purpose is to obtain a contradiction with the condition $\bar{\omega}(a_0) < \varphi(a_0)$. Let us define

$$\varphi_0(t) = \max\{\varphi(s) : \; s \in [0, t]\} \text{ for } t \in [0, a_0].$$

Then

$$D^+\varphi_0(0) = 0, \quad D^+\varphi_0(t) \le \sigma_0(t, \varphi_0)$$

and φ_0 is a nondecreasing function. Define also

$$\kappa(p) = \inf\{\varphi_0^{-1}(p) : \; p \in (0, \varphi_0(a_0))\}.$$

Since $p \to 0+$ iff $\kappa(p) \to 0+$ we obtain

$$\lim_{p \to 0+} \frac{\kappa(p)}{p} = \lim_{p \to 0+} \frac{\kappa(p)}{\varphi_0(\kappa(p))} = \lim_{t \to 0+} \frac{t}{\varphi_0(t)} = +\infty.$$

It follows from (1.56) and from the condition $D^+\varphi_0(0) = 0$ that there exists $t_0 \in (0, a_0)$ such that

$$\bar{\omega}(t_0) \ge \bar{a}t_0 > \varphi_0(t_0), \quad \kappa(\bar{a}t_0) > \frac{t_0}{K},$$

and

$$\bar{\omega}(t) - \varphi_0(t) > \frac{L_0}{1-L_0}\varphi_0(t_0) \text{ for } t \in [t_0, t_0\frac{1}{K}].$$

As $\varphi_0(s) < \varphi_0(\kappa(p))$ for $s < \kappa(p)$ we have from the above estimates

$$\bar{\omega}(t) \geq \bar{\omega}(t_0) \geq \bar{a}t_0 = \varphi_0(\kappa(\bar{a}t_0)) > \varphi_0(t) \text{ for } t \in [t_0, \kappa(\bar{a}t_0)),$$

so we obtain for $t_1 = t_0K^{-1}$ that $\bar{\omega}(t) \geq \varphi_0(t)$ for $t \in [Kt_1, t_1]$. Let us define

$$T^* = \{t \in (t_1, a_0) : \bar{\omega}(t) - \varphi_0(t) \leq \frac{L_0}{1-L_0}\varphi_0(t_0)\}.$$

Since $\bar{\omega}(a_0) < \varphi_0(a_0)$, we have $T^* \neq \emptyset$. If $t_2 = \inf T^*$ then

$$\bar{\omega}(t) > \varphi_0(t) \text{ for } t \in [Ks, s],\ s \in [t_1, t_2]. \tag{1.57}$$

Moreover, since

$$\left(\int_0^t h(t,s)ds\right)^{-1} \int_0^{t_0} h(t,s)ds \leq L_0 \text{ for } t \in [t_1, a_0],$$

we have

$$\int_{t_0}^t h(t,s)ds \geq \frac{1-L_0}{L_0}\int_0^{t_0} h(t,s)ds$$

and

$$\begin{aligned}
\int_0^t h(t,s)\bar{\omega}(s)ds &\geq \int_{t_0}^t h(t,s)\left(\varphi_0(s) + \frac{L_0}{1-L_0}\varphi_0(t_0)\right)ds \\
&= \int_{t_0}^t h(t,s)\,\varphi_0(s)\,ds + \frac{L_0}{1-L_0}\varphi_0(t_0)\int_{t_0}^t h(t,s)\,ds \\
&\geq \int_{t_0}^t h(t,s)\,\varphi_0(s)\,ds + \frac{L_0}{1-L_0}\varphi_0(t_0)\frac{1-L_0}{L_0}\int_0^{t_0} h(t,s)\,ds \\
&\geq \int_{t_0}^t h(t,s)\varphi_0(s)\,ds + \int_0^{t_0} h(t,s)\varphi_0(s)ds \\
&= \int_0^t h(t,s)\varphi_0(s)\,ds, \quad t \in [t_1, t_2].
\end{aligned}$$

It follows from the above estimates and from (1.57) that $\sigma_0(t,\bar{\omega}) \geq \sigma_0(t,\varphi_0)$, $t \in [t_1, t_2]$, and consequently $\bar{\omega}'(t) = \sigma_0(t,\bar{\omega}) \geq \sigma_0(t,\varphi_0) \geq D^+\varphi_0(t)$ for $t \in [t_1, t_2]$, so the function $\bar{\omega} - \varphi_0$ is nondecreasing on $[t_1, t_2]$ and

$$\bar{\omega}(t_2) - \varphi_0(t_2) \geq \bar{\omega}(t_1) - \varphi_0(t_1) > \frac{L_0}{1-L_0}\varphi_0(t_0).$$

This implies that $t_2 \notin T^*$. Since the set T^* is closed we obtain a contradiction with the condition $T^* \neq \emptyset$. The proof is complete.

Remark 1.31 *It is obvious that in the case* $\sigma_0(t, w) = g_1(t, w(t))$ *the assertion of Theorem 1.30 is satisfied. Moreover, in this case Theorem 1.29 holds true without the assumption about monotonicity of the function* $\sigma_0(t, \cdot)$ *and with the function* $\bar{\omega}$ *such that* $\bar{\omega}(t) \leq \varphi(t)$ *for* $t \in [0, a_0]$.

Remark 1.32 *It is easy to verify that the assumptions of Theorem 1.30 are satisfied if*

$$\sigma_0(t, w) = g_0(t, w(\alpha(t)), \int_0^t h(t,s)w(s)ds)$$

and there exist constants $0 < K_1, K_2, K_3, K_4 \leq 1$, $K_3 < K_4$, *and functions* $h_1, h_2 : (0, a_0) \to R_+$ *such that for* $t \in (0, a_0)$ *we have*

$$K_1 t \leq \alpha(t) \leq t, \quad h(t,s) \leq h_1(t) \quad for\ s \in (0, K_2 t),$$

$$h(t,s) \geq h_2(t) \quad for\ s \in (K_3 t,\ K_4 t), \qquad \sup\left\{ \frac{h_1(t)}{h_2(t)} \ : \ t \in (0, a_0) \right\} < +\infty.$$

Assumption H $[\sigma, K]$. Suppose that the function σ satisfies the conditions:

1) $\sigma : (0, a) \times C([0, a), R_+) \to R_+$, $a > 0$ and $\sigma(t, \theta) = 0$ for $t \in (0, a)$ where $\theta(s) = 0$ for $s \in [0, a)$,

2) the function σ satisfies the Volterra condition and is nondecreasing with respect to the functional variable,

3) $\sigma \in C((0, a) \times U, R_+)$ where

$$U = \{\omega \in C([0, a), R_+) : \ \omega(t) \leq L \ \text{ for } t \in [0, a)\}, \ L > 1,$$

and $\sigma(s, w) \leq \tilde{M}(t)$ for $s \in [t, a)$, $w \in U$, where $\tilde{M}$ is a locally bounded function on $(0, a)$,

4) for each $s \in (0, a)$ if

$$w'(t) = \sigma(t, w) \ \text{ for } t \in (0, s) \ \text{ and } w(0) = 0$$

then $w(t) = 0$ for $t \in [0, s)$.

Now we prove the following comparison lemma.

Lemma 1.33 *Suppose that Assumption H* $[\sigma, K]$ *is satisfied and*

1) *the function* $\psi : [0, a) \to R_+$ *is continuous and* $\psi(0) = 0$,

2) $I_+ = \{t \in (0, a) : \ \psi(t) > 0\}$ *and the functional differential inequality*

$$D_-\psi(t) \leq \sigma(t, \psi), \quad t \in I_+, \tag{1.58}$$

is satisfied.

Under these assumptions we have $\psi(t) = 0$ *for* $t \in [0, a)$.

PROOF. Assume that there exists $t^* \in (0, a)$ such that $\psi(t^*) > 0$. It follows from conditions 2), 3) of Assumption H $[\sigma, K]$ and from the continuity of the function σ that the differential inequality (1.58) is fulfilled for $t \in (0, a)$. From Theorem 1.29 it follows that (see also Remark 1.28) there exists a function $\varphi_0 : [0, t^*] \to R_+$ such that

$$\varphi_0(t^*) > 0, \quad \varphi_0(0) = 0, \quad \varphi_0'(t) = \sigma(t, \varphi_0) \ \text{ for } t \in (0, t^*].$$

This contradicts the assumption 4) from H $[\sigma, K]$. The Lemma 1.33 is proved.

1.7 Uniqueness of classical solutions

We formulate next assumptions on comparison problems.

Assumption H $[\sigma, g]$. Suppose that

1) the function σ satisfies the conditions 1) – 3) of Assumption H $[\sigma, K]$ and there exists $K \in (0, 1]$ such that

$$\sigma(t, w) = g(t, w|_{[Kt,t]}, \int_0^t h(t, s)w(s)ds), \quad t \in (0, a), \; w \in C([0, a), R_+),$$

where the function g is nondecreasing with respect to the second and third variables, $h : [0, a) \times [0, a) \to R_+$, $h(t, \cdot)$ is an integrable function on $(0, t)$ for each $t \in (0, a)$ and

$$\sup \left\{ \left[\int_0^\tau h(\tau, t)dt \right]^{-1} \int_0^{K\tau} h(\tau, t)dt \; : \; \tau \in (0, a] \right\} < 1,$$

2) for each $s \in (0, a)$ if

$$\omega'(t) = \sigma(t, \omega) \text{ for } t \in (0, s) \text{ and } \omega(0) = 0, \quad \lim_{t \to 0+} \frac{\omega(t)}{t} = 0,$$

then $\omega(t) = 0$ for $t \in [0, s)$.

Lemma 1.34 *Suppose that*

1) *Assumption* $H[\sigma, g]$ *is satisfied and* $\psi \in C([0, a), R_+)$, $\psi(0) = 0$, $D^+\psi(0) = 0$,

2) $I_+ = \{t \in (0, a) : \psi(t) > 0\}$ *and for* $t \in I_+$ *we have* $D_-\psi(t) \leq \sigma(t, \psi)$.

Then $\psi(t) = 0$ *for* $t \in [0, a)$.

This Lemma is an immediate consequence of Theorem 1.30. We omit details. The function σ in Lemmas 1.33 and 1.34 fulfils the Kamke type condition on $(0, a) \times C([0, a), R_+)$. For comparison functions of the Perron type we have the following result.

Lemma 1.35 *Suppose that the function* σ *satisfies the conditions:*

1) $\sigma : [0, a) \times C([0, a), R_+) \to R_+$ *and* $\sigma(t, \theta) = 0$ *for* $t \in [0, a)$,

2) σ *is a continuous function on* $[0, a) \times C([0, a), R_+)$, *it fulfils the Volterra condition and is nondecreasing with respect to the functional variable,*

3) *for every* $s \in (0, a)$ *the function* $\bar{\omega}(t) = 0$ *is the unique solution on* $[0, s)$ *of the problem*

$$\omega'(t) = \sigma(t, \omega), \quad \omega(0) = 0,$$

3) *the function* $\psi : [0, a) \to R_+$ *is continuous,* $\psi(0) = 0$ *and the differential inequality* $D_-\psi(t) \leq \sigma(t, \psi)$ *is fulfilled for* $t \in I_+$ *where* $I_+ = \{t \in (0, a) : \psi(t) > 0\}$.

Then $\psi(t) = 0$ *for* $t \in [0, a)$.

We omit the simple proof of this Lemma. Now we prove a theorem on uniqueness of solutions of problem (1.53), (1.54).

Theorem 1.36 *Assume that $f : E \times C(\bar{E}, R) \times R^n \to R$, $\phi \in C(E_0, R)$ and*

1) the function f satisfies the Volterra condition and there exists a function $\sigma : (0, a) \times C([0, a), R_+) \to R_+$ such that

$$| f(x, y, z, q) - f(x, y, \bar{z}, \bar{q}) | \le \sigma(x, V(u - v)) + \sum_{k=1}^{n} M_k |q_k - \bar{q}_k|, \tag{1.59}$$

where $(x, y) \in E$, $q, \bar{q} \in R^n$, $z, \bar{z} \in C(\bar{E}, R)$ and V is defined by (1.20) with $r_0 = 0$.

2) the function σ fulfils Assumption $H\,[\sigma, K\,]$.

Then problem (1.53), (1.54) has at most one solution of class D on $\bar{E}$.

PROOF. Suppose that u and v are solutions of (1.53), (1.54) and they are of class D on $\bar{E}$. Let

$$\psi(x) = \max\{ |u(x, y) - v(x, y)| : y \in S_x \}, \quad x \in [0, a). \tag{1.60}$$

We prove that the function ψ fulfils the assumptions of Lemma 1.33. It follows from (1.60) that $\psi \in C([0, a), R_+)$ and $\psi(0) = 0$. We prove that

$$D_-\psi(x) \le \sigma(x, \psi) \text{ if } \psi(x) > 0, \quad x \in (0, a).$$

Suppose that $\bar{x} \in (0, a)$ and $\psi(\bar{x}) > 0$. Let $\bar{y} \in S_{\bar{x}}$ be such a point that $\psi(\bar{x}) = |u(\bar{x}, \bar{y}) - v(\bar{x}, \bar{y})|$. It follows from (1.60) that either (i) $\psi(\bar{x}) = (u - v)(\bar{x}, \bar{y})$ or (ii) $\psi(\bar{x}) = -(u - v)(\bar{x}, \bar{y})$. Consider the case (i). Let I_0, I_+, I_- be the functions defined in Section 1.2 for E given by (1.16). Then we have

$$D_{y_i}(u - v)(\bar{x}, \bar{y}) \le 0 \text{ for } i \in I_-[\bar{x}, \bar{y}], \quad D_{y_i}(u - v)(\bar{x}, \bar{y}) \ge 0 \text{ for } i \in I_+[\bar{x}, \bar{y}]$$

and

$$D_{y_i}(u - v)(\bar{x}, \bar{y}) = 0 \text{ for } i \in I_0[\bar{x}, \bar{y}].$$

Let $\tilde{q}(x) = (\tilde{q}_1(x), \ldots, \tilde{q}_n(x))$ where

$$\tilde{q}_i(x) = -b_i + M_i x \text{ for } i \in I_-[\bar{x}, \bar{y}], \quad \tilde{q}_i(x) = b_i - Mx \text{ for } i \in I_+[\bar{x}, \bar{y}],$$

and

$$\tilde{q}_i(x) = \bar{y}_i \text{ for } i \in I_0[\bar{x}, \bar{y}].$$

Then the function $\tilde{\omega}(x) = (u - v)(x, \tilde{q}(x))$, $x \in [0, \bar{x}]$, satisfies the following conditions: $\tilde{\omega}(\bar{x}) = \psi(\bar{x})$ and $\tilde{\omega}(x) \le \psi(x)$ for $x \in [0, \bar{x}]$. Therefore $D_-\psi(\bar{x}) \le D_-\tilde{\omega}(\bar{x})$. It follows from the last inequality and from (1.60) that

$$D_-\psi(\bar{x}) \le D_x(u-v)(\bar{x}, \bar{y}) + \sum_{i \in I_-[\bar{x}, \bar{y}]} M_i D_{y_i}(u-v)(\bar{x}, \bar{y}) - \sum_{i \in I_+[\bar{x}, \bar{y}]} M_i D_{y_i}(u-v)(\bar{x}, \bar{y})$$

$$\le \sigma(\bar{x}, \psi) + \sum_{i=1}^{n} M_i D_{y_i}(u - v)(\bar{x}, \bar{y})$$

$$+ \sum_{i\in I_-[\bar{x},\bar{y}]} M_i D_{y_i}(u-v)(\bar{x},\bar{y}) - \sum_{i\in I_+[\bar{x},\bar{y}]} M_i D_{y_i}(u-v)(\bar{x},\bar{y}) = \sigma(\bar{x},\psi).$$

In a similar way we prove that $D_-\psi(\bar{x}, \leq \sigma(\bar{x},\psi)$ in the case (ii). It follows from Lemma 1.33 that $\psi(x) = 0$ for $x \in [0,a)$. This completes the proof.

Theorem 1.37 *Assume that*

1) $\phi \in C(E_0, R)$ and the function $f : E \times C(\bar{E}, R) \times R^n \to R$ satisfies the Volterra condition,

2) the function $\sigma : (0,a) \times C([0,a), R_+) \to R_+$ satisfies Assumption H $[\sigma, g]$ and estimation (1.59) is satisfied for $(x,y) \in E$, $q, \bar{q} \in R^n$, $z, \bar{z} \in C(E,R)$.

Then problem (1.53), (1.54) has at most one solution of class D_0 on E.

PROOF. If u and v are solutions of (1.53, (1.54) of class D_0 then ψ defined by (1.60) is continuous on $[0,a)$, $\psi(0) = 0$ and

$$D_-\psi(x) \leq \sigma(x,\psi) \text{ if } \psi(x) > 0.$$

Let $\{t_\nu\}$ be a sequence such that $t_\nu > 0$, $\nu \geq 1$, and

$$\lim_{\nu\to\infty} t_\nu = 0, \quad D^+\psi(0) = \lim_{\nu\to\infty} \frac{\psi(t_\nu) - \psi(0)}{t_\nu}.$$

Let $\{y^{(\nu)}\}$ be a sequence such that

$$y^{(\nu)} \in S_{t_\nu} \text{ and } \psi(t_\nu) = |u(t_\nu, y^{(\nu)}) - v(t_\nu, y^{(\nu)})|, \quad \nu \geq 1.$$

Then there exist points $x_\nu, \bar{x}_\nu \in (0, t_\nu)$ such that

$$\frac{\psi(t_\nu) - \psi(0)}{t_\nu} = |D_x(x_\nu, y^{(\nu)}) - D_x v(\bar{x}_\nu, y^{(\nu)})|.$$

Since u and v are of class D_0, we get $D^+\psi(0) = 0$. It follows from Lemma 1.34 that $\psi(x) = 0$ for $x \in [0,a)$. This completes the proof.

Suppose that $f : \bar{E} \times C(\bar{E}, R) \times R^n \to R$. We have not assumed the continuity of f on $\bar{E} \times C(\bar{E}, R) \times R^n$ in the uniqueness theorems. Suppose that f is continuous now. We prove that the Kamke type uniqueness theorem for (1.53), (1.54) is equivalent to the Perron type theorem in this case. It is easy to see that the Perron type comparison function satisfies the assumptions of Theorem 1.36. Suppose that f satisfies the inequality (1.59) where σ fulfils Assumption H $[\sigma, K]$. We prove that there exists a function $\tilde{\sigma}$ such that estimation (1.59) holds with $\sigma = \tilde{\sigma}$ and $\tilde{\sigma}$ satisfies the assumptions 1) – 3) of Lemma 1.35.

Theorem 1.38 *Suppose that*

1) the function $f : \bar{E} \times C(\bar{E}, R) \times R^n \to R$ is continuous and fulfils the Volterra condition,

2) there exists a function $\sigma : (0,a) \times C([0,a), R_+) \to R_+$ and constants M_i, $i = 1, \ldots, n$, such that the inequality (1.59) is fulfilled for $(x,y) \in E$, $q, \bar{q} \in R^n$, $z, \bar{z} \in C(\bar{E}, R)$,

3) *the function* σ *satisfies Assumption* $H\,[\sigma, K\,]$.

Then there exists a function $\tilde{\sigma} : [0,a) \times C([0,a), R_+) \to R_+$ *such that*

(i) $\tilde{\sigma}$ *is continuous and satisfies the Volterra condition,*

(ii) *estimation (1.59) is fulfilled for* $\sigma = \tilde{\sigma}$ *and for* $(x,y) \in \bar{E}$, $q, \bar{q} \in R^n$, $z, \bar{z} \in C(\bar{E}, R)$,

(iii) $\tilde{\sigma}(t,\theta) = 0$ *for* $t \in [0,a)$ *where* $\theta(s) = 0$ *for* $s \in [0,a)$, *and the function* $\omega = 0$ *is the only solution of the problem*

$$\omega'(t) = \tilde{\sigma}(t,\omega), \quad \omega(0) = 0. \tag{1.61}$$

Proof. Let $x \in [0,a)$, $y \in S_x$, $w \in C([0,a), R_+)$, $\eta = (\eta_1, \ldots, \eta_n)$, $\eta \in R^n_+$. We define

$$h(x,y,w,\eta) = \sup\ \{|f(x,y,z,q) - f(x,y,\bar{z},\bar{q})| : \tag{1.62}$$

$$|q_i - \bar{q}_i| \le \eta_i \ \text{for}\ i = 1,\ldots,n,\ \ \max\{\,|z(x,\tilde{y}) - \bar{z}(x,\tilde{y})| :\ \tilde{y} \in S_x\,\} \le w(x)\}$$

and

$$\tilde{\sigma}(t,w) = \sup\{\, h(t,y,w,0) :\ y \in S_t\,\}.$$

The function $\tilde{\sigma}$ is continuous on $[0,a) \times C([0,a), R_+)$, satisfies the Volterra condition and $\tilde{\sigma}(x,\theta) =$ for $x \in [0,a)$. It follows from (1.59) and (1.62) that

$$|f(x,y,z,q) - f(x,y,\bar{z},\bar{q})| \le \tilde{\sigma}(x, V(z-\bar{z})) + \sum_{i=1}^{n} M_i|q_i - \bar{q}_i|.$$

We see at once that $\tilde{\sigma}$ is nondecreasing with respect to the functional variable. We prove that $w = 0$ is the only solution of (1.61). Suppose that there exists a function φ satisfying (1.61) and such that $\varphi(a_0) > 0$ for some $a_0 \in (0,a)$. Since $\tilde{\sigma}(x,w) \le \sigma(x,w)$ for $x \in (0,a)$, $w \in C([0,a), R_+)$, we obtain $\varphi'(x) \le \sigma(x,\varphi)$, $x \in (0,a)$. It follows from this inequality and from Theorem 1.29 that there exists a solution φ_0 of the equation $\omega'(x) = \sigma(x,\omega)$ on $(0,a_0]$ such that $\varphi_0(a_0) > 0$ and $\varphi(0) = 0$. This contradicts assumption 3) of the theorem. Thus we see that $\tilde{\sigma}$ satisfies all the conditions of Theorem 1.38.

1.8 Nonlinear systems

Let E and E_0 be the sets given by (1.16) and (1.15) respectively. Write $\Xi = E \times R^k \times C(E_0 \cup E, R^k) \times R^n$ and suppose that

$$f = (f_1, \ldots, f_k) : \Xi \to R^k, \quad \varphi : E_0 \to R^k$$

are given functions. In this Section we consider functional differential inequalities generated by the Cauchy problem

$$D_x z_i(x,y) = f_i(x,y,z(x,y),z\,,D_y z_i(x,y)), \quad i = 1,\ldots,k, \tag{1.63}$$

$$z(x,y) = \varphi(x,y) \ \ \text{on}\ \ E_0, \tag{1.64}$$

where $z = (z_1, \ldots, z_k)$. We will say that the function $f : \Xi \to R^k$ satisfies the Volterra condition if for each $(x,y) \in E$ there is a set $E[x,y] \subset E_x$ such that if $z, \bar{z} \in C(E_0 \cup E, R^k)$ and $z(t,s) = \bar{z}(t,s)$ for $(t,x) \in E[x,y]$ then $f(x,y,p,z,q) = f(x,y,p,\bar{z},q)$. The Volterra condition for f means that the value of f at the point (x,y,p,z,q) depends on (x,y,p,q) and on the restriction of z to the set $E[x,y]$. Let us denote by F the Niemycki operator corresponding to (1.63) i.e.

$$F[\,z\,](x,y) = (\,F_1[\,z\,](x,y), \ldots, F_k[\,z\,](x,y)\,)$$

and

$$F_i[\,z\,](x,y) = f_i(\,x,y,z(x,y),z,\,D_y z_i(x,y)\,), \quad i = 1, \ldots, k.$$

Write $D_x z = (\,D_x z_1, \ldots, D_x z_k\,)$. A function $z = (z_1, \ldots, z_k) : E_0 \cup E \to R^k$ will be called the function of class D if the components $z_i : E_0 \cup E \to R$ are of of class D for $1 \le i \le k$. We start with a theorem on strong inequalities.

Theorem 1.39 *Suppose that*

1) the function $f : \Xi \to R^k$ *of the variables* (x,y,p,z,q), $p = (p_1, \ldots, p_k)$, *satisfies the Volterra condition and is nondecreasing with respect to the functional variable,*

2) the following quasi-monotonicity condition is satisfied: for each i, $1 \le i \le k$, *if* $p \le \bar{p}$, $p = (p_1, \ldots, p_k)$, $\bar{p} = (\bar{p}_1, \ldots, \bar{p}_k)$ *and* $p_i = \bar{p}_i$ *then*

$$f_i(x,y,p,z,q) \le f_i(x,y,\bar{p},z,q)$$

where $(x,y,z,q) \in E \times C(E_0 \cup E, R^k) \times R^n$,

3) the Lipschitz condition

$$|\,f_i(x,y,p,z,q) - f_i(x,y,p,z,\bar{q})\,| \le \sum_{j=1}^{n} M_j\,|\,q_j - \bar{q}_j\,|, \quad i = 1, \ldots, k, \tag{1.65}$$

is satisfied on Ξ *where the vector* $M = (M_1, \ldots, M_n)$ *appears in (1.16),*

4) the functions $u, v : E_0 \cup E \to R^k$, $u = (u_1, \ldots, u_k)$, $v = (v_1, \ldots, v_k)$, *are of class* D *and*

$$u(x,y) \le v(x,y) \;\mathit{for}\; (x,y) \in E_0, \quad u(0,y) < v(0,y) \text{ for } y \in [-b,b], \tag{1.66}$$

5) the functional differential inequality

$$D_x u(x,y) - F[\,u\,](x,y) < D_x v(x,y) - F[\,v\,](x,y) \tag{1.67}$$

is satisfied on Ξ.

Under these assumptions

$$u(x,y) < v(x,y) \quad \mathit{for} \quad E. \tag{1.68}$$

PROOF. It follows from the initial inequalities (1.66) and from the continuity of u and v that the set

$$J^{(+)} = \{ x^* > 0 : \ u(x,y) < v(x,y) \ \text{for} \ (x,y) \in E \cap ((0,x^*) \times R^n) \}$$

is not empty. Defining $\bar{x} = \sup J^{(+)}$ we prove that $\bar{x} = a$. Suppose it is not true and hence $\bar{x} < a$. Then there exists an index j, $1 \le j \le k$, and a point $\bar{y} \in S_{\bar{x}}$ such that

$$u(x,y) < v(x,y) \ \text{for} \ (x,y) \in E \cap ((0,\bar{x}) \times R^n)$$

and

$$u(\bar{x},\bar{y}) \le v(\bar{x},\bar{y}), \quad u_j(\bar{x},\bar{y}) = v_j(\bar{x},\bar{y}).$$

Then

$$D_{y_i}(u_j - v_j)(\bar{x},\bar{y}) \ge 0 \ \text{for} \ i \in I_+[\bar{x},\bar{y}], \tag{1.69}$$

$$D_{y_i}(u_j - v_j)(\bar{x},\bar{y}) \le 0 \ \text{for} \ i \in I_-[\bar{x},\bar{y}], \tag{1.70}$$

and

$$D_{y_i}(u_j - v_j)(\bar{x},\bar{y}) = 0 \ \text{for} \ i \in I_0[\bar{x},\bar{y}]. \tag{1.71}$$

Now, for $0 \le x \le \bar{x}$ put $\eta(x) = (\eta_1(x), \ldots, \eta_n(x))$ where

$$\eta_i(x) = -b_i + M_i x \text{ for } i \in I_-[\bar{x},\bar{y}], \quad \eta_i(x) = b_i - M_i x \text{ for } i \in I_+[\bar{x},\bar{y}],$$

$$\eta_i(x) = \bar{y}_i \quad \text{for} \ i \in I_0[\bar{x},\bar{y}],$$

and consider the composite function $\gamma_j(x) = (u_j - v_j)(x,\eta(x))$, $x \in [0,\bar{x}]$. It attains maximum at $\bar{x}$. Since u_j and v_j are of class D on $E_0 \cup E$ and $\gamma_j'(\bar{x}) \ge 0$ then we have

$$D_x(u_j - v_j)(\bar{x},\bar{y}) + \sum_{i \in I_-[\bar{x},\bar{y}]} M_i \, D_{y_i}(u_j - v_j)(\bar{x},\bar{y}) \tag{1.72}$$
$$- \sum_{i \in I_+[\bar{x},\bar{y}]} M_i \, D_{y_i}(u_j - v_j)(\bar{x},\bar{y}) \ge 0.$$

It follows from the monotonicity of f with respect to the functional variable and from the quasi-monotone property with respect to p and from (1.65), (1.67), (1.69) - (1.71) that

$$D_x(u_j - v_j)(\bar{x},\bar{y}) < F_j[u](\bar{x},\bar{y}) - F_j[v](\bar{x},\bar{y})$$
$$\le f_j(\bar{x},\bar{y},v(\bar{x},\bar{y}),v\,,D_y u_j(\bar{x},\bar{y})) - f_j(\bar{x},\bar{y},v(\bar{x},\bar{y}),v\,,D_y v_j(\bar{x},\bar{y}))$$
$$\le \sum_{i=1}^{n} M_i \, | D_{y_i}(u_j - v_j)(\bar{x},\bar{y}) |$$
$$= \sum_{i \in I_+[\bar{x},\bar{y}]} M_i \, D_{y_i}(u_j - v_j)(\bar{x},\bar{y}) - \sum_{i \in I_-[\bar{x},\bar{y}]} M_i \, D_{y_i}(u_j - v_j)(\bar{x},\bar{y}),$$

which contradicts (1.72). Hence $\bar{x} = a$ and the statement (1.68) follows.

Remark 1.40 *In Theorem 1.39 we can assume instead of (1.67) on E that*

$$D_x u_j(x,y) - F_j[u](x,y) < D_x v_j(x,y) - F_j[v](x,y)$$

for $(x,y) \in \Sigma_j$ *where*

$$\Sigma_j = \{ (x,y) \in E : \ u(t,s) < v(t,s) \ \ for \ \ (t,s) \in E \cap ((0,x) \times R^n) \\ \text{and } \ u_j(x,y) = v_j(x,y) \}$$

and $1 \leq j \leq k$.

Assumption H $[f, \sigma_0]$. Suppose that the function $\sigma_0 : [0,a) \times R_- \to R_+$ satisfies Assumption H $[\sigma_0]$ and for each i, $1 \leq i \leq k$, we have

$$f_i(x,y,p,z,q) - f_i(x,y,p_1,\ldots,p_{i-1},\bar{p}_i,p_{i+1},\ldots,p_k,z,q) \\ \leq \sigma_0(x, p_i - \bar{p}_i) \ \text{ on } \ \Xi,$$

where $p_i \leq \bar{p}_i$.

Theorem 1.41 *Suppose that Assumption H* $[f, \sigma_0]$ *is satisfied and*

1) the assumptions 1) - 4) of Theorem 1.39 hold,

2) the functional differential inequality

$$D_x u(x,y) - F[u](x,y) \leq D_x v(x,y) - F[v](x,y) \tag{1.73}$$

is satisfied on E.

Then

$$u(x,y) < v(x,y) \tag{1.74}$$

for $(x,y) \in E$.

PROOF. Let $0 < a_0 < a$ be fixed. At first we prove (1.74) for $(x,y) \in E \cap ((0,a_0) \times R^n)$. Put

$$0 < \tilde{p} < \min_{1 \leq i \leq k} \{ \min \{ [v_i(0,y) - u_i(0,y)] : \ y \in [-b,b] \} \}.$$

For $\delta > 0$ denote by $\tilde{\omega}$ the right-hand minimum solution of the problem

$$\omega'(t) = -\sigma_0(t, -\omega(t)) - \delta, \quad \omega(0) = \tilde{p}.$$

If $\tilde{p} > 0$ is fixed then to every $a_0 \in (0,a)$ there corresponds $\delta_0 > 0$ such that for $0 < \delta < \delta_0$ the solution $\tilde{\omega}$ exists and is positive on $[0,a_0)$. Let $\delta > 0$ be a such small constant that the function $\tilde{\omega}$ satisfies the above conditions. Denote by $\tilde{z} = (\tilde{z}_1 \ldots, \tilde{z}_k) : E_0 \to R^k$ a continuous function such that

$$u(x,y) \leq \tilde{z}(x,y) \leq v(x,y) \ \text{ on } \ E_0$$

and $\tilde{z}_i(0,y) = u_i(0,y) + \tilde{p}$, $1 \leq i \leq k$, for $y \in [-b,b]$. Let $\tilde{u} = (\tilde{u}_1, \ldots, \tilde{u}_k)$ be defined by

$$\tilde{u}(x,y) = \tilde{z}(x,y) \ \text{ for } \ (x,y) \in E_0,$$

$$\tilde{u}_i(x,y) = u_i(x,y) + \tilde{\omega}(x), \quad i = 1,\ldots,k, \quad \text{for} \quad (x,y) \in E \cap (\,(0,a_0) \times R^n\,)\,.$$

It follows that $\tilde{u}$ is of class D on $(E_0 \cup E) \cap (\,[-r_0,a_0) \times R^n\,)$ and

$$\tilde{u}(x,y) \leq v(x,y) \ \text{ on } \ E_0, \quad \tilde{u}(0,y) < v(0,y) \ \text{ on } \ [-b,b].$$

Now we prove that

$$D_x\tilde{u}(x,y) - F[\,\tilde{u}\,](x,y) < D_x v(x,y) - F[\,v\,](x,y) \tag{1.75}$$

for $(x,y) \in E \cap (\,(0,a_0) \times R^n\,)$. Let $1 \leq i \leq k$ be fixed. Write

$$(\,\Gamma_i u\,)(x,y) = (\,u_1(x,y),\ldots,u_{i-1}(x,y), \tilde{u}_i(x,y), u_{i+1}(x,y),\ldots,u_k(x,y)\,)\,.$$

Then

$$\begin{aligned} &D_x\tilde{u}_i(x,y) - F_i[\,\tilde{u}\,](x,y) - [\,D_x v_i(x,y) - F_i[\,v\,](x,y)\,] \\ &= D_x u_i(x,y) - F_i[\,u\,](x,y) - [\,D_x v_i(x,y) - F_i[\,v\,](x,y)\,] \\ &\quad + \tilde{\omega}'(x) - F_i[\,\tilde{u}\,](x,y) + F_i[\,u\,](x,y) \\ &\leq f_i(\,x,y,u(x,y),\tilde{u},D_y u_i(x,y)\,) - f_i(\,x,y,(\,\Gamma_i u\,)(x,y),\tilde{u},D_y u_i(x,y)\,) \\ &+ [f_i(\,x,y,(\,\Gamma_i u\,)(x,y),\tilde{u},D_y u_i(x,y)\,) - f_i(\,x,y,\tilde{u}(x,y),\tilde{u},D_y u_i(x,y)\,)\,] + \tilde{\omega}'(x) \\ &\leq \sigma_0(\,x,-\tilde{\omega}(x)\,) + \tilde{\omega}'(x) = -\delta, \end{aligned}$$

where $(x,y) \in E \cap (\,(0,a_0) \times R^n\,)$. Then condition (1.75) is proved. It follows from Theorem 1.39 that $\tilde{u}(x,y) < v(x,y)$ on $E \cap (\,(0,a_0) \times R^n\,)$, and consequently $u(x,y) < v(x,y)$ on $E \cap (\,(0,a_0) \times R^n\,)$. Since $a_0 \in (0,a)$ is arbitrary, inequality (1.74) follows.

Now we prove a theorem on weak functional differential inequalities. For a function $z = (z_1,\ldots,z_k) \in C(E_0 \cup E, R^k)$ and for a point $x \in [0,a)$ we write

$$[\,\|z\|_x\,] = (\,\|z_1\|_x,\ldots,\|z_k\|_x\,)\,.$$

In this Section we will denote by $V\,:\,C(E_0 \cup R, R^k) \to C(\,[-r_0,a), R^k_+\,)$ the operator given by

$$\begin{aligned} (Vz)(x) = (\,\max\{\,|z_1(x,y)|\,:\,y \in S_x\,\},\ldots, \\ \max\{\,|z_k(x,y)|\,:\,y \in S_x\,\}\,), \end{aligned} \tag{1.76}$$

where $-r_0 \leq x < a$. Note that continuity of the function Vz given by (1.76) follows from Theorem 33.1 in [195].

Let $\sigma = (\,\sigma_1,\ldots,\sigma_k\,) : [0,a) \times R^k \times C(\,[-r_0,a), R^k_+\,) \to R^k$ be a given function.

Assumption $\tilde{H}$ $[\,\sigma\,]$. Suppose that

1) σ is continuous and $\sigma(\,x,0,\theta\,) = 0$ for $x \in [0,a)$ where $\theta : [-r_0,a) \to R^n_+$ is given by $\theta(t) = 0$, $t \in [-r_0,a)$,

2) σ satisfies the Volterra condition, it is nondecreasing with respect to the functional variable and the following quasi-monotonicity condition is satisfied: for each i, $1 \leq i \leq k$, if $p \leq \bar{p}$ and $p_i = \bar{p}_i$ then $\sigma_i(\,t,p,w\,) \leq \sigma_i(\,x,\bar{p},w\,)$, where $(t,w) \in [0,a) \times C(\,[-r_0,a), R^k_+\,)$,

3) the function $\tilde{\omega}(x) = 0$, $x \in [-r_0,a)$, is the right-hand maximum solution of the problem

$$\omega'(x) = \sigma(\,x,\omega(x),\omega\,), \quad \omega(x) = 0 \ \text{ on } \ [-r_0,0]. \tag{1.77}$$

Theorem 1.42 *Suppose that*

1) the function $f : \Xi \to R^k$ *satisfies the conditions 1) - 3) of Theorem 1.39,*

2) there is a function $\sigma : [0,a) \times R^k \times C([-r_0,a), R^k_+) \to R^k$ *satisfying Assumption* $\tilde{H}[\sigma]$ *and such that*

$$f(x,y,p,z,q) - f(x,y,\bar{p},\bar{z},q) \leq \sigma(x, p-\bar{p}, V(z-\bar{z})) \text{ on } \Xi$$

where $p \geq \bar{p}$, $z \geq \bar{z}$,

3) the functions $u, v : E_0 \cup E \to R^k$ *are of class D and* $u(x,y) \leq v(x,y)$ *on* E_0,

4) the functional differential inequality

$$D_x u(x,y) - F[u](x,y) \leq D_x v(x,y) - F[v](x,y)$$

is satisfied on E.

Then we have

$$u(x,y) \leq v(x,y) \text{ on } E. \tag{1.78}$$

PROOF. Let $0 < a_0 < a$ be fixed. We prove that $u(x,y) \leq v(x,y)$ for $x \in (0,a_0)$, $(x,y) \in E$. For every $\varepsilon = (\varepsilon_1, \ldots, \varepsilon_k)$ where $\varepsilon_i > 0$ for $1 \leq i \leq k$, denote by $\omega(\cdot,\varepsilon) = (\omega_1(\cdot,\varepsilon), \ldots, \omega_k(\cdot,\varepsilon))$ the right-hand maximum solution of the problem

$$\omega'(x) = \sigma(x, \omega(x), \omega) + \varepsilon, \quad \omega(x) = \varepsilon \text{ on } [-r_0, 0].$$

There is $\tilde{\varepsilon} > 0$ such that for $0 < \varepsilon_i < \tilde{\varepsilon}$, $1 \leq i \leq k$, the solution $\omega(\cdot,\varepsilon)$ is defined on $[0,a_0)$ and $\lim_{\varepsilon\to 0}\omega(x,\varepsilon) = 0$ uniformly on $[-r_0,a_0)$. Let $u_\varepsilon = (u_{\varepsilon.1}, \ldots, u_{\varepsilon.k})$ be the function given by

$$u_\varepsilon(x,y) = u(x,y) - \varepsilon \text{ on } E_0,$$

$$u_\varepsilon(x,y) = u(x,y) - \omega(x,\varepsilon) \text{ on } E \cap ((0,a_0) \times R^n).$$

We prove that $u_\varepsilon(x,y) < v(x,y)$ on $E \cap ((0,a_0) \times R^n)$. It follows that the strong initial inequality $u_\varepsilon(x,y) < v(x,y)$ is satisfied on E_0. We infer, by Assumption $\tilde{H}[\sigma]$ that

$$\begin{aligned} & D_x u_{\varepsilon.i}(x,y) - F_i[u_\varepsilon](x,y) - [D_x v_i(x,y) - F_i[v](x,y)] \\ &= D_x u_i(x,y) - F_i[u](x,y) - [D_x v_i(x,y) - F_i[v](x,y)] - \omega_i'(x,\varepsilon) \\ &+ f_i(x,y,u(x,y),u,D_y u_i(x,y)) - f_i(x,y,u_\varepsilon(x,y),u_\varepsilon,D_y u_i(x,y)) \\ &\leq -\omega_i'(x,\varepsilon) + \sigma(x,\omega(x,\varepsilon),\omega(\cdot,\varepsilon)) = -\varepsilon_i, \end{aligned}$$

where $1 \leq i \leq k$, $(x,y) \in E \cap ((0,a_0) \times R^n)$. Then we have

$$D_x u_\varepsilon(x,y) - F[u_\varepsilon](x,y) < D_x v(x,y) - F[v](x,y) \text{ on } E \cap ((0,a_0) \times R^n).$$

It follows from Theorem 1.39 that

$$u_\varepsilon(x,y) < v(x,y) \text{ on } E \cap ((0,a_0) \times R^n).$$

From the above inequality we obtain in the limit, letting ε tend to 0, inequality (1.78) on $E \cap ((0,a_0) \times R^n)$. By the arbitrariness of $a_0 \in (0,a)$, the assertion follows.

Remark 1.43 *Consider the following example of condition 2) of Theorem 1.42. Suppose that there is a function*

$$\tilde{\sigma} = (\tilde{\sigma}_1, \ldots, \tilde{\sigma}_k) : [0, a) \times R^k_+ \times R^k_+ \to R^k_+$$

such that

1) $\tilde{\sigma}$ is continuous and $\tilde{\sigma}(x, 0, 0) = 0$ for $x \in [0, a)$,

2) the function $\tilde{\sigma}$ is nondecreasing with respect to the last variable and the following quasi-monotonicity condition is satisfied: for each i, $1 \le i \le k$, if $p \le \bar{p}$ and $p_i = \bar{p}_i$ then $\tilde{\sigma}_i(x, p, s) \le \tilde{\sigma}_i(x, \bar{p}, s)$ where $(x, s) \in [0, a) \times R^k_+$,

3) the estimate

$$f(x, y, p, z, q) - f(x, y, \bar{p}, \bar{z}, q) \le \tilde{\sigma}(x, p - \bar{p}, [\,\|z - \bar{z}\|_x\,])$$

is satisfied on Ξ for $p \ge \bar{p}$, $z \ge \bar{z}$.

Then the comparison problem (1.77) has the form

$$\omega'(x) = \tilde{\sigma}(x, \omega(x), \omega(x)), \quad \omega(0) = 0.$$

We assume in this case that the function $\bar{\omega}(x) = 0$, $x \in [0, a)$, is the maximum solution of the above Cauchy problem.

Theorem 1.44 *Suppose that the assumption 1) - 3) of Theorem 1.42 are satisfied and*

$$D_x u(x, y) - F[u](x, y) < D_x v(x, y) - F[v](x, y) \text{ on } E.$$

Then $u(x, y) < v(x, y)$ on E.

This theorem can be proved by applying Theorem 1.42 and then be repeating the arguments used in the proof of Theorem 1.39.

1.9 Haar inequality for nonlinear systems

An important technique in the theory of functional differential equations is concerned with estimating of a function satisfying a system of functional differential inequalities by the extremal solution of corresponding comparison problem. Let E_0, E and Ξ be the sets given in Section 1.8. For a function $z = (z_1, \ldots, z_k) : E_0 \cup E \to R^k$ we write $[|z(x, y)|] = (|z_1(x, y)|, \ldots, |z_k(x, y)|)$. Suppose that there exist the derivatives

$$D_x z(x, y) = (D_x z_1(x, y), \ldots, D_x z_k(x, y)),$$
$$D_y z(x, y) = [D_{y_i} z_j(x, y)]_{i=1,\ldots,n,\ j=1,\ldots,k}.$$

We put

$$[|D_x z(x, y)|] = (|D_x z_1(x, y)|, \ldots, |D_x z_k(x, y)|),$$
$$[|D_y z(x, y)|] = [|D_{y_i} z_j(x, y)|]_{i=1,\ldots,n,\ j=1,\ldots,k}$$

In this Section we denote by $A \circ B$ the product of two matrices. Suppose that

$$f = (f_1, \ldots, f_k) : \Xi \to R^k, \quad \varphi = (\varphi_1, \ldots, \varphi_k) : E_0 \to R^k$$

are given functions. We consider comparison theorems generated by the Cauchy problem (1.63), (1.64). We assume that the operator f satisfies the Volterra condition. We consider solutions of class D to problem (1.63), (1.64). In this Section we will denote by $V : C(E_0 \cup E, R^k) \to C([-r_0, a), R^k_+)$ the operator given by (1.76) Let us state now a lemma on strong functional differential inequalities for functions of one variable.

Assumption $\bar{H}[\sigma]$. Suppose that the function

$$\sigma = (\sigma_1, \ldots, \sigma_k) : [0,a) \times R^k_+ \times C([-r_0, a), R^k_+) \to R^k_+$$

of the variables (x, p, ω) satisfies the conditions:

1) σ is continuous, satisfies the Volterra condition and it is nondecreasing with respect to the functional variable,

2) for each i, $1 \le i \le k$, the function σ_i is nondecreasing with respect to variables

$$(p_1, \ldots, p_{i-1}, p_{i+1}, \ldots, p_k),$$

3) for each function $\lambda \in C([-r_0, 0], R^k_+)$ the maximum solution $\omega(\,\cdot\,, \lambda)$ of the problem

$$\omega'(x) = \sigma(x, \omega(x), \omega), \quad \omega(x) = \lambda(x) \text{ for } x \in [-r_0, 0], \tag{1.79}$$

is defined on $[-r_0, a)$.

Lemma 1.45 *Suppose that*

1) *the function σ satisfies conditions* 1), 2) *of Assumption $\bar{H}[\sigma]$ and*

2) *the functions $\bar{\alpha}$, $\bar{\beta} : [-r_0, a) \to R^k$, $\bar{\alpha} = (\bar{\alpha}_1, \ldots, \bar{\alpha}_k)$, $\bar{\beta} = (\bar{\beta}_1, \ldots, \bar{\beta}_k)$ are continuous and satisfy the strong initial inequality $\bar{\alpha}(x) < \bar{\beta}(x)$ for $x \in [-r_0, 0]$,*

4) *denoted by*

$$J_i^+ = \{x \in (0,a) : \ \bar{\alpha}(t) < \bar{\beta}(t) \ for \ t \in [-r_0, x) \ and \ \bar{\alpha}_i(x) = \bar{\beta}_i(x)\},$$

suppose that

$$D_-\bar{\alpha}_i(x) - \sigma_i(x, \bar{\alpha}(x), \bar{\alpha}) < D_-\bar{\beta}_i(x) - \sigma_i(x, \bar{\beta}(x), \bar{\beta}), \quad x \in J_i^+.$$

Then we have $\bar{\alpha}(x) < \bar{\beta}(x)$ for $x \in [0, a)$.

Proof. Suppose by contradiction, that the above assertion is fails to be true. Then there are a point $\bar{x} > 0$ and an index i, $1 \le i \le k$, such that $\bar{\alpha}(t) < \bar{\beta}(t)$ for $t \in [0, x)$ and $\bar{\alpha}_i(\bar{x} = \bar{\beta}_i(\bar{x})$. Then $\bar{x} \in J_i^+$ and we have

$$D_-\bar{\alpha}(\bar{x}) - D_-\bar{\beta}_i(\bar{x}) < \sigma_i(\bar{x}, \bar{\alpha}(\bar{x}), \bar{\alpha}) - \sigma_i(\bar{x}, \bar{\beta}(\bar{x}), \bar{\beta}) < 0. \tag{1.80}$$

On the other hand, it follows that

$$\frac{\bar{\alpha}_i(t) - \bar{\alpha}_i(\bar{x}}{t - x} > \frac{\bar{\beta}_i(t) - \bar{\beta}_i(\bar{x})}{t - x}, \quad 0 < t < x,$$

and consequently $D_-\bar{\alpha}(\bar{x}) \ge D_-\bar{\beta}(\bar{x})$, which contradicts (1.80). The Lemma 1.45 is proved.

Now we present a functional extension of the classical Haar inequality.

Theorem 1.46 *Suppose that Assumption $\bar{H}\,[\sigma]$ is satisfied and*

1) *the function $u = (u_1, \dots, u_k) : E_0 \cup E \to R^k$ is of class D and the initial estimation is satisfied*

$$[|u(x,y)|] \leq \lambda(x), \quad (x,y) \in E_0, \tag{1.81}$$

where $\lambda \in C([-r_0, 0], R^k_+)$,

2) *the functional differential inequality*

$$[|D_x u(x,y)|] \leq \sigma(\,x, [|u(x,y)|], Vu\,) + M \circ [|\,D_y u(x,y)\,|] \tag{1.82}$$

is satisfied on E.

Then

$$[|u(x,y)|] \leq \omega(x,\,\lambda) \;\; on \;\; E_0 \cup E. \tag{1.83}$$

PROOF. Let us define the function $\zeta = (\zeta_1, \dots, \zeta_k)$ by

$$\zeta_i(x) = \max\{\,|u_i(x,y)| : \; y \in S_x\,\}, \;\; -r_0 \leq x < a, \;\; 1 \leq i \leq k.$$

Then $\zeta \in C([-r_0, a), R^k_+)$ and the assertion (1.83) is equivalent to

$$\zeta(x) \leq \omega(x,\,\lambda), \;\; x \in [-r_0, a). \tag{1.84}$$

Let $0 < \tilde{a} < a$ be fixed. Consider the Cauchy problem

$$\omega'(x) = \sigma(\,x, \omega(x), \omega\,) + \bar{\varepsilon}, \quad \omega(x) = \lambda(x) + \bar{\varepsilon} \;\text{ for }\; x \in [-r_0, 0] \tag{1.85}$$

where $\bar{\varepsilon} = (\varepsilon, \dots, \varepsilon) \in R^k$ and $\varepsilon > 0$. There exists $\varepsilon_0 > 0$ such that for $0 < \varepsilon < \varepsilon_0$ the maximum solution

$$\omega(\,\cdot\,, \lambda, \bar{\varepsilon}) = (\,\omega_1(\,\cdot\,, \lambda, \bar{\varepsilon}), \dots, \omega_k(\,\cdot\,, \lambda, \bar{\varepsilon})\,)$$

of problem (1.85) is defined on $[-r_0, \tilde{a})$ and

$$\lim_{\bar{\varepsilon} \to 0} \omega(x\,, \lambda, \bar{\varepsilon}) = \omega(x, \lambda) \text{ uniformly with respect to } x \in [-r_0, \tilde{a}). \tag{1.86}$$

We prove that

$$\zeta(x) < \omega(x\,, \lambda, \bar{\varepsilon}) \;\text{ for } x \in [-r_0, \tilde{a}) \tag{1.87}$$

where $0 < \varepsilon < \varepsilon_0$. We apply Lemma 1.45 to prove (1.87). It follows from (1.81) that the initial inequality $\zeta(x) < \lambda(x) + \bar{\varepsilon}$, $x \in [-r_0, \tilde{a}]$ is satisfied. Put

$$J_i^+ = \{\, x \in (0, a) : \;\; \zeta(t) < \omega(t,, \lambda, \bar{\varepsilon}) \text{ for } t \in [-r_0, x) \;\text{ and }\; \zeta_i(x) = \omega_i(x,\,\lambda, \bar{\varepsilon})\}.$$

We prove that

$$D_- \zeta_i(x) < \sigma_i(\,x, \zeta(x), \zeta\,) + \varepsilon \;\text{ for } x \in J_i^+. \tag{1.88}$$

Let $\bar{x} \in J_i^+$ be fixed. Then $\zeta_i(\bar{x}) > 0$ and there exists a point $\bar{y} = (\bar{y}_1, \dots, \bar{y}_n) \in S_{\bar{x}}$ such that $\zeta_i(\bar{x}) = u_i(\bar{x}, \bar{y})$ or $\zeta_i(\bar{x} = -u_i(\bar{x}, \bar{y})$. Let us consider the first case. We have

$$D_{y_j} u_i(\bar{x}, \bar{y}) \geq 0 \;\text{ for } j \in I_+[\bar{x}, \bar{y}],$$

$$D_{y_j}u_i(\bar{x},\bar{y}) \le 0 \text{ for } j \in I_-[\bar{x},\bar{y}], \quad D_{y_j}u_i(\bar{x},\bar{y}) = 0 \text{ for } j \in I_0[\bar{x},\bar{y}].$$

Let the function $\eta = (\eta_1, \dots, \eta_n) : [0,\bar{x}] \to R^n$ be given by

$$\eta_j(x) = -b_j + M_j x \text{ for } j \in I_-[\bar{x},\bar{y}],$$

$$\eta_j(x) = b_j - M_j x \text{ for } j \in I_+[\bar{x},\bar{y}], \quad \eta_j(x) = \bar{y}_j \text{ for } j \in I_0[\bar{x},\bar{y}],$$

and $\gamma(x) = u_i(x,\eta(x))$ for $x \in [0,\bar{x}]$. We have $\gamma(x) \le \zeta_i(x)$ for $x \in [0,\bar{x}]$ and $\gamma(\bar{x}) = \zeta(\bar{x})$. Therefore $D_-\zeta_i(\bar{x}) \le \gamma'(\bar{x})$ and consequently

$$D_-\zeta_i(\bar{x}) \le D_x u_i(\bar{x},\bar{y}) + \sum_{j=1}^{n} D_{y_j}u_i(\bar{x},\bar{y})\,\eta_j'(\bar{x})$$

$$\le \sigma_i\left(\bar{x}, [|u(\bar{x},\bar{y})|], Vu\right) + \sum_{j=1}^{n} M_j\,|D_{y_j}u_i(\bar{x},\bar{y})|$$

$$+ \sum_{j\in I_-[\bar{x},\bar{y}]} M_j\, D_{y_j}u_i(\bar{x},\bar{y}) - \sum_{j\in I_+[\bar{x},\bar{y}]} M_j\, D_{y_j}u_i(\bar{x},\bar{y}) \le \sigma_i\left(\bar{x}, \zeta(x), \zeta\right),$$

which proves (1.88). In a similar way we prove functional differential inequality (1.88) in the case $\zeta_i(\bar{x}) = -u_i(\bar{x},\bar{y})$. Thus, an application of Lemma 1.16 gives (1.87). From (1.86), (1.87) we obtain in the limit letting ε tend to 0, inequality $\zeta(x) \le \omega(x,\lambda)$ for $x \in [0,\tilde{a})$. By the arbitrariness of $0 < \tilde{a} < a$ estimate (1.84) follows, which completes the proof of (1.83).

1.10 Uniqueness and continuous dependence

A simple criterion that implies the uniqueness of solutions on the Haar pyramid is the following Perron type condition.

Theorem 1.47 *Suppose that the function $f : \Xi \to R^k$ satisfies the Volterra condition and*

1) there is a function $\sigma : [0,a) \times R^k_+ \times C([-r_0,a), R^k_+) \to R^k_+$ such that Assumption $\bar{H}\,[\sigma]$ is satisfied and the maximum solution $\omega(\,\cdot\,,\lambda)$ of problem (1.79) with $\lambda(x) = 0$ for $x \in [-r_0, 0]$ is $\bar{\omega}(x) = 0$, $x \in [-r_0, a)$,

2) the function f satisfies the Lipschitz condition

$$\|\, f(x,y,p,z,q) - f(x,y,p,z,\bar{q})\,\| \le \sum_{i=1}^{n} M_i |q_i - \bar{q}_i| \tag{1.89}$$

and the inequality

$$[|\, f(x,y,p,z,q) - f(x,y,\bar{p},\bar{z},q)\,|] \le \sigma\left(x, [|p-\bar{p}|], V(z-\bar{z})\right) \tag{1.90}$$

is satisfied on Ξ.

Under these assumptions the Cauchy problem (1.63), (1.64) admits at most one solution $v : E_0 \cup E \to R$ of class D.

PROOF. If v and $\bar{v}$ are solutions of problem (1.63), (1.64) and they are of class D on $E_0 \cup E$ then the function $u = v - \bar{v}$ satisfies all the assumptions of Theorem 1.46 with $\lambda(x) = 0$, $x \in [-r_0, 0]$. Since $\omega(x,0) = 0$ for $x \in [-r_0, a)$ then the assertion follows.

Remark 1.48 *Consider the following example of condition (1.90). Suppose that there is a function*

$$\tilde{\sigma} = (\tilde{\sigma}_1, \ldots, \tilde{\sigma}_k) : [0,a) \times R_+^k \times R_+^k \to R_+^k$$

such that

1) $\tilde{\sigma}$ is continuous and $\tilde{\sigma}(x,0,0) = 0$ for $x \in [0,a)$,

2) the function $\tilde{\sigma}$ is nondecreasing with respect to the last variable and the following quasi-monotonicity condition is satisfied: for each i, $1 \le i \le k$, if $p \le \bar{p}$ and $p_i = \bar{p}_i$ then $\tilde{\sigma}_i(x,p,s) \le \tilde{\sigma}_i(x,\bar{p},s)$ where $(x,s) \in [0,a) \times R_+^k$,

3) the estimate

$$[\,| f(x,y,p,z,q) - f(x,y,\bar{p},\bar{z},q) |\,] \le \tilde{\sigma}(x, p - \bar{p}, [\,\|z - \bar{z}\|_x\,])$$

is satisfied on Ξ.
Then the comparison problem (1.77) has the form

$$\omega'(x) = \tilde{\sigma}(x, \omega(x), \omega(x)), \quad \omega(0) = 0.$$

We assume in this case that the function $\bar{\omega}(x) = 0$, $x \in [0,a)$, is the maximum solution of the above Cauchy problem.

Let us consider now two problems, (1.63), (1.64) and the problem

$$D_x z(x,y) = \tilde{f}\,(x, y, z(x,y), z, D_y z(x,y)), \tag{1.91}$$

$$z(x,y) = \tilde{\varphi}(x,y) \quad \text{on} \quad E_0. \tag{1.92}$$

where $\tilde{f} : \Xi \to R^k$ and $\tilde{\varphi} : E_0 \to R^k$ are given functions. Now we prove a theorem on the continuous dependence of solutions on initial data and on the right-hand sides of systems.

Theorem 1.49 *Suppose that*

1) *the function σ satisfies Assumption $\bar{H}[\sigma]$ and the maximum solution $\omega(\cdot, \lambda)$ of problem (1.79) with $\lambda(x) = 0$ for $x \in [-r_0, 0]$ is $\bar{\omega}(x) = 0$ for $x \in [-r_0, a)$,*

2) *the functions f, $\tilde{f} : \Xi \to R^k$ satisfy the Volterra condition and estimates (1.89), (1.90) hold on Ξ,*

3) *the functions v, $\tilde{v} : E_0 \cup E \to R^k$ are solutions of problems (1.63), (1.64) and (1.91), (1.92) respectively, v and $\tilde{v}$ are of class D on $E_0 \cup E$.*

Let $0 < \tilde{a} < a$. Then for every $\varepsilon > 0$ there corresponds a $\delta > 0$ such that if

$$\| f(x,y,p,z,q) - \tilde{f}(x,y,p,z,q) \| < \delta \;\; on \;\; \Xi \tag{1.93}$$

and

$$\| \varphi(x,y) - \tilde{\varphi}(x,y) \| < \delta \;\; on \;\; E_0 \tag{1.94}$$

then

$$\| v(x,y) - \tilde{v}(x,y) \| < \varepsilon \;\; for \;\; 0 \le x < \tilde{a}, \;\; (x,y) \in E. \tag{1.95}$$

PROOF. Suppose that $0 < \tilde{a} < a$ is fixed. Let the function $\lambda = (\lambda_1, \ldots, \lambda_k) : [-r_0, 0] \to R_+^k$ be given by

$$\lambda_i(x) = \max \{ |\varphi_i(x,y) - \tilde{\varphi}_i(x,y)| : \ y \in S_x \}, \ -r_0 \le x \le 0, \ 1 \le i \le k.$$

For $\varepsilon > 0$ we can choose $\delta > 0$ such that the right-hand maximum solution $\omega(\,\cdot\,, \lambda, \bar{\delta})$ of the problem

$$\omega'(x) = \sigma\,(x, \omega(x), \omega\,) + \bar{\delta}, \ \ \omega(x) = \lambda(x) \ \text{ for } x \in [-r_0, 0],$$

where $\bar{\delta} = (\delta, \ldots, \delta) \in R^k$ and $\lambda(x) \le \bar{\delta}$ on $[-r_0, 0]$ is defined on $[-r_0, \tilde{a})$ and

$$\|\,\omega(x,\, \lambda, \delta)\,\| < \varepsilon \ \text{ for } \ x \in [-r_0, \tilde{a}). \tag{1.96}$$

Suppose that estimates (1.93), (1.94) hold true with the above chosen δ. Then we have

$$[|D_x(v - \tilde{v})(x,y)|] \le \sigma\,(x, [|v(x,y) - \tilde{v}(x,y)|], V(v - \tilde{v})\,)$$
$$+\bar{\delta} + M \circ [|D_y(v - \tilde{v})(x,y)|] \ \text{ on } \ E.$$

Hence, by Theorem 1.46 we obtain $[|\,v(x,y) - \tilde{v}(x,y)\,|] \le \omega(x,\, \lambda, \bar{\delta})$ for $0 \le x < \tilde{a}$, $(x,y) \in E$. From (1.96) follows (1.95). This completes the proof of the theorem.

1.11 Chaplygin method for initial problems

Let E_0, E be the sets defined by (1.15), (1.16) and $\Omega_0 = E \times R \times C(E_0 \cup E, R)$. Suppose that

$$f = (\,f_1, \ldots, f_n\,) : E \to R^n, \ \ F : \Omega_0 \to R, \ \ \phi : E_0 \to R,$$

are given functions. We consider the initial problem for the almost linear equation

$$D_x z(x,y) = F(\,x, y, z(x,y), z\,) + \sum_{i=1}^{n} f_i(x,y)\, D_{y_i} z(x,y), \tag{1.97}$$

$$z(x,y) = \phi(x,y) \ \text{ on } \ E_0. \tag{1.98}$$

We are interested in working out a method of approximation of a solution of problem (1.97), (1.98) by solutions of associated linear equations and in estimating of the difference between the exact and approximate solutions. This is precisely what the Chaplygin method accomplishes (see [63, 104, 158, 166, 170]).

Assumption H $[\,F, f\,]$. Suppose that

1) the function F of the variables $(x,y,p,z) \in \Omega_0$ is continuous, satisfies the Volterra condition and it is nondecreasing with respect to the functional variable,

2) the derivative D_pF exists and is continuous on Ω_0,

3) D_pF is nondecreasing with respect to p and there exist constants M, $N \ge 0$ such that $|\,D_pF(\,x,y,p,z\,)\,| \le N$ and

$$|\,F(\,x,y,p,z\,) - F(\,x,y,p,\bar{z}\,)\,| \le M\,\|\,z - \bar{z}\,\|_x \text{ on } \Omega_0$$

where $\|\cdot\|_x$ is the supremum norm in the space $C(E_x, R)$,

4) $f \in C(E, R^n)$ and

$$\sum_{i=1}^{n} f_i(x,y)\,q_i + \sum_{i\in I_-[x,y]} \alpha_i(x)\,q_i + \sum_{i\in I_+[x,y]} \beta_i(x)\,q_i \le 0,\ (x,y)\in E,\ q\in R^n,$$

where $q_i \le 0$ for $i \in I_-[x,y]$, $q_i \ge 0$ for $i \in I_+[x,y]$, $q_i = 0$ for $i \in I_0[x,y]$ and I_0, I_-, $I_+$are the functions defined in Section 1.2.

Now we formulate a monotone iterative method for problem (1.97), (1.98). For given function u, $v \in C(E_0 \cup E, R)$ we define the functions G, $H : E\times R\times R^n \to R$ as follows

$$G(x,y,p,q;u) = F(x,y,u(x,y),u)$$
$$+D_pF(x,y,u(x,y),u)(p-u(x,y)) + \sum_{i=1}^{n} f_i(x,y)\,q_i,$$

and

$$H(x,y,p,q;u,v) = F(x,y,v(x,y),v)$$
$$+\frac{F(x,y,u(x,y),v) - F(x,y,v(x,y),v)}{u(x,y)-v(x,y)}\,(p - v(x,y))$$
$$+\sum_{i=1}^{n} f_i(x,y)\,q_i \ \text{ if } \ u(x,y)-v(x,y)\neq 0,$$

and

$$H(x,y,p,q;u,v) = F(x,y,v(x,y),v) + D_pF(x,y,v(x,y),v)\,(p-v(x,y))$$
$$+\sum_{i=1}^{n} f_i(x,y)\,q_i \ \text{ if } \ u(x,y)-v(x,y) = 0.$$

Suppose that we have a sequence

$$\left\{u^{(m)}, v^{(m)}\right\},\quad u^{(m)},\ v^{(m)} : E_0\cup E \to R, \tag{1.99}$$

where $u^{(m)}$, $v^{(m)}$ are of class D and

(i) for every $m \ge 1$ the function $u^{(m)}$ is a solution on E of the linear equation

$$D_x z(x,y) = G(x,y,z(x,y),D_y z(x,y);u^{(m-1)}), \tag{1.100}$$

(ii) for every $m \ge 1$ the function $v^{(m)}$ satisfies on E the linear equation

$$D_x z(x,y) = H(x,y,z(x,y),D_y z(x,y);u^{(m-1)},v^{(m-1)}). \tag{1.101}$$

The sequence (1.99) which satisfies the above conditions will be called the Chaplygin sequence. Let F^* be the Niemycki operator corresponding to equation (1.97):

$$F^*[z](x,y) = F(x,y,z(x,y),z) + \sum_{i=1}^{n} f_i(x,y)\,D_{y_i} z(x,y).$$

Now we prove a theorem on the monotonicity of the Chaplygin sequence.

Theorem 1.50 *Suppose that Assumption $H\,[\,F, f\,]$ is satisfied and*

1) $\bar{z} : E_0 \cup E \to R$ is a solution of problem (1.97), (1.98) and $\bar{z}$ is of class D,

2) the functions $u^{(0)}$, $v^{(0)} : E_0 \cup E \to R$ are of class D and the differential inequalities

$$D_x u^{(0)}(x,y) \le F^*[\,u^{(0)}\,](x,y), \tag{1.102}$$

$$D_x v^{(0)}(x,y) \ge F^*[\,v^{(0)}\,](x,y) \tag{1.103}$$

are satisfied on E,

3) the Chaplygin sequence (1.99) satisfies the initial estimates

$$u^{(m-1)}(x,y) \le u^{(m)}(x,y) \le \bar{z}(x,y) \le v^{(m)}(x,y) \le v^{(m-1)}(x,y) \;\; on \;\; E_0$$

for $m \ge 1$.

Under these assumptions we have

$$u^{(m-1)}(x,y) \le u^{(m)}(x,y) \le \bar{z}(x,y) \le v^{(m)}(x,y) \le v^{(m-1)}(x,y) \;\; on \;\; E \tag{1.104}$$

and

$$D_x u^{(m)}(x,y) \le F^*[\,u^{(m)}\,](x,y) \;\; on \;\; E, \tag{1.105}$$

$$D_x v^{(m)}(x,y) \ge F^*[\,v^{(m)}\,](x,y) \;\; on \;\; E \tag{1.106}$$

for $m \ge 1$.

Proof. From (1.100), (1.102) it follow that

$$D_x u^{(0)}(x,y) \le G(\,x, y, u^{(0)}(x,y), D_y u^{(0)}(x,y); u^{(0)}\,) \text{ on } E. \tag{1.107}$$

Since $(u^{(0)} - u^{(1)})(x,y) \le 0$ on E_0, we obtain by a theorem on differential inequalities (see [195], Theorem 59.1) that

$$u^{(0)}(x,y) \le u^{(1)}(x,y) \text{ on } E. \tag{1.108}$$

In a similar way we prove that

$$v^{(1)}(x,y) \le v^{(0)}(x,y) \text{ on } E. \tag{1.109}$$

Now we prove the differential inequality

$$D_x u^{(1)}(x,y) \le F^*[\,u^{(1)}\,](x,y) \text{ on } E. \tag{1.110}$$

Inequality (1.108) together with the monotonicity condition for F and assumption 2) of our theorem imply

$$F^*[\,u^{(1)}\,] - D_x u^{(1)}(x,y) \ge F(\,x, y, u^{(1)}(x,y), u^{(0)}\,) + \sum_{i=1}^{n} f_i(x,y)\, D_{y_i} u^{(1)}(x,y)$$

$$-F(\,x, y, u^{(0)}(x,y), u^{(0)}\,) - D_p F(\,x, y, u^{(0)}(x,y), u^{(0)}\,)\left(u^{(1)}(x,y) - u^{(0)}(x,y)\right)$$

$$-\sum_{i=1}^{n} f_i(x,y) D_{y_i} u^{(1)}(x,y) \geq 0, \ (x,y) \in E,$$

which proves (1.110). The function $u^{(1)}$ and $\bar{z}$ fulfil the assumptions of Theorem 1.6. Then we obtain

$$u^{(1)}(x,y) \leq \bar{z}(x,y) \text{ on } E. \tag{1.111}$$

In a similar way we get the inequality

$$u^{(0)}(x,y) \leq v^{(0)}(x,y) \text{ on } E. \tag{1.112}$$

Hence,

$$D_x u^{(0)}(x,y) \leq F^*[u^{(0)}](x,y) \leq H(x,y,u^{(0)}(x,y), D_y u^{(0)}(x,y); u^{(0)}, v^{(0)}) \text{ on } E,$$

and consequently, we have, applying the theorem on partial differential inequalities ([195], Theorem 59.1) that

$$u^{(0)}(x,y) \leq v^{(1)}(x,y) \text{ on } E. \tag{1.113}$$

The function F is convex with respect to p and it is nondecreasing with respect to the functional variable. It follows from (1.108), (1.109), (1.111) that

$$D_x v^{(1)}(x,y) \geq F^*[v^{(1)}](x,y) \text{ on } E. \tag{1.114}$$

From the initial inequalities for $m = 1$ and from (1.114) we obtain

$$\bar{z}(x,y) \leq v^{(1)}(x,y) \text{ on } E. \tag{1.115}$$

It follows from (1.108) - (1.111), (1.114), (1.115) that assertions (1.104) are satisfied for $m = 1$. The proof of (1.104) for $m \geq 2$ is simple, it runs by induction. This proves Theorem 1.50.

In the next theorem we take up the problem of the convergence of the Chaplygin sequence to the exact solution. We also give estimates of the difference between the exact and approximate solutions.

Theorem 1.51 *Suppose that all the assumptions of Theorem 1.50 are satisfied and there are the sequences* $\{\varepsilon_m\}$, $\{\bar{\varepsilon}_m\}$ *such that*

(i) for $(x,y) \in E_0 \cup E$ *we have*

$$(\bar{z} - u^{(0)})(x,y) \leq \varepsilon_0 \ \text{and} \ (v^{(0)} - \bar{z})(x,y) \leq \bar{\varepsilon}_0, \tag{1.116}$$

(ii) for $(x,y) \in E_0$ *and* $m \geq 1$ *we have*

$$(\bar{z} - u^{(m)})(x,y) \leq \varepsilon_m \ \text{and} \ (v^{(m)} - \bar{z})(x,y) \leq \bar{\varepsilon}_m,$$

and

$$\lim_{m\to\infty} \varepsilon_m = 0, \quad \lim_{m\to\infty} \bar{\varepsilon}_m = 0.$$

Under these assumptions

$$\lim_{m\to\infty} u^{(m)}(x,y) = \bar{z}(x,y) \ \text{and} \ \lim_{m\to\infty} v^{(m)}(x,y) = \bar{z}(x,y) \tag{1.117}$$

uniformly on E and

$$(\bar{z} - u^{(m)})(x,y) \leq h_m(x), \quad (v^{(m)} - \bar{z})(x,y) \leq \bar{h}_m(x) \tag{1.118}$$

where $(x,y) \in E$ and

$$h_m(x) = \exp(Nx) \sum_{i=0}^{m} \frac{(2N+M)^i x^i}{i!} \varepsilon_{m-i},$$

$$\bar{h}_m(x) = \exp(Nx) \sum_{i=0}^{m} \frac{(2N+M)^i x^i}{i!} \bar{\varepsilon}_{m-i}.$$

PROOF. We prove (1.118) by induction on m. It follows that the function $\bar{z} - u^{(1)}$ satisfies on E the differential inequality

$$D_x(\bar{z} - u^{(1)})(x,y) \leq N(\bar{z} - u^{(1)})(x,y) + \varepsilon_0(2N+M)\exp(Nx) + \sum_{i=1}^{n} f_i(x,y)\, D_{y_i}(\bar{z} - u^{(1)})(x,y),$$

and $(\bar{z}-u^{(1)})(0,y) \leq \varepsilon_1$ for $y \in [-b,b]$. Then we obtain from a comparison theorem for partial differential inequalities that $(\bar{z}-u^{(1)})(x,y) \leq h_1(x)$ on E. Suppose now that inequality

$$(\bar{z} - u^{(m)})(x,y) \leq h_m(x) \text{ on } E \tag{1.119}$$

is true for fixed $m \geq 1$. Then the function $\bar{z} - u^{(m+1)}$ satisfies on E the differential inequality

$$D_x(\bar{z} - u^{(m+1)})(x,y) \leq N(\bar{z} - u^{(m+1)})(x,y) + (2N+M)h_m(x) + \sum_{i=1}^{n} f_i(x,y)\, D_{y_i}(\bar{z} - u^{(m+1)})(x,y),$$

and the initial estimate $(\bar{z} - u^{(m+1)})(0,y) \leq \varepsilon_{m+1}$ holds for $y \in [-b,b]$. The above relations together with the comparison theorem, imply

$$(\bar{z} - u^{(m+1)})(x,y) \leq \tilde{h}_{m+1}(x) \text{ on } E,$$

where $\tilde{h}_{m+1}$ is the solution of

$$\omega'(x) = N\,\omega(x) + (2N+M)\,h_m(x), \quad \omega(0) = \varepsilon_{m+1}.$$

Because $\tilde{h}_{m+1}(x) = h_{m+1}(x)$, we obtain estimate (1.119) for $m+1$. Thus , by induction, relation (1.119) is true for all $m \in \mathbf{N}$. In the same way we prove the second inequality in (1.118). Since $\lim_{m\to\infty} h_m(x) = 0$, $\lim_{m\to\infty} \bar{h}_m(x) = 0$ uniformly on $[0,a)$, we get (1.117. This proves Theorem 1.51.

Remark 1.52 *If $(v^{(0)} - u^{(0)})(x,y) \leq K$ on E then estimates (1.116) are satisfied with $\varepsilon_0 = \bar{\varepsilon}_0 = K$.*

Remark 1.53 *The right-hand sides of equations (1.100), (1.101) are obtained by linearization of F with respect to the classical argument denoted by p. We consider the Chaplygin method with the linearization with respect to the functional variable in Chapter V.*

Chapter 2

Existence of Solutions on the Haar Pyramid

2.1 Introduction

Let E be the Haar pyramid

$$E = \{ (x, y) \in R^{1+n} : x \in [0, a], \quad -b + M x \le y \le b - M x \},$$

where b, $M \in R^n_+$, $b = (b_1, \ldots, b_n)$, $M = (M_1, \ldots, M_n)$ and $b - M a > 0$. Write

$$E_0 = [-r_0, 0] \times [-b, b] \text{ where } r_0 \in R_+, \quad \Omega = E \times C(E_0 \cup E, R) \times R^n$$

and

$$\tilde{E}_x = E \cap ([0, x] \times R^n), \quad E_x = (E_0 \cup E) \cap ([-r_0, x] \times R^n) \quad 0 \le x \le a.$$

Suppose that $f : \Omega \to R$ and $\varphi : E_0 \to R$ are given functions. Consider the Cauchy problem

$$D_x z(x, y) = f(x, y, z, D_y z(x, y)), \tag{2.1}$$

$$z(x, y) = \varphi(x, y) \text{ on } E_0, \tag{2.2}$$

where $D_y z = (D_{y_1} z, \ldots, D_{y_n} z)$. In this Chapter we consider classical solutions of problem (2.1), (2.2). We assume that the operator f satisfies the Volterra condition.

Now we present relations between local and global (with respect to y) initial problems for differential and functional differential equations. Let E be the Haar pyramid and suppose that $\tilde{F} : E \times R \times R^n \to R$ and $\tilde{\varphi} : [-b, b] \to R$ are given functions. Consider the Cauchy problem without the functional dependence

$$D_x z(x, y) = \tilde{F} (x, y, z(x, y), D_y z(x, y)), \quad z(0, y) = \tilde{\varphi}(y) \text{ for } y \in [-b, b]. \tag{2.3}$$

We formulate now the following assumption on $\tilde{F}$.

Assumption $\tilde{H}$. Suppose that

1) the function $\tilde{F}$ of the variables (x, y, p, q) and its partial derivatives

$$D_y\tilde{F} = (D_{y_1}\tilde{F}, \ldots, D_{y_n}\tilde{F}), \quad D_p\tilde{F}, \quad D_q\tilde{F} = (D_{q_1}\tilde{F}, \ldots, D_{q_n}\tilde{F})$$

are continuous on $E \times R^{1+n}$ and

$$|D_{q_i}\tilde{F}(x, y, p, q)| \leq M_i \text{ on } E \times R^{1+n} \text{ for } 1 \leq i \leq n,$$

where the constants $(M_1, \ldots, M_n)$ appear in the definition of E,

2) the functions $\tilde{F}$, $D_y\tilde{F}$, $D_p\tilde{F}$ are bounded and $D_y\tilde{F}$, $D_p\tilde{F}$, $D_q\tilde{F}$ satisfy the Lipschitz condition with respect to (y, p, q) on $E \times R^{1+n}$,

3) the function $\tilde{\varphi}$ together with its derivatives $(D_{y_1}\tilde{\varphi}, \ldots, D_{y_n}\tilde{\varphi}) = D_y\tilde{\varphi}$ are continuous on $[-b, b]$ and $D_y\tilde{\varphi}$ satisfies the Lipschitz condition.

Lemma 2.1 *If Assumption $\tilde{H}$ is satisfied then there exists exactly one solution $\tilde{u}$ of problem (2.3) on $\tilde{E}_c$ for sufficiently small $c \in (0, a]$. The solution $\tilde{u}$ is of class C^1 and $D_y\tilde{u}$ satisfies the Lipschitz condition with respect to y on $\tilde{E}_c$. Moreover, the solution $\tilde{u}$ depends continuously on given functions.*

We only give the main ideas of the proof of the Lemma. The proof of the existence of a solution of (2.3) is divided into two steps.

I. Assume additionally that $\tilde{F}$ and $\tilde{\varphi}$ are of class C^2 on $E \times R^{1+n}$ and $[-b, b]$ respectively. Consider the characteristic system corresponding to (2.3)

$$y'(x) = -D_q\tilde{F}(x, y(x), p(x), q(x)),$$

$$p'(x) = \tilde{F}(x, y(x), p(x), q(x)) - \sum_{i=1}^{n} q_i(x) D_{q_i}\tilde{F}(x, y(x), p(x), q(x)),$$

$$q'(x) = D_y\tilde{F}(x, y(x), p(x), q(x)) + q(x) D_p\tilde{F}(x, y(x), p(x), q(x))$$

and its solutions $\tilde{y}(\cdot, \eta)$, $\tilde{p}(\cdot, \eta)$, $\tilde{q}(\cdot, \eta)$, $\eta \in [-b, b]$, satisfying the initial conditions

$$y(0) = \eta, \; p(0) = \tilde{\varphi}(\eta), \; q(0) = D_y\tilde{\varphi}(\eta), \quad \eta \in [-b, b].$$

Let us denote by

$$\tilde{t}(\cdot, \eta) = [\tilde{t}_{ij}(\cdot, \eta)]_{i,j=1,\ldots,n}$$

the characteristics of the second order corresponding to (2.3). They are solutions of a system of ordinary differential equations and satisfy the initial conditions

$$t_{ij}(0) = D_{y_i}D_{y_j}\tilde{\varphi}(\eta), \quad 1 \leq i, j \leq n, \quad \eta \in [-b, b].$$

The right-hand sides of the system are polynomials of the second order with respect to t_{ij}. Details can be found in [177, 193]. We next claim that the equation $y = \tilde{y}(x, \eta)$ can be solved with respect to η. Let $\eta = \tilde{\eta}(x, y)$ be this solution. Define $\tilde{u}(x, y) = \tilde{p}(x, \tilde{\eta}(x, y))$. Then there is $c \in (0, a]$ such that

(i) $\tilde{u}$ is of class C^2 on $\tilde{E}_c$,

(ii) $D_y\tilde{u}(x, y) = \tilde{q}(x, \tilde{\eta}(x, y))$ and $D_{y_i}D_{y_j}\tilde{u}(x, y) = \tilde{t}_{ij}(x, \tilde{\eta}(x, y))$, $1 \leq i, j \leq n$, on $\tilde{E}_c$,

(iii) $\tilde{u}$ is the solution of (2.3) on $\tilde{E}_c$ and the functions $\tilde{q}(\cdot, \eta)$, $\tilde{t}(\cdot, \eta)$ are bounded.

II. Now we consider the original assumptions on $\tilde{F}$ and $\tilde{\varphi}$. Let $\{\tilde{F}^{(k)}\}$ and $\{\tilde{\varphi}^{(k)}\}$ be the sequences of functions uniformly convergent to $\tilde{F}$ and $\tilde{\varphi}$ and satisfying **I**. Let us denote be $\{\tilde{u}^{(k)}\}$ the sequence of solutions of corresponding initial problems. All these solutions are given on $\tilde{E}_c$ with $c \in (0, a]$ sufficiently small and independent on k. There exists a subsequence $\{\tilde{u}^{(k_i)}\}$ and a function $\tilde{u}$ such that

$$\tilde{u} = \lim_{i\to\infty} \tilde{u}^{(k_i)}, \quad D_x\tilde{u} = \lim_{i\to\infty} D_x\tilde{u}^{(k_i)}, \quad D_y\tilde{u} = \lim_{i\to\infty} D_y\tilde{u}^{(k_i)}$$

uniformly on $\tilde{E}_c$. This $\tilde{u}$ satisfies all the conditions of the Lemma. Uniqueness and continuous dependence of the solution $\tilde{u}$ on given functions can be proved using classical theorems on differential inequalities.

Thus we see that initial problems for nonlinear equations have the following property: the proof of the existence of classical solutions of problem (2.3) (or adequate hyperbolic systems) and the existence results for the Cauchy problem which is global with respect to y ([118, 177]) are based on the same ideas.

Let us note that generalized solutions in the Cinquini Cibrario sense also have the same property ([44, 51, 52]).

The situation is completely different for equations (or systems) with the functional dependence. We discuss the problem. We will consider equations with the generalized Hale operator. This model of the functional dependence is the most suitable in existence theorems for the global Cauchy problem. Let $B = [-r_0, 0] \times [-r, r]$ where $r_0 \in R_+$ and $r = (r_1, \ldots, r_n) \in R^n_+$. For a function $z : [-r_0, a] \times R^n \to R$ and for a point $(x, y) \in [0, a] \times R^n$ we define a function $z_{(x,y)} : B \to R$ by

$$z_{(x,y)}(t, s) = z(x + t, y + s), \quad (t, s) \in B. \tag{2.4}$$

Suppose that $f : [0, a] \times R^n \times C(B, R) \times R^n$ and $\psi : [-r_0, 0] \times R^n$ are given functions. Consider the initial problem global with respect to y

$$D_x z(x, y) = f(\, x, y, z_{(x,y)}, D_y z(x, y)\,) \tag{2.5}$$

$$z(x, y) = \psi(x, y) \quad \text{on} \quad [-r_0, 0] \times R^n. \tag{2.6}$$

Classical differential equations, differential equations with a deviated argument, differential integral equations can be derived from (2.5) by specializing the operator f. The papers [109, 110] contain a survey of results for problem (2.5), (2.6), see also [72].

It is evident from (2.4) that the formulation (2.5), (2.6) is not suitable for the local Cauchy problem considered in the Haar pyramid and consequently, the results of papers [98, 146, 149] are not applicable to problem (2.1), (2.2). It is easy to see that results of the paper [200] can not be applicable to equations with a deviated argument and to differential integral equations of the Volterra type. Until now there are not any results on the existence of classical or generalized solutions to problem (2.1), (2.2). The aim of the next Sections is to prove a theorem on the existence of classical solutions of the problem.

The proof of the existence of a solution to problem (2.1), (2.2) is based on the following idea. We construct the set X_c which is the closed subset of the Banach space consisting of all functions $z \in C(\tilde{E}_c, R)$, $0 < c \leq a$. For $u \in X_c$ consider the initial problem

$$D_x z(x,y) = f(x,y,\tilde{u},D_y z(x,y)), \quad z(0,y) = \varphi(0,y) \text{ for } y \in [-b,b], \tag{2.7}$$

where $\tilde{u}(x,y) = u(x,y)$ on $\tilde{E}_c$ and $\tilde{u}(x,y) = \varphi(x,y)$ on E_0. Let $\tilde{v}(\,\cdot\,;u)$ denote the solution of (2.7). We formulate sufficient conditions for the existence and uniqueness of the solution $\tilde{v}(\,\cdot\,;u)$ of the above problem. We consider the operator U defined on X_c as follows: $Uu = \tilde{v}(\,\cdot\,;u)$. We prove that under suitable assumptions on f and φ there is $0 < c \leq a$ such that $U : X_c \to X_c$ and U has exactly one fixed point $\bar{u}$ on X_c. Let $\bar{v} : E_c \to R$ be a function given by $\bar{v} = \bar{u}$ on $\tilde{E}_c$ and $\bar{v} = \varphi$ on E_0. This $\bar{v}$ is the classical solution of (2.1), (2.2). We use the ideas introduced in [193] for hyperbolic systems and generalized Cauchy problems.

Write $\Omega_0 = E \times R^n$ and suppose that $G : \Omega_0 \to R$ and $\omega : [-b,b] \to R$ are given functions. Consider the nonlinear equation

$$D_x z(x,y) = G(\,x,y,D_y z(x,y)\,) \tag{2.8}$$

with the initial condition

$$z(0,y) = \omega(y) \text{ for } y \in [-b,b]. \tag{2.9}$$

The above method of the proof of the existence result for (2.1), (2.2) is based on the theorem on the existence of the solution of problem (2.8), (2.9) and on the estimates of the partial derivatives of the solution and the Lipschitz coefficients for the derivatives. Now we state the auxiliary theorem.

Theorem 2.2 *Suppose that*

1) the function G of the variables (x,y,q) is continuous and bounded on Ω_0,

2) the derivatives D_yG, D_qG exist on Ω_0 and $D_yG, D_qG \in C(\Omega_0, R^n)$,

3) there are $\bar{P}$, $P_0 \in R_+$ such that

$$\|\, D_yG(x,y,q)\,\| \leq \bar{P},$$

$$\|\, D_yG(x,y,q) - D_yG(x,\bar{y},\bar{q})\,\| \leq P_0\,[\,\|y-\bar{y}\| + \|q-\bar{q}\|\,],$$

$$\|\, D_qG(x,y,q) - D_qG(x,\bar{y},\bar{q})\,\| \leq P_0\,[\,\|y-\bar{y}\| + \|q-\bar{q}\|\,]$$

and

$$|\, D_{q_i}G(x,y,q)\,| \leq M_i, \quad i = 1,\ldots,n,$$

on Ω_0,

4) the function $\omega : [-b,b] \to R$ is of class C^1 on R^n and

$$\|\, D_y\omega(y)\,\| \leq \bar{J}, \quad \|\, D_y\omega(y) - D_y\omega(\bar{y})\,\| \leq J\,\|y-\bar{y}\| \text{ on } [-b,b].$$

Under these assumptions there exists the unique solution $v : \tilde{E}_\delta \to R$ of problem (2.8), (2.9) where

$$\delta = \min\left\{\, a, \ \frac{1}{P_0(1+J)}\,\right\}.$$

Moreover, the solution v satisfies the conditions

$$\| D_y v(x,y) - D_y v(x,\bar{y}) \| \leq \Gamma(x) \, \|y - \bar{y}\|,$$

$$\| D_y v(x,y) - D_y v(\bar{x},y) \| \leq \left[\bar{P} + P^* \Gamma(x) \right] |x - \bar{x}|$$

and

$$\| D_y v(x,y) \| \leq \bar{J} + \bar{P} x$$

on $\tilde{E}_\delta$ where

$$\Gamma(x) = \frac{P_0(1+J)x + J}{1 - P_0(1+J)x}, \qquad P^* = \sum_{j=1}^{n} M_j.$$

We only give the main ideas of the proof of the theorem.

I. We first assume additionally that G and ω are of class C^2 on Ω_0 and $[-b, b]$ respectively. Then the assumptions on the Lipschitz condition for D_yG, D_qG and $D_y\omega$ mean that adequate derivatives of the second order are bounded. Let us denote by

$$\tilde{y}(\,\cdot\,,\eta), \;\; \tilde{p}(\,\cdot\,,\eta), \;\; \tilde{q}(\,\cdot\,,\eta), \;\; \tilde{t}(\,\cdot\,,\eta) = [\, t_{ij}(\,\cdot\,,\eta) \,]_{i,j=1,\dots,n}$$

the solution of the system of ordinary differential equations consisting of the characteristic system

$$y'(x) = -D_q G(\, Q(x)\,),$$

$$p'(x) = G(\,Q(x)\,) - \sum_{i=1}^{n} q_i(x) D_{q_i} G(\,Q(x)\,), \;\; q'(x) = D_y G(\,Q(x)\,)$$

where $Q(x) = (x, y(x), q(x))$ and the system

$$t'_{ij}(x) = \sum_{k,r=1}^{n} D_{q_k} D_{q_r} G(\,Q(x)\,)\, t_{ki}(x)\, t_{rj}(x) + D_{y_i} D_{y_j} G(\,Q(x)\,) \tag{2.10}$$

$$+ \sum_{k=1}^{n} \left[D_{q_k} D_{y_i} G(\,Q(x)\,) t_{kj}(x) + D_{q_k} D_{y_j} G(\,Q(x)\,) t_{ki}(x) \right], \;\; 1 \leq i, j \leq n,$$

satisfying the initial conditions

$$y(0) = \eta, \; p(0) = \omega(\eta), \;\; q(0) = D_y\omega(\eta),$$

$$t(0) = [\, t_{ij}(0) \,]_{i,j=1,\dots,n} = \left[\, D_{y_i} D_{y_j} \omega(\eta) \,\right]_{i,j=1,\dots,n},$$

where $\eta \in [-b, b]$. It follows that the equation $y = \tilde{y}(x,\eta)$ has the unique solution with respect to η: $\eta = \tilde{\eta}(x,y)$ and $\tilde{\eta}$ is of class C^1 on $\tilde{E}_\delta$. Write $v(x,y) = \tilde{p}(x,\tilde{\eta}(x,y))$, $(x,y) \in \tilde{E}_\delta$. Then $D_y v(x,y) = \tilde{q}(x,\tilde{\eta}(x,y))$ and

$$D_{yy} v(x,y) \;=\; \left[\, D_{y_i} D_{y_j} v(x,y) \,\right]_{i,j=1,\dots,n} = \tilde{t}(x,\tilde{\eta}(x,y)) \;\text{ on }\; \tilde{E}_\delta$$

and v is the solution of problem (2.8), (2.9). Moreover v is of class C^2 on $\tilde{E}_\delta$. Write

$$\beta(x,\eta) = \max_{1\leq j\leq n} \sum_{k=1}^{n} |\, \tilde{t}_{kj}(x,\eta)\,|.$$

It follows from (2.10) that β satisfies the differential inequality

$$D_-\beta(x,\eta) \leq P_0\ [\,\beta(x,\eta)+1\,]^2 \quad \text{and} \quad \beta(0,\eta)\leq J,$$

where D_- is the left-hand lower Dini derivative with respect to x. Let $\tilde{\omega}$ denote the solution of the problem

$$\zeta'(x) = P_0\ [\,\zeta(x)+1\,]^2, \quad \zeta(0)=J.$$

Then $\beta(x,\eta)\leq \tilde{\omega}(x)$. Since $\tilde{\omega}(x)=\Gamma(x)$ it follows that $\beta(x,\eta)\leq\Gamma(x)$ and consequently

$$\|\, D_{yy}v(x,y)\,\| \leq \Gamma(x) \quad \text{on} \quad \tilde{E}_\delta. \tag{2.11}$$

In a similar way we prove that

$$\|\, D_{y}v(x,y)\,\| \leq \bar{J}+\bar{P}x \quad \text{on} \quad \tilde{E}_\delta. \tag{2.12}$$

It is easily seen that

$$\|\, D_yD_xv(x,y)\,\| \leq \|\, D_yG(x,y,D_yv(x,y))\,\|$$

$$+\sum_{i,k=1}^{n} |\, D_{q_k}G(x,y,D_yv(x,y))\, D_{y_i}D_{y_k}v(x,y)\,|\,.$$

According to (2.11), we have

$$\|\, D_yD_xv(x,y)\,\| \leq \bar{P}+P^*\,\Gamma(x) \quad \text{on} \quad \tilde{E}_\delta. \tag{2.13}$$

II. Consider now the original assumptions on G and ω. There are sequences $\{\, G^{(k)}\,\}$ and $\{\,\omega^{(k)}\,\}$ such that

(i) $F^{(k)}$ and $\omega^{(k)}$ satisfy all the assumption of Theorem 2.2,

(ii) $F^{(k)}$ and $\omega^{(k)}$ are of class C^2 on Ω_0 and $[-b,\, b]$ respectively,

(iii) $\lim_{k\to\infty} G^{(k)} = G$ uniformly on Ω_0 and $\lim_{k\to\infty}\omega^{(k)} = \omega$ uniformly on $[-b,\, b]$.

Let $\{\, v^{(k)}\,\}$ be the sequence of solutions of adequate differential problems. The functions $v^{(k)}$ are defined on $\tilde{E}_\delta$ and satisfy (2.11) - (2.13). There is a subsequence $\{\, v^{(k_i)}\,\}$ and a function v such that

$$\lim_{i\to\infty} v^{(k_i)} = v, \quad \lim_{i\to\infty} D_xv^{(k_i)} = D_xv, \quad \lim_{i\to\infty} D_yv^{(k_i)} = D_yv$$

uniformly on $\tilde{E}_\delta$. This v satisfies all the conditions of Theorem 2.2.

Lemma 2.3 *Suppose that all the assumptions of the Theorem 2.2 are satisfied and*

1) $|G(0,y,q)| \leq \bar{J}$ *on* $[-b,b] \times R^n$,

2) the derivative $D_x G$ *exists on* Ω_0 *and*

$$|D_x G(x,y,q)| \leq P_0 \quad on \ \Omega_0.$$

Then the solution v *of problem (2.8), (2.9) satisfies the conditions*

$$|D_x v(x,y)| \leq \bar{J} + \bar{P}x$$

and

$$|D_x v(x,y) - D_x v(x,\bar{y})| \leq \big[\,\bar{P} + P^*\Gamma(x)\,\big]\ \|y - \bar{y}\|,$$

$$|D_x v(x,y) - D_x v(\bar{x},y)| \leq \big[\,\bar{P} + P^*\left(\bar{P} + P^*\Gamma(x)\right)\big]\ |x - \bar{x}|$$

on $\tilde{E}_\delta$.

We give comments on the proof of the lemma. Let us first consider a reduced form of the lemma. Suppose that G and ω are of class C^2. Write $u = D_x v$. Then

$$|\,D_x u(x,y)\,| \leq P_0 + \sum_{i=1}^{n} M_i\ |\,D_{y_i} u(x,y)\,| \quad \text{on} \ \ \tilde{E}_\delta$$

and $|\,u(0,y)\,| \leq \bar{J}$ for $y \in [-b,\,b]$. It follows from a theorem on partial differential inequalities on the Haar pyramid ([195], Th. 73.1, see also [140], Th. 9.2.1) that

$$|\,D_x v(x,y)\,| \leq \bar{J} + \bar{P}x \ \ \text{on} \ \ \tilde{E}_\delta. \tag{2.14}$$

It is easily seen from (2.8) that

$$\|\,D_y D_x v(x,y)\,\| \leq \bar{P} + P^*\,\Gamma(x) \tag{2.15}$$

and

$$|\,D_x D_x v(x,y)\,| \leq \bar{P} + P^*(\,\bar{P} + P^*\,\Gamma(x)\,) \tag{2.16}$$

on $\tilde{E}_\delta$. We can now proceed analogously to the step II of the proof of Theorem 2.2. Details are omitted.

2.2 Function spaces

We will denote by $\|\ \|_{(x;0)}$ the supremum norm in the space $C(E_x,R)$. For any $x \in [0,a]$ we consider the following subspaces of the space $C(E_x,R)$. Let $C^{0.L}(E_x,R)$ be the class of all $z \in C(E_x,R)$ such that

$$[\|\,z\,\|]_{(x;L)} = \sup\left\{ \frac{|z(t,s) - z(\bar{t},\bar{s})|}{|t-\bar{t}| + \|s-\bar{s}\|} \ : \ (t,s),(t,\bar{s}) \in E_x \right\}.$$

We will use the symbol $\|\ \|_{(x;0.L)}$ to denote the norm in the space $C^{0.L}(E_x,R)$ and we put

$$\|\,z\,\|_{(x;0.L)} = \|\,z\,\|_{(x;0)} + [\|\,z\,\|]_{(x;L)}.$$

Let us denote by $C^1(E_x, R)$ the class of all continuous functions $z : E_x \to R$ such that the derivatives $D_x z$, $D_y z$ exist on E_x and $D_x z \in C(E_x, R)$, $D_y z \in C(E_x, R^n)$. For $z \in C^1(E_x, R)$ we put

$$\| z \|_{(x;1)} = \| z \|_{(x;0)} + \|D_x z\|_{(x;0)} + \| D_y z \|_{(x;0)},$$

where

$$\| D_y z \|_{(x;0)} = \max \{ \| D_y z(t,s) \| : \ (t,s) \in E_x \} .$$

Let $C^{1.L}(E_x, R)$ denote the class of all functions $z \in C^1(E_x, R)$ such that

$$\| z \|_{(x;1.L)} < +\infty$$

where

$$\| z \|_{(x;1.L)} = \| z \|_{(x;1)} + [|D_x z|]_{(x;L)} + [| D_y z |]_{(x;L)}$$

and

$$[| D_y z |]_{(x;L)} = \sup \left\{ \frac{\| D_y z(t,s) - D_y z(\bar{t},\bar{s}) \|}{|t-\bar{t}| + \| s - \bar{s} \|} : \ (t,s), \ (\bar{t},\bar{s}) \in E_x \right\}.$$

If $x = 0$ then $C(E_x, R)$ is the space of all initial functions for problem (2.1), (2.2). For simplicity of notation we use the symbols $\| \cdot \|_0$, $\| \cdot \|_{0.L}$, $\| \cdot \|_1$, $\| \cdot \|_{1.L}$ to denote the norms in the spaces

$$C(E_0, R), \ \ C^{0.L}(E_0, R), \ \ C^1(E_0, R), \ \ C^{1.L}(E_0, R)$$

respectively. The symbol $\| \cdot \|_0$ will also denote the supremum norm in the space $C(E_0, R^n)$. Let us denote by $CL(E_x, R)$ the set of all linear and continuous operators defined on $C(E_x, R)$ and taking values in R. The norm in the space $CL(E_x, R)$ will be denoted by $\| \cdot \|_{*;x}$. We will prove that under suitable assumptions on f and φ and for sufficiently small $c \in (0, a]$ there exists a solution $\bar{z}$ of problem (2.1), (2.2) such that $\bar{z} \in C^{1.L}(E_c, R)$.

2.3 Existence of classical solutions

We start with the formulation of assumptions on f and φ.

Assumption H $[f]$. Suppose that

1) the function $f : \Omega \to R$ is continuous and there exists a nondecreasing function $\alpha : R_+ \to R_+$ such that

$$\|f(x,y,z,q)\| \le \alpha(\|z\|_{(x;0)}) \ \text{ on } \Omega,$$

2) for each $P = (x, y, w, q) \in E \times C^1(E_x, R) \times R^n$ there exist the derivatives

$$D_x f(P), \ D_y f(P) = (D_{y_1} f(P), \ldots, D_{y_n} f(P)),$$

$$D_q f(P) = (D_{q_1} f(P), \ldots, D_{q_n} f(P))$$

and the functions $D_x f$, $D_y f$, $D_q f$ are continuous on $E \times C^1(E_0 \cup E, R) \times R^n$,

3) for each $P \in E \times C^1(E_x, R) \times R^n$ there exists the Fréchet derivative $D_z f(P)$ and $D_z f(P) \in CL(E_x, R)$,

4) there exist positive constants d_0, d_1, d such that

$$|D_x f(P)|, \ \|D_y f(P)\| \leq d_0 + d_1 \|z\|_{(x;1)}, \ \|D_z f(P)\|_{*;x} \leq d,$$

and

$$|\, D_{q_i} f(P) \,| \leq M_i \quad \text{for} \ \ i = 1, \ldots, n,$$

where $P \in E \times C^1(E_x, R) \times R^n$,

5) there exists a nondecreasing function $\beta : R_+ \to R_+$ such that the terms

$$\|D_y f(x, y, z, q) - D_y f(x, \bar{y}, z + h, \bar{q})\|, \ \|D_q f(x, y, z, q) - D_q f(x, \bar{y}, z + h, \bar{q})\|,$$

$$\|D_z f(x, y, z, q) - D_z f(x, \bar{y}, z + h, \bar{q})\|_{*;x}$$

are bounded from above by $\beta(\|z\|_{(x;1.L)}) \left[\, \|y - \bar{y}\| + \|h\|_{(x;1)} + \|q - \bar{q}\| \,\right]$ where

$$(x, y, q), \ (x, \bar{y}, \bar{q}) \in E \times R^n, \ z \in C^{1.L}(E_x, R), \ h \in C^1(E_x, R).$$

Assumption H $[\varphi]$. Suppose that

1) the function $\varphi : E_0 \to R$ is of class C^1 and $\|D_x \varphi\|_0 \leq b_1$, $\|D_y \varphi\|_0 \leq b_1$,

2) there is $b_2 \in R_+$ such that $[|\, D_x \varphi \,|] \leq b_2$, $[|\, D_y \varphi \,|] \leq b_2$.

Assumption H $[\varphi, f]$ **.** If $r_0 > 0$ then the consistency condition

$$D_x \varphi(0, y) = f\,(\, 0, y, \varphi, D_y \varphi(0, y)\,)$$

is satisfied for $y \in [-b, b]$.

Theorem 2.4 *If Assumptions $H\,[\,f\,]$, $H\,[\varphi]$, $H\,[\varphi, f\,]$ are satisfied then there is $c \in (0, a]$ such that problem (2.1), (2.2) has exactly one classical solution $\bar{u}$ on E_c.*

PROOF. Let us denote by $C(\tilde{E}_c, R; \lambda)$, $0 < c \leq a$, the Banach space of all continuous functions from $\tilde{E}_c$ into R with the norm

$$\|z\|_{[\lambda]} = \sup\left\{\, |\, z(x, y) \,| \exp\left(\, -\lambda x \,\right) : \ \ (x, y) \in E_0 \cup E_c \,\right\}$$

where $\lambda > d$. Let W be the set of all functions $u : E_c \to R$ such that

(i) u is of class C^1,

(ii) $u(x, y) = \varphi(x, y)$ on E_0.

For $u \in W$ denote by $u\,|_{\tilde{E}_c}$ the restriction of u to the set $\tilde{E}_c$. Let W^* denotes the set of all functions $u\,|_{\tilde{E}_c}$ where $u \in W$. Let X be the class of all functions z belonging to W^* and satisfying the conditions

$$\|\, D_y z(x, y) \,\| \leq \tilde{m}, \quad |\, D_x z(x, y) \,| \leq \tilde{m}, \tag{2.17}$$

$$\|\, D_y z(x, y) - D_y z(\bar{x}, \bar{y}) \,\| \leq s_0 |x - \bar{x}| + (2b_2 + 1)\, \|y - \bar{y}\|, \tag{2.18}$$

$$\|D_x z(x, y) - D_x z(\bar{x}, \bar{y})\| \leq s_1 |x - \bar{x}| + s_0\, \|y - \bar{y}\|, \tag{2.19}$$

on $\tilde{E}_c$ where

$$\tilde{m} = 2b_1 + \frac{d_0 + d_1 b_0}{d_1 a + d + d_1}, \quad p = d_0 + d_1(\tilde{m}\, a + b_0 + 2\tilde{m}),$$

$$s_0 = d\tilde{m} + 2p(b_2 + 1), \quad s_1 = (\, p + d\tilde{m}\,)\,(p+1) + p^2(2b_2 + 1).$$

Let c denote the constant defined by

$$c = \min\left\{ a,\ \frac{1}{2K(1+b_2)},\ \frac{1}{2(d_1 a + d + 2d_1)} \right\}.$$

where

$$K = \max\left\{ (\,\tilde{m} + 1\,)\beta(s)\,[\,\tilde{m} + 1 + s_0 + 2b_2 + 1\,],\ d(2b_2+1) \right\},$$

$$s = \tilde{m}\, a + b_0 + 2\tilde{m} + 2b_2 + 1 + 2s_0 + s_1.$$

The set X is the closed subset of the Banach space $C(\tilde{E}_c, R; \lambda)$. Let u be an arbitrary element of X. Consider the initial problem (2.1), (2.2) where

$$G(x,y,q) = f(x,y,\tilde{u},q), \quad \text{on } \Omega_0 \quad \text{and } \omega(y) = \varphi(0,y) \text{ on } [-b,b] \tag{2.20}$$

and $\tilde{u}(x,y) = u(x,y)$ on $\tilde{E}_c$, $\tilde{u}(x,y) = \varphi(x,y)$ on E_0. We will prove that there exists on E_c a unique solution $z(\,\cdot\,;u)$ of problem (2.1), (2.2), (2.20) and this solution satisfies (2.17) - (2.19). We will use Theorem 2.2 and Lemma 2.3. Since $u \in X$ then we have $\|\tilde{u}\|_{(x;1)} \le Ma + b_0 + 2\tilde{m}$, $\|\tilde{u}\|_{(x;1.L)} \le s$. For the function G given by (2.20) we have

$$D_{y_j}G(x,y,q) = D_{y_j}f(x,y,\tilde{u},q) + D_z f(x,y,\tilde{u},q)\,(D_{y_j}\tilde{u}),$$

where $j = 1,\ldots,n$, and

$$D_x G(x,y,q) = D_x f(x,y,\tilde{u},q) + D_z f(x,y,\tilde{u},q)\ (D_x\tilde{u}),$$

$$D_q G(x,y,q) = D_q f(x,y,\tilde{u},q).$$

It follows from Assumption H $[\,f\,]$ that

$$|D_x G(x,y,q)|, \quad \|D_y G(x,y,q)\| \le p + d\tilde{m},$$

$$|\,D_{q_i} G(x,y,q)\,| \le M_i, \quad i = 1,\ldots,n,$$

and

$$\|\,D_y G(x,y,q) - D_y G(x,\bar{y},\bar{q})\,\| \le K\,[\,\|y - \bar{y}\| + \|q - \bar{q}\|\,].$$

Let

$$\tilde{\Gamma}(x) = \frac{K(1+b_2)x + b_2}{1 - K(1+b_2)x}.$$

It follows from Theorem 2.2 and Lemma 2.3 that

(i) there exists the solution $z(\,\cdot\,;u)$ of problem (2.1), (2.2), (2.20) on $\tilde{E}_c$,

(ii) the solution satisfies the conditions:

$$\|D_y z(x,y;u) - D_y z(\bar{x},\bar{y};u)\| \le \left[p + d\tilde{m} + p\tilde{\Gamma}(x)\right] |x-\bar{x}| + \tilde{\Gamma}(x)\,\|y-\bar{y}\|, \quad (2.21)$$

$$|D_x z(x,y;u) - D_x z(\bar{x},\bar{y};u)| \le \left[(p+d\tilde{m})(p+1) + p^2\tilde{\Gamma}(x)\right] |x-\bar{x}| \quad (2.22)$$

$$+ \left[p + d\tilde{m} + p\tilde{\Gamma}(x)\right] \|y-\bar{y}\|,$$

and

$$|D_x z(x,y;u)|, \ \|D_y z(x,y;u)\| \le b_1 + (p+d\tilde{m})x \quad (2.23)$$

on $\tilde{E}_c$. Estimates (2.21), (2.22) and the condition $\tilde{\Gamma}(c) = 2b_2 + 1$ imply

$$\|D_y z(x,y;u) - D_y z(\bar{x},\bar{y};u)\| \le s_0|x-\bar{x}| + (2b_2+1)\,\|y-\bar{y}\|, \quad (2.24)$$

and

$$|D_x z(x,y;u) - D_x z(\bar{x},\bar{y};u)| \le s_1|x-\bar{x}| + s_0\,\|y-\bar{y}\|. \quad (2.25)$$

For all $x \in [0,c]$ we have

$$b_1 + (p+d\tilde{m})x \le b_1 + (d_0 + d_1 b_0)c + \tilde{m}\,[d_1 a + 2d_1 + d]\,c \le \tilde{m}.$$

Then by (2.23) we get

$$|D_x z(x,y;u)|, \quad \|D_y z(x,y;u)\| \le \tilde{m} \ \text{ on } \ \tilde{E}_c. \quad (2.26)$$

Let U be the operator defined on X in the following way: for $u \in X$ we put $(Uu)(x,y) = z(x,y;u)$ on $\tilde{E}_c$. It follows from (2.24) - (2.26) that the function Uu satisfies (2.17) - (2.19) and therefore $U : X \to X$. Now we prove that U is a contraction. Let u, $v \in X$ and $\tilde{v}(x,y) = v(x,y)$ on $\tilde{E}_c$, $\tilde{v}(x,y) = \varphi(x,y)$ on E_0. It follows from Assumption H $[f]$ that

$$|\,D_x[z(x,y;u) - z(x,y;v)]\,| \le d\|u-v\|_{[\lambda]}\exp(\,\lambda x\,)$$

$$+\sum_{j=1}^{n} M_j\,|\,D_{y_j}[z(x,y;u) - z(x,y;v)]\,|$$

on $\tilde{E}_c$ and $z(0,y;u) - z(0,y;v) = 0$ on $[-b,b]$. By the comparison theorem for hyperbolic differential inequalities ([140, 195]) we get

$$|\,z(x,y;u) - z(x,y;v)\,| \le \frac{d}{\lambda}\|u-v\|_{[\lambda]}\exp(\,\lambda x\,) \ \text{ on } \ \tilde{E}_c,$$

and hence

$$\|\,Uu - Uv\,\|_{[\lambda]} \le \frac{d}{\lambda}\|\,u-v\,\|_{[\lambda]}.$$

Since $d < \lambda$ then by the Banach fixed point theorem it follows that there exists $\bar{z} \in X$ such that $\bar{z} = U\bar{z}$. Let $\bar{u}(x,y) = \bar{z}(x,y)$ on $\tilde{E}_c$, $\bar{u} = \varphi(x,y)$ on E_0. This $\bar{u}$ is the solution of problem (2.1), (2.2) satisfying all the conditions of our Theorem.

2.4 Examples

It is important in Assumption H $[f]$ that we have assumed the local Lipschitz condition for the derivatives $D_y f$, $D_q f$, $D_z f$ on some special function spaces. Let us consider the simplest assumptions. Suppose that there is $\bar{L}$ such that

$$\| D_y f(x,y,z,q) - D_y f(x,\bar{y},z+h,\bar{q}) \| \leq \bar{L} \; [\, \|y-\bar{y}\| + \|h\|_{(x;0)} + \|q-\bar{q}\| \,] \tag{2.27}$$

and that suitable inequalities for the derivatives $D_q f$, $D_z f$ are satisfied. Of course, our result are true under the above stronger assumptions. Now we show that the formulation 5) of Assumption H $[f]$ is important. More precisely, we show that there is a class of nonlinear equations satisfying condition 5) of Assumption H $[f]$ and do not satisfying (2.27).

Example 2.5 *Suppose that $F : E \times R \times R^n \to R$, $\alpha : E \to R$, $\beta : E \to R^n$ are given function and consider the equation with a deviated argument*

$$D_x z(x,y) = F(\, x, y, z(\alpha(x,y), \beta(x,y)), D_y z(x,y)\,). \tag{2.28}$$

Assume that the functions F, α, β are continuous and

(i) there are the derivatives $D_y F$, $D_p F, D_q F$, the functions $D_y F$, $D_p F$, $D_q F$ are continuous and there are L_0, $\tilde{L} \in R_+$ such that

$$\| D_y F(P) \|, \quad | D_p F(P) |, \quad \| D_q F(P) \| \leq L_0$$

for $P = (x,y,p,q) \in E \times R \times R^n$ and that these derivatives satisfy the Lipschitz condition with respect to (y,p,q) with the constant $\tilde{L}$,

(ii) for $(x,y) \in E$ we have $(\, \alpha(x,y), \beta(x,y)\,) \in E_0 \cup E$ and $\alpha(x,y) \leq x$,

(iii) there are the derivatives $D_y\alpha$, $D_y\beta$, the functions $D_y\alpha$, $D_y\beta$ are continuous and there are $\tilde{d}$, $\bar{d} \in R_+$ such that

$$\| D_y\alpha(x,y) \| \leq \tilde{d}, \quad \| D_y\beta(x,y)\| \leq \tilde{d}$$

where

$$D_y\beta(x,y) = [\, D_{y_i}\beta_j(x,y)\,]_{i,j=1,\ldots,n}\,,$$

and the derivatives $D_y\alpha$, $D_y\beta$ satisfy the Lipschitz condition with respect to y with the constant $\bar{d}$. Put $f(x,y,z,q) = F(\, x,y,z(\alpha(x,y),\beta(x,y)\,), q)$. Then equation (2.1) is equivalent to (2.28).

From now we consider the function $D_y f$ only. It follows that for $z \in C^1(E_x, R)$ we have

$$D_{y_i} f(x,y,z,q) = D_{y_i} F(Q) +$$

$$+ D_p F(Q) \left[D_x z(Q_0) D_{y_i}\alpha(x,y) + \sum_{j=1}^{n} D_{y_j} z(Q_0)\, D_{y_i}\beta_j(x,y) \right]$$

where $Q = (\, x,y,z(\alpha(x,y),\beta(x,y)), q\,)$ and $Q_0 = (\alpha(x,y), \beta(x,y))$.
For (x,y,q), $(x,\bar{y},\bar{q}) \in E \times R^n$ and $z \in C^{1.L}(E_x, R)$, $h \in C^1(E_x, R)$ we get

$$\| D_y f(x,y,z,q) - D_y f(x,\bar{y},z+h,\bar{q}) \|$$

$$\leq \tilde{L}\left[1+\tilde{d}\left(\| D_x z \|_{(x;0)} + \| D_y z \|_{(x;0)}\right)\right]^2 \|y-\bar{y}\|$$
$$+\tilde{L}\left[1+\tilde{d}\left(\| D_x z \|_{(x;0)} + \| D_y z \|_{(x;0)}\right)\right]\left(\|h\|_{(x;0)} + \|q-\bar{q}\|\right)$$
$$+L_0\left[2\tilde{d}^2\left([|D_x z|]_{(x;L)} + [|D_y z|]_{(x;L)}\right) + \tilde{d}\left(\|D_x\|_{(x;0)} + \|D_y z\|_{(x;0)}\right)\right]\|y-\bar{y}\|$$
$$+L_0\tilde{d}\left[\|D_x h\|_{(x;0)} + \|D_y h\|_{(x;0)}\right].$$

It follows from the above considerations that condition 5) of Assumption $H[f]$ is satisfied. We see at once that the function $D_y f$ does not satisfy the global Lipschitz condition (2.27).

Example 2.6 *As particular cases of (2.1), (2.2) we obtain the Cauchy problem for differential integral equations*

$$D_x z(x,y) = F\left(x, y, z(x,y), \int_{E_x} z(t,s)\,dt\,ds, D_y z(x,y)\right),$$

where $F : E \times R^2 \times R^n \to R$, $\alpha : E \to R$, $\beta : E \to R^n$ and $(\alpha(x,y), \beta(x,y)) \in E_0 \cup E$, $\alpha(x,y) \leq x$ for $(x,y) \in E$. It is easy to see that Theorem 2.4 can be extended on the following systems of functional differential equations

$$D_x z_i(x,y) = f_i\left(x, y, z, D_y z_i(x,y)\right), \quad i = 1, \ldots, k,$$

with the initial condition $z(x,y) = \phi(x,y)$ on E_0 where $f = (f_1, \ldots, f_k) : E \times C(E_0 \cup E, R^k) \times R^n \to R^k$, $\phi : E_0 \to R^k$ and $z = (z_1, \ldots, z_k)$.

2.5 Quasi - linear systems

Let $L([0,a], R^n_+)$ be the class of all functions $\gamma : [0,a] \to R^n_+$ which are integrable on $[0,a]$. We denote by $M[k,n]$ the space of real $k \times n$ matrices. For the matrix

$$U \in M[k,n], \quad U = [\,u_{ij}\,]_{i=1,\ldots,k, j=1\ldots,n}$$

we define the norm

$$\|U\| = \max\left\{\sum_{j=1}^{n} |\,u_{ij}\,| : \ 1 \leq i \leq k\right\}.$$

Suppose that the function $M = (M_1, \ldots, M_n) \in C([0,a], R^n_+)$, $a > 0$, is non-decreasing and $M(0) = 0$. Let E be the generalized Haar pyramid

$$E = \{(x,y) \in R^{1+n} : \ x \in [0,a], \quad -b + M(x) \leq y \leq b - M(x)\},$$

where $b = (b_1, \ldots, b_n) \in R^n_+$ and $b - M(a) > 0$. Write $E_0 = [-r_0, 0] \times [-b, b]$ with $r_0 \in R_+$ and

$$E_x = (E_0 \cup E) \cap ([-r_0, x] \times R^n), \ x \in [0,a].$$

Put

$$S_x = [-b, b] \text{ for } x \in [-r_0, 0] \text{ and } S_x = [-b + M(x),\, b - M(x)] \text{ for } x \in [0, a].$$

Let $\Omega_0 = E \times C(E_0 \cup E, R^k)$ and assume that

$$\varrho : \Omega_0 \to M[k, n], \quad \varrho = [\varrho_{ij}]_{i=1,\dots,k,\ j=1,\dots,n}, \qquad f : \Omega_0 \to R^k, \quad f = (f_1, \dots, f_k),$$

are given functions of the variables (x, y, z), $z = (z_1, \dots, z_k)$. Given a function $\phi = (\phi_1, \dots, \phi_k) : E_0 \to R^k$, we consider the Cauchy problem

$$D_x z_i(x, y) + \sum_{j=1}^{n} \varrho_{ij}(x, y, z) D_{y_j} z_i(x, y) = f_i(x, y, z), \quad i = 1, \dots, k, \tag{2.29}$$

$$z(x, y) = \phi(x, y) \quad \text{on } E_0. \tag{2.30}$$

We will consider weak solutions of problem (2.29), (2.30). For $(x, y) \in [0, a] \times [-b, b]$ we put $I[x, y] = \{t : (t, y) \in E_x \setminus E_0\}$. A function $u : E_c \to R^k$ where $0 < c \le a$, is the Carathéodory solution of (2.29), (2.30) provides

(i) $u \in C(E_c, R^k)$ and the derivatives

$$D_x u_i u(x, y),\ D_y u_i(x, y) = (D_{y_1} u_i(x, y), \dots, D_{y_n} u_i(x, y)),\ 1 \le i \le k,$$

exist for almost all $(x, y) \in E_c \setminus E_0$,

(ii) u satisfies (2.29) almost everywhere on $E_c \setminus E_0$ and condition (2.30) holds. We assume that f and $\varrho_i = (\varrho_{i1}, \dots, \varrho_{in})$, $1 \le i \le k$, satisfy the Volterra condition. We will discuss the question of the existence and uniqueness of solutions to problem (2.29), (2.30). Our results are based on the method of bicharacteristics. The Cauchy problem will be transformed into integral functional equations. The existence of solutions of this system will be proved by using integral inequalities and by the Banach fixed point theorem. We list below a few examples of systems which can be derived from (2.29) by specializing given functions.

Example 2.7 *Assume that*

$$\tilde\varrho : E \times R^k \times R^k \to M[k, n], \quad \tilde\varrho = [\tilde\varrho_{ij}]_{i=1,\dots,k,\ j=1,\dots,n},$$

$$\tilde f = (\tilde f_1, \dots, \tilde f_k) : E \times R^k \times R^k \to R^k$$

and

$$\alpha,\ \beta : E \to R^{1+n}, \quad \alpha = (\alpha_0, \alpha_1, \dots, \alpha_n),\ \beta = (\beta_0, \beta_1, \dots, \beta_n)$$

are given functions and $\alpha(t, s)$, $\beta(t, s) \in E_x$ for $(t, s) \in E_x$. Put

$$\varrho(x, y, z) = \tilde\varrho\,(x, y, z(x, y), z(\alpha(x, y))),$$

$$f(x, y, z) = \tilde f\,(x, y, z(x, y), z(\beta(x, y)))$$

on Ω_0. Then (2.29) reduces to the system with a deviated argument

$$D_x z_i(x, y) + \sum_{j=1}^{n} \tilde\varrho_{ij}(x, y, z(x, y), z(\alpha(x, y)))\, D_{y_j} z_i(x, y)$$

$$= \tilde f_i(x, y, z(x, y), z(\beta(x, y))), \quad i = 1, \dots, k.$$

Example 2.8 *For the above $\tilde{\varrho}$ and $\tilde{f}$ we define*

$$\varrho(x,y,z) = \tilde{\varrho}(x,y,z(x,y), \int_{E_x} z(t,s)dt\,ds),$$

$$f(x,y,z) = \tilde{f}(x,y,z(x,y), \int_{E_x} z(t,s)dt\,ds).$$

Then (2.29) is the integral differential system

$$D_x z_i(x,y) + \sum_{j=1}^{n} \tilde{\varrho}_{ij}(x,y,z(x,y), \int_{E_x} z(t,s)dt\,ds)\, D_{y_j} z_i(x,y)$$

$$= \tilde{f}_i(x,y,z(x,y), \int_{E_x} z(t,s)dt\,ds), \quad i = 1,\ldots,k.$$

Remark 2.9 *In the same way we can prove that the functional differential equations considered in [214, 215] are particular cases of (2.29) under suitable assumptions on Volterra operators .*

Now we present relations between local and global (with respect to y) solutions of differential and functional differential systems. Suppose that the function $\tilde{M} = (\tilde{M}_1,\ldots,\tilde{M}_n) \in C([0,a], R^n_+)$ is nondecreasing and $\tilde{M}(0) = 0$, $b - \tilde{M}(a) > 0$. Let

$$\tilde{E} = \{ (x,y) : \; x \in [0,a], \quad -b + \tilde{M}(x) \le y \le b - \tilde{M}(x) \}.$$

Assume that $\tilde{\phi} : [-b,\, b] \to R^k$ and

$$\tilde{\varrho} : \tilde{E} \times R^k \to M[k,n], \quad \tilde{\varrho} = [\, \tilde{\varrho}_{ij} \,]_{i=1,\ldots,k,\, j=1,\ldots,n}\,,$$

$$\tilde{f} : \tilde{E} \times R^k \to R^k, \; \tilde{f} = (\tilde{f}_1,\ldots,\tilde{f}_k)$$

are given functions. Consider the quasilinear system without the functional dependence

$$D_x z_i(x,y) + \sum_{j=1}^{n} \tilde{\varrho}_{ij}(x,y,z(x,y))\, D_{y_j} z_i(x,y) \tag{2.31}$$

$$= \tilde{f}_i(x,y,z(x,y)), \quad i = 1,\ldots,k,$$

with the initial condition

$$z(0,y) = \tilde{\phi}(y) \quad \text{on } [-b,\, b]. \tag{2.32}$$

We formulate now the following assumption on $\tilde{\varrho}$.

Assumption $\tilde{H}$. Suppose that the function $\tilde{\varrho}$ of the variables (x,y,p) is measurable in x for every (y,p) and it is continuous in (y,p) for almost all x. Assume that there exists a function $\gamma = (\gamma_1,\ldots,\gamma_n) \in L([0,a], R^n_+)$ such that for almost all $x \in [0,a]$ we have

$$|\, \tilde{\varrho}_{ij}(x,y,p)\,| \le \gamma_j(x), \;\; 1 \le j \le n, \;\; 1 \le i \le k,$$

for y, p such that $(x,y,p) \in \tilde{E} \times R^k$, and

$$\tilde{M}(x) \geq \int_0^x \gamma(\tau)\, d\tau \ \text{ for } \ x \in [0,a].$$

Initial problems for quasilinear systems without the functional dependence have the following property: existence and uniqueness results for the Cauchy problem (2.31), (2.32) can be deduced from known results for the global Cauchy problem ([45]). More precisely, suppose that

1) Assumption $\tilde{H}$ is satisfied,

2) the functions $\tilde{\varrho}$, $\tilde{f}$ satisfy all the assumptions of Theorem 1 in [45] on the set $\tilde{E} \times R^k$ (instead of $[0,a] \times R^n \times [-d,d]$, $[-d,d] \subset R^k$),

3) $\tilde{\phi}$ satisfies all the assumptions of Theorem 1 in [45] on $[-b,b]$ (instead of R^n).

Then there exists exactly one generalized solution $\tilde{u}$ of problem (2.31), (2.32). The solution $\tilde{u}$ is defined on the set $\tilde{E} \cap ([0,c] \times R^n)$ with $c \in (0,a]$ sufficiently small and it depends continuously on given functions.

This result can be proved by exactly the same methods as in [45], see also [13 – 15]. The situation is completely different for systems with the functional dependence. We discuss the problem. Several authors introduced various hereditary structures for description different situations in partial differential equations. Let' s recall some of the main settings. For simplicity, let $k = 1$ and consider the nonlinear equation

$$D_x z(x,y) = F(x, y, T(z; x, y), D_y z(x,y)) \tag{2.33}$$

where $D_y z = (D_{y_1} z, \ldots, D_{y_n} z)$ and T is a delay operator. If T is given by $T(z;x,y) = z(x,y)$ then (2.33) reduces to a classical equation. There are a lot of papers concerning equation (2.33) with T defined by

$$T(z;x,y) = z_{(x,y)} \tag{2.34}$$

where the function $(x,y) \to z_{(x,y)}$ is a natural extension of the Hale operator given by (2.4). Consider the equation (2.33), (2.34) with the initial condition

$$z(x,y) = \phi_0(x,y) \quad \text{on} \quad [-r_0, 0] \times R^n. \tag{2.35}$$

This formulation is natural and suitable for initial problems which are global with respect to y. The paper [110] contains a survey of existence results for nonlinear equations and quasilinear systems in the second canonical form. It is evident from (2.4) that the formulation (2.33), (2.34) is not suitable for the local Cauchy problems considered on the Haar pyramid. The second group of papers is connected with the initial problems for the equation

$$D_x z(x,y) = G(x, y, z, D_y z(x,y)) \tag{2.36}$$

where G is an operator of the Volterra type. If we assume that $G : \tilde{E} \times C(E_0 \cup \tilde{E}, R) \times R^n \to R$ then we can consider the initial problem consisting of equation (2.36) and the condition

$$z(x,y) = \phi_0(x,y) \ \text{ on } \ E_0. \tag{2.37}$$

Quasilinear system (2.29) is generated by equation (2.36).

2.6 Bicharacteristics of quasilinear systems

We start with definitions of some function spaces. Let us denote by $\|\cdot\|_x$ the supremum norm in the space $C(E_x, R^k)$ where $0 \le x \le a$. We will use the symbol $C_{0.L}(E_x, R^k)$ to denote the class of all functions $z \in C(E_x, R^k)$ such that

$$\|z\|_{(x;L)} = \sup\left\{ \frac{\|z(t,s) - z(t,\bar{s})\|}{\|s-\bar{s}\|} : \ (t,s),\ (t,\bar{s}) \in E_x \right\} < \infty.$$

For $z \in C_{0.L}(E_x, R^k)$ we write

$$\|z\|_{(x;0,L)} = \|z\|_x + \|z\|_{(x;L)}$$

where $0 \le x \le a$. Let

$$C(E_x, R^k; \kappa) = \{z \in C(E_x, R^k) : \ \|z\|_x \le \kappa\},$$

$$C_{0.L}(E_x, R^k; \kappa) = \{z \in C_{0.L}(E_x, R^k) : \ \|z\|_{(x;0,L)} \le \kappa\},$$

where $\kappa \in R_+$ and $0 \le x \le a$. Let $\omega_0 \in L([-r_0, 0], R_+)$, $p = (p_0, p_1) \in R_+^2$. Denote by $J[\omega_0, p]$ the set of all functions $\psi \in C(E_0, R^k)$ such that $\|\psi(x,y)\| \le p_0$ and

$$\|\psi(x,y) - \psi(\bar{x},\bar{y})\| \le \left| \int_x^{\bar{x}} \omega_0(\tau)d\tau \right| + p_1\|y-\bar{y}\| \ \text{ on } E_0.$$

Suppose that

$$c \in (0,a],\ d = (d_0, d_1) \in R_+^2,\ d_0 \ge p_0,\ d_1 \ge p_1$$

and $\omega \in L([-r_0, c], R_+)$, $\omega(t) \ge \omega_0(t)$ for almost all $t \in [-r_0, 0]$. Let $K_{c.\phi}[\omega, d]$ be the class of all functions $z \in C(E_c, R^k)$ such that

(i) $z(x,y) = \phi(x,y)$ on E_0,

(ii) $\|z(x,y)\| \le d_0$ and

$$\|z(x,y) - z(\bar{x},\bar{y})\| \le \left| \int_x^{\bar{x}} \omega(\tau)d\tau \right| + d_1\|y-\bar{y}\| \ \text{ on } E_c.$$

Put $|d| = d_0 + d_1$. Denote by Θ the class of all functions $\delta : [0,a] \times R_+ \to R_+$ such that $\delta(\cdot, t) \in L([0,a], R_+)$ for $t \in R_+$ and the function $\delta(t, \cdot)$ is continuous and nondecreasing on R_+ for almost all $t \in [0,a]$.

Assumption H $[\varrho]$. Suppose that

1) the function $\varrho(\cdot, y, z) : I[a,y] \to M[k,n]$ is measurable for $(y,z) \in [-b,b] \times C(E_0 \cup E, R^k)$, $0 \le x \le a$, and $\varrho(x, \cdot) : S_x \times C(E_x, R^k) \to M[k,n]$ is continuous for almost all $x \in [0,a]$,

2) there exists $\gamma = (\gamma_1, \ldots, \gamma_n) \in L([0,a], R_+^n)$ such that

$$|\varrho_{ij}(x,y,z)| \le \gamma_j(x),\ \ 1 \le j \le n,\ 1 \le i \le k,$$

for $(y,z) \in S_x \times C(E_x, R^k)$ and for almost all $x \in [0,a]$,

3) there exists $\beta_1 \in \Theta$ such that

$$\|\varrho(x,y,z) - \varrho(x,\bar{y},\bar{z})\| \le \beta_1(x,\kappa)\,[\,\|y-\bar{y}\| + \|z-\bar{z}\|_x\,],$$

for (y,z), $(\bar{y},\bar{z}) \in S_x \times C_{0.L}(E_x, R^k;\kappa)$ almost everywhere on $[0,a]$,

4) for $x \in [0,a]$ we have

$$M(x) \ge \int_0^x \gamma(\tau)d\tau.$$

We give now the notion of bicharacteristics to system (2.29). For the matrix $\varrho \in M[k,n]$ we consider the vectors $\varrho_i = (\,\varrho_{i1},\dots,\varrho_{in}\,)$, $1 \le i \le k$. Suppose that $\phi \in J[\omega,p]$, $c \in (0,a]$ and $z \in K_{c.\phi}[\omega,d]$. Consider the Cauchy problem

$$\eta'(t) = \varrho_i(\,t,\eta(t),z\,), \quad \eta(x) = y, \tag{2.38}$$

where $(x,y) \in E_c \setminus E_0$. Let $g_i[z](\,\cdot\,,x,y)$ denote the solution of problem (2.38). The function $g_i[z]$ is called the i - th bicharacteristic of system (2.29) corresponding to $z \in K_{c.\phi}[\omega,p]$.

Lemma 2.10 *Suppose that Assumption H [ϱ] is satisfied and $c \in (0,a]$, ϕ, $\bar{\phi} \in J[\omega_0,p]$, $z \in K_{c.\phi}[\omega,d]$, $\bar{z} \in K_{c.\bar{\phi}}[\omega,d]$.*

Then for each i, $1 \le i \le k$, the solutions $g_i[z](\,\cdot\,,x,y)$ and $g_i[\bar{z}](\,\cdot\,,x,y)$ are defined on such intervals $[0,c_i(x,y)]$ and $[0,\bar{c}_i(x,y)]$ that

$$(c_i(x,y),\ g_i[z](c_i(x,y),x,y)) \in \partial E_c \ \text{ and } \ (\bar{c}_i(x,y),\ g_i[z](\bar{c}_i(x,y),x,y)) \in \partial E_c$$

where ∂E_c is the boundary of E_c. Moreover we have the estimates

$$\|g_i[z](t,x,y) - g_i[z](t,\bar{x},\bar{y})\| \tag{2.39}$$

$$\le \left[\,\|y-\bar{y}\| + \left|\int_x^{\bar{x}} \|\gamma(\tau)\|d\tau\right|\,\right] \exp\left[\,\left|\int_x^t \beta_1(\tau,|d|)\,d\tau\right|\,\right]$$

where (x,y), $(\bar{x},\bar{y}) \in E_c \setminus E_0$, $t \in [\,0,\min\{c_i(x,y),\ c_i(\bar{x},\bar{y}\}\,]$ and

$$\|g_i[z](t,x,y) - g_i[\bar{z}](t,x,y)\| \tag{2.40}$$

$$\le \left|\int_x^t \beta_1(\tau,|d|)\,\|z-\bar{z}\|_\tau\,d\tau\right| \exp\left[\,\left|\int_x^t \beta_1(\tau,|d|)d\tau\right|\,\right],$$

where $(x,y) \in E_c \setminus E_0$, $t \in [\,0,\ \min\{c_i(x,y),\bar{c}_i(x,y)\}\,]$.

PROOF. The existence and uniqueness of the solutions of (2.38) follows from classical theorems. Note that the right-hand side of the differential system satisfies the Carathéodory conditions and the Lipschitz estimate holds

$$\|\varrho_i(t,\nu,z) - \varrho_i(t,\bar{\nu},z)\| \le \beta_1(t,|d|)\,\|\nu-\bar{\nu}\|,\ (t,\nu),\ (t,\bar{\nu}) \in E_c \setminus E_0.$$

The function $g_i[z](\,\cdot\,,x,y)$ satisfies the integral equation

$$g_i[z](t,x,y) = y + \int_x^t \varrho_i(\,\tau, g_i[z](\tau,x,y),z\,)\,d\tau, \ \ 1 \le i \le k, \tag{2.41}$$

where $(x,y) \in E_c \setminus E_0$, $t \in [0, c_i(x,y)]$. It follows from Assumption H $[\varrho]$ that the integral inequality is satisfied

$$\|g_i[z](t,x,y) - g_i[z](t,\bar{x},\bar{y})\| \le \|y - \bar{y}\| + \left| \int_x^{\bar{x}} \|\gamma(\tau)\| d\tau \right|$$

$$+ \left| \int_x^t \beta_1(\tau, |d|) \, \|g_i[z](\tau,x,y) - g_i[z](\tau,\bar{x},\bar{y})\| d\tau \right|,$$

where (x,y), $(\bar{x},\bar{y}) \in E_c \setminus E_0$, $t \in [0, \min\{c_i(x,y), c_i(\bar{x},\bar{y})\}]$. Then we obtain (2.39) by Gronwall inequality. For $z \in K_{c.\phi}[\omega, d]$, $\bar{z} \in K_{c.\bar{\phi}}[\omega, d]$ we have the estimate

$$\|g_i[z](t,x,y) - g_i[\bar{z}](t,x,y)\| \le \left| \int_x^t \beta_1(\tau, |d|) \, \|z - \bar{z}\|_\tau \, d\tau \right|$$

$$+ \left| \int_x^t \beta_1(\tau, |d|) \, \|g_i[z](\tau,x,y) - g_i[\bar{z}](\tau,x,y)\| d\tau \right|,$$

where $(x,y) \in E_c \setminus E_0$, $t \in [0, \min\{c_i(x,y), \bar{c}_i(x,y)\}]$. Now we obtain (2.40) by the Gronwall inequality. This completes the proof of the Lemma.

2.7 Integral operators for initial problems

Suppose that

$$\phi \in J[\omega_0, p], \; c \in (0, a], \; z \in K_{c.\phi}[\omega, d]$$

and

$$(g_1[z](\cdot, x, y), \ldots, g_k[z](\cdot, x, y)) = g[z](\cdot, x, y)$$

are bicharacteristics of system (2.29). Write

$$f[g,z](t,x,y) = (\, f_1(t, g_1[z](t,x,y), z), \ldots, f_k(t, g_k[z](t,x,y), z)\,)$$

and

$$\phi[g,z](t,x,y) = (\, \phi_1(t, g_1[z](t,x,y)), \ldots, \phi_k(t, g_k[z](t,x,y))\,).$$

For $z \in K_{c.\phi}[\omega, d]$ we define the function $U = (U_1, \ldots, U_k) = U_\phi z$ by

$$U(x,y) = \phi[g,z](0,x,y) + \int_0^x f[g,z](\tau,x,y) d\tau, \quad (x,y) \in E_c \setminus E_0,$$

$$U(x,y) = \phi(x,y) \;\text{ on }\; E_0.$$

We prove that the system of functional integral equations consisting of (2.41) and the equation $z = U_\phi z$ has exactly one solution.

Remark 2.11 *The operator U_ϕ is obtained by integrating system (2.29) along bicharacteristics. The relations*

$$D_x z_i(t, g_i[z](t,x,y)) + \sum_{j=1}^{n} \varrho_{ij}\,(\,t, g_i[z](t,x,y), z\,)\, D_{y_j} z_i(t, g_i[z](t,x,y))$$

$$= f_i\,(t, g_i[z](t,x,y), z)$$

where $1 \le i \le k$, lead to

$$\frac{d}{dt} z_i\,(t, g_i[z](t,x,y)) = f_i\,(\,t, g_i[z](t,x,y), z\,), \quad 1 \le i \le k. \tag{2.42}$$

Integration of (2.42) with respect to t from 0 to x leads to

$$z_i(x,y) = \phi_i(0, g_i[z](0,x,y)) + \int_0^x f_i\,(\,\tau, g_i[z](\tau,x,y), z\,)\, d\tau, \quad 1 \le i \le k, \tag{2.43}$$

and $z = \phi$ on E_0. Then we have obtained the functional integral equation $z = U_\phi z$.

Assumption H $[\,f\,]$. Suppose that

1) the function $f(\,\cdot\,, y, z) : I[a,y] \to R^k$ is measurable for $(y,z) \in [-b,b] \times C(E_0 \cup E, R^k)$ and $f(x,\,\cdot\,) : S_x \times C(E_x, R^k) \to R^k$ is continuous for almost all $x \in [0,a]$,

2) there exist functions $\alpha_2,\ \beta_2 \in \Theta$ such that

$$\|f(x,y,z)\| \le \alpha_2(x,\kappa) \ \text{ for } \ (y,z) \in S_x \times C(E_x, R^k, \kappa)$$

almost everywhere on $[0,a]$ and

$$\|f(x,y,z) - f(x,\bar{y},\bar{z})\| \le \beta_2(x,\kappa)\,[\,\|y-\bar{y}\| + \|z-\bar{z}\|_x\,]$$

for (y,z), $(\bar{y},\bar{z}) \in S_x \times C_{0.L}(E_x, R^k; \kappa)$ almost everywhere on $[0,a]$.

Assumption H $[\omega, d]$. Suppose that the constants $(d_0, d_1) = d,\ c \in (0,a]$ and the function $\omega : [-r_0, c] \to R_+$ satisfy the conditions:

$$d_0 \ge p_0 + \int_0^c \alpha_2(\tau, d_0)d\tau,$$

$$d_1 \ge \left[p_1 + \int_0^c \beta_2(\tau, |d|)d\tau\right] \exp\left[\int_0^c \beta_1(\tau, |d|)d\tau\right]$$

and

$$\omega(t) \ge \|\gamma(t)\| \left[p_1 + \int_0^c \beta_2(\tau, |d|)d\tau\right] \exp\left[\int_0^c \beta_1(\tau, |d|)d\tau\right]$$

for almost all $t \in [0,c]$.

Remark 2.12 *If $d_0 > p_0$ and $d_1 > p_1$ then there is $c \in (0,a]$ such that (d_0, d_1) satisfy Assumption H $[\omega, d\,]$.*

Lemma 2.13 *If $\phi \in J[\omega_0, p]$ and Assumptions $H[\varrho]$, $H[f]$, $H[\omega, d]$ are satisfied then*

$$U_\phi : K_{c.\phi}[\omega, d] \to K_{c.\phi}[\omega, d].$$

PROOF. Suppose that $z \in K_{c.\phi}[\omega, d]$. It follows from Assumption H $[f]$ that

$$\|U(x,y)\| \le p_0 + \int_0^c \alpha_2(\tau, |d|)d\tau \le d_0 \text{ on } E_c \setminus E_0. \tag{2.44}$$

If (x,y), $(\bar{x},\bar{y}) \in E_c \setminus E_0$ then we have

$$\|U(x,y) - U(\bar{x},\bar{y})\| \le \|\phi[g,z](0,x,y) - \phi[g,z](0,\bar{x},\bar{y})\|$$

$$+ \int_0^x \|f[g,z](\tau,x,y) - f[g,z](\tau,\bar{x},\bar{y})\|d\tau \le p_1 \max_{1\le i\le k} \|g_i[z](0,x,y) - g_i[z](0,\bar{x},\bar{y})\|$$

$$+ \int_0^x \beta_2(\tau,|d|) \max_{1\le i\le k} \|g_i[z](\tau,x,y) - g_i[z](\tau,\bar{x},\bar{y})\|\, d\tau.$$

The above inequality and Lemma 2.10 imply

$$\|U(x,y) - U(\bar{x},\bar{y})\|$$

$$\le \left[\|y-\bar{y}\| + \left|\int_x^{\bar{x}} \|\gamma(\tau)\|d\tau\right|\right] \exp\left[\int_0^c \beta_1(\tau,|d|)d\tau\right] \left[p_1 + \int_0^c \beta_2(\tau,|d|)d\tau\right],$$

and consequently

$$\|U(x,y) - U(\bar{x},\bar{y})\| \le \left|\int_x^{\bar{x}} \omega(\tau)d\tau\right| + d_1\|y-\bar{y}\| \text{ on } E_c \setminus E_0. \tag{2.45}$$

It follows from (2.44), (2.45) that $U \in K_{c.\phi}[\omega, d]$. This proves the lemma.

2.8 Existence of Carathéodory solutions

Put

$$\Gamma_c = \int_0^c \beta_1(\tau,|d|)d\tau \exp\left[\int_0^c \beta_1(\tau,|d|)d\tau\right].$$

Theorem 2.14 *Suppose that $\phi \in J[\omega_0, p]$ and Assumptions $H[\varrho]$, $H[f]$, $H[\omega, d]$ are satisfied. Let $c \in (0, a]$ be such a constant that*

$$\Lambda_c = p_1\Gamma_c + (1+\Gamma_c)\int_0^c \beta_2(\tau,|d|)d\tau < 1.$$

Then there exists exactly one solution u of problem (2.29), (2.30) on E_c. Moreover, if $\bar{\phi} \in J[\omega_0, p]$ and $\bar{u}$ is a solution of system (2.29) with the initial condition

$$z(x,y) = \bar{\phi}(x,y) \text{ on } E_0, \tag{2.46}$$

then there exists $\tilde{\Lambda}_c \in R_+$ such that

$$\|u - \bar{u}\|_x \le \tilde{\Lambda}_c \|\phi - \bar{\phi}\|_0, \quad x \in [0,c]. \tag{2.47}$$

PROOF. Lemma 2.13 shows that $U_\phi : K_{c.\phi}[\omega, d] \to K_{c.\phi}[\omega, d]$. Now we prove that U_ϕ is a contraction on $K_{c.\phi}[\omega, d]$. If z, $\tilde{z} \in K_{c.\phi}[\omega, d]$ then

$$||(U_\phi z)(x,y) - (U_\phi \tilde{z})(x,y)|| \leq ||\phi[g,z](0,x,y) - \phi[\tilde{g},\tilde{z}](0,x,y)||$$

$$+ \int_0^x ||f[g,z](\tau,x,y) - f[\tilde{g},\tilde{z}](\tau,x,y)||\, d\tau$$

$$\leq ||z - \tilde{z}||_x \left[p_1 \Gamma_c + (1+\Gamma_c) \int_0^c \beta_2(\tau, |d|) d\tau \right], \quad (x,y) \in E_c \setminus E_0,$$

where $\tilde{g} = (\tilde{g}_1, \ldots, \tilde{g}_k)$ is the set of bicharacteristics corresponding to $\tilde{u}$. The above estimates implies $||U_\phi z - U_\phi \tilde{z}||_c \leq \Lambda_c ||z - \tilde{z}||_c$. By the Banach fixed point theorem there exists a unique solution $u \in K_{c.\phi}[\omega, d]$ of equation $z = U_\phi z$. Now we prove that u is a solution of system (2.29). We have proved that

$$u_i(x,y) = \phi_i(0, g_i[u](0,x,y)) + \int_0^x f_i\left(\tau, g_i[u](\tau,x,y), u\right) d\tau, \tag{2.48}$$

where $1 \leq i \leq k$, $(x,y) \in E_c \setminus E_0$. For given $(x,y) \in E_c \setminus E_0$, $1 \leq i \leq k$, let us put $\eta^{(i)} = g_i[u](0,x,y)$. It follows that

$$g_i[u](t,x,y) = g_i[u](t,0,\eta^{(i)}) \quad \text{for} \quad t \in [0, c_i(x,y)],$$

and $y = g_i[u](x,0,\eta^{(i)})$. It follows from (2.48) that

$$u_i(x, g_i[u](x,0,\eta^{(i)})) = \phi_i(0,\eta^{(i)}) + \int_0^x f_i(\tau, g_i[u](\tau,0,\eta^{(i)}), u)\, d\tau, \tag{2.49}$$

where $i = 1, \ldots, k$. By differentiating (2.49) with respect to x and by putting again $\eta^{(i)} = g_i[u](0,x,y)$ we obtain that u satisfies system (2.29) for almost all $(x,y) \in E_c \setminus E_0$. Now we prove relation (2.47). If $u = U_\phi u$ and $\bar{u} = U_{\bar{\phi}} \bar{u}$ then

$$||u(x,y) - \bar{u}(x,y)|| \leq ||\phi[g,u](0,x,y) - \bar{\phi}[\bar{g},\bar{u}](0,x,y)|| \tag{2.50}$$

$$+ \int_0^x ||f[g,u](\tau,x,y) - f[\bar{g},\bar{u}](\tau,x,y)|| d\tau, \quad (x,y) \in E_c \setminus E_0,$$

where $\bar{g} = (\bar{g}_1, \ldots, \bar{g}_k)$ are bicharacteristics corresponding to $\bar{u}$. It follows from Lemma 2.10 that

$$||g_i[u](t,x,y) - g_i[\bar{u}](t,x,y)|| \leq B_c \int_0^x \beta_1(\tau, |d|)\, ||u - \bar{u}||_\tau\, d\tau, \tag{2.51}$$

where $0 \leq t \leq x$, $(x,y) \in E_c \setminus E_0$ and

$$B_c = \exp\left[\int_0^c \beta_1(\tau, |d|) d\tau \right].$$

Assumption H $[\,f\,]$ and estimates (2.50), (2.51) imply

$$||u(x,y) - \bar{u}(x,y)|| \leq ||\phi - \bar{\phi}||_0 + p_1 B_c \int_0^x \beta_1(\tau, |d|)\, ||u - \bar{u}||_\tau\, d\tau$$

$$+(1 + B_c) \int_0^x \beta_2(\tau, |d|) \, \|u - \bar{u}\|_\tau \, d\tau, \quad (x, y) \in E_c \setminus E_0.$$

The right-hand side of the above inequality is a nondecreasing function of x, therefore we have the integral inequality

$$\|u - \bar{u}\|_x \leq \|\phi - \bar{\phi}\|_0 + \int_0^x [\, p_1 B_c \beta_1(\tau, |d|) + (1 + B_c)\beta_2(\tau, |d|) \,] \, \|u - \bar{u}\|_\tau \, d\tau,$$

where $x \in [0, c]$. Using the Gronwall inequality we obtain

$$\|u - \bar{u}\|_x \leq \tilde{\Gamma}(x) \, \|\phi - \bar{\phi}\|_0, \quad x \in [0, c],$$

with

$$\tilde{\Gamma}(x) = \exp\left\{ \int_0^x [\, p_1 B_c \beta_1(\tau, |d|) + (1 + B_c)\beta_2(\tau, |d|) \,] \, d\tau \right\}.$$

Then we have (2.47) with $\tilde{\Lambda}_c = \tilde{\Gamma}(c)$. This proves the Theorem.

2.9 Uniqueness of Generalized Solutions

Now we prove a general theorem on the uniqueness of solutions of the Cauchy problem. Given the functions $\sigma : [0, a] \times C([-r_0, a], R_+) \to R_+$, $\omega : [0, a] \to R_+$ and $\theta : [-r_0, 0] \to R_+$, we consider the Cauchy problem

$$\eta'(x) = \omega(x) \, \max\{\eta(t) : t \in [-r_0, x] \} + \sigma(x, \eta), \tag{2.52}$$

$$\eta(x) = \theta(x) \;\text{ on }\; [-r_0, 0]. \tag{2.53}$$

The above system will be considered as a comparison problem for quasilinear hyperbolic equations. The function σ is said to satisfy Carathéodory conditions if

(i) $\sigma(x, \cdot) : C([-r_0, a], R_+) \to R_+$ is continuous for almost all $x \in [0, a]$,

(ii) $\sigma(\cdot, \eta) : [0, a] \to R_+$ is measurable for every $\eta \in C([-r_0, a], R_+)$ and there exists $m_\sigma \in L([0, a], R_+)$ such that $\sigma(x, \eta) \leq m_\sigma(x)$ for $\eta \in C([-r_0, a], R_+)$ and for almost all $x \in [0, a]$.

We consider Cauchy problem (2.52), (2.53) under the assumptions that σ satisfies Carathéodory conditions, $\theta \in C([-r_0, 0], R_+)$ and $\omega \in L([0, a], R_+)$. We assume also that σ fulfils the Volterra condition. We will deal with Carathéodory solutions of problem (2.52), (2.53).

Assumption H $[\sigma]$. Suppose that

1) σ satisfies Carathéodory conditions and is nondecreasing with respect to the functional variable,

2) σ fulfils the Volterra condition and for every

$$\theta \in C([-r_0, 0], R_+), \quad \omega \in L([0, a], R_+)$$

there exists on $[-r_0, a]$ the maximum solution of problem (2.52), (2.53).

Lemma 2.15 *Suppose that Assumption H [σ] is satisfied and*

1) $\alpha \in C([-r_0, a], R_+)$ *satisfies the initial inequality* $\alpha(x) \leq \theta(x)$ *for* $x \in [-r_0, 0]$,

2) *for* $x \in [0, c]$ *we have*

$$\alpha(x) \leq \theta(0) + \int_0^x \omega(\tau)\ \max\{\alpha(t) : t \in [-r_0, \tau]\}d\tau + \int_0^x \sigma(\tau, \alpha)\, d\tau,$$

3) $\beta : [-r_0, a] \to R_+$ *is the maximum solution of (2.52), (2.53).*

Then $\alpha(x) \leq \beta(x)$ *on* $[0, a]$.

PROOF. Let us consider the function $\tilde{\alpha} : [-r_0, a] \to R_+$ defined by

$$\tilde{\alpha}(x) = \theta(x) \ \text{ on } \ [-r_0, 0],$$

$$\tilde{\alpha}(x) = \theta(0) + \int_0^x \omega(\tau)\ \max\{\alpha(t) : t \in [-r_0, \tau]\}\, d\tau + \int_0^x \sigma(\tau, \alpha)\, d\tau$$

on $[0, a]$. Then $\tilde{\alpha}$ satisfies the functional differential inequality

$$\tilde{\alpha}'(x) \leq \omega(x) \max\{\tilde{\alpha}(t) : t \in [-r_0, x]\} + \sigma(x, \tilde{\alpha}) \ \text{ almost everywhere on } \ [0, a].$$

The above relation and the theorem on functional differential inequalities in the Carathéodory sense (Theorem 5 from [31], see also Theorem 1 in [32]) imply the thesis.

Let $V : C(E_0 \cup E, R^k) \to C([-r_0, a], R_+)$ be the operator defined by

$$(Vz)(x) = \max\{\|z(x, y)\| : y \in S_x\}, \quad -r_0 \leq x \leq a.$$

Note that the continuity of Vz follows from Theorem 33.1 in the monograph [195].

Assumption $\tilde{H}[f]$. Suppose that condition 1) of Assumption H [f] is satisfied and there exist functions $\beta_2 \in \Theta$ and $\sigma : [0, a] \times C([-r_0, a], R_+) \to R_+$ such that

(i) Assumption H [σ] is satisfied,

(ii) for (y, z), $(\bar{y}, \bar{z}) \in S_x \times C_{0.L}(E_x, R^k, \kappa)$ we have

$$\|f(x, y, z) - f(x, \bar{y}, \bar{z})\| \leq \beta_2(x, \kappa)\, \|y - \bar{y}\| + \sigma(x, V(z - \bar{z}))$$

almost everywhere on $[0, a]$,

(iii) the maximum solution of problem (2.52), (2.53) with $\theta(x) = 0$ on $[-r_0, 0]$ is $\bar{\eta}(x) = 0$ on $[-r_0, a]$.

Theorem 2.16 *Suppose that Assumptions H [ϱ], $\tilde{H}[f]$ are satisfied and* $\phi \in J[\omega_0, p]$. *Then problem (2.29), (2.30) admits one solution of class* $K_{a.\phi}[\omega, d]$ *at the most.*

PROOF. Suppose that u and $\tilde{u}$ are two solutions of (2.29), (2.30) which are of class $K_{a.\phi}[\omega, d]$. Then u and $\tilde{u}$ satisfy integral functional equation (2.43) and we have the estimates

$$|u_j(x, y) - \tilde{u}_j(x, y)| \leq |\phi_j(0, g_j[u](0, x, y)) - \phi_j(0, g_j[\tilde{u}](0, x, y))|$$

$$+\int_0^x |\, f_j\,(\,\tau, g_j[u](\tau,x,y), u\,) - f_j\,(\,\tau, g_j[\tilde{u}](\tau,x,y), \tilde{u}\,)\,|\; d\tau$$

$$\leq p_1 \max_{1\leq i\leq k} \|g_i[u](0,x,y) - g_i[\tilde{u}](0,x,y)\|$$

$$+\int_0^x \beta_2(\tau,|d|)\left[\max_{1\leq i\leq k} \|g_i[u](\tau,x,y) - g_i[\tilde{u}](\tau,x,y)\|\right] d\tau$$

$$+\int_0^x \sigma(\,\tau, V(u-\tilde{u})\,)\, d\tau,$$

where $1 \leq j \leq k$. It follows from Lemma 2.10 that

$$\|g_i[u](\tau,x,y) - g_i[\tilde{u}](\tau,x,y)\| \leq \int_0^x \beta_1(\tau,|d|)\;\|u-\tilde{u}\|_\tau\, d\tau \exp\left[\int_0^a \beta_1(\tau,|d|)\, d\tau\right]$$

where $(x,y) \in E$, $0 \leq t \leq x$. Put

$$A = p_1 \exp\left[\int_0^a \beta_1(\tau,|d|)d\tau\right], \quad B = \int_0^a \beta_2(\tau,|d|)d\tau\, \exp\left[\int_0^a \beta_1(\tau,|d|)d\tau\right].$$

The above estimates imply

$$\|u(x,y) - \tilde{u}(x,y)\| \leq (A+B)\int_0^x \beta_1(\tau,|d|)\;\|u-\tilde{u}\|_\tau\, d\tau + \int_0^x \sigma(\,\tau, V(u-\tilde{u})\,)d\tau$$

where $(x,y) \in E$. Let $\alpha = V(u-\tilde{u})$. Then α satisfies the functional integral inequality

$$\alpha(x) \leq (A+B)\int_0^x \beta_1(\tau,|d|)\;\max\{\alpha(t):\; t\in[-r_0,\tau]\,\}\, d\tau$$

$$+\int_0^x \sigma(\,\tau,\alpha\,)\, d\tau, \quad x\in[0,a].$$

Since $\alpha(x) = 0$ on $[-r_0, 0]$ then Lemma 2.15 and the condition (iii) of Assumption $\tilde{H}\,[\,f\,]$ imply $\alpha(x) = 0$ on $[0,a]$ which completes the proof of the theorem.

Remark 2.17 *Suppose that the function ϱ, f and φ are defined by*

$$\varrho(x,y,z) = \tilde{\varrho}(x,y,z(x,y)), \quad f(x,y,z) = \tilde{f}(x,y,z(x,y)), \quad \varphi(x,y) = \tilde{\varphi}(y).$$

where $\tilde{\varrho}$, $\tilde{f}$, $\tilde{\varphi}$ are given functions. Then problem (2.29), (2.30) reduces to the Cauchy problem without the functional dependence. Note that in this case our assumptions on given functions are identical with adequate conditions in [45] where the systems without functional dependence were considered, see also [13 – 15].

We have been working under the assumption that given functions satisfy the Lipschitz condition with respect to (y,z). The following examples show that this assumption is essential.

Example 2.18 *Let $n = 1$ and*

$$F(p) = 0 \ \ for \ \ p < 0, \quad F(p) = \sqrt{p} \ \ for \ \ p \geq 0,$$

$$\varphi(y) = 0 \ \ for \ \ y < 0, \quad \varphi(y) = y \ \ for \ \ y \geq 0.$$

Consider the differential integral equation

$$D_x z(x,y) = \left[F(z(x,y)) + F\left(\int_0^y z(x,s)\, ds \right) \right] D_y z(x,y) \tag{2.54}$$

$$-\sqrt{f(x,y)} D_y z(x,y)$$

with the initial condition

$$z(0,y) = \varphi(y) \ \ for \ \ y \in R, \tag{2.55}$$

where

$$f(x,y) = \frac{1}{2}x^2 y + \frac{1}{2}y^2 + \frac{x}{12}\left[\sqrt{(x^2+4y)^3} - x^3 \right] \ \ for \ \ (x,y) \in [0,a] \times R_+,$$

$$f(x,y) = 0 \ \ for \ \ (x,y) \in [0,a] \times R_-$$

and $R_- = (-\infty, 0]$. It is easily seen that the functions

$$u_+(x,y) = \frac{1}{4}\left(x + \sqrt{x^2+4y} \right)^2, \quad (x,y) \in [0,a] \times (0,+\infty),$$

and

$$u_-(x,y) = 0, \quad (x,y) \in [0,a] \times (-\infty, 0)$$

are unique classical solutions of problem (2.54), (2.55) on $[0,a] \times (0,+\infty)$ and $[0,a] \times (-\infty,0)$ respectively. Let

$$v(x,y) = u_+(x,y) \ \ on \ \ [0,a] \times R_+, \quad v(x,y) = u_-(x,y) \ \ on \ \ [0,a] \times (-\infty,0).$$

The function v is not continuous at points $(x,0)$ for $x \in (0,a]$. It follows that problem (2.54), (2.55) has not the generalized solution on the set $[0,\varepsilon] \times [-b,b$ with $\varepsilon > 0$, $b > 0$.

Example 2.19 *Let $n = 1$ and*

$$\varphi(y) = 0 \ \ for \ \ y < 0, \qquad \varphi(y) = \sqrt{y} \ \ for \ \ y \geq 0.$$

Consider the Cauchy problem

$$D_x z(x,y) = \left[z(x,y) + \int_0^y z(x,s)\, ds \right] D_y z(x,y) - f(x,y)\, D_y z(x,y), \tag{2.56}$$

$$z(0,y) = \varphi(y) \ \ for \ \ y \in R, \tag{2.57}$$

where

$$f(x,y) = \frac{xy}{2} + \frac{1}{2}\left[\sqrt{(x^2+4y)^3} - x^3\right] \quad for \quad (x,y) \in [0,a] \times R_+,$$

$$f(x,y) = 0 \quad for \quad (x,y) \in R_-.$$

It is easy to check that the functions

$$u_+(x,y) = \frac{1}{2}\left(x + \sqrt{x^2+4y}\right) \quad for \quad (x,y) \in [0,a] \times (0,+\infty),$$

and

$$u_-(x,y) = 0 \quad for \quad (x,y) \in [0,a] \times (-\infty,0)$$

are unique solutions of problem (2.56), (2.57) on $[0,a] \times (0,+\infty)$ *and* $[0,a] \times (-\infty,0)$ *respectively. Let*

$$v(x,y) = u_+(x,y) \quad on \quad [0,a] \times R_+, \qquad v(x,y) = u_-(x,y) \quad on \quad [0,a] \times (-\infty,0).$$

The function v *is not continuous at points* $(x,0)$ *for* $x \in (0,a]$. *It follows that the problem (2.56), (2.57) has not the generalized solution on* $[0,\varepsilon] \times [-b,b]$ *with* $\varepsilon > 0,\ b > 0$.

Remark 2.20 *Differential systems with a deviated argument and integral differential problems can be obtained from (2.29) by specializing* ϱ *and* f. *Functional differential systems considered in [210] are particular cases of (2.29) under suitable assumptions on Volterra operators.*

Remark 2.21 *The theorem on the existence and uniqueness of Carathéodory solutions can be extended on quasilinear systems in the second canonical form*

$$\sum_{j=1}^{k} A_{ij}(x,y,z)\left[D_x z_j(x,y) + \sum_{l=1}^{n} \varrho_{il}(x,y,z)\, D_{y_l} z_j(x,y)\right] = f_i(x,y,z), \tag{2.58}$$

where $i = 1,\ldots,k$, *with the initial condition (2.30) and*

$$A : \Omega_0 \to M[k,k], \ \ A = [A_{ij}]_{i,j=1,\ldots,k}, \qquad \varrho : \Omega_0 \to M[k,n], \ \ f : \Omega_0 \to R^k$$

are given functions. We construct an integral operator corresponding to initial problem (2.31), (2.58). Suppose that there is $\nu > 0$ *such that* $\det A(x,y,z) > \nu$ *on* Ω_0. *Let* $(g_1[z](\cdot,x,y),\ldots,g_k[z](\cdot,x,y)) = g[z](\cdot,x,y)$ *be the bicharacteristics of (2.58) corresponding to* z. *The functions* $g_i[z](\cdot,x,y)$, $1 \le i \le k$, *are solutions of 2.38. Write*

$$f[g,z](t,x,y) = (\, f_1(t,g_1[z](t,x,y),z),\ldots,f_k(t,g_k[z](t,x,y),z)\,),$$

$$A[g,z](t,x,y) = [\, A_{ij}(t,g_i[z](t,x,y),z)\,]_{i,j=1,\ldots,k},$$

$$\phi[g,z](t,x,y) = [\, \phi_i(0,g_j[z](t,x,y))\,]_{i,j=1,\ldots,k},$$

$$Z[g,z](t,x,y) = [\, z_i(t, g_j[z](t,x,y)) \,]_{i,j=1,\dots,k} \,.$$

For matrices

$$U, V \in M[k,k], \quad U = [\, u_{ij} \,]_{i,j=1,\dots,k} \,, V = [\, v_{ij} \,]_{i,j=1,\dots,k} \,,$$

we denote by $U * V$ *the vector* $(d_1, \dots, d_k)$ *where*

$$d_i = \sum_{j=1}^{k} u_{ij} v_{ji}, \quad i = 1, \dots, k. \tag{2.59}$$

Let us define the operator T_ϕ *for all* $z \in K_{c.\phi}[\omega, d]$ *by the formula*

$$T_\phi z(x,y) = A^{-1}(x,y,z)\,\{A[g,z](0,x,y) * \phi[g,z](0,x,y)\}$$

$$+A^{-1}(x,y,z) \int_0^x \{D_t A[g,z](t,x,y) * Z[g,z](t,x,y) + f[g,z](t,x,y)\}\, dt$$

for $(x,y) \in E_c \setminus E_0$ *and* $T_\phi z(x,y) = \phi(x,y)$ *for* $(x,y) \in E_0$. *Under natural assumptions on* A, ϱ, f, ϕ *there exists exactly one fixed point* $\bar{z} \in K_{c.\phi}[\omega, d]$ *of the operator* T_ϕ *and this* $\bar{z}$ *is the Carathéodory solution of problem (2.31), (2.58). Details can be found in [73].*

Chapter 3

Numerical Methods for Initial Problems

3.1 Introduction

Let E be the Haar pyramid

$$E = \{(x, y) \in R^{1+n} : \ x \in [0, a), \ y \in [-b + Mx, \ b - Mx]\}$$

where $a > 0$, $b = (b_1, \ldots, b_n)$, $M = (M_1, \ldots, M_n) \in R^n_+$ and $b - Ma > 0$. Write

$$E_0 = [-r_0, 0] \times [-b, b] \ \text{ and } \ \Sigma = E \times R \times C(E_0 \cup E, R) \times R^n,$$

where $r_0 \in R_+$. Suppose that $f : \Sigma \to R$ and $\varphi : E_0 \to R$ are given functions. In this Section we consider functional difference inequalities generated by the Cauchy problem

$$D_x z(x, y) = f(\, x, y, z(x, y), z, \ D_y z(x, y)\,), \tag{3.1}$$

$$z(x, y) = \varphi(x, y) \ \text{ for } \ (x, y) \in E_0. \tag{3.2}$$

We assume that f satisfies the Volterra condition. We define a mesh on the set $E_0 \cup E$ in the following way. Let $\mathbf{N}$ and $\mathbf{Z}$ be the sets of natural numbers and integers respectively. For $y, \bar{y} \in R^n$, $y = (y_1, \ldots, y_n)$, $\bar{y} = (\bar{y}_1, \ldots, \bar{y}_n)$, we write $y * \bar{y} = (y_1\bar{y}_1, \ldots, y_n\bar{y}_n) \in R^n$. Suppose that (h_0, h') where $h' = (h_1, \ldots, h_n)$ stand for steps of the mesh. For $h = (h_0, h')$ and $(i, m) \in \mathbf{Z}^{1+n}$ where $m = (m_1, \ldots, m_n)$ we define nodal points as follows:

$$x^{(i)} = ih_0, \ \ y^{(m)} = m * h', \ \ y^{(m)} = (y_1^{(m_1)}, \ldots, y_n^{(m_n)}).$$

Denote by Δ the set of all $h = (h_0, h')$ such that there are $\tilde{N}_0 \in \mathbf{Z}$, $N = (N_1, \ldots, N_n) \in \mathbf{N}^n$ with the properties: $\tilde{N}_0 h_0 = r_0$ and $N * h' = b$. We assume that $\Delta \neq \emptyset$ and that there is a sequence $\{h^{(j)}\}$, $h^{(j)} \in \Delta$, and $\lim_{j\to\infty} h^{(j)} = 0$. For $h \in \Delta$ we put $|h| = h_0 + h_1 + \ldots + h_n$. There is $N_0 \in \mathbf{N}$ such that $N_0 h_0 < a \leq (N_0 + 1)h_0$. Let

$$R^{1+n}_h = \{\, (x^{(i)}, y^{(m)}) : \ (i, m) \in \mathbf{Z}^{1+n}\,\}$$

and

$$E_{0.h} = E_0 \cap R_h^{1+n}, \quad E_h = E \cap R_h^{1+n}.$$

In this Chapter we assume that $h' \le h_0 M$. For a function $z : E_{0.h} \cup E_h \to R$ we write $z^{(i,m)} = z(x^{(i)}, y^{(m)})$. Put

$$E_{j.h} = \{ (x^{(i)}, y^{(m)}) \in E_{0.h} \cup E_h : j \le i \},$$

$$\|z\|_{j.h} = \max \{ |z^{(i,m)}| : (x^{(i)}, y^{(m)}) \in E_{j.h} \},$$

where $0 \le j \le N_0$, and

$$E_h' = \{ (x^{(i)}, y^{(m)}) \in E_h : (x^{(i)} + h_0, y^{(m)}) \in E_h \}.$$

Let

$$S = \{ s = (s_1, \ldots, s_n) : s_i \in \{ -1, 0, 1 \} \text{ for } 1 \le i \le n \}.$$

We define the operators A, δ_0, $\delta = (\delta_1, \ldots, \delta_n)$ in the following way. For $z : E_{0.h} \cup E_h \to R$ we put

$$Az^{(i,m)} = \sum_{s \in S} a_s z^{(i,m+s)}, \quad \delta_0 z^{(i,m)} = \frac{1}{h_0} \left[z^{(i+1,m)} - Az^{(i,m)} \right], \tag{3.3}$$

and

$$\delta_j z^{(i,m)} = \frac{1}{h_j} \sum_{s \in S} c_{s.j} z^{(i,m+s)}, \quad j = 1, \ldots, n, \tag{3.4}$$

where a_s, $c_{s.j}$ are given numbers. Write $\delta z^{(i,m)} = (\delta_1 z^{(i,m)}, \ldots, \delta_n z^{(i,m)})$. We will approximate the derivatives $D_x z$ and $D_y z$ by the difference operators $\delta_0 z$ and δz respectively. Let $\Sigma_h = E_h' \times R \times \mathbf{F}(E_{0.h} \cup E_h, R) \times R^n$ and suppose that $f_h : \Sigma_h \to R$ and $\varphi_h : E_{0.h} \to R$, $h \in \Delta$, are given functions. We will approximate solutions of problem (3.1), (3.2) by means of solutions of the problem

$$\delta_0 z^{(i,m)} = f_h(x^{(i)}, y^{(m)}, Az^{(i,m)}, z, \delta z^{(i,m)}), \tag{3.5}$$

$$z^{(i,m)} = \varphi_h^{(i,m)} \quad \text{on } E_{0.h}. \tag{3.6}$$

The function f_h is said to satisfy the Volterra condition if for each $(x^{(i)}, y^{(m)}) \in E_h'$ there is a set $E[i,m]$ such that

(i) $E[i,m] \subset E_{i.h}$,

(ii) for $z, \bar{z} \in \mathbf{F}(E_{0.h} \cup E_h, R)$ such that $z = \bar{z}$ on $E[i,m]$ we have

$$f_h(x^{(i)}, y^{(m)}, p, z, q) = f_h(x^{(i)}, y^{(m)}, p, \bar{z}, q), \quad (p, q) \in R^{1+n}.$$

If f_h satisfies the Volterra condition then relation $h' \le h_0 M$ implies that there exists exactly one solution $u_h : E_{0.h} \cup E_h \to R$ of problem (3.5), (3.6). The motivation for the definition of the set E_h' is the following. Approximate solutions of problem (3.1), (3.2) are functions u_h defined on E_h. We will write a difference equation generated by (3.1) at each point of the set E_h'. It follows from condition $h' \le h_0 M$ that we calculate all the values of u_h on E_h.

Results of this Chapter are based on the papers [28, 29, 105, 115, 120, 121, 145]. Functional difference problems were considered also in [159 – 163] and [179 – 181]. Classical results concerning difference problems generated by first-order partial differential equations can be found in [133 – 135] and [178]. Nonlinear parabolic difference problems problems were considered in [22, 87, 88].

3.2 FUNCTIONAL DIFFERENCE INEQUALITIES

In this Section we show that the results on functional difference inequalities are consequences of a simple result concerned with the recurrent inequalities for monotone operators. To be more precise, the result we mean reads: if X is a partially ordered set, $F : X \to X$ is a nondecreasing operator and $\{x_i\}$, $\{y_i\}$ are sequences of elements of X then conditions $x_0 \le y_0$ and $x_{i+1} \le Fx_i$, $y_{i+1} \ge Fy_i$ for $i = 0, 1, \ldots$, imply that $x_i \le y_i$ for $i \ge 0$. Moreover, if X is also a linear space then assumptions $x_0 \le y_0$ and $x_{i+1} - Fx_i \le y_{i+1} - Fy_i$ for $i = 0, 1, \ldots$, imply $x_i \le y_i$ for $i \ge 0$. It means that the operator $\mathbf{\Phi}$ defined by the formula

$$\mathbf{\Phi}x = \begin{pmatrix} x_0 \\ x_{i+1} - Fx_i,\ i \ge 0 \end{pmatrix}$$

is inverse monotone, i.e. the $\mathbf{\Phi}x \le \mathbf{\Phi}y$ implies $x \le y$. Having this in mind we see that in order to establish a functional difference inequality result it is sufficient to check that the corresponding operator is monotone (or inverse monotone). This is the point which we will use in the Section. We adopt the ideas mentioned to the situation we have in considerations concerned with functional difference inequalities related to some classes of partial functional differential equations. We use general ideas presented in [115].

Suppose that $F_h : E'_h \times \mathbf{F}(E_{0.h} \cup E_h, R) \to R$ is a given function. For $z \in \mathbf{F}(E_{0.h} \cup E_h, R)$ and $(x^{(i)}, y^{(m)}) \in E'_h$ we write $F_h[(i,m), z] = F_h(x^{(i)}, y^{(m)}, z)$. We say that F_h satisfies the Volterra condition if for each $(x^{(i)}, y^{(m)}) \in E'_h$ there is a set $E[i,m]$ such that

(i) $E[i,m] \subset E_{i.h}$,

(ii) for $z, \bar{z} \in \mathbf{F}(E_{0.h} \cup E_h, R)$ such that $z = \bar{z}$ on $E[i,m]$ we have $F_h[(i,m), z] = F_h[(i,m), \bar{z}]$.

For a given function $\varphi_h \in \mathbf{F}(E_{0.h}, R)$ we consider the problem

$$z^{(i+1,m)} = F_h[(i,m), z], \quad z^{(i,m)} = \varphi_h^{(i,m)} \text{ on } E_{0.h}. \tag{3.7}$$

If F_h satisfies the Volterra condition then relation $h' \le h_0 M$ implies that there exists exactly one solution $u_h : E_{0.h} \cup E_h \to R$ of problem (3.7). Now we formulate a theorem on recurrent inequalities corresponding to (3.7).

Theorem 3.1 *Suppose that*

1) *the function F_h satisfies the Volterra condition and it is nondecreasing with respect to the functional variable,*

2) *the functions $u, v : E_{0.h} \cup E_h \to R$ satisfy the initial inequality*

$$u^{(i,m)} \le v^{(i,m)} \text{ on } E_{0.h} \tag{3.8}$$

and

$$u^{(i+1,m)} - F_h[(i,m), u] \le v^{(i+1,m)} - F_h[(i,m), v] \tag{3.9}$$

where $(x^{(i)}, y^{(m)}) \in E'_h$.

Then we have

$$u^{(i,m)} \le v^{(i,m)} \text{ on } E_h. \tag{3.10}$$

PROOF. We prove (3.10) by induction on i. It follows from (3.8) that assertion (3.10) is satisfied for $i = 0$ and for $(x^{(i)}, y^{(m)}) \in E_h$. Suppose now that $u^{(j,m)} \leq v^{(j,m)}$ for $(x^{(j)}, y^{(m)}) \in E_{i.h}$ where $0 \leq i < N_0$. Then

$$u^{(i+1,m)} - v^{(i+1,m)} \leq F_h[(i,m),u] - F_h[(i,m),v] \leq 0, \quad (x^{(i+1)}, y^{(m)}) \in E_h,$$

which completes the proof.

We formulate the main assumptions on the function f_h in (3.5).

Assumption H $[f_h]$. Suppose that

1) the function $f_h : \Sigma_h \to R$ of the variables (x, y, p, z, q) satisfies the Volterra condition and is nondecreasing with respect to the functional variable,

2) for each $P = (x, y, p, z, q) \in \Sigma_h$ there exist the derivatives

$$D_p f_h(P), \quad D_q f_h(P) = (D_{q_1} f_h(P), \ldots, D_{q_n} f_h(P))$$

and the functions

$$D_p f_h(x, y, \cdot, z, \cdot) : R^{1+n} \to R, \quad D_q f_h(x, y, \cdot, z, \cdot) : R^{1+n} \to R^n$$

are continuous,

3) for each $P \in \Sigma_h$ we have

$$a_s + h_0 a_s D_p f_h(P) + h_0 \sum_{j=1}^{n} \frac{1}{h_j} c_{s.j} D_{q_j} f_h(P) \geq 0, \quad s \in S. \tag{3.11}$$

Now we prove a theorem on functional difference inequalities generated by problem (3.5), (3.6).

Theorem 3.2 *Suppose that Assumption H* $[f_h]$ *is satisfied and*

1) *the functions* $u, v : E_{0.h} \cup E_h \to R$ *satisfy the initial inequality (3.8),*

2) *for* $(x^{(i)}, y^{(m)}) \in E'_h$ *we have*

$$\delta_0 u^{(i,m)} - f_h(x^{(i)}, y^{(m)}, Au^{(i,m)}, u, \delta u^{(i,m)}) \tag{3.12}$$

$$\leq \delta_0 v^{(i,m)} - f_h(x^{(i)}, y^{(m)}, Av^{(i,m)}, v, \delta v^{(i,m)}).$$

Under these assumptions we have $u^{(i,m)} \leq v^{(i,m)}$ on E_h.

PROOF. We apply Theorem 3.1 to prove the above assertion. We define the operator $F_h : E'_h \times \mathbf{F}(E_{0.h} \cup E_h, R) \to R$ by

$$F_h[(i,m), z] = Az^{(i,m)} + h_0 f_h(x^{(i)}, y^{(m)}, Az^{(i,m)}, z, \delta z^{(i,m)}).$$

Then the functions u, v satisfy inequality (3.9). It follows that F_h fulfils the Volterra condition. Now we prove that F_h is nondecreasing with respect to the functional variable. Suppose that $z \leq \bar{z}$. Let $w = z - \bar{z}$. Then we have

$$F_h[(i,m), z] - F_h[(i,m), \bar{z}] \leq Aw^{(i,m)}$$

$$+h_0\left[f_h(x^{(i)},y^{(m)},Az^{(i,m)},\bar{z},\delta z^{(i,m)})-f_h(x^{(i)},y^{(m)},A\bar{z}^{(i,m)},\bar{z},\delta\bar{z}^{(i,m)})\right]$$

$$=Aw^{(i,m)}+h_0D_pf_h(\bar{P})\,Aw^{(i,m)}+h_0\sum_{j=1}^{n}D_{q_j}f_h(\bar{P})\,\delta_jw^{(i,m)},$$

where $\bar{P}\in\Sigma_h$ is an intermediate point. It follows from (3.3), (3.4) that

$$F_h[(i,m),z]-F_h[(i,m),\bar{z}]$$

$$\leq\sum_{s\in S}w^{(i,m+s)}\left[a_s+h_0a_sD_pf_h(\bar{P})+h_0\sum_{j=1}^{n}\frac{1}{h_j}c_{s.j}\,D_{q_j}f_h(\bar{P})\right].$$

The above estimates and assumption (3.11) imply $F_h[(i,m),z]\leq F_h[(i,m),\bar{z}]$ which completes the proof of the monotonicity of the operator F_h. Now, Theorem 3.2 follows from Theorem 3.1.

We give examples of operators δ_0 and δ which are important in our theory. For $1\leq j\leq n$ we write $e_j=(0,\ldots,0,1,0,\ldots,0)\in R^n$, 1 standing on j - th place. Let $Az^{(i,m)}=z^{(i,m)}$. Then

$$\delta_0z^{(i,m)}=\frac{1}{h_0}\left[z^{(i+1,m)}-z^{(i,m)}\right]. \tag{3.13}$$

Suppose that the operator $\delta=(\delta_1,\ldots,\delta_n)$ is given by

$$\delta_jz^{(i,m)}=\frac{1}{h_j}\left[z^{(i,m)}-z^{(i,m-e_j)}\right]\ \text{ for }\ 1\leq j\leq\kappa, \tag{3.14}$$

$$\delta_jz^{(i,m)}=\frac{1}{h_j}\left[z^{(i,m+e_j)}-z^{(i,m)}\right]\ \text{ for }\ \kappa+1\leq j\leq n, \tag{3.15}$$

where κ is fixed natural number. Consider the functional difference equation

$$\delta_0z^{(i,m)}=f_h(x^{(i)},y^{(m)},z^{(i,m)},z,\delta z^{(i,m)}) \tag{3.16}$$

with δ_0, δ given by (3.13) - (3.15) and initial condition (3.6).

Remark 3.3 *Condition (3.11) for equation (3.16) with δ_0, δ given by (3.13) - (3.15) is equivalent to*

$$1+h_0D_pf_h(P)-h_0\sum_{j=1}^{n}\frac{1}{h_j}\,|D_{q_j}f_h(P)|\geq0, \tag{3.17}$$

and

$$D_{q_j}f_h(P)\leq0\ \ for\ \ 1\leq j\leq\kappa,\quad D_{q_j}f_h(P)\geq0\ \ for\ \ \kappa+1\leq j\leq n \tag{3.18}$$

where $P=(x,y,p,z,q)\in\Sigma_h$.

Let the operators A and δ_0 be given by

$$\delta_0 z^{(i,m)} = \frac{1}{h_0}\left[z^{(i+1,m)} - Az^{(i,m)}\right], \tag{3.19}$$

where

$$Az^{(i,m)} = \frac{1}{2n}\sum_{j=1}^{n}\left(z^{(i,m+e_j)} + z^{(i,m-e_j)}\right).$$

Write

$$\delta_j z^{(i,m)} = \frac{1}{2h_j}\left[z^{(i,m+e_j)} - z^{(i,m-e_j)}\right], \quad j = 1,\ldots,n. \tag{3.20}$$

and $\delta z^{(i,m)} = (\delta_1 z^{(i,m)},\ldots,\delta_n z^{(i,m)})$. Consider the functional difference equation

$$\delta_0 z^{(i,m)} = f_h(x^{(i)}, y^{(m)}, Az^{(i,m)}, z, \delta z^{(i,m)}) \tag{3.21}$$

with the above given δ_0, δ and initial condition (3.6).

Remark 3.4 *Condition (3.11) for equation (3.21) with δ_0 and δ given by (3.19), (3.20) is equivalent to*

$$1 + h_0 D_p f_h(P) - \frac{nh_0}{h_j}\,|D_{q_j} f_h(P)| \geq 0, \quad 1 \leq j \leq n, \tag{3.22}$$

where $P \in \Sigma_h$.

Remark 3.5 *All the theorems of this Section can be extended on the weakly coupled system of functional difference equations*

$$\delta_0 z_j^{(i,m)} = f_{h.j}(x^{(i)}, y^{(m)}, z^{(i,m)}, z,\ \delta z_j^{(i,m)}), \quad j = 1,\ldots,k,$$

where $f_h = (f_{h.1},\ldots,f_{h.k}) : E_h' \times R^k \times F(E_{0.h} \cup E_h, R^k) \times R^n \to R^k$ and $z = (z_1,\ldots,z_k)$. Some quasi-monotonicity conditions for f_h are assumed in this case.

3.3 Applications of functional difference inequalities

In this Section we consider initial problem (3.1), (3.2) and difference method (3.5), (3.6). We give sufficient conditions for the convergence of the sequence $\{u_h\}$, where u_h are solutions of problem (3.5), (3.6) to a solution of (3.1), (3.2). An error estimates implying the convergence of difference schemes are obtained by functional difference inequalities theorems.

Assumption H $[\sigma, C]$. Suppose that the function $\sigma : [0, a_0) \times R_+ \to R_+$ where $a_0 > a$, satisfies the conditions:

1) σ is continuous and $\sigma(x, 0) = 0$ for $x \in [0, a_0)$,

2) for each $C \in R_+$ the unique solution of the Cauchy problem

$$\omega'(x) = C\,\omega(x) + \sigma(x, \omega(x)), \quad \omega(0) = 0, \tag{3.23}$$

is $\omega(x) = 0$ for $x \in [0, a_0)$,

3) the function σ is nondecreasing with respect to both variables,

Let

$$I_{0.h} = \{x^{(i)} : -\tilde{N}_0 \le i \le 0\}, \quad I_h = \{x^{(i)} : 0 \le i \le N_0\}, \quad I'_h = I_h \setminus \{x^{(N_0)}\}.$$

For $\omega \in \mathbf{F}(I_{0.h} \cup I_h, R)$ we write $\omega^{(i)} = \omega(x^{(i)})$ and

$$\|\omega\|_{i.h} = \max\{|\omega^{(j)}| : -\tilde{N}_0 \le j \le i\}, \quad i = 0, \dots, N_0.$$

Assumption H $[f_h, \sigma, L_0]$. Suppose that

1) the function σ satisfies Assumption H $[\sigma, C]$ and

$$f_h(x^{(i)}, y^{(m)}, p, z, q) - f_h(x^{(i)}, y^{(m)}, p, \bar{z}, q) \le \sigma(x, \|z - \bar{z}\|_{i.h}) \tag{3.24}$$

on Σ_h, where $z \ge \bar{z}$ on $E_{0.h} \cup E_h$,

2) there exists $L_0 \in R_+$ such that $|D_p f_h(x, y, p, z, q)| \le L_0$ on Σ_h.

Assumption H $[A, \delta]$. Suppose that

1) the operators A and δ satisfy the conditions:

$$\sum_{s \in S} a_s = 1, \qquad \sum_{s \in S} c_{s.j} = 0 \ \text{ for } \ 1 \le j \le n \tag{3.25}$$

and

$$\sum_{s \in S} s_j\, a_s = 0, \quad \sum_{s \in S} s_j c_{s.i} = \delta_{ij}, \quad i, j = 1, \dots, n, \tag{3.26}$$

where δ_{ij} is the Kronecker symbol,

2) there exists $C_0 > 0$ such that $h_i h_j^{-1} \le C_0$ for $i, j = 1, \dots, n$.

Remark 3.6 *Suppose that Assumption H $[A, \delta]$ is satisfied. Let $v \in C(E_0 \cup E, R)$. If the function $v|_E$ is of class C^2 and the derivatives of the second order of v are bounded on E then there is $C > 0$ such that*

$$|\delta_0 v^{(i,m)} - D_x v^{(i,m)}| \le C|h|, \quad \|\delta v^{(i,m)} - D_y v^{(i,m)}\| \le C|h|,$$

$$|A v^{(i,m)} - v^{(i,m)}| \le C|h|$$

on E'_h.

We omit the simple proof of the above properties of the operators δ_0, δ, A. Now we give an estimate between the exact and approximate solutions of the initial problem (3.5), (3.6).

Theorem 3.7 *Suppose that Assumptions H $[f_h]$ and H $[f_h, \sigma, L_0]$ are satisfied and*

1) the operators A and δ satisfy condition (3.25) of Assumption H $[A, \delta]$,

2) the function $u_h : E_{0.h} \cup E_h \to R$ is a solution of problem (3.5), (3.6),

3) the function $v_h : E_{0.h} \cup E_h \to R$ is such that there are functions γ_0, γ : $\Delta \to R_+$ satisfying

$$|u_h^{(i,m)} - v_h^{(i,m)}| \le \gamma_0(h) \ \text{ on } \ E_{0.h}, \tag{3.27}$$

$$\left| \delta_0 v_h^{(i,m)} - f_h(x^{(i)}, y^{(m)}, Av_h^{(i,m)}, v_h, \delta v_h^{(i,m)}) \right| \leq \gamma(h) \text{ on } E'_h \tag{3.28}$$

and

$$\lim_{h \to 0} \gamma_0(h) = 0, \quad \lim_{h \to 0} \gamma(h) = 0. \tag{3.29}$$

Under these assumptions there exists $\varepsilon_0 > 0$ *and a function* $\omega_h : [0, a) \to R_+$, $h \in \Delta$, *such that for* $|h| < \varepsilon_0$ *we have*

$$\|u_h - v_h\|_{i.h} \leq \omega_h^{(i)}, \quad i = 0, 1, \ldots, N_0, \tag{3.30}$$

and

$$\lim_{h \to 0} \omega_h(x) = 0 \ \ uniformly\ with\ respect\ to\ x \in [0, a). \tag{3.31}$$

PROOF. Denote by ω_h the solution of the problem

$$\omega'(x) = L_0\, \omega(x) + \sigma(\, x, \omega(x)\,) + \gamma(h), \quad \omega(0) = \gamma_0(h). \tag{3.32}$$

It follows from Assumption H $[\sigma, C]$ that there is $\varepsilon_0 > 0$ such that for $|h| < \varepsilon_0$ the solution ω_h exists on $[0, a)$ and satisfies condition (3.31). Let the function $\tilde{v}_h : E_{0.h} \cup E_h \to R$ be defined by

$$\tilde{v}_h^{(i,m)} = v_h^{(i,m)} + \omega_h^{(i)} \text{ on } E_h, \qquad \tilde{v}_h^{(i,m)} = v_h^{(i,m)} + \omega_h^{(0)} \text{ on } E_{0.h}.$$

We prove that

$$u_h^{(i,m)} \leq \tilde{v}_h^{(i,m)} \quad \text{on} \quad E_h. \tag{3.33}$$

We apply Theorem 3.2 to prove (3.33). It follows from the definition of $\tilde{v}_h$ and from (3.27), (3.32) that $u_h^{(i,m)} \leq \tilde{v}_h^{(i,m)}$ on $E_{0.h}$. We prove that

$$\delta_0 \tilde{v}_h^{(i,m)} \geq f_h(x^{(i)}, y^{(m)}, A\tilde{v}_h^{(i,m)}, \tilde{v}_h, \delta\, \tilde{v}_h^{(i,m)}) \text{ on } E'_h. \tag{3.34}$$

It follows from the definition of the function $\tilde{v}_h$ and from Assumption H $[\,f_h, \sigma, L_0\,]$ that

$$\delta_0 \tilde{v}_h^{(i,m)} - f_h(x^{(i)}, y^{(m)}, A\tilde{v}_h^{(i,m)}, \tilde{v}_h, \delta\, \tilde{v}_h^{(i,m)}) \tag{3.35}$$

$$= \delta_0 v_h^{(i,m)} - f_h(x^{(i)}, y^{(m)}, Av_h^{(i,m)}, v_h, \delta\, v_h^{(i,m)}) + \frac{1}{h_0}\left[\omega_h^{(i+1)} - \omega_h^{(i)}\right]$$

$$+ \left[f_h(x^{(i)}, y^{(m)}, Av_h^{(i,m)}, v_h, \delta\, v_h^{(i,m)}) - f_h(x^{(i)}, y^{(m)}, A\tilde{v}_h^{(i,m)}, \tilde{v}_h, \delta\, v_h^{(i,m)}) \right]$$

$$\geq -\gamma(h) - L_0\, \omega_h^{(i)} - \sigma(\, x^{(i)}, \omega_h^{(i)}\,) + \frac{1}{h_0}\left[\omega_h^{(i+1)} - \omega_h^{(i)}\right],$$

where $(x^{(i)}, y^{(m)}) \in E'_h$. It follows from the condition 3) of Assumption H $[\sigma, C]$ that ω_h is a convex function and therefore it satisfies the difference inequality

$$\frac{1}{h_0}\left[\omega_h^{(i+1)} - \omega_h^{(i)}\right] \geq L_0\, \omega_h^{(i)} + \sigma(\, x^{(i)}, \omega_h^{(i)}\,) + \gamma(h), \ \ x^{(i)} \in I'_h.$$

Now we obtain (3.34) from (3.35) and from the above inequality. Using Theorem 3.2 we get estimate (3.33). In a similar way we prove that

$$\tilde{u}_h^{(i,m)} \leq u_h^{(i,m)} \text{ on } E_h \tag{3.36}$$

where

$$\tilde{u}_h^{(i,m)} = v_h^{(i,m)} - \omega_h^{(i)} \text{ on } E_h, \quad \tilde{u}_h^{(i,m)} = v_h^{(i,m)} - \omega_h^{(0)} \text{ on } E_{0.h}.$$

Since the function ω_h is nondecreasing on $[0, a)$, we obtain inequality (3.30) from estimates (3.33), (3.36). Condition (3.31) follows from Assumption H $[\sigma, C]$ and from (3.29). This completes the proof of the Theorem.

Remark 3.8 *Suppose that all the assumptions of Theorem 3.7 are satisfied with $\sigma(x, p) = L_1 p$ where $L_1 \in R_+$. Then condition (3.24) takes the form*

$$f_h(x^{(i)}, y^{(m)}, p, z, q) - f_h(x^{(i)}, y^{(m)}, p, \bar{z}, q) \leq L_1 \|z - \bar{z}\|_{i.h}$$

on Σ_h where $z \geq \bar{z}$ on $E_{0.h} \cup E_h$. Let $L = L_0 + L_1$. In this case we have the error estimates

$$\|u_h - v_h\|_{i.h} \leq \gamma_0(h) \exp\left(Lx^{(i)}\right) + \gamma(h) \frac{\exp\left(Lx^{(i)}\right) - 1}{L} \quad \text{if } L > 0$$

and

$$\|u_h - v_h\|_{i.h} \leq \gamma_0(h) + x^{(i)} \gamma(h) \text{ if } L = 0$$

where $0 \leq i \leq N_0$.

Theorem 3.9 *Suppose that Assumptions H $[f_h]$, H $[f_h, \sigma, L_0]$, H $[A, \delta]$ are satisfied and*

1) the function $v : E_0 \cup E \to R$ is a solution of problem (3.1), (3.2), the function $v|_E$ is of class C^2 and the derivatives of the second order of the function $v|_E$ are bounded,

2) the functions $\varphi : E_0 \to R$ and $f : \Sigma \to R$ are continuous and $u_h : E_{0.h} \cup E_h \to R$ is a solution of problem (3.5), (3.6),

3) the function v_h is the restriction of v to the set $E_{0.h} \cup E_h$ and there exists a function $\beta_0 : \Delta \to R_+$ such that the following compatibility condition is satisfied :

$$\left| f_h(x^{(i)}, y^{(m)}, v_h^{(i,m)}, v_h, \delta v_h^{(m)}) - f(x^{(i)}, y^{(m)}, v^{(i,m)}, v, \delta v^{(i,m)}) \right| \leq \beta_0(h) \tag{3.37}$$

on E'_h and

$$\lim_{h \to 0} \beta_0(h) = 0, \tag{3.38}$$

4) there exists a function $\gamma_0 : \Delta \to R_+$ such that

$$|\varphi^{(i,m)} - \varphi_h^{(i,m)}| \leq \gamma_0(h) \text{ on } E_{0.h} \text{ and } \lim_{h \to 0} \gamma_0(h) = 0. \tag{3.39}$$

Then there exist a number $\varepsilon_0 > 0$ and a function $\alpha : \Delta \to R_+$ such that for $h \in \Delta$, $|h| < \varepsilon_0$, we have

$$\|u_h - v_h\|_{i.h} \leq \alpha(h) \text{ for } 0 \leq i \leq N_0 \text{ and } \lim_{h \to 0} \alpha(h) = 0. \tag{3.40}$$

PROOF. It follows from the compatibility condition that there exists a function $\gamma : \Delta \to E_+$ such that $\lim_{h\to 0}\gamma(h) = 0$ and

$$\left| \delta_0 v_h^{(i,m)} - f_h(x^{(i)}, y^{(m)}, Av_h^{(i,m)}, v_h, \delta v_h^{(i,m)}) \right| \le \gamma(h) \text{ on } E'_h.$$

Using Theorem 3.11 we get (3.40) for $\alpha(h) = \omega_h(a)$ where ω_h is the solution of problem (3.32).

Remark 3.10 *It is easy to see that the operator A given by $Az^{(m)} = z^{(i,m)}$ and δ given by (3.14), (3.15) satisfy conditions (3.25), (3.26). The operators A, δ given by (3.19), (3.20) satisfy conditions (3.25), (3.26).*

3.4 Almost linear problems

Let $g_h : E'_h \times R \times \mathbf{F}(E_{0.h} \cup E_h) \to R$ and $G = (g_1, \dots, g_n) : E'_h \to R^n$ be given functions. Consider the functional difference equation

$$\delta_0 z^{(i,m)} = g_h(x^{(i)}, y^{(m)}, z^{(i,m)}, z) + \sum_{j=1}^{n} g_j(x^{(i)}, y^{(m)})\, \delta_j z^{(i,m)} \tag{3.41}$$

with initial condition (3.6) where δ_0 and δ are given by (3.13) - (3.15). Write $g_j^{(i,m)} = g_j(x^{(i)}, y^{(m)})$ for $1 \le j \le n$. If we apply Theorem 3.2 to equation (3.41) then we need the following assumption on G (see (3.18)): for each j, $1 \le j \le n$, the function

$$\tilde{g}_j(x, y) = \text{sign}\, g_j(x, y), \quad (x, y) \in E'_h,$$

is constant on E'_h. Now we prove that this assumption can be omitted if we define

$$\delta_j z^{(i,m)} = \frac{1}{h_j}\left[z^{(i,m)} - z^{(i,m-e_j)}\right] \text{ if } g_j^{(i,m)} \le 0, \tag{3.42}$$

$$\delta_j z^{(i,m)} = \frac{1}{h_j}\left[z^{(i,m+e_j)} - z^{(i,m)}\right] \text{ if } g_j^{(i,m)} > 0. \tag{3.43}$$

Assumption H $[g_h, G]$. Suppose that

1) the function g_h satisfies the Volterra condition and it is nondecreasing with respect to the functional variable,

2) the derivative $D_p g_h$ exists on $E'_h \times R \times \mathbf{F}(E_{0.h} \cup E_h, R)$,

3) the following estimate is satisfied on $E'_h \times R \times \mathbf{F}(E_{0.h} \cup E_h, R)$

$$1 + h_0 g_h(x, y, p, z) - h_0 \sum_{j=1}^{n} \frac{1}{h_j} |\, g_j(x, y)\,| \ge 0. \tag{3.44}$$

Theorem 3.11 *Suppose that Assumption H $[g_h, G]$ holds true and*

1) the functions u, $v : E_{0.h} \cup E_h \to R$ satisfy initial inequality (3.8),

2) for $(x^{(i)}, y^{(m)}) \in E'_h$ we have

$$\delta_0 u^{(i,m)} - g_h(x^{(i)}, y^{(m)}, z^{(i,m)}, z) - \sum_{j=1}^{n} g_j^{(i,m)} \delta_j u^{(i,m)} \tag{3.45}$$

$$\leq \delta_0 v^{(i,m)} - g_h(x^{(i)}, y^{(m)}, v^{(i,m)}, v) - \sum_{j=1}^{n} g_j^{(i,m)} \delta_j v^{(i,m)}$$

where δ_0 and δ are given by (3.13) and (3.42), (3.43) respectively.

Under these assumptions we have

$$u^{(i,m)} \leq v^{(i,m)} \text{ on } E_h. \tag{3.46}$$

PROOF. We define $F_h : E'_h \times \mathbf{F}(E_{0.h} \cup E_h, R) \to R$ as follows

$$F_h[(i,m), z] = z^{(i,m)} + h_0 g_h(x^{(i)}, y^{(m)}, z^{(i,m)}, z) + h_0 \sum_{j=1}^{n} g_j^{(i,m)} \delta_j z^{(i,m)}.$$

Then the functions u and v satisfy inequality (3.9). Now we prove that the operator F_h is nondecreasing with respect to z. Suppose that $z \leq \bar{z}$. Let $w = z - \bar{z}$ and $(x^{(i)}, y^{(m)}) \in E'_h$. Write

$$I^-[i,m] = \{j : 1 \leq j \leq n,\ g_j(x^{(i)}, y^{(m)}) \leq 0\},$$

$$I^+[i,m] = \{1, \ldots, n\} \setminus I^-[i,m].$$

Then there exists $\tilde{P} = (x^{(i)}, y^{(m)}, \tilde{p}, z) \in E'_h \times R \times \mathbf{F}(E_{0.h} \cup E_h, R)$ such that

$$F_h[(i,m), z] - F_h[(i,m), \bar{z}]$$

$$\leq w^{(i,m)} \left[1 + h_0 D_p g_h(\tilde{P}) + h_0 \sum_{j \in I^-[i,m]} \frac{1}{h_j} g_j^{(i,m)} - h_0 \sum_{j \in I^+[i,m]} \frac{1}{h_j} g_j^{(i,m)} \right]$$

$$- h_0 \sum_{j \in I^-[i,m]} \frac{1}{h_j} g_j^{(i,m)} w^{(i,m-e_j)} + h_0 \sum_{j \in I^+[i,m]} \frac{1}{h_j} g_j^{(i,m)} w^{(i,m+e_j)}$$

$$\leq w^{(i,m)} \left[1 + h_0 D_p g_h(\tilde{P}) - h_0 \sum_{j=1}^{n} \frac{1}{h_j} | g_j^{(i,m)} | \right].$$

It follows from assumption (3.44) that $F_h[(i,m), z] - F_h[(i,m), \bar{z}] \leq 0$. Now, we obtain inequality (3.46) from Theorem 3.1.

3.5 Error estimates of approximate solutions

Let E, E_0, E_h, $E_{0.h}$ and E'_h, $E_{i.h}$ be the sets given in Section 3.1. We assume that the step h of the mesh satisfies the condition: $h' \leq Mh_0$. Let $F_h : E'_h \times \mathbf{F}(E_{0.h} \cup E_h, R) \to R$ and $\varphi_h : E_{0.h} \to R$ be given functions. We consider the initial problem

$$z^{(i+1,m)} = F_h[\,(i,m), z\,], \qquad z^{(i,m)} = \varphi_h^{(i,m)} \text{ on } E_{0.h}. \tag{3.47}$$

We assume that F_h satisfies the Volterra condition. Then there exists exactly one solution $u_h : E_{0.h} \cup E_h \to R$ of problem (3.47). Let I_h, $I_{0.h}$ and I'_h be the sets defined in Section 3.3. Very frequently we need to know what is the relation between the solution $u_h : E_{0.h} \cup E_h \to R$ of problem (3.47) and a function $v_h : E_{0.h} \cup E_h \to R$ satisfying the relations

$$|\,v_h^{(i+1,m)} - F_h[\,(i,m), v_h\,]\,| \leq \gamma_h^{(i)} \text{ for } (x^{(i)}, y^{(m)}) \in E'_h \tag{3.48}$$

and

$$|\,v_h^{(i,m)} - u_h^{(i,m)}\,| \leq \gamma_{0.h}^{(i)} \text{ on } E_{0.h} \tag{3.49}$$

for some functions $\gamma_h \in \mathbf{F}(I'_h, R_+)$ and $\gamma_{0.h} \in \mathbf{F}(I_{0.h}, R_+)$. The function v_h satisfying the above relations is considered as an approximate solution of problem (3.47). Notice that v_h can be a solution of the recurrent equation with the initial condition

$$z^{(i+1,m)} = \tilde{F}_h[\,(i,m), z\,], \qquad z^{(i,m)} = \tilde{\varphi}_h^{(i,m)} \text{ on } E_{0.h}$$

with the mappings $\tilde{F}_h : E'_h \times \mathbf{F}(E_{0.h} \cup E_h, R) \to R$ and $\tilde{\varphi}_h : E_{0.h} \to R$ satisfying

$$|\,F_h[\,(i,m), v_h\,] - \tilde{F}_h[\,(i,m), v_h\,]\,| \leq \gamma_h^{(i)} \text{ on } E'_h$$

and

$$|\,\varphi_h^{(i,m)} - \tilde{\varphi}_h^{(i,m)}\,| \leq \gamma_{0.h}^{(i)} \text{ on } E_{0.h}.$$

We prove a theorem on the estimation of the difference between the exact and approximate solutions of (3.47). Suppose that we have a function

$$\sigma_h : I'_h \times \mathbf{F}(I_{0.h} \cup I_h, R_+) \to R_+.$$

If $(x^{(i)}, \omega) \in I'_h \times \mathbf{F}(I_{0.h} \cup I_h, R_+)$ then we write $\sigma_h[i, \omega] = \sigma_h(x^{(i)}, \omega)$. The function σ_h is said to satisfy the Volterra condition if for each i, $0 \leq i \leq N_0 - 1$, and for $\omega, \bar{\omega} \in \mathbf{F}(I_{0.h} \cup I_h, R)$ such that $\omega^{(j)} = \bar{\omega}^{(j)}$ where $-\tilde{N}_0 \leq j \leq i$, we have $\sigma_h[i, \omega] = \sigma_h[i, \bar{\omega}]$. In the sequel we will need the following operator

$$V_h : \mathbf{F}(E_{0.h} \cup E_h, R) \to \mathbf{F}(I_{0.h} \cup I_h, R_+).$$

If $z : E_{0.h} \cup E_h \to R$ then $V_h z$ is given by

$$(V_h z)(x^{(i)}) = \max\{\,|\,z^{(i,m)}\,| : \ (x^{(i)}, y^{(m)}) \in E_{0.h} \cup E_h\,\}, \quad -\tilde{N}_0 \leq i \leq N_0.$$

Theorem 3.12 *Suppose that*

1) *the function F_h satisfies the Volterra condition and there exists a function*

$$\sigma_h : I'_h \times \mathbf{F}(I_{0.h} \cup I_h, R_+) \to R_+$$

such that

(i) *σ_h is nondecreasing with respect to the functional variable and satisfies the Volterra condition,*

(ii) *for $z, \bar{z} \in \mathbf{F}(E_{0.h} \cup E_h, R)$ we have on E'_h*

$$|\, F_h[\,(i,m), z\,] - F_h[\,(i,m), \bar{z}\,]\,| \leq \sigma_h[\,i, V_h(z - \bar{z})\,], \tag{3.50}$$

2) *the function $u_h : E_{0.h} \cup E_h \to R$ is a solution of problem (3.47),*

3) *the functions $v_h : E_{0.h} \cup E_h \to R$ and $\gamma_h : I'_h \to R_+$, $\gamma_{0.h} : I_{0.h} \to R_+$ are such that estimation (3.48) is satisfied and initial inequality (3.49) holds,*

4) *the function $\beta_h : I_{0.h} \cup I_h \to R_+$ satisfies the recurrent inequality*

$$\beta_h^{(i+1)} \geq \sigma_h[\,i, \beta_h\,] + \gamma_h^{(i)} \quad for \quad 0 \leq i \leq N_0 - 1, \tag{3.51}$$

and the initial condition holds

$$\beta_h^{(i)} = \gamma_{0.h}^{(i)} \quad on \quad I_{0.h}. \tag{3.52}$$

Under these assumptions we have

$$|\, u_h^{(i,m)} - v_h^{(i,m)}\,| \leq \beta_h^{(i)} \tag{3.53}$$

on E_h.

PROOF. We prove the assertion (3.53) by induction on i. It follows from the initial estimation (3.49) that inequality (3.53) is satisfied for $i = 0$. Suppose that (3.53) holds true on $E_{i.h}$. Then we have

$$V_h(u_h - v_h)^{(j)} \leq \beta_h^{(j)} \quad \text{for} \quad -\tilde{N}_0 \leq j \leq i.$$

It follows from (3.50) - (3.52) that

$$|\, u_h^{(i+1,m)} - v_h^{(i+1,m)}\,| \leq |\, F_h[\,(i,m), u_h\,] - F_h[\,(i,m), v_h\,]\,|$$

$$+|\, F_h[\,(i,m), v_h\,] - v_h^{(i+1,m)}\,|$$

$$\leq \sigma_h[\,i, V_h(u_h - v_h)\,] + \gamma_h^{(i)} \leq \sigma_h[\,i, \beta_h\,] + \gamma_h^{(i)} \leq \beta_h^{(i+1)}$$

where $(x^{(i+1)}, y^{(m)}) \in E_h$. Hence, the proof is completed by induction.

Remark 3.13 *If all the assumptions of Theorem 3.12 are satisfied and the function β_h is nondecreasing on $I_{0.h} \cup I_h$ then $\|\, u_h - v_h\,\|_{i.h} \leq \beta_h^{(i)}$ for $0 \leq i \leq N_0$.*

Remark 3.14 *In applications we consider the function* $\beta_h : I_{0.h} \cup I_h \to R$ *satisfying (3.51), (3.52) such that* $\lim_{h\to 0} \beta_h^{(i)} = 0$ *for* $0 \le i \le N_0$. *This implies* $\lim_{h\to 0} |\, u_h^{(i,m)} - v_h^{(i,m)} \,| = 0$ *on* E_h.

Remark 3.15 *Theorem 3.12 enables us to get estimates of the function* $u_h - v_h$ *of the variables* $(x, y) \in E_{0.h} \cup E_h$ *by means of the function* β_h *of one variable. This estimate is convenient for applications in difference methods for hyperbolic functional differential problems. It is easy to obtain a more general estimate of the function* $u_h - v_h$ *by means of a function depending on* $(x, y) \in E_{0.h} \cup E_h$.

For a function $z : E_{0.h} \cup E_h \to R$ we define a function $W_h z : E_{0.h} \cup E_h \to R_+$ as follows: $W_h z^{(i,m)} = |\, z^{(i,m)} \,|$, $(x^i, y^{(m)}) \in E_{0.h} \cup E_h$.

Lemma 3.16 *Suppose that*

1) *the function* $F_h : E'_h \times \mathbf{F}(E_{0.h} \cup E_h, R) \to R$ *satisfies the Volterra condition and there exists a function*

$$Q_h : E'_h \times \mathbf{F}(E_{0.h} \cup E_h, R_+) \to R_+$$

such that

(i) Q_h *is nondecreasing with respect to the functional variable and satisfies the Volterra condition,*

(ii) *for* z, $\bar{z} \in \mathbf{F}(E_{0.h} \cup E_h, R)$ *we have on* E'_h

$$|\, F_h[\,(i,m), z\,] - F_h[\,(i,m), \bar{z}\,] \,| \le Q_h[\,(i,m), W_h(z - \bar{z})\,],$$

2) *the function* $u_h : E_{0.h} \cup E_h \to R$ *is a solution of problem (3.47),*

2) *the functions* $v_h : E_{0.h} \cup E_h \to R$ *and* $\tilde{z}_h : E_{0.h} \cup E_h \to R_+$, $\tilde{\eta}_h : E'_h \to R_+$ *satisfy the conditions:*

(i) *the initial estimate is satisfied*

$$|\, u_h^{(i,m)} - v_h^{(i,m)} \,| \le \tilde{z}_h^{(i,m)} \quad \text{on } E_{0.h},$$

(ii) *for* $(x^{(i)}, y^{(m)}) \in E'_h$ *we have*

$$|\, v_h^{(i+1,m)} - F_h[\,(i,m), v_h\,] \,| \le \tilde{\eta}_h^{(i,m)}$$

and

$$\tilde{z}_h^{(i+1,m)} \ge Q_h[\,(i,m), \tilde{z}_h\,] + \tilde{\eta}_h^{(i,m)}, \quad (x^{(i)}, y^{(m)}) \in E'_h.$$

Under these assumptions we have $|\, u_h^{(i,m)} - v_h^{(i,m)} \,| \le \tilde{z}_h^{(i,m)}$ *on* E_h.

We omit the simple proof of Lemma 3.16.

3.6 Difference methods for nonlinear equations

Let E and E_0 be the sets defined in Section 3.1 and $\Omega = E \times C(E_0 \cup E, R) \times R^n$. Suppose that $f : \Omega \to R$ and $\varphi : E_0 \to R$ are given functions We consider the Cauchy problem

$$D_x z(x, y) = f(\,x, y, z,\, D_y z(x, y)\,), \tag{3.54}$$

$$z(x,y) = \varphi(x,y) \text{ for } (x,y) \in E_0. \tag{3.55}$$

We assume that the function f satisfies the Volterra condition. We give sufficient conditions for the convergence of a sequence $\{u_h\}$ where u_h are solutions of difference problems generated by (3.54), (3.55) to a classical solution of the original problem. Suppose that E_h, $E_{0.h}$, E'_h and $E_{i.h}$, S are the sets defined in Section 3.1. We assume that $h' \le h_0 M$. Put $\Omega_h = E'_h \times \mathbf{F}(E_{0.h} \cup E_h, R) \times R^n$ and assume that the operators A, δ_0, δ are given by (3.3), (3.4). Suppose that $f_h : \Omega_h \to R$ and $\varphi_h : E_{0.h} \to R$ are given functions. We will approximate solutions of problem (3.54), (3.55) by solutions of the problem

$$\delta_0 z^{(i,m)} = f_h(x^{(i)}, y^{(m)}, z, \delta z^{(i,m)}), \ (x^{(i)}, y^{(m)}) \in E'_h, \tag{3.56}$$

$$z^{(i,m)} = \varphi_h^{(i,m)} \text{ on } E_{0.h}. \tag{3.57}$$

We will assume the same conditions for f and f_h with respect to the functional variable z and the real variable p. Therefore we omit the variable $z(x,y)$ in the functional differential problem and in the difference equation. We assume that the operator f_h satisfies the Volterra condition. Then there exists exactly one solution $u_h : E_{0.h} \cup E_h \to R$ of problem (3.56), (3.57). We omit the assumption on the monotonicity of the function f_h with respect to the functional variable.

Assumption H $[\sigma_h]$. Suppose that the function $\sigma_h : I'_h \times \mathbf{F}(I_{0.h} \cup I_h, R_+) \to R_+$ satisfies the conditions:

1) σ_h is nondecreasing with respect to the functional variable and fulfils the Volterra condition,

2) $\sigma_h[i, \theta_h] = 0$ for $0 \le i \le N_0 - 1$ where $\theta_h^{(j)} = 0$ for $-\tilde{N}_0 \le j \le N_0$ and the difference problem

$$\eta^{(i+1)} = \eta^{(i)} + h_0\, \sigma_h[i, \eta] \text{ for } 0 \le i \le N_0 - 1, \tag{3.58}$$

$$\eta^{(i)} = 0 \text{ for } -\tilde{N}_0 \le i \le 0, \tag{3.59}$$

is stable in the following sense: if η_h is a solution of the problem

$$\eta^{(i+1)} = \eta^{(i)} + h_0\, \sigma_h[i, \eta] + h_0 \gamma(h) \text{ for } 0 \le i \le N_0 - 1, \tag{3.60}$$

$$\eta^{(i)} = \alpha_0(h) \text{ for } -\tilde{N}_0 \le i \le 0, \tag{3.61}$$

where γ, $\alpha_0 : \Delta \to R_+$ and $\lim_{h\to 0} \gamma(h) = 0$, $\lim_{h \to 0} \alpha_0(h) = 0$, then there exists a function $\beta : \Delta \to R_+$ such that $\eta_h^{(i)} \le \beta(h)$ for $0 \le i \le N_0$ and $\lim_{h\to 0} \beta(h) = 0$.

Assumption H $[f_h, \sigma_h]$. Suppose that the function f_h satisfies the Volterra condition and

1) the derivatives

$$D_q f_h = (D_{q_1} f_h, \ldots, D_{q_n} f_h)$$

exist on Ω_h and

$$D_q f_h(x, y, z, \cdot) \in C(R^n, R^n),$$

2) for each $s \in S$ we have

$$a_s + h_0 \sum_{i=1}^{n} \frac{1}{h_i} c_{s.i}\, D_{q_i} f_h(x, y, z, q) \ge 0 \text{ on } \Omega_h, \tag{3.62}$$

3) Assumption H $[\sigma_h]$ is satisfied and

$$|f_h(x,y,z,q) - f_h(x,y,\bar{z})| \le \sigma_h(x, V_h(z-\bar{z})) \text{ on } \Omega_h. \tag{3.63}$$

Now we prove a theorem on the convergence of the method (3.56), (3.57).

Theorem 3.17 *Suppose that Assumptions H $[f_h, \sigma_h]$, H $[A,\delta]$ are satisfied and*

1) the function $u_h : E_{0.h} \cup E_h \to R$ is a solution of problem (3.56), (3.57) and there exists $\alpha_0 : \Delta \to R_+$ such that $|\varphi^{(i,m)} - \varphi_h^{(i,m)}| \le \alpha_0(h)$ on $E_{0.h}$ and $\lim_{h\to 0}\alpha_0(h) = 0$,

2) the function $v \in C(E_0 \cup E, R)$ is a solution of problem (3.54), (3.55), the function $v|_E$ is of class C^2 and the derivatives of the second order of $v|_E$ are bounded,

3) there exists a function $\tilde{\beta} : \Delta \to R_+$ such that

$$\left| f_h(x^{(i)}, y^{(m)}, v_h, \delta v_h^{(i,m)}) - f(x^{(i)}, y^{(m)}, v, \delta v^{(i,m)}) \right| \le \tilde{\beta}(h) \text{ on } E_h' \tag{3.64}$$

and $\lim_{h\to 0}\tilde{\beta}(h) = 0$ where v_h is the restriction of the function v to the set $E_{0.h} \cup E_h$.

Then there exists a function $\tilde{\gamma} : \Delta \to R_+$ such that

$$|u_h^{(i,m)} - v_h^{(i,m)}| \le \tilde{\gamma}(h) \text{ on } E_h \text{ and } \lim_{h\to 0}\tilde{\gamma}(h) = 0. \tag{3.65}$$

PROOF. We apply Theorem 3.12 to prove (3.65). Let the function $\Gamma_h : E_h' \to R$ be defined by

$$v_h^{(i+1,m)} = A v_h^{(i,m)} + h_0 f_h(x^{(i)}, y^{(m)}, v_h, \delta v_h^{(i,m)}) + h_0 \Gamma_h^{(i,m)} \text{ on } E_h'.$$

It follows from (3.64) and from Assumption H $[A,\delta]$ that there exists a function $\gamma : \Delta \to R_+$ such that $|\Gamma_h^{(i,m)}| \le \gamma(h)$ on E_h' and $\lim_{h\to 0}\gamma(h) = 0$. We define

$$F_h[(i,m), z] = A z^{(i,m)} + h_0 f_h(x^{(i)}, y^{(m)}, z, \delta z^{(i,m)})$$

on $E_h' \times \mathbf{F}(E_{0.h} \cup E_h, R)$. Then u_h satisfies (3.47) and

$$|v_h^{(i+1,m)} - F_h[(i,m), v_h]| \le h_0\gamma(h) \text{ on } E_h',$$

$$|u_h^{(i,m)} - v_h^{(i,m)}| \le \alpha_0(h) \text{ on } E_h'.$$

Now we estimate the difference $F_h[(i,m),z] - F_h[(i,m),\bar{z}]$ on $E_h' \times \mathbf{F}(E_{0.h} \cup E_h, R)$. It follows from Assumption H $[f_h, \sigma_h]$ that

$$|F_h[(i,m),z] - F_h[(i,m),\bar{z}]|$$

$$\le \left| A z^{(i,m)} - A\bar{z}^{(i,m)} + h_0 \left[f_h(x^{(i)}, y^{(m)}, z, \delta z^{(i,m)}) - f_h(x^{(i)}, y^{(m)}, z, \delta\bar{z}^{(i,m)}) \right] \right|$$

$$+ h_0 \left| f_h(x^{(i)}, y^{(m)}, z, \delta\bar{z}^{(i,m)}) - f_h(x^{(i)}, y^{(m)}, \bar{z}, \delta\bar{z}^{(i,m)}) \right|$$

$$\leq \left| \sum_{s \in S} \left[a_s + h_0 \sum_{j=1}^{n} \frac{1}{h_j} c_{s.j} \, D_{q_j} f_h(Q) \right] \left(z^{(i,m+s)} - \bar{z}^{(i,m+s)} \right) \right|$$

$$+ h_0 \, \sigma_h [\, i, \, V_h(z - \bar{z}) \,]$$

where $Q \in E'_h \times \mathbf{F}(E_{0.h} \cup E_h, R) \times R^n$ is an intermediate point. We conclude from (3.62) and from Assumption H $[\, A, \delta \,]$ that we have on E'_h :

$$|\, F_h[\,(i,m), z\,] - F_h[\,(i,m), \bar{z}\,]\,| \leq V_h(z - \bar{z})^{(i)} + h_0 \sigma_h [\, i, \, V_h(z - \bar{z})\,].$$

Let $\bar{\eta}_h : I_{0.h} \cup I_h \to R_+$ be the solution of problem (3.60), (3.61). It follows from Theorem 3.12 that $|\, u_h^{(i,m)} - v_h^{(i,m)} \,| \leq \bar{\eta}_h^{(i)}$ on E_h. Now we obtain the assertion of Theorem 3.17 from the Assumption H $[\, \sigma_h \,]$ on the stability of problem (3.58), (3.59).

3.7 Interpolating operators on Haar pyramid

Now we consider a special class of difference methods for problem (3.54), (3.55). We give examples of function f_h corresponding to f and we give an error estimate for the method. We adopt additional assumptions for the mesh $E_{0.h} \cup E_h$. We assume that the step of the mesh satisfies the condition: $h' = M h_0$. Then we can write the definitions of the sets $E_{0.h}$ and E_h in the following way:

$$E_{0.h} = \{ (x^{(i)}, y^{(m)}) : \; -\tilde{N}_0 \leq i \leq 0, \; -N \leq m \leq N \},$$

$$E_h = \{ (x^{(i)}, y^{(m)}) : \; 0 \leq i \leq N_0, \; |m_i| \leq N_i - i \; \text{ for } i = 1, \ldots, n \}$$

where $\tilde{N} \in \mathbf{Z}$, $N = (N_1, \ldots, N_n) \in \mathbf{N}^n$ and $N_0 \in \mathbf{N}$ satisfy the conditions: $\tilde{N}_0 h_0 = r_0$, $N * h' = b$, $N_0 h_0 < a \leq (N_0 + 1) h_0$.

Put $B = [\, -b, b \,]$ and $B_{h'} = \{ y^{(m)} \; : \; -N \leq m \leq N \}$. For a function $w : B_{h'} \to R$ and a point $y^{(m)} \in B_{h'}$ we write $w^{(m)} = w(y^{(m)})$. We define now the operator $T_{h'} : \mathbf{F}(B_{h'}, R) \to \mathbf{F}(B, R)$ as follows. Put $S_+ = \{ s = (s_1, \ldots, s_n) : \; s_i \in \{0, 1\} \text{ for } 1 \leq i \leq n \}$. Let $w \in \mathbf{F}(B_{h'}, R)$ and $y \in B$. There exists $m \in \mathbf{Z}^n$ such that $y^{(m)}, \; y^{(m+1)} \in B_{h'}$ where $m+1 = (m_1+1, \ldots, m_n+1)$ and $y^{(m)} \leq y \leq y^{(m+1)}$. We define

$$(T_{h'} w)(y) = \sum_{s \in S_+} w^{(m+s)} \left(\frac{y - y^{(m)}}{h'} \right)^s \left(1 - \frac{y - y^{(m)}}{h'} \right)^{1-s}$$

where

$$\left(\frac{y - y^{(m)}}{h'} \right)^s = \prod_{i=1}^{n} \left(\frac{y_i - y_i^{(m_i)}}{h_i} \right)^{s_i},$$

$$\left(1 - \frac{y - y^{(m)}}{h'} \right)^{1-s} = \prod_{i=1}^{n} \left(1 - \frac{y_i - y_i^{(m_i)}}{h_i} \right)^{1-s_i}$$

and we take $0^0 = 1$ in the above formulas. It is easy to see that the function $T_{h'}w$ is continuous on B. If $z : [-r_0, a) \times B_{h'} \to R$ then we write

$$(T_{h'}z)(x,y) = \sum_{s \in S_+} z(x, y^{(m+s)}) \left(\frac{y - y^{(m)}}{h'} \right)^s \left(1 - \frac{y - y^{(m)}}{h'} \right)^{1-s}.$$

We define now the interpolating operator $T_h : \mathbf{F}(E_{0.h} \cup E_h, R) \to \mathbf{F}(E_0 \cup E, R)$ in the following way. Suppose that $z : E_{0.h} \cup E_h \to R$ and $(x,y) \in E_0 \cup E$, $-r_0 \leq x \leq N_0 h_0$. Three cases will be distinguished.

I. Suppose that $(x,y) \in E_0$. Then there is $(i,m) \in \mathbf{Z}^{1+n}$ such that

$$(x^{(i)}, y^{(m)}),\ (x^{(i+1)}, y^{(m+1)}) \in E_{0.h} \ \text{ and } \ x^{(i)} \leq x \leq x^{(i+1)},\ y^{(m)} \leq y \leq y^{(m+1)}.$$

We define

$$(T_h z)(x,y) = \left(1 - \frac{x - x^{(i)}}{h_0} \right) (T_{h'}z)(x^{(i)}, y) + \frac{x - x^{(i)}}{h_0} (T_{h'}z)(x^{(i+1)}, y). \tag{3.66}$$

II. Suppose that $(x,y) \in E$ and there is $(i,m) \in \mathbf{Z}^{1+n}$ such that

$$[x^{(i)}, x^{(i+1)}] \times [y^{(m)}, y^{(m+1)}] \subset E$$

and $x^{(i)} \leq x < x^{(i+1)}$, $y^{(m)} \leq y \leq y^{(m+1)}$. Then we define $(T_h z)(x,y)$ by formula (3.66).

III. Suppose that $(x,y) \in E$ and

(a) $x^{(i)} \leq x < x^{(i+1)}$ for some i, $0 \leq i \leq N_0 - 1$, and $y^{(m)} \leq y \leq y^{(m+1)}$ for some $m \in \mathbf{Z}^n$,

(b) $(x^{(i)}, y^{(m)})$, $(x^{(i)}, y^{(m+1)}) \in E$ and

$$(x^{(i)}, y^{(m)}) \in \partial_0 E \ \text{ or } \ (x^{(i)}, y^{(m+1)}) \in \partial_0 E.$$

Define the sets of integers $I_+[i,m]$, $I_-[i,m]$, $I_0[i,m]$ (possibly empty) as follows :

$$I_+[i,m] = \{ j : 1 \leq j \leq n \ \text{ and } \ y_j^{(m_j+1)} = b_j - M_j x^{(i)} \}$$

$$I_-[i,m] = \{ j : 1 \leq j \leq n \ \text{ and } \ y_j^{(m_j)} = -b_j + M_j x^{(i)} \},$$
$$I_0[i,m] = \{1, \ldots, n\} \setminus (I_+[i,m] \cup I_-[i,m]).$$

Then we have

$$-b_j + M_j x \leq y_j^{(m_j)} \leq y_j \leq b_j - M_j x < y_j^{(m_j+1)} \ \text{ for } \ j \in I_+[i,m], \tag{3.67}$$

$$y_j^{(m_j)} < -b_j + M_j x \leq y_j \leq y_j^{(m_j+1)} \leq b_j - M_j x \ \text{ for } \ j \in I_-[i,m], \tag{3.68}$$

$$-b_j - M_j x \leq y_j^{(m_j)} \leq y_j \leq y_j^{(m_j+1)} \leq b_j - M_j x \ \text{ for } \ j \in I_0[i,m]. \tag{3.69}$$

We define $Uy = (\bar{y}_1, \ldots, \bar{y}_n)$ and $Wy = (\tilde{y}_1, \ldots, \tilde{y}_n)$ as follows:

$$\bar{y}_j = y_j^{(m_j)} + \frac{h_0}{x^{(i)} + h_0 - x} \left(y_j - y_j^{(m_j)} \right) \ \text{ and } \ \tilde{y}_j = y_j^{(m_j)} \tag{3.70}$$

for $j \in I_+[i,m]$, and

$$\bar{y}_j = y_j^{(m_j+1)} + \frac{h_0}{x^{(i)} + h_0 - x}\left(y_j - y_j^{(m_j+1)}\right) \text{ and } \tilde{y}_j = y_j^{(m_j+1)} \tag{3.71}$$

for $j \in I_-[i,m]$, and

$$\bar{y}_j = \tilde{y}_j = y_j \text{ for } j \in I_0[i,m]. \tag{3.72}$$

Then we define $(T_h z)(x,y)$ by the formula:

$$(T_h z)(x,y) = \left(1 - \frac{x - x^{(i)}}{h_0}\right)(T_{h'} z)(x^{(i)}, Uy) \tag{3.73}$$

$$+\frac{x - x^{(i)}}{h_0}(T_{h'} z)(x^{(i+1)}, Wy).$$

If $(x,y) \in E_0 \cup E$ and $N_0 h_0 < x < a$ then we put $(T_h z)(x,y) = (T_h z)(N_0 h_0, y)$. Then we have defined $T_h z : E_0 \cup E \to R$. It is easy to see that $T_h z$ is a continuous function on $E_0 \cup E$. The function $T_h z$ is obtained by interpolation by splines of z.

Theorem 3.18 *Suppose that the function $v : E_0 \cup E \to R$ is of class C^2 and denote by v_h the restriction of v to the set $E_{0.h} \cup E_h$. Then there is $C \in R_+$ such that*

$$\|T_h v_h - v\|_x \le C h_0^2, \quad x \in [-r_0, N_0 h_0], \tag{3.74}$$

where $\|\cdot\|_x$ is the supremum norm in the space $C(E_x, R)$.

PROOF. Let $y^{(m)} \le y \le y^{(m+1)}$. It is easy to prove by induction with respect to n that

$$\sum_{s \in S_+} \left(\frac{y - y^{(m)}}{h'}\right)^s \left(1 - \frac{y - y^{(m)}}{h'}\right)^{1-s} = 1 \tag{3.75}$$

and

$$\sum_{s \in S_+} \left(\frac{y - y^{(m)}}{h'}\right)^s \left(1 - \frac{y - y^{(m)}}{h'}\right)^{1-s} h_j s_j = y_j - y_j^{(m_j)}, \tag{3.76}$$

where $j = 1, \ldots, n$. Suppose that $(x,y) \in E_0 \cup E$ and $x^{(i)} \le x < x^{(i+1)}$, $y^{(m)} \le y \le m^{(m+1)}$. Consider the case when $[x^{(i)}, x^{(i+1)}] \times [y^{(m)}), y^{(m+1)}] \subset E_0 \cup E$. Then $(T_h v_h)(x,y)$ is given by formula (3.66). Put

$$A[s,P] = v^{(i,m)} + \sum_{j=1}^{n} h_j s_j D_{y_j} v^{(i,m)} + \frac{1}{2}\sum_{j,k=1}^{n} h_j s_j h_k s_k D_{y_j y_k}(P),$$

$$B[s,P] = v^{(i,m)} + h_0 D_x v^{(i,m)} + \sum_{j=1}^{n} h_j s_j D_{y_j} v^{(i,m)} + \frac{1}{2} h_0^2 D_{xx} v(P)$$

$$+h_0 \sum_{j=1}^{n} s_j h_j D_{y_j x} v(P) + \frac{1}{2}\sum_{j,k=1}^{n} h_j s_j h_k s_k D_{y_j y_k} v(P),$$

and

$$C[(x,y),P] = v^{(i,m)} + (x - x^{(i)})D_x v^{(i,m)} + \sum_{j=1}^{n} (y_j - y_j^{(m_j)}) D_{y_j} v^{(i,m)}$$

$$+\frac{1}{2}\left(x - x^{(i)}\right)^2 D_{xx}v(P) + \left(x - x^{(i)}\right)\sum_{j=1}^{n}\left(y_j - y_j^{(m_j)}\right) D_{xy_j}v(P)$$

$$+\frac{1}{2}\sum_{j,k=1}^{N}\left(y_j - y_j^{(m_j)}\right)\left(y_k - y_k^{(m_k)}\right) D_{y_j y_k}v(P).$$

There are $\bar{P}$, $\tilde{P}$, $Q \in E_0 \cup E$ such that

$$(T_h v_h)(x,y) - v(x,y) \tag{3.77}$$

$$= \left(1 - \frac{x - x^{(i)}}{h_0}\right) \sum_{s \in S_+} \left(\frac{y - y^{(m)}}{h'}\right)^s \left(1 - \frac{y - y^{(m)}}{h'}\right)^{1-s} A[s, \bar{P}]$$

$$+\frac{x - x^{(i)}}{h_0} \sum_{s \in S_+} \left(\frac{y - y^{(m)}}{h'}\right)^s \left(1 - \frac{y - y^{(m)}}{h'}\right)^{1-s} B[s, \tilde{P}] - C[(x,y),Q].$$

Let $\tilde{C}$ be such a constant that

$$|D_{xx}v(x,y)| \le \tilde{C}, \ \ |D_{y_j x}v(x,y)| \le \tilde{C}, \ \ |D_{y_j y_k}v(x,y)| \le \tilde{C} \tag{3.78}$$

on $E_0 \cup E$ where $j,k = 1,\ldots,n$. It follows from (3.75) - (3.78) that

$$\|(T_h v_h)(x,y) - v(x,y)\| \le C h_0^2 \tag{3.79}$$

for $(x,y) \in [\, x^{(i)}, x^{(i+1)},\,] \times [\, y^{(m)}, y^{(m+1)}\,]$ where

$$C = \frac{\tilde{C}}{2}\left[1 + 2\sum_{j=1}^{n} M_j + \sum_{j,k=1}^{n} M_j M_k\right]. \tag{3.80}$$

Suppose that $(x,y) \in E$ and there is $(i,m) \in \mathbf{Z}^{1+n}$ such that relations (3.67) - (3.69) hold. There are $\bar{P}$, $\tilde{P}$, Q such that

$$(T_h v_h)(x,y) - v(x,y) \tag{3.81}$$

$$= \left(1 - \frac{x - x^{(i)}}{h_0}\right) \sum_{s \in S_+} \left(\frac{\bar{y} - y^{(m)}}{h'}\right)^s \left(1 - \frac{\bar{y} - y^{(m)}}{h'}\right)^{1-s} A[s, \bar{P}]$$

$$+\frac{x - x^{(i)}}{h_0} \sum_{s \in S_+} \left(\frac{\tilde{y} - y^{(m)}}{h'}\right)^s \left(1 - \frac{\tilde{y} - y^{(m)}}{h'}\right)^{1-s} B[s, \tilde{P}] - C[(x,y),Q].$$

Put

$$\Gamma_j(x,y) = \left(1 - \frac{x - x^{(i)}}{h_0}\right) \sum_{s \in S_+} \left(\frac{\bar{y} - y^{(m)}}{h'}\right)^s \left(1 - \frac{\bar{y} - y^{(m)}}{h'}\right)^{1-s} s_j h_j$$

$$+ \frac{x - x^{(i)}}{h_0} \sum_{s \in S_+} \left(\frac{\tilde{y} - y^{(m)}}{h'}\right)^s \left(1 - \frac{\tilde{y} - y^{(m)}}{h'}\right)^{1-s} h_j s_j$$

where $1 \le j \le n$ and $\bar{y}, \tilde{y}$ are given by (3.70) - (3.72). We prove that

$$\Gamma_j(x,y) = y_j - y_j^{(m_j)} \text{ for } 1 \le j \le n. \tag{3.82}$$

It follows from (3.70) - (3.72) that $y^{(m)} \le \bar{y} \le y^{(m+1)}$ and $y^{(m)} \le \tilde{y} \le y^{(m+1)}$. Therefore relations (3.75) and (3.76 are satisfied for $\bar{y}$ and $\tilde{y}$.

If $j \in I_0[i,m]$ then we have from (3.72), (3.76)

$$\Gamma_j(x,y) = \left(1 - \frac{x - x^{(i)}}{h_0}\right)(y_j - y_j^{(m_j)}) + \frac{x - x^{(i)}}{h_0}(y_j - y_j^{(m_j)}) = y_j - y_j^{(m_j)}.$$

Suppose that $j \in I_+[i,m]$. Relation (3.76) and (3.70) imply

$$\Gamma_j(x,y) = \left(1 - \frac{x - x^{(i)}}{h_0}\right)(\bar{y}_j - y_j^{(m_j)}) + \frac{x - x^{(i)}}{h_0}(\tilde{y}_j - y_j^{(m_j)}) = y_j - y_j^{(m_j)}.$$

Suppose that $j \in I_-[i,m]$. Then relations (3.76) and (3.71) imply

$$\Gamma_j(x,y) = \left(1 - \frac{x - x^{(i)}}{h_0}\right)\left[y_j^{(m_j+1)} - y_j^{(m_j)} + \frac{h_0}{x^{(i)} + h_0 - x}\left(y_j - y_j^{(m_j+1)}\right)\right]$$

$$+ \frac{x - x^{(i)}}{h_0}\left(y_j^{(m_j+1)} - y_j^{(m_j)}\right) = y_j - y_j^{(m_j)}$$

which completes the proof of (3.82). It follows from (3.75), (3.81), (3.82) that estimate (3.79) is satisfied with C given by (3.80). Then we have proved inequality (3.79) on $E_0 \cup E$ with C given by (3.80) and assertion (3.74) follows. This completes the proof of the Theorem.

Lemma 3.19 *Suppose that the function $v : E_0 \cup E \to R$ is of class C^1 and the derivatives $D_x v$, $D_y v$ are bounded on $E_0 \cup E$. Let v_h be the restriction of v to the set $E_{0.h} \cup E_h$.*

Then there is $C^* \in R_+$ such that $\|T_h v_h - v\|_x \le C^* |h|$ for $x \in [-r_0, a)$.

We omit the simple proof of the lemma.

Let $L[h_0, \cdot] : \mathbf{F}(I_{0.h} \cup I_h, R) \to \mathbf{F}([-r_0, a), R)$ be the operator given by

$$L[h_0, \eta](x) = \eta^{(i+1)} \frac{x - x^{(i)}}{h_0} + \eta^{(i)}\left(1 - \frac{x - x^{(i)}}{h_0}\right) \text{ for } x^{(i)} \le x \le x^{(i+1)},$$

where $-\tilde{N}_0 \le i \le N_0 - 1$ and

$$L[h_0, \eta](x) = L[h_0, \eta](x^{(N_0)}) \text{ for } x^{(N_0)} < x < a.$$

3.8 The Euler method for the Cauchy problem

Now we consider functional differential problem (3.54), (3.55) and the difference equation

$$\delta_0 z^{(i,m)} = f(x^{(i)}, y^{(m)}, T_h z, \delta z^{(i,m)}) \tag{3.83}$$

with the initial condition

$$z^{(i,m)} = \varphi_h^{(i,m)} \text{ on } E_{0.h}. \tag{3.84}$$

The operators δ_0, δ are defined by (3.13) - (3.15), where $1 \le \kappa \le n$ is given integer.

Assumption H $[f, \sigma]$**.** Suppose that the function $f : \Omega \to R$ is continuous, it satisfies the Volterra condition and

1) there exists a continuous function $\sigma : R_+ \times R_+ \to R_+$ such that

(i) σ is nondecreasing with respect to both variables and $\sigma(x, 0) = 0$ for $x \in R_+$,

(ii) the maximum solution of the problem

$$\omega'(x) = \sigma(x, \omega(x)), \quad \omega(0) = 0$$

is $\bar{\omega}(x) = 0$ for $x \in R_+$,

2) for $(x, y, q) \in E \times R^n$, $z, \bar{z} \in C(E_0 \cup E, R)$ we have

$$|f(x, y, z, q) - f(x, y, \bar{z}, q)| \le \sigma(x, \|z - \bar{z}\|_x), \tag{3.85}$$

3) the derivatives $D_q f = (D_{q_1} f, \ldots, D_{q_n} f)$ exist on Ω and $D_q f(x, y, z, \cdot) \in C(R^n, R^n)$,

4) the estimates

$$1 - h_0 \sum_{j=1}^{n} \frac{1}{h_j} |D_{q_j} f(P)| \ge 0 \tag{3.86}$$

and

$$D_{q_j} f(P) \le 0 \text{ for } 1 \le j \le \kappa, \tag{3.87}$$

$$D_{q_j} f(P) \ge 0 \text{ for } \kappa + 1 \le j \le n \tag{3.88}$$

are satisfied for $P = (x, y, z, q) \in \Omega$.

Theorem 3.20 *Suppose that Assumption H $[f, \sigma]$ is satisfied and*

1) *the function $u_h : E_{0.h} \cup E_h \to R$ is a solution of problem (3.83), (3.84) with δ_0, δ given by (3.13) - (3.15),*

2) *$v : E_0 \cup E \to R$ is s solution of (3.54), (3.55), v is of class C^1 on $E_0 \cup E$ and it is of class C^2 on E,*

3) *the derivatives of the second order of v are bounded on E,*

4) *there exists a function $\alpha_0 : \Delta \to R_+$ such that $|\varphi_h^{(i,m)} - \varphi^{(i,m)}| \le \alpha_0(h)$ on $E_{0.h}$ and $\lim_{h \to 0} \alpha_0(h) = 0$.*

Then there is a number $\varepsilon_0 > 0$ and a function $\tilde{\gamma} : \Delta \to R_+$ such that we have for $|h| < \varepsilon_0$

$$|u_h^{(i,m)} - v_h^{(i,m)}| \le \tilde{\gamma}(h) \quad \text{on } E_h \tag{3.89}$$

where $v_h = v|_{E_{0.h} \cup E_h}$ and $\lim_{h \to 0} \tilde{\gamma}(h) = 0$.

PROOF. We prove that the function

$$f_h(x,y,z,q) = f(x,y,T_h z,q), \quad (x,y,z,q) \in \Omega_h,$$

satisfies all the assumptions of Theorem 3.17. Let $\sigma_h : I'_h \times \mathbf{F}(I_{0.h} \cup I_h, R_+) \to R_+$ be defined by

$$\sigma_h(\,x,\omega\,) = \sigma(\,x,\ \|L[h_0,\omega]\|_x\,). \tag{3.90}$$

Assumption (3.85) and relation (3.75) imply the estimates

$$|\,f_h(x,y,z,q) - f_h(x,y,\bar z,q)\,| = |\,f(x,y,T_h z,q) - f(x,y,T_h\bar z,q)\,|$$

$$\le \sigma(\,x, \|T_h z - T_h \bar z\|_x\,) = \sigma(\,x, \|L[\,h_0, V_h(z-\bar z)\,]\,\|_x\,)$$

$$= \sigma_h(\,x, V_h(z-\bar z)\,) \ \text{ on } \ \Omega_h,$$

which proves condition (3.63). It follows from Lemma 3.19 that the consistency condition (3.64) is satisfied. We prove that the difference problem

$$\eta^{(i+1)} = \eta^{(i)} + h_0\sigma_h[\,i,\eta\,] \ \text{ for } \ 0 \le i \le N_0 - 1, \tag{3.91}$$

$$\eta^{(i)} = 0 \ \text{ for } \ -\tilde N_0 \le i \le 0, \tag{3.92}$$

is stable in the sense of Assumption H $[\,\sigma_h\,]$. Let $\eta_h : I_{0.h} \cup I_h \to R_+$ be a solution of problem

$$\eta^{(i+1)} = \eta^{(i)} + h_0\sigma_h[\,i,\eta\,] + h_0\gamma(h) \ \text{ for } 0 \le i \le N_0 - 1, \tag{3.93}$$

$$\eta^{(i)} = \alpha_0(h) \ \text{ for } \ -\tilde N_0 \le i \le 0, \tag{3.94}$$

where $\gamma,\ \alpha_0 : \Delta \to R_+$ and $\lim_{h\to 0}\gamma(h) = 0$, $\lim_{h\to 0}\alpha_0(h) = 0$. The above problem is equivalent to

$$\eta^{(i+1)} = \eta^{(i)} + h_0\sigma(\,x,\ \eta^{(i)}\,) + h_0\gamma(h) \ \text{ for } \ 0 \le i \le N_0 - 1, \tag{3.95}$$

$$\eta^{(0)} = \alpha_0(h). \tag{3.96}$$

Let $\omega_h : [0,a) \to R_+$ be the maximum solution of the problem

$$\omega'(x) = \sigma(\,x,\omega(x)\,) + \gamma(h), \quad \omega(0) = \alpha_0(h).$$

There exists $\varepsilon_0 > 0$ such that the solution ω_h is defined on $[0,a)$ for $|h| < \varepsilon_0$ and

$$\lim_{h\to 0}\omega_h(x) = 0 \ \text{ uniformly on } [0,a).$$

The function ω_h is convex on $[0,a)$, therefore we have

$$\omega_h^{(i+1)} \ge \omega_h^{(i)} + h_0\,\sigma(\,x^{(i)},\omega_h^{(i)}\,) + h_0\gamma(h) \ \text{ for } 0 \le i \le N_0 - 1.$$

Since η_h satisfies conditions (3.95), (3.96) and $\eta_h(0) = \omega_h(0)$ then we have $\eta_h^{(i)} \le \omega_h^{(i)}$ for $0 \le i \le N_0$, which completes the proof of the stability of problem (3.91), (3.92). The assumption (3.86) - (3.88) for the method (3.83), (3.84) is equivalent to condition (3.62). It follows from Theorem 3.17 that there is $\tilde{\gamma} : \Delta \to R_+$ such that estimation (3.89) is satisfied and $\lim_{h\to 0} \tilde{\gamma}(h) = 0$. This proves the Theorem.

Now we consider the equation (3.83) with the initial condition (3.84) where the operators δ_0, $\delta = (\delta_1, \ldots, \delta_n)$ are given by (3.19), (3.20).

Assumption H' $[f, \sigma]$. Suppose that conditions 1) - 3) of Assumption H $[f, \sigma]$ are satisfied and the estimates

$$1 - n\, h_0 \frac{1}{h_i} |D_{q_i} f(x, y, z, q)| \geq 0, \quad \text{for } i = 1, \ldots, n, \tag{3.97}$$

hold on Ω.

Theorem 3.21 *Suppose that Assumption H' $[f, \sigma]$ is satisfied and*

1) the function $u_h : E_{0.h} \cup E_h \to R$ is a solution of problem (3.83), (3.84) with δ_0, δ given by (3.19), (3.20),

2) conditions 2) - 4) of Theorem 3.20 hold.

Then there is a number $\varepsilon_0 > 0$ and a function $\tilde{\gamma} : \Delta \to R_+$ such that we have for $|h| < \varepsilon_0$

$$|\, u_h^{(i,m)} - v_h^{(i,m)} \,| \leq \tilde{\gamma}(h) \;\; \text{on } E_h$$

where v_h is the restriction of v to the set $E_{0.h} \cup E_h$ and $\lim_{h \to 0} \tilde{\gamma}(h) = 0$.

The proof of the above Theorem is analogous to Theorem 3.20. Details are omitted.

3.9 Error estimates for the Euler method

Now we give an error estimate for the method (3.83), (3.84) with difference operators given by (3.19), (3.20). In a general case we have the estimate $\|u_h - v_h\|_{i.h} \leq \tilde{\omega}_h^{(i)}$, $i = 0, 1, \ldots, N_0$, where $\tilde{\omega}_h : I'_h \to R_+$ satisfies the difference inequality

$$\frac{1}{h_0} \left(\tilde{\omega}_h^{(i+1)} - \tilde{\omega}_h^{(i)} \right) \geq \sigma(x^{(i)}, \tilde{\omega}_h^{(i)}) + \gamma(h) \;\; \text{for } \; 0 \leq i \leq N_0 - 1,$$

and the initial estimate $\alpha_0(h) \leq \tilde{\omega}_h^{(0)}$. The functions γ, $\alpha_0 : \Delta \to R_+$ are defined by the relations

$$\left| \delta_0 v_h^{(i,m)} - f(x^{(i)}, y^{(m)}, T_h v_h, \delta v_h^{(i,m)}) \right| \leq \gamma(h) \tag{3.98}$$

where $(x^{(i)}, y^{(m)}) \in E'_h$ and

$$|\varphi_h^{(i,m)} - \varphi^{(i,m)}| \leq \alpha_0(h) \quad \text{on } E_{0.h} \tag{3.99}$$

and the functions f, σ satisfy Assumption H $[f, \sigma]$ or H' $[f, \sigma]$.

Theorem 3.22 *Suppose that*

1) the assumptions of Theorem 3.21 are satisfied with $\sigma(x, p) = Lp$, $L \in R_+$, (i.e. the function f satisfies the Lipschitz condition with respect to the functional variable),

2) the function v is of class C^3 on E and

$$|D_{xx}v(x,y)| \le \tilde{C}, \ |D_{xy_j}v(x,y)| \le \tilde{C}, \ |D_{y_jy_k}v(x,y)| \le \tilde{C} \tag{3.100}$$

on $E_0 \cup E$, $1 \le j,k \le n$, and

$$|D_{y_jy_jy_j}v(x,y)| \le \bar{C} \text{ on } E, \ 1 \le j \le n. \tag{3.101}$$

Then

$$\|u_h - v_h\|_{i.h} \le \tilde{\eta}_h^{(i)} \text{ for } 0 \le i \le N_0, \tag{3.102}$$

where

$$\tilde{\eta}_h^{(i)} = \alpha_0(h)\,(1+Lh_0)^i + h_0\gamma(h_0)\left[1+(1+Lh_0)+\ldots+(1+Lh_0)^{i-1}\right] \tag{3.103}$$

and

$$\gamma(h_0) = \frac{h_0}{2}\tilde{C}\left[1+\frac{1}{n}\sum_{j=1}^{n}M_j^2\right] + h_0LC + \frac{h_0^2}{6n}\bar{C}\sum_{j=1}^{n}M_j^3 \tag{3.104}$$

where

$$C = \frac{\tilde{C}}{2}\left[1+2\sum_{j=1}^{n}M_j + \sum_{j,k=1}^{n}M_jM_k\right]. \tag{3.105}$$

PROOF. I follows from (3.19), (3.20) and (3.100), (3.101) that v satisfies

$$\left|\delta_0 v_v^{(i,m)} - D_x v^{(i,m)}\right| \le \frac{1}{2}\tilde{C}h_0\left[1+\frac{1}{n}\sum_{j=1}^{n}M_j^2\right]$$

and

$$\left|\delta_j v_h^{(i,m)} - D_{y_j}v^{(i,m)}\right| \le \frac{1}{6}\bar{C}h_j^2, \quad j=1,\ldots,n.$$

Then we have relation (3.98) with $\gamma(h_0)$ given by (3.104), (3.105). The function $\tilde{\eta}_h$ defined by (3.103) is the solution of the problem

$$\frac{1}{h_0}\left(\eta^{(i+1)}-\eta^{(i)}\right) = L\eta^{(i)} + \gamma(h_0), \ i=0,1,\ldots,N_0-1, \qquad \eta^{(0)} = \alpha_0(h).$$

Then assertion (3.102) follows from Theorem 3.12.

Theorem 3.23 *Suppose that all the assumptions of Theorem 3.20 are satisfied with $\sigma(x,p) = Lp$ where $L \in R_+$. Assume that the solution v is of class C^2 on $E_0 \cup E$ and there is $\tilde{C}$ such that estimates (3.100) hold.*
Then we have $\|u_h - v_h\|_{i.h} \le \tilde{\eta}_h^{(i)}$, $0 \le i \le N_0$, where $\tilde{\eta}_h$ is given by (3.103) with

$$\gamma(h_0) = \frac{h_0}{2}\tilde{C} + LCh_0^2 + \frac{h_0}{2}\tilde{C}\|M\|^2$$

and C is given by (3.105).

The proof of the above theorem is similar to the proof of Theorem 3.22. Details are omitted.

3.10 Difference methods for almost linear equations

Now we consider difference methods for almost linear functional differential problems. Suppose that the functions

$$g : E \times C(E_0 \cup E, R) \to R, \quad G : E \to R^n, \quad G = (g_1, \ldots, g_n),$$

and $\varphi : E \to R$ are given. We consider the Cauchy problem

$$D_x z(x,y) = g(x,y,z) + \sum_{j=1}^{n} g_j(x,y) D_{y_j} z(x,y), \tag{3.106}$$

$$z(x,y) = \varphi(x,y) \text{ for } (x,y) \in E_0 \tag{3.107}$$

and the difference method

$$z^{(i+1,m)} = z^{(i,m)} + h_0 g(x^{(i)}, y^{(m)}, T_h z) + h_0 \sum_{j=1}^{n} g_j^{(i,m)} \delta_j z^{(i,m)}, \tag{3.108}$$

$$z^{(i,m)} = \varphi_h^{(i,m)} \text{ on } E'_h, \tag{3.109}$$

where $\varphi_h : E_{0.h} \to R$ is a given function and $\delta = (\delta_1, \ldots, \delta_n)$ is defined by (3.42), (3.43). We assume that the step of the mesh satisfies the condition: $h' = M h_0$.

Theorem 3.24 *Suppose that the functions g and G are continuous, g satisfies the Volterra condition, φ is of class C^1 and*

1) *there exists a function $\sigma : R_+ \times R_+ \to R_+$ satisfying condition 1) of Assumption $H[f, \sigma]$ and such that*

$$|g(x,y,z) - g(x,y,\bar{z})| \le \sigma(x, \|z - \bar{z}\|_x) \text{ on } E \times C(E_0 \cup E, R), \tag{3.110}$$

2) *the estimation*

$$1 - h_0 \sum_{j=1}^{n} \frac{1}{h_j} |g_j(x,y)| \ge 0 \tag{3.111}$$

is satisfied on E,

3) *the function $v : E_0 \cup E \to R$ is a solution of problem (3.106), (3.107), v is of class C^1 on $E_0 \cup E$ and the function $v|_E$ is of class C^2,*

4) *the derivatives of the second order of v are bounded on E,*

5) *the function $u_h : E_{0.h} \cup E_h \to R$ is a solution of problem (3.108), (3.109) with difference operators given by (3.42), (3.43) and there is $\alpha_0 : \Delta \to R_+$ such that $|\varphi_h^{(i,m)} - \varphi^{(i,m)}| \le \alpha_0(h)$ on $E_{0.h}$ and $\lim_{h \to 0} \alpha_0(h) = 0$.*

Then there exist a number $\varepsilon_0 > 0$ and a function $\tilde{\gamma} : \Delta \to R_+$ such that for $|h| < \varepsilon$ we have

$$|u_h^{(i,m)} - v^{(i,m)}| \le \tilde{\gamma}(h) \text{ on } E_h \tag{3.112}$$

and $\lim_{h \to 0} \tilde{\gamma}(h) = 0$.

PROOF. We apply Theorem 3.12 to prove (3.112). Let the function $\Gamma_h : E'_h \to R$ be defined by

$$\frac{1}{h_0}\left(v_h^{(i+1,m)} - v_h^{(i,m)}\right) = g(\,x^{(i)}, y^{(m)}, T_h v_h\,) + \sum_{j=1}^{n} g_j^{(i,m)}\,\delta_j v_h^{(i,m)} + \Gamma_h^{(i,m)}$$

where $v_h = v\,|_{E_{0.h}\cup E_h}$. It follows from Lemma 3.19 that there exists a function $\gamma : \Delta \to R_+$ such that $|\,\Gamma_h^{(i,m)}\,| \le \gamma(h)$ on E'_h and $\lim_{h\to 0}\gamma(h) = 0$. We define the operator F_h by

$$F_h[(i,m),z] = z^{(i,m)} + h_0 g(x^{(i)}, y^{(m)}, T_h z) + h_0 \sum_{j=1}^{n} g_j^{(i,m)}\,\delta_j z^{(i,m)}.$$

Then the function u_h satisfies (3.47) and

$$|\,v_h^{(i+1,m)} - F_h[\,(i,m), v_h\,]\,| \le h_0\gamma(h) \;\text{ on } E'_h,$$

$$|\,u_h^{(i,m)} - v_h^{(i,m)}\,| \le \alpha_0(h) \;\text{ on } E_{0.h}.$$

Now we estimate the difference $F_h[\,(i,m),z\,] - F_h[\,(i,m),\bar z\,]$ on $E'_h \times \mathbf{F}(E_{0.h}\cup E_h, R)$. Let $(x^{(i)}, y^{(m)}) \in E'_h$ and

$$I^+[i,m] = \{j :\; g_j(x^{(i)}, y^{(m)}) \ge 0\},\quad I^-[i,m] = \{1,\dots,n\} \setminus I^+[i,m].$$

It follows from (3.110), (3.111) that

$$|\,F_h[\,(i,m),z\,] - F_h[\,(i,m),\bar z\,]\,| \le h_0\sigma\,(\,x, \|L\,[h_0, V_h(z-\bar z)]\,\|_x\,)$$

$$+|\,(z-\bar z)^{(i,m)}\,|\left[1 - h_0\sum_{j=1}^{n}\frac{1}{h_j}|\,g_j^{(i,m)}\,|\right]$$

$$+h_0\left|\sum_{j\in I^+[i,m]}\frac{1}{h_j}g_j^{(i,m)}(z-\bar z)^{(i,m+e_j)} - \sum_{j\in I^-[i,m]}\frac{1}{h_j}g_j^{(i,m)}(z-\bar z)^{(i,m-e_j)}\right|$$

$$\le h_0\sigma_h[\,i, V_h(z-\bar z)\,] + V_h(z-\bar z)^{(i)},$$

where σ_h is defined by (3.90). Thus we see that all the assumptions of Theorem 3.12 are satisfied. Then we have the estimate $|\,u_h^{(i,m)} - v_h^{(i,m)}\,| \le \eta_h^{(i)}$ on E_h where $\eta_h : I_{0.h}\cup I_h \to R_+$ is the solution of problem (3.93), (3.94). Difference problem (3.91), (3.92 is stable. Therefore there exists $\varepsilon > 0$ and $\tilde\gamma : \Delta \to R_+$ such that assertion (3.112) is satisfied for $|h| < \varepsilon_0$ and $\lim_{h\to 0}\tilde\gamma(h) = 0$. This completes the proof of the theorem.

Now we give an error estimate for the method (3.108), (3.109) with δ given by (3.42), (3.43).

Theorem 3.25 *Suppose that all the assumptions of Theorem 3.24 are satisfied with*

$$\sigma(s,p) = Lp, \quad L \in R_+.$$

Assume that the solution $v : E_0 \cup E \to R$ of problem (3.106), (3.107) is of class C^2 and there is $\tilde{C} \in R_+$ such that estimates (3.100) are satisfied.

The we have

$$\|u_h - v_h\|_{i.h} \leq \tilde{\eta}_h^{(i)}, \quad 0 \leq i \leq N_0,$$

where $\tilde{\eta}_h$ is given by (3.103) with

$$\gamma(h_0) = \frac{h_0}{2}\tilde{C} + LCh_0^2 + \frac{h_0}{2}\tilde{C}\|M\|^2$$

and C is given by (3.105).

The proof is based on Theorem 3.12. We omit details.

Remark 3.26 *The function $\tilde{\eta}_h$ given by (3.103) satisfies the conditions:*

$$\tilde{\eta}_h^{(i)} \leq \alpha_0(h)\exp[La] + \gamma(h_0)\frac{\exp[La]-1}{L} \quad \textit{if } L > 0$$

and

$$\tilde{\eta}_h^{(i)} \leq \alpha_0(h) + a\gamma(h_0) \quad \textit{if } L = 0.$$

where $0 \leq i \leq N_0$.

Remark 3.27 *All the theorems of difference methods can be extended on weakly coupled functional differential systems.*

Chapter 4

Initial Problems on Unbounded Domains

4.1 Introduction

Let $E_0 = [-r_0, 0] \times R^n$ and $B = [-r_0, 0] \times [-r, r] \subset R^{1+n}$ where $r_0 \in R_+$, $r = (r_1, \ldots, r_n) \in R^n_+$. Given a function $z = (z_1, \ldots, z_k) : [-r_0, a] \times R^n \to R^k$ with $a > 0$ and a point $(x, y) \in [0, a] \times R^n$, we consider the function $z_{(x,y)} : B \to R^k$ defined by

$$z_{(x,y)}(t, s) = z(x + t, y + s), \quad (t, s) \in B. \tag{4.1}$$

The function $z_{(x,y)}$ is the restriction of z to the set $[x - r_0, x] \times [y - r, y + r]$ and this restriction is shifted to the set B. Put $\Omega = [0, a] \times R^n \times C(B, R^k) \times R^n$ and assume that $f = (f_1, \ldots, f_k) : \Omega \to R^k$ is a given function of the variables (x, y, w, q), where $w = (w_1, \ldots, w_k)$, $q = (q_1, \ldots, q_n)$. For a function $z : [-r_0, a] \times R^n \to R^k$ and a point $(x, y) \in [0, a] \times R^n$ we write

$$f(\, x, y, z_{(x,y)}, D_y z(x, y)\,)$$

$$= (\; f_1(x, y, z_{(x,y)}, D_y z_1(x, y)), \ldots, f_k(x, y, z_{(x,y)}, D_y z_k(x, y))\;)\,.$$

Given the function $\phi = (\phi_1, \ldots, \phi_k) : E_0 \to R^k$, we consider the Cauchy problem

$$D_x z(x, y) = f(\, x, y, z_{(x,y)}, D_y z(x, y)\,), \tag{4.2}$$

$$z(x, y) = \phi(x, y) \quad \text{on } E_0. \tag{4.3}$$

We will consider weak solutions of problem (4.2), (4.3). More precisely, a function $u : [-r_0, c] \times R^n \to R^k$ where $0 < c \le a$, is a solution of (4.2), (4.3) provides

(i) $u \in C([-r_0, c] \times R^n, R^k)$ and

$$D_y u_i(x, y) = (\, D_{y_1} u_i(x, y), \ldots, D_{y_n} u_i(x, y)\,)\,, \; i = 1, \ldots, k,$$

exist on $[0, c] \times R^n$,

(ii) the function $u(\cdot, y) : [-r_0, c] \to R^k$ is absolutely continuous on $[0, c]$ for each $y \in R^n$,

(iii) for each $y \in R^n$ system (4.2) is satisfied for almost all $x \in [0, c]$ and condition (4.3) holds.

We will discuss the question of the existence and uniqueness of solutions of problem (4.2), (4.3) and adequate quasilinear systems. Existence results are based on the method of bicharacteristics. The Cauchy problem will be transformed into integral functional equations. The existence of solutions of this system will be proved by using integral inequalities theorems and by the Banach fixed point principle. A theorem on uniqueness is based on a result on functional differential inequalities in Carathéodory sense. We list below a few examples of systems which can be derived from (4.2) by specializing the operator f.

Example 4.1 *Assume that*

$$H = (H_1, \ldots, H_k) : [0, a] \times R^n \times R^k \times R^k \times R^n \to R^k$$

and

$$\alpha = (\alpha_0, \alpha') = (\alpha_0, \alpha_1, \ldots, \alpha_n) : [0, a] \times R^n \to R^{1+n}$$

are given functions and $-r_0 \le \alpha_0(x, y) - x \le 0, \quad -r \le \alpha'(x, y) - y \le r$ *on* $[0, a] \times R^n$. *Write* $f(x, y, w, q) = H\left(x, y, w(0, 0), w(\alpha_0(x, y) - x, \alpha'(x, y) - y), q\right)$ *on* $[0, a] \times R^n \times C(B, R^k) \times R^n$. *Then system (4.2) reduces to the system with a deviated argument*

$$D_x z(x, y) = H\left(x, y, z(x, y), z(\alpha(x, y)), D_y z(x, y)\right).$$

Example 4.2 *For the above H we put*

$$f(x, y, w, q) = H(x, y, w(0, 0), \int_B w(t, s) dt\, ds, q).$$

Then (4.2) is the integral differential system

$$D_x z(x, y) = H(x, y, z(x, y), \int_B z(x + t, y + s)\, dt\, ds, D_y z(x, y)).$$

Example 4.3 *There are a lot of papers concerning an initial value problem for the systems*

$$D_x z_i(x, y) = F_i\left(x, y, (Vz)(x, y), D_y z_i(x, y)\right), \quad i = 1, \ldots, k, \tag{4.4}$$

where

$$V : C\left([-r_0, a] \times R^n, R^k\right) \to C\left([0, a] \times R^n, R^k\right)$$

is an operator of the Volterra type and $F = (F_1, \ldots, F_k)$ *is defined on finite dimension Euclidean space. The main assumptions in existence theorems for (4.4) concern the operator V. They are formulated in a form of inequalities for norms in some function spaces [20, 106, 212]. These inequalities are linear and it is the main shortcoming of this theory. In [110] it is proved that system (4.4) is a particular case of (4.2).*

Let us mention some of the methods that are more frequently adopted for proving the existence and uniqueness of classical or generalized solutions to initial or mixed problems for system (4.2). For simplicity suppose that $k = 1$.

Method of successive approximations is based on the idea which was first introduced by T. Ważewski for systems without functional dependence ([231]). We construct a sequence $\{z^{(m)}\}$, $z^{(m)} : [-r_0, c] \times R^n \to R$ such that $z^{(0)}$ is an arbitrary function satisfying an adequate initial condition and $z^{(m+1)}$ is a solution of the Cauchy problem for the equation

$$D_x z(x,y) = f(\,x, y, z^{(m)}_{(x,y)}, D_y z(x,y)\,).$$

Under suitable assumptions on an initial function and on f, the sequence $\{z^{(m)}\}$ exists and is convergent to a classical solution of (4.2). By means of this process, the first results on C^1 solutions for functional differential problems were obtained ([27, 106]). This method can be used also for proving existence theorems for weakly coupled functional differential systems ([146, 201]).

The fixed point method is based on the well-known Banach fixed point theorem. The closed subset X of the Banach space consisting of all functions $z \in C([-r_0, c] \times R^n, R)$ which are bounded, is considered. For $u \in X$ we study the equation

$$D_x z(x,y) = f(\,x, y, u_{(x,y)}, D_y z(x,y)\,),$$

with initial condition generated by (4.3) with $k = 1$. Let Wu be the solution of the above problem. Then, under suitable assumptions on f we have:

(i) $Wu \in X$,

(ii) the operator $W : X \to X$ is a contraction.

The fixed point of W is a solution of the original problem.

The Schauder fixed point theorem was used in [167] for generalized solutions of almost linear problems with unknown function of two variables, see also [102]. Difference method was used in [127] for discussing the existence of Carathéodory solutions for equations with a deviated argument and in [59] for general class of functional differential problems with unknown function of two variables. The method of bicharacteristics for quasilinear problems was introduced and widely studied in nonfunctional setting by S. Cinquini and M. Cinquini Cibrario ([52, 53]). It was also adopted by L. Cesari ([45 – 47]) and P. Bassanini ([13 – 15]) for quasilinear systems in the second canonical form. An application of the Cesari process to functional setting is due to T. Człapiński ([56 – 58] and [60]). Existence results given by P. Besala in [24] are based on the theory of bicharacteristics.

Method of linearization for generalized solutions of nonlinear problems was introduced in nonfunctional setting by S. Cinquini and M. Cinquini Cibrario ([51, 54, 55]). It consists on linearization of the right-hand side of equation with respect to unknown function. In the second step a quasilinear system is constructed for unknown function and for their spatial derivatives. The system thus obtained is equivalent to a system of integral equations of the Volterra type. The method reveals very sharp since the results obtained by these authors for generalized solutions in nonfunctional setting are deeper than those proved by means of other approaches. For this reason, we decided to extend this last process to functional setting for initial problems in Chapter IV and for mixed problems in Chapter V. The method of linearization for nonlinear functional differential problems was initiated in the papers [30, 146, 149].

This Chapter is organized as follows. In the first part (Sections 4.2 - 4.4) we prove existence results for weakly coupled quasilinear systems. The second part (Sections 4.5 and 4.6) concerns with quasilinear systems in the second canonical form. In the last part we investigate nonlinear problem (4.2), (4.3). We prove a theorem on uniqueness of generalized solutions and a result on existence of solutions.

4.2 Bicharacteristics for quasilinear systems

Put $\Omega_0 = [0,a] \times R^n \times C(B,R^k)$ and assume that

$$\varrho : \Omega_0 \to M[k,n], \quad \varrho = [\varrho_{ij}]_{i=1,\dots,k,\; j=1,\dots,n},$$

and $f : \Omega_0 \to M[k,1]$, $f = (f_1,\dots,f_k)^T$, are given functions of the variables (x,y,w), $w = (w_1,\dots,w_k)$, and T means transposition of a vector or a matrix. Write $\varrho_i = (\varrho_{i1},\dots,\varrho_{in})$ for $1 \le i \le k$ and suppose that $\phi : E_0 \to R^k$ is a given function. We consider the quasilinear functional differential system

$$D_x z_i(x,y) + \sum_{j=1}^{n} \varrho_{ij}(x,y,z_{(x,y)}) D_{y_j} z_i(x,y) = f_i(x,y,z_{(x,y)}), \quad i = 1,\dots,k, \tag{4.5}$$

with the initial condition

$$z(x,y) = \phi(x,y) \quad \text{on } E_0. \tag{4.6}$$

A function $u : [-r_0,c] \times R^n \to R^k$, $0 < c \le a$, is the Carathéodory solution of problem (4.5), (4.6) if

(i) u is continuous and the derivatives $D_x u$, $D_y u_i = (D_{y_1} u_i,\dots,D_{y_n} u_i)$, $1 \le i \le k$, exist almost everywhere on $[0,c] \times R^n$,

(ii) u satisfies (4.5) almost everywhere on $[0,c] \times R^n$ and condition (4.6) holds.

Let $C_{0.L}(B,R^k)$ be the class of all functions $w \in C(B,R^k)$ such that

$$\|w\|_L = \sup\left\{ \frac{\|w(t,s) - w(t,\bar{s})\|}{\|s - \bar{s}\|} : \; (t,s),\, (t,\bar{s}) \in B \right\} < \infty.$$

Let us denote by $\|\cdot\|_0$ the supremum norm in the space $C(B,R^k)$ and $\|w\|_{0.L} = \|w\|_0 + \|w\|_L$ for $w \in C_{0.L}(B,R^k)$. Write

$$C(B,R^k;\kappa) = \{ w \in C(B,R^k) : \; \|w\|_0 \le \kappa \}$$

and

$$C_{0.L}(B,R^k;\kappa) = \{ w \in C_{0.L}(B,R^k) : \; \|w\|_{0.L} \le \kappa \}.$$

where $\kappa \in R_+$. Let $\omega_0 \in L([-r_0,0],R_+)$, $p=(p_0,p_1) \in R_+^2$. Denote by $J[\omega_0,p]$ the set of all functions $\phi \in C(E_0,R^k)$ such that $\|\phi(x,y)\| \le p_0$ on E_0 and

$$\|\phi(x,y)-\phi(\bar{x},\bar{y})\| \le \left| \int_x^{\bar{x}} \omega_0(t)dt \right| + p_1\|y-\bar{y}\| \text{ on } E_0.$$

Put $|p| = p_0 + p_1$. Suppose that $c \in (0,a]$, $d=(d_0,d_1) \in R_+^2$, $d_0 \ge p_0$, $d_1 \ge p_1$ and $\omega \in L([-r_0,c],R_+)$, $\omega(t) \ge \omega_0(t)$ for almost all $t \in [-r_0,0]$. Let $K_{c.\phi}[\omega,d]$ be the class of all functions $z \in C([-r_0,c]\times R^n,R^k)$ such that

(i) $z(x,y)=\phi(x,y)$ on E_0,

(ii) $\|z(x,y)\| \le d_0$ and

$$\|z(x,y)-z(\bar{x},\bar{y})\| \le \left| \int_x^{\bar{x}} \omega(t)dt \right| + d_1\|y-\bar{y}\| \text{ on } [0,c]\times R^n.$$

Put $|d| = d_0 + d_1$. Denote by Θ the set of all functions $\alpha : [0,a]\times R_+ \to R_+$ such that $\alpha(\cdot,t) \in L([0,a],R_+)$ for $t \in R_+$ and the function $\alpha(t,\cdot) \in C(R_+,R_+)$ is nondecreasing on R_+ for almost all $t \in [0,a]$ and $\alpha(t,0)=0$.

Assumption H $[\varrho]$. Suppose that

1) the function $\varrho(\cdot,y,w) : [0,a] \to M[k,n]$ is measurable for $(y,w) \in R^n \times C(B,R^k)$ and $\varrho(x,\cdot) : R^n \times C(B,R^k) \to M[k,n]$ is continuous for almost all $x \in [0,a]$,

2) there exist α_1, $\beta_1 \in \Theta$ such that

$$\|\varrho(x,y,w)\| \le \alpha_1(x,\kappa) \text{ for } (y,w) \in R^n \times C(B,R^k;\kappa) \text{ almost everywhere on } [0,a],$$

and

$$\|\varrho(x,y,w)-\varrho(x,\bar{y},\bar{w}\| \le \beta_1(x,\kappa)\,[\,\|y-\bar{y}\| + \|w-\bar{w}\|_0\,]$$

for (y,w), $(\bar{y},\bar{w}) \in R^n \times C_{0.L}(B,R^k;\kappa)$ almost everywhere on $[0,a]$.

Suppose that $\phi \in J[\omega,p]$, $c \in (0,a]$ and $z \in K_{c.\phi}[\omega,d]$. Consider the Cauchy problem

$$\eta'(t) = \varrho_i(\,t,\eta(t),z_{(t,\eta(t))}\,), \quad \eta(x)=y, \tag{4.7}$$

where $(x,y) \in [0,c]\times R^n$ and $1 \le i \le k$ is fixed. Denote by $g_i[z](\cdot,x,y)$ the solution of (4.7). The function $g_i[z]$ is the i-th bicharacteristic of system (4.5) Let $\Delta_c = [0,c]\times[0,c]\times R^n$. For $z \in C([-r_0,a]\times R^n,R^k)$ and $x \in [0,a]$ we put

$$\|z\|_x = \sup\{\|z(t,s)\| : (t,s) \in [-r_0,x]\times R^n\,\}.$$

Lemma 4.4 *Suppose that Assumption H $[\varrho]$ is satisfied and $c \in (0,a]$, ϕ, $\bar{\phi} \in J[\omega_0,p]$, $z \in K_{c.\phi}[\omega,d]$, $\bar{z} \in K_{c.\bar{\phi}}[\omega,d]$.*

Then for each i, $1 \le i \le k$, the solutions $g_i[z](\cdot,x,y)$ and $g_i[\bar{z}](\cdot,x,y)$ are defined on $[0,c]$. Furthermore we have the estimates

$$\|g_i[z](t,x,y) - g_i[z](t,\bar{x},\bar{y})\| \tag{4.8}$$

$$\le \left[\|y-\bar{y}\| + \left|\int_x^{\bar{x}} \alpha_1(\tau,d_0)d\tau\right|\right] \exp\left[(1+d_1)\left|\int_x^t \beta_1(\tau,|d|)d\tau\right|\right] \text{ on } \Delta_c$$

and

$$\|g_i[z](t,x,y) - g_i[\bar{z}](t,x,y)\| \tag{4.9}$$

$$\leq \left| \int_x^t \beta_1(\tau, |d|) \, \|z - \bar{z}\|_\tau \, d\tau \right| \exp \left[(1 + d_1) \left| \int_x^t \beta_1(\tau, |d|) d\tau \right| \right] \quad on \;\; \Delta_c.$$

PROOF. The existence and uniqueness of the solution of (4.7) follows from classical theorems. On this purpose, note that the right-hand side of the differential system satisfies the Carathéodory assumptions and the following Lipschitz condition

$$\|\varrho_i(t, \mu, z_{(t,\mu)}) - \varrho_i(t, \bar{\mu}, z_{(t,\bar{\mu})})\| \leq \beta_1(t, |d|)(1 + d_1) \, \|\mu - \bar{\mu}\| \;\text{ on }\; [0, c] \times R^n$$

holds. The function $g_i[z](\,\cdot\,, x, y)$ satisfies the integral equation

$$g_i[z](t,x,y) = y + \int_x^t \varrho_i \left(\tau, g_i[z](\tau, x, y), z_{(\tau, g_i[z](\tau,x,y))} \right) d\tau.$$

For (t,x,y), $(t,\bar{x},\bar{y}) \in \Delta_c$ we have $\|z_{(t,g_i[z](t,x,y))}\|_0 \leq d_0$, and

$$\|z_{(t,g_i[z](t,x,y))} - z_{(t,g_i[z](t,\bar{x},\bar{y}))}\|_0 \leq d_1 \|g_i[z](t,x,y) - g_i[z](t,\bar{x},\bar{y})\|.$$

It follows from Assumption H $[\,\varrho\,]$ and from the above estimates that the following integral inequality is satisfied

$$\|g_i[z](t,x,y) - g_i[z](t,\bar{x},\bar{y})\| \leq \|y - \bar{y}\| + \left| \int_x^{\bar{x}} \alpha_1(\tau, d_0) d\tau \right|$$

$$+ (1 + d_1) \left| \int_x^t \beta_1(\tau, |d|) \, \|g_i[z](\tau,x,y) - g_i[z](\tau,\bar{x},\bar{y})\| d\tau \right|.$$

Now we obtain (4.8) by the Gronwall inequality. For $z \in K_{c.\phi}[\omega, d]$, $\bar{z} \in K_{c.\bar{\phi}}[\omega, d]$ we have the estimate

$$\|z_{(t,g_i[z](t,x,y))} - \bar{z}_{(t,g_i[\bar{z}](t,x,y))}\|_0 \leq d_1 \|g_i[z](t,x,y) - g_i[\bar{z}](t,x,y)\| + \|z - \bar{z}\|_t.$$

It follows that the function $g_i[z](\,\cdot\,,x,y) - g_i[\bar{z}](\,\cdot\,,x,y)$ satisfies the integral inequality

$$\|g_i[z](t,x,y) - g_i[\bar{z}](t,x,y)\| \leq \left| \int_x^t \beta_1(\tau, |d|) \|z - \bar{z}\|_\tau d\tau \right|$$

$$+ (1 + d_1) \left| \int_x^t \beta_1(\tau, |d|) \, \|g_i[z](\tau,x,y) - g_i[\bar{z}](\tau,x,y)\| d\tau \right|.$$

Now we obtain (4.9) by the Gronwall inequality. This proves the Lemma.

4.3 OPERATOR U_φ AND ITS PROPERTIES

Suppose that $\phi \in J[\omega_0, p]$, $c \in (0, a]$, $z \in K_{c.\phi}[\omega, d]$ and

$$(g_1[z](\cdot, x, y), \dots, g_k[z](\cdot, x, y)) = g[z](\cdot, x, y)$$

are bicharacteristics of system (4.5). Write

$$f^*\left(t, g[z](t,x,y), z_{(t,g[z](t,x,y))}\right)$$

$$= \left(f_1\left(t, g_1[z](t,x,y), z_{(t,g_1[z](t,x,y))}\right), \dots, f_k\left(t, g_k[z](t,x,y), z_{(t,g_k[z](t,x,y))}\right)\right),$$

and

$$\phi^*(t, g[z](t,x,y)) = \left(\phi_1(t, g_1[z](t,x,y)), \dots, \phi_k(t, g_k[z](t,x,y))\right).$$

For $z \in K_{c.\phi}[\omega, d]$ we define the function $Z = (Z_1, \dots, Z_k) = U_\phi z$ by

$$Z(x,y) = \phi^*(0, g[z](0,x,y))$$

$$+ \int_0^x f^*\left(\tau, g[z](\tau,x,y), z_{(\tau,g[z](\tau,x,y))}\right) d\tau \text{ on } [0,c] \times R^n,$$

and

$$Z(x,y) = \phi(x,y) \quad \text{on } E_0.$$

Remark 4.5 *The operator U_ϕ is obtained in the following way. Consider equations (4.5) along bicharacteristics*

$$D_x z_i(t, g_i[z](t,x,y)) + \sum_{j=1}^n \varrho_{ij}\left(t, g_i[z](t,x,y), z_{(t,g_i[z](t,x,y))}\right) D_{y_j} z_i(t, g_i[z](t,x,y))$$

$$= f_i\left(t, g_i[z](t,x,y), z_{(t,g_i[z](t,x,y))}\right), \quad i = 1, \dots, k.$$

The above relations imply

$$\frac{d}{dt} z_i(t, g_i[z](t,x,y)) = f_i\left(t, g_i[z](t,x,y), z_{(t,g_i[z](t,x,y))}\right), \quad i = 1, \dots, k,$$

and consequently, by integrating, we have

$$z_i(x,y) = \phi_i(0, g_i[z](0,x,y))$$

$$+ \int_0^x f_i\left(\tau, g_i[z](\tau,x,y), z_{(\tau,g_i[z](\tau,x,y))}\right) d\tau, \quad i = 1, \dots, k.$$

Then we have obtained the integral equation $z = U_\phi z$ with the initial condition $z = \phi$ on E_0.

Assumption H $[f]$. Suppose that

1) the function $f(\cdot, y, w) : [0,a] \to R^k$ is measurable for $(y,w) \in R^n \times C(B, R^k)$ and $f(x, \cdot) : R^n \times C(B, R^k) \to R^k$ is continuous for almost all $x \in [0,a]$,

2) there exist functions $\alpha_2, \beta_2 \in \Theta$ such that $\|f(x,y,w)\| \le \alpha_2(x,\kappa)$ on $R^n \times C(B, R^k; \kappa)$ almost everywhere on $[0,a]$ and

$$\|f(x,y,w) - f(x,\bar{y},\bar{w})\| \le \beta_2(x,\kappa)[\,\|y-\bar{y}\| + \|w-\bar{w}\|_0\,]$$

where (y,w), $(\bar{y},\bar{w}) \in R^n \times C_{0.L}(B, R^k; \kappa)$ almost everywhere on $[0,a]$.

Lemma 4.6 *If $\phi \in J[\omega_0, p]$ and Assumptions $H\ [\varrho]$, $H\ [f]$ are satisfied then there are $(d_0, d_1) = d \in R_+^2$, $c \in (0, a]$ and $\omega \in L([-r_0, c], R_+)$ such that*

$$U_\phi : K_{c.\phi}[\omega, d] \to K_{c.\phi}[\omega, d].$$

PROOF. Suppose that the constants $(d_0, d_1) = d \in R_+^2$, $c \in (0, a]$ and the function $\omega \in L([-r_0, c], R_+)$ satisfy the conditions:

$$d_0 \geq p_0 + \int_0^c \alpha_2(\tau, d_0) d\tau,$$

$$d_1 \geq \left[p_1 + (1 + d_1) \int_0^c \beta_2(\tau, |d|) d\tau\right] \exp\left[(1 + d_1) \int_0^c \beta_1(\tau, |d|)\, d\tau\right],$$

and

$$\omega(t) \geq s_c \alpha_1(t, d_0) + \alpha_2(t, d_0) \quad \text{for almost all } t \in [0, c]$$

where

$$s_c = \left[p_1 + (1 + d_1) \int_0^c \beta_2(\tau, |d|)\, d\tau\right] \exp\left[(1 + d_1) \int_0^c \beta_1(\tau, |d|)\, d\tau\right].$$

Suppose that $z \in K_{c.\phi}[\omega, d]$. Then we have

$$\|Z(x, y)\| \leq p_0 + \int_0^c \alpha_2(\tau, d_0)\, d\tau \leq d_0 \quad \text{on } [0, c] \times R^n. \tag{4.10}$$

If $(x, y),\ (x, \bar{y}) \in [0, c] \times R^n$ then using Lemma 4.4 we obtain

$$\|Z(x, y) - Z(x, \bar{y})\| \leq \|\phi^*(0, g[z](0, x, y)) - \phi^*(0, g[z](0, x, \bar{y}))\|$$

$$+ \int_0^x \|f^*\left(\tau, g[z](\tau, x, y), z_{(\tau, g[z](\tau, x, y))}\right) - f^*\left(\tau, g[z](\tau, x, \bar{y}), z_{(\tau, g[z](\tau, x, \bar{y}))}\right)\| d\tau$$

$$\leq p_1 \max_{1 \leq i \leq k} \|g_i[z](0, x, y) - g_i[z](0, x, \bar{y})\|$$

$$+ (1 + d_1) \int_0^x \beta_2(\tau, |d|) \max_{1 \leq i \leq k} \|g_i[z](\tau, x, y) - g_i[z](\tau, x, \bar{y})\|\, d\tau.$$

It follows from Assumption H $[f]$ and Lemma 4.4 that

$$\|Z(x, y) - Z(x, \bar{y})\| \leq d_1 \|y - \bar{y}\| \quad \text{on } [0, c] \times R^n. \tag{4.11}$$

For $(x, y),\ (\bar{x}, y) \in [0, c] \times R^n$ we have the estimates

$$\|Z(x, y) - Z(\bar{x}, y)\| \leq \|\phi^*(0, g[z](0, x, y)) - \phi^*(0, g[z](0, \bar{x}, y))\|$$

$$+ \int_0^x \|f^*\left(\tau, g[z](\tau, x, y), z_{(\tau, g[z](\tau, x, y))}\right) - f^*\left(\tau, g[z](\tau, \bar{x}, y), z_{(\tau, g[z](\tau, \bar{x}, y))}\right)\|\, d\tau$$

$$+\left|\int_x^{\bar{x}} \|f^*\left(\tau, g[z](\tau,\bar{x},y), z_{(\tau,g[z](\tau,\bar{x},y))}\right)\| \, d\tau\right|.$$

It follows from Assumption H [f] and Lemma 4.4 that

$$\|Z(x,y) - Z(\bar{x},y)\|$$

$$\leq p_1 \max_{1\leq i\leq k} \|g_i[z](0,x,y) - g_i[z](0,\bar{x},y)\| + \left|\int_x^{\bar{x}} \alpha_2(\tau,d_0)\, d\tau\right|$$

$$+(1+d_1)\int_0^c \beta_2(\tau,|d|) \max_{1\leq i\leq k} \|g_i[z](\tau,x,y) - g_i[z](\tau,\bar{x},y)\|\, d\tau$$

$$\leq \left|\int_x^{\bar{x}} \alpha_1(\tau,d_0)d\tau\right| \left[p_1 + (1+d_1)\int_0^c \beta_2(\tau,|d|)\, d\tau\right] \exp\left[(1+d_1)\int_0^c \beta_1(\tau,|d|)\, d\tau\right]$$

$$+\left|\int_x^{\bar{x}} \alpha_2(\tau,d_0)\, d\tau\right|.$$

Thus we see that

$$\|Z(x,y) - Z(\bar{x},y\| \leq \left|\int_x^{\bar{x}} \omega(\tau)d\tau\right| \quad \text{on } [0,c]\times R^n. \tag{4.12}$$

It follows from (4.10) - (4.12) that $Z \in K_{c.\phi}[\omega,d]$, which completes the proof of the Lemma.

4.4 Existence of weak solutions

Next we will show that under Assumptions H [ϱ] , H [f] there exists exactly one solution of (4.5), (4.6). The solution is local with respect to x.

Theorem 4.7 *Suppose that $\phi \in J[\omega_0,p]$ and Assumptions H [ϱ], H [f] are satisfied. Then there are $(d_0,d_1) = d \in R_+^2$, $c \in (0,a]$ and $\omega \in L([-r_0,c],R_+)$ such that there exists exactly one solution $u \in K_{c.\phi}[\omega,d]$ of problem (4.5), (4.6). If $\bar{\phi} \in J[\omega_0,p]$ and $\bar{u} \in K_{c.\bar{\phi}}[\omega,d]$ is a solution of system (4.5) with the initial condition $z(x,y) = \bar{\phi}(x,y)$ on E_0 then there is $\tilde{\Lambda}_c \in R_+$ such that*

$$\|u-\bar{u}\|_x \leq \tilde{\Lambda}_c\|\phi-\bar{\phi}\|_0, \quad x\in[0,c]. \tag{4.13}$$

Proof. Lemma 4.6 shows that there are

$$(d_0,d_1) = d \in R_+^2, \quad c \in (0,a] \text{ and } \omega \in L([-r_0,c],R_+)$$

such that $U_\phi : K_{c.\phi}[\omega,d] \to K_{c.\phi}[\omega,d]$. Write

$$\Gamma_c = \int_0^c \beta_1(\tau,|d|)d\tau \, \exp\left[(1+d_1)\int_0^c \beta_1(\tau,|d|)d\tau\right].$$

Let $c \in (0, a]$ be such a constant that

$$\Lambda_c = p_1\Gamma_c + [\Gamma_c(1+d_1)+1]\int_0^c \beta_2(\tau, |d|)d\tau < 1.$$

Now we prove that U_ϕ is a contraction. If $z,\ \tilde{z} \in K_{c.\phi}[\omega, d]$ then we have

$$||(U_\phi z)(x,y) - (U_\phi \tilde{z})(x,y)|| \le p_1 \max_{1\le i\le k} ||g_i[z](0,x,y) - g_i[\tilde{z}](0,x,y)||$$

$$+\int_0^x ||f^*(\tau, g[z](\tau,x,y), z_{(\tau,g[z](\tau,x,y))}) - f^*(\tau, g[\tilde{z}](\tau,x,y), \tilde{z}_{(\tau,g[\tilde{z}](\tau,x,y))})||d\tau$$

$$\le p_1 \exp\left[(1+d_1)\int_0^x \beta_1(\tau,|d|)\,d\tau\right]\int_0^x \beta_1(\tau,|d|)\,||z-\tilde{z}||_\tau\,d\tau$$

$$+\int_0^x \beta_2(\tau,|d|)\left[\max_{1\le i\le k}||g_i[z](\tau,x,y) - g_i[\tilde{z}](\tau,x,y)||\right.$$

$$\left.+\max_{1\le i\le k}||z_{(\tau,g_i[z](\tau,x,y))} - \tilde{z}_{(\tau,g_i[\tilde{z}](\tau,x,y))}||_0\right]d\tau.$$

The estimates

$$||z_{(t,g_i[z](t,x,y))} - \tilde{z}_{(t,g_i[\tilde{z}](t,x,y))}||_0 \le ||z-\tilde{z}||_t + d_1||g_i[z](t,x,y) - g_i[\tilde{z}](t,x,y)||,$$

where $1 \le i \le k$, and Lemma 4.4 imply

$$||(U_\phi z)(x,y) - (U_\phi \tilde{z})(x,y)||$$

$$\le \int_0^x \beta_2(\tau,|d|)\,d\tau\,||z-\tilde{z}||_x\,[\,1+\Gamma_c(1+d_1)\,] + p_1\Gamma_c\,||z-\tilde{z}||_x,$$

and consequently $||U_\phi z - U_\phi\tilde{z}||_c \le \Lambda_c||z-\tilde{z}||_c$. By the Banach fixed point theorem there exists a unique solution $u \in K_{c.\phi}[\omega, d]$ of the equation $z = U_\phi z$. Now we prove that u is a solution of the system (4.5). We have proved that

$$u_i(x,y) = \phi_i\,(0, g_i[u](0,x,y)) \tag{4.14}$$

$$+\int_0^x f_i\left(\tau, g_i[u](\tau,x,y), u_{(\tau,g_i[u](\tau,x,y))}\right)d\tau,\quad i=1,\ldots,k,\quad \text{on}\quad [0,c]\times R^n.$$

For given $y \in R^n$, $1 \le i \le k$, let us put $\eta^{(i)} = g_i[u](0,x,y)$. It follows that $g_i[u](t,x,y) = g_i[u](t,0,\eta^{(i)})$ for $t \in [0,c]$ and $y = g_i[u](x,0,\eta^{(i)})$. For each $1 \le i \le k$ the relations

$$\eta^{(i)} = g_i[u](0,x,y) \ \text{ and } \ y = g_i[u](x,0,\eta^{(i)})$$

are equivalent for $y,\ \eta^{(i)} \in R^n$. It follows from (4.14) that

$$u_i(x, g_i[u](x,0,\eta^{(i)})) = \phi_i(0,\eta^{(i)}) \tag{4.15}$$

$$+\int_0^x f_i(\,\tau, g_i[u](\tau,0,\eta^{(i)}), u_{(\tau,g_i[u](\tau,0,\eta^{(i)}))}\,)d\tau,\quad i=1,\ldots,k.$$

By differentiating (4.15) with respect to x and by using the transformations $\eta^{(i)} = g_i[u](0,x,y)$ which preserve the sets of measure zero, we obtain that u satisfies (4.5) for almost all $(x,y) \in [0,c] \times R^n$. Now we prove relation (4.13). If $u = U_\phi u$ and $\bar{u} = U_{\bar{\phi}}\bar{u}$ then

$$\|u(x,y) - \bar{u}(x,y)\| \le \|\phi^*\left(0, g[u](0,x,y)\right) - \bar{\phi}^*\left(0, g[\bar{u}](0,x,y)\right)\|$$

$$+ \int_0^x \|f^*\left(\tau, g[u](\tau,x,y), u_{(\tau,g[u](\tau,x,y))}\right) - f^*\left(\tau, g[\bar{u}](\tau,x,y), \bar{u}_{(\tau,g[\bar{u}](\tau,x,y))}\right)\| d\tau$$

$$\le \|\phi - \bar{\phi}\|_0 + p_1 \max_{1\le i\le k} \|g_i[u](0,x,y) - g_i[\bar{u}](0,x,y)\|$$

$$+ \int_0^x \beta_2(\tau,|d|) \left[\max_{1\le i\le k} \|g_i[u](\tau,x,y) - g_i[\bar{u}](\tau,x,y)\|\right.$$

$$\left. + \max_{1\le i\le k} \|u_{(\tau,g_i[u](\tau,x,y))} - \bar{u}_{(\tau,g_i[\bar{u}(\tau,x,y))}\|_0\right] d\tau.$$

It follows from Lemma 4.4 that

$$\|g_i[u](t,x,y) - g_i[\bar{u}](t,x,y)\| \le C_c \int_0^x \beta_1(\tau,|d|)\|u-\bar{u}\|_\tau \, d\tau, \;\; 0 \le t \le x,$$

with

$$C_c = \exp\left[(1+d_1)\int_0^c \beta_1(\tau,|d|)d\tau\right]$$

and

$$\|u_{(t,g_i[u](t,x,y))} - \bar{u}_{(t,g_i[\bar{u}](t,x,y))}\|_0 \le \|u-\bar{u}\|_t + d_1\|g_i[u](t,x,y) - g_i[\bar{u}](t,x,y)\|$$

$$\le \|u-\bar{u}\|_t + d_1 C_c \int_0^x \beta_1(\tau,|d|)\|u-\bar{u}\|_\tau \, d\tau, \;\; 0 \le t \le x.$$

Then we have

$$\|u(x,y) - \bar{u}(x,y)\| \le \|\phi - \bar{\phi}\|_0$$

$$+p_1 C_c \int_0^x \beta_1(\tau,|d|)\|u-\bar{u}\|_\tau \, d\tau + C_c \int_0^x \beta_1(\tau,|d|)\, \|u-\bar{u}\|_\tau \, d\tau \int_0^c \beta_2(\tau,|d|)d\tau$$

$$+ \int_0^x \beta_2(\tau,|d|)\|u-\bar{u}\|_\tau \, d\tau + C_c d_1 \int_0^x \beta_1(\tau,|d|)\, \|u-\bar{u}\|_\tau \, d\tau \int_0^c \beta_2(\tau,|d|)\, d\tau.$$

The right-hand side of the above inequality is a nondecreasing function of x, therefore we have the integral inequality

$$\|u-\bar{u}\|_x \le \|\phi - \bar{\phi}\|_0$$

$$+ \int_0^x \left[\beta_1(\tau,|d|)\left(p_1 C_c + D_c + d_1 D_c\right) + \beta_2(\tau,|d|)\right] \|u-\bar{u}\|_\tau \, d\tau, \;\; x \in [0,c],$$

with

$$D_c = C_c \int_0^c \beta_2(\tau,|d|)d\tau.$$

Using the Gronwall inequality we obtain $\|u-\bar{u}\|_x \le \tilde{\Gamma}(x)\|\phi-\bar{\phi}\|_0$, $x \in [0,c]$ with

$$\tilde{\Gamma}(x) = \exp\left\{\int_0^x [\beta_1(\tau,|d|)(p_1C_c + D_c + d_1D_c) + \beta_2(\tau,|d|)]\, d\tau\right\}.$$

Then we have estimate (4.13) for $\tilde{\Lambda}_c = \tilde{\Gamma}(c)$. This proves the theorem.

It is important in our consideration that we have assumed the Lipschitz condition for given functions on some special function spaces. We have assumed that the functions $\varrho(x,\cdot)$ and $f(x,\cdot)$ satisfy the Lipschitz condition on the space $R^n \times C_{0.L}(B,R^k;\kappa)$ for almost all $x \in [0,a]$ and this condition is local with respect to the functional variable. Let us consider simplest assumptions on ϱ and f. Suppose that there is $\tilde{L}$ such that for almost all $x \in [0,a]$

$$\|\varrho(x,y,w) - \varrho(x,\bar{y},\bar{w})\| \le \tilde{L}[\,\|y-\bar{y}\| + \|w-\bar{w}\|_0\,] \tag{4.16}$$

$$\|f(x,y,w) - f(x,\bar{y},\bar{w})\| \le \tilde{L}[\,\|y-\bar{y}\| + \|w-\bar{w}\|_0\,] \tag{4.17}$$

on $R^n \times C(B,R^k)$. Of course, our results are true if we assume (4.16), (4.17) instead of suitable estimates in Assumptions H $[\varrho]$ and H $[f]$. Now we show that there is a class of quasilinear systems satisfying Assumptions H $[\varrho]$ and H $[f]$ and do not satisfying (4.16), (4.17). Consider the system with a deviated argument

$$D_x z_i(x,y) = \sum_{j=1}^{n} \tilde{\varrho}_{ij}(\,x,y,z(\psi_0(x),\psi(x,y))\,)\, D_{y_j} z_i(x,y) \tag{4.18}$$

$$= \tilde{f}_i(\,x,y,z(\psi_0(x),\psi(x,y))\,),\quad i = 1,\ldots,k,$$

where $\psi_0 : [0,a] \to R$, $\psi = (\psi_1,\ldots,\psi_n) : [0,a]\times R^n \to R^n$ and

$$\tilde{\varrho} : [0,a]\times R^n \times R^k \to M[k,n],\quad \tilde{\varrho} = [\,\tilde{\varrho}_{ij}\,]_{i=1,\ldots,k\; j=1,\ldots,n},$$

$$\tilde{f} : [0,a]\times R^n \times R^k \to M[k,1],\quad \tilde{f} = (\,f_1,\ldots,\tilde{f}_k\,)^T,$$

are given functions. We get system (4.18) by putting in (4.5)

$$f(x,y,w) = \tilde{f}(\,x,y,w(\psi_0(x)-x,\psi(x,y)-y)\,) \tag{4.19}$$

$$\varrho(x,y,w) = \tilde{\varrho}(\,x,y,w(\psi_0(x)-x,\psi(x,y)-y)\,) \tag{4.20}$$

Assumption H $[\tilde{f},\tilde{g}]$. Suppose that

1) the functions $\tilde{\varrho}(\cdot,y,\zeta) : [0,a] \to M[k,n]$ and $\tilde{f}(\cdot,y,\zeta) : [0,a] \to M[k,1]$ are measurable for $(y,\zeta) \in R^n \times R^k$,

2) the functions $\tilde{\varrho}(x,\cdot) : R^n \times R^k \to M[k,n]$ and $\tilde{f}(x,\cdot) : R^n \times R^k \to M[k,1]$ are continuous for almost all $x \in [0,a]$,

3) there exist $\alpha,\ \beta \in L(\,[0,a],R_+\,)$ such that

$$\|\tilde{\varrho}(x,y,\zeta)\| \le \alpha(x),\quad \|\tilde{f}(x,y,\zeta)\| \le \alpha(x),$$

for $(y,\zeta) \in R^n \times R^k$ almost everywhere on $[0,a]$ and

$$\|\tilde{\varrho}(x,y,\zeta) - \tilde{\varrho}(x,\bar{y},\bar{\zeta})\| \le \beta(x)\ [\,\|y-\bar{y}\| + \|\zeta-\bar{\zeta}\|\,],$$

$$\|\tilde{f}(x,y,\zeta) - \tilde{f}(x,\bar{y},\bar{\zeta})\| \le \beta(x)\ \left[\,\|y-\bar{y}\| + \|\zeta - \bar{\zeta}\|\,\right],$$

for $(y,\zeta) \in R^n \times R^k$ almost everywhere on $[0,a]$.

Assumption H $[\psi_0, \psi]$**.** Suppose that

1) $\psi_0 \in C([0,a],R)$ and $-r_0 \le \psi_0(x) - x \le 0$ on $[0,a]$,
2) $\psi \in C([0,a]\times R^n, R^n)$ and $-r \le \psi(x,y) - y \le r$ on $[0,a]\times R^n$,
3) there is $\bar{L} \in R_+$ such that

$$\|\psi(x,y) - \psi(x,\bar{y})\| \le \bar{L}\,\|y-\bar{y}\| \text{ on } [0,a]\times R^n.$$

Theorem 4.8 *Suppose that Assumptions $H\,[\tilde{f},\tilde{\varrho}]$ and $H\,[\psi_0,\psi]$ are satisfied and $\phi \in J[\omega,p]$. Then there is $c \in (0,a]$ such that the Cauchy problem (4.6), (4.18) has exactly one solution $u \in K_{c.\phi}[\omega,d]$ for suitable ω and d. If $\bar{\phi} \in J[\omega_0,p]$ and $\bar{u} \in K_{c.\bar{\phi}}[\omega,d]$ is a solution of system (4.18) with the initial condition $z(x,y) = \bar{\phi}(x,y)$ on E_0 then there is $\tilde{\Lambda}_c \in R_+$ such that $\|u-\bar{u}\|_x \le \tilde{\Lambda}_c\|\phi - \bar{\phi}\|_0$, $x \in [0,c]$.*

The above result follows from Theorem 4.7.

Remark 4.9 *Note that the function f, ϱ given by (4.19), (4.20) do not satisfy the Lipschitz condition (4.16), (4.17).*

4.5 Integral operators for quasilinear systems

Let B, E_0, Δ_c and Ω_0 be the sets defined in Sections 4.1 and 4.2. Assume that

$$A : \Omega_0 \to M[k,k], \qquad A = [A_{i,j}]_{i,j=1,\dots,k},$$

$$\varrho : \Omega_0 \to M[k,n],, \qquad \varrho = [\varrho_{ij}]_{i=1,\dots,k,\ j=1,\dots,n},$$

$$f : \Omega_0 \to M[k,1], \quad f = (f_1,\dots,f_k)^T, \quad \phi : E_0 \to R^k,$$

are given functions. In this Section we consider the functional differential system in the second canonical form

$$\sum_{j=1}^{k} A_{ij}(x,y,z_{(x,y)}) \left[D_x z_j(x,y) + \sum_{l=1}^{n} \varrho_{il}(x,y,z_{(x,y)}) D_{y_l} z_j(x,y) \right] \tag{4.21}$$

$$= f_i(x,y,z_{(x,y)}), \quad i = 1,\dots,k,$$

with the initial condition

$$z(x,y) = \phi(x,y) \text{ on } E_0. \tag{4.22}$$

Note that system (4.21) is a strongly coupled system. We consider Carathéodory solutions of problem (4.21), (4.22). If A is the identity matrix then (4.21) reduces to the weakly coupled system (4.5). We will discuss the problem of the existence of solutions of problem (4.21), (4.22). We start with definitions of some function spaces. Let $C_L(B,R^k)$ be the class of all functions $w \in C(B,R^k)$ such that

$$\|w\|_L = \sup\left\{ \frac{\|w(t,s) - w(\bar{t},\bar{s})\|}{|t-\bar{t}| + \|s-\bar{s}\|} :\ (t,s),\ (\bar{t},\bar{s}) \in B \right\} < \infty.$$

For $w \in C_L(B, R^k)$ we define $\|w\|_{0.L} = \|w\|_0 + \|w\|_L$ where $\|\cdot\|_0$ is the supremum norm in the space $C(B, R^k)$. Write

$$C(B, R^k; \kappa) = \{w \in C(B, R^k) : \|w\|_0 \le \kappa\},$$

$$C_L(B, R^k; \kappa) = \{w \in C_L(B, R^k) : \|w\|_{0.L} \le \kappa\},$$

where $\kappa \in R_+$. Let $\tilde{\Theta}$ be the class of all functions $\alpha : R_+ \to R_+$ which are nondecreasing on R_+. For $U, V \in M[k,k]$ we denote by $U * V = (d_1, \dots, d_k)$ the vector given by (2.59). Now we formulate a functional integral system corresponding to (4.21), (4.22). For $z : [-r_0, c] \times R^n \to R^k$ and $g : \Delta_c \to M[k,n]$, $g = [g_{ij}]_{i=1,\dots,k,\ j=1,\dots,n}$ where $0 < c \le a$ we put

$$g_i(t,x,y) = (g_{i1}(t,x,y), \dots, g_{in}(t,x,y)), \quad i = 1, \dots, k,$$

$$A[g,z](t,x,y) = \left[\, A_{ij}\left(t, g_i(t,x,y), z_{(t,g_i(t,x,y))}\right) \right]_{i,j=1,\dots,k},$$

$$\phi[g](t,x,y) = \left[\, \phi_i\left(0, g_j(t,x,y)\right) \right]_{i,j=1,\dots,k},$$

$$Z[g,z] = \left[\, z_i(t, g_j(t,x,y)) \right]_{i,j=1,\dots,k},$$

$$f[g,z](t,x,y) = \left(f_1(t, g_1(t,x,y), z_{(t,g_1(t,x,y))}), \dots, f_k(t, g_k(t,x,y), z_{(t,g_k(t,x,y))}) \right)^T.$$

Consider the system of functional integral equations

$$g_i(t,x,y) = y + \int_x^t \varrho_i\left(\tau, g_i(\tau,x,y), z_{(\tau,g_i(\tau,x,y))}\right) d\tau, \quad i = 1, \dots, k, \tag{4.23}$$

$$z(x,y) = A^{-1}(x,y,z_{(x,y)}) \left\{ A[g,z](0,x,y) * \phi[g](0,x,y) \right\} \tag{4.24}$$

$$+ A^{-1}(x,y,z_{(x,y)}) \left\{ \int_0^x \left[\frac{d}{d\tau} A[g,z](\tau,x,y) * Z[g,z](\tau,x,y) + f[g,z](\tau,x,y) \right] d\tau \right\},$$

$$z(x,y) = \phi(x,y) \quad \text{on } E_0. \tag{4.25}$$

We give sufficient conditions for the existence and uniqueness of solutions to problem (4.23) - (4.25). We establish relationships between solutions of (4.21), (4.22) and (4.23) - (4.25). For $\lambda = (\lambda_0, \lambda_1, \lambda_2) \in R_+^3$ we put

$$J[\lambda] = \{\, \psi \in C(E_0, R^k) : \|\psi(x,y)\| \le \lambda_0$$

$$\text{and } \|\psi(x,y) - \psi(\bar{x},\bar{y}\| \le \lambda_1 |x - \bar{x}| + \lambda_2 \|y - \bar{y}\| \text{ on } E_0 \,\}.$$

Suppose that $\phi \in J[\lambda]$, $0 < c \le a$ and $d = (d_0, d_1, d_2) \in R_+^3$, $d_i \ge \lambda_i$, $i = 0, 1, 2$. We denote by $K_{c.\phi}[d]$ the class of all functions $z \in C([-r_0, c] \times R^n, R^k)$ such that

(i) $z(x,y) = \phi(x,y)$ on E_0,

(ii) $\|z(x,y)\| \le d_0$ and

$$\|z(x,y) - z(\bar{x},\bar{y})\| \le d_1 |x - \bar{x}| + d_2 \|y - \bar{y}\| \quad \text{on } [0,c] \times R^n.$$

Assumption $\tilde{H}\,[\varrho]$. Suppose that

1) $\varrho(\cdot, y, w) : [0, a] \to M[k, n]$ is measurable for $(y, w) \in R^n \times C(B, R^k)$ and $\varrho(x, \cdot) \in C(R^n \times C(B, R^k), M[k, n])$ for almost all $x \in [0, a]$,

2) there are $\alpha, \beta \in \tilde{\Theta}$ such that $\|\varrho(x, y, w)\| \leq \alpha(\kappa)$ for $(y, w) \in R^n \times C(B, R^k; \kappa)$ for almost all $x \in [0, a]$ and

$$\|\varrho(x, y, w) - \varrho(x, \bar{y}, \bar{w})\| \leq \beta(\kappa)[\,\|y - \bar{y}\| + \|w - \bar{w}\|_0\,] \text{ on } R^n \times C_L(B, R^k; \kappa)$$

almost everywhere on $[0, a]$.

Let $c \in (0, a]$, $p \in (0, 1)$, $d_0 \in R_+$. Denote by $\tilde{K}_c[p, d_0]$ the set of all functions

$$h \in C(\,\Delta_c, M[k, n]\,), \quad h = [h_{ij}]_{i=1,\dots,k,\; j=1,\dots,n}\,,$$

such that $h_i(x, x, y) = 0$ and

$$\|h_i(t, x, y) - h_i(\bar{t}, x, \bar{y})\| \leq \alpha_1(d_0)\,|t - \bar{t}| + p\,\|y - \bar{y}\| \text{ on } \Delta_c, \quad i = 1, \dots, k,$$

where $h_i = (h_{i1}, \dots, h_{in})$.

Remark 4.10 *If $h \in \tilde{K}_c[p, d_0]$ and $(t, x, y) \in \Delta_c$ then*

$$\|h_i(t, x, y)\| \leq \|h_i(t, x, y) - h_i(x, x, y)\| \leq c\,\alpha(d_0).$$

Thus functions $h \in \tilde{K}_c[p, d_0]$ are equibounded and $\tilde{K}_c[p, d_0]$ is a closed subset of the Banach space of continuous and bounded functions $h : \Delta_c \to M[k, n]$ with the norm

$$\|h\|_c = \sup\{\|h(t, x, y)\| : \ (t, x, y) \in \Delta_c\}\,.$$

Suppose that $c \in (0, a]$, $\phi \in J[\lambda]$ and $z \in K_{c.\phi}[d]$. Consider the operator

$$H = T_z h, \quad H = [h_{ij}]_{i=1,\dots,k, j=1,\dots,n}\,, \quad H_i = (H_{i1}, \dots, H_{in}),$$

defined for $h \in \tilde{K}_c[p, d_0]$ by

$$H_i(t, x, y) = \int_x^t \varrho_i\left(\tau, h_i(\tau, x, y) + y, z_{(\tau, h_i(\tau, x, y) + y)}\right) d\tau,$$

where $(t, x, y) \in \Delta_c$, $i = 1, \dots, k$.

Lemma 4.11 *Suppose that Assumption $\tilde{H}\,[\,\varrho\,]$ is satisfied, $c \in (0, a]$ and $\phi \in J[\lambda]$, $z \in K_{c.\phi}[d]$. Assume that*

$$c\,\beta(|d|)(1 + p)(1 + d_2) \leq p, \quad p \in (0, 1), \tag{4.26}$$

where $|d| = d_0 + d_1 + d_2$. Then $T_z : \tilde{K}_c[p, d_0] \to \tilde{K}_c[p, d_0]$ and T_z has exactly one fixed point on $\tilde{K}_c[p, d_0]$. Furthermore, if

$$\phi, \bar{\phi} \in J[\lambda], \ z \in K_{c.\phi}[d], \ \ \bar{z} \in K_{c.\bar{\phi}}[d], \quad h, \ \bar{h} \in \tilde{K}_c[p, d_0]$$

and $h = T_z h$, $\bar{h} = T_{\bar{z}}\bar{h}$ then

$$\|h - \bar{h}\|_c \leq N_c \|z - \bar{z}\|_c \tag{4.27}$$

where

$$N_c = c\beta(|d|) \exp\left[c\beta(|d|)(1 + d_2)\right]. \tag{4.28}$$

PROOF. Theorems on integral inequalities and the Banach fixed point theorem in spaces with Bielecki norms are used in the proof of the Lemma. We omit details.

Let $K_c[p,d_0]$ be the class of all functions

$$g:\Delta_c \to M[k,n],\quad g=[g_{ij}]_{i=1,\dots,k,\ j=1,\dots,n},\quad g_i=(g_{i1},\dots,g_{in}),\ 1\le i\le k,$$

such that $g_i(t,x,y)=h_i(t,x,y)+y,\ i=1,\dots,k,$ and $h\in\tilde{K}_c[p,d_0]$. Suppose that $h\in\tilde{K}_c[p,d_0]$ is the fixed point of T_z on $\tilde{K}_c[p,d_0]$ with $z\in K_{c.\phi}[d]$ and $g_i(t,x,y)=h_i(t,x,y)+y,\ 1\le i\le k$. Then $g_i(\cdot,x,y)$ is a solution in the almost everywhere sense of the Cauchy problem

$$\eta'(t)=\varrho_i(t,\eta(t),z_{(t,\eta(t))}),\quad \eta(x)=y. \tag{4.29}$$

Functions g_i are called bicharacteristics of system (4.21). Note that for given $z\in K_{c.\phi}[d]$ integral system (4.23) is equivalent to problem (4.29).

Remark 4.12 *Under the assumptions of Lemma 4.11 the fixed point h of the operator T_z satisfies*

$$\|h_i(t,x,y)-h_i(t,\bar{x},y)\|\le\lambda_c|x-\bar{x}|\ \text{ on }\ \Delta_c, \tag{4.30}$$

where $i=1,\dots,k,$ and $\lambda_c=\alpha(d_0)\exp\left[c\beta(|d|)(1+d_2)\right]$.

Indeed, we have

$$\|h_i(t,x,y)-h_i(t,\bar{x},y)\|$$
$$\le\alpha(d_0)|x-\bar{x}|+\beta(|d|)(1+d_2)\left|\int_x^t\|h_i(\tau,x,y)-h_i(\tau,\bar{x},y)\|d\tau\right|\ \text{ on }\ \Delta_c.$$

Now, we obtain (4.30) by the Gronwall inequality.

4.6 QUASILINEAR SYSTEMS IN THE SECOND CANONICAL FORM

Now we consider system (4.24), (4.25) with $g[z]$ satisfying (4.23). We will need the following assumptions.

Assumption $\tilde{H}[A]$. Suppose that $A\in C(\Omega_0,M[k,k])$ and there are $\nu>0,\ \alpha,\beta\in\Theta$ such that

1) $\det A(x,y,w)\ge\nu$ on Ω_0 and

$$\|A(x,y,w)\|\le\alpha(\kappa)\ \text{ on }\ [0,a]\times R^n\times C(B,R^k;\kappa),$$

2) for $(x,y,w),(\bar{x},\bar{y},\bar{w})\in[0,a]\times R^n\times C_L(B,R^k;\kappa)$ we have

$$\|A(x,y,w)-A(\bar{x},\bar{y},\bar{w})\|\le\beta(\kappa)\left[\,|x-\bar{x}|+\|y-\bar{y}\|+\|w-\bar{w}\|_0\right].$$

Assumption $\tilde{H}[f]$. Suppose that

1) $f(\cdot,y,w):[0,a]\to M[k,1]$ is measurable for $(y,w)\in R^n\times C(B,R^k)$ and $f(x,\cdot)\in C\left(R^n\times C(B,R^k),\ M[k,1]\right)$ for almost all $x\in[0,a]$,

2) for $(y, w), \in R^n \times C(B, R^k; \kappa)$ we have $\|f(x, y, w)\| \leq \alpha(\kappa)$ almost everywhere on $[0, a]$,

3) for $(y, w), (\bar{y}, \bar{w}) \in R^n \times C_L(B, R^k; \kappa)$ we have

$$\|f(x, y, w) - f(x, \bar{y}, \bar{w})\| \leq \beta(\kappa)\, [\, \|y - \bar{y}\| + \|w - \bar{w}\|_0\,]$$

almost everywhere on $[0, a]$.

Remark 4.13 *For simplicity of notation, we have assumed the same estimation for ϱ, A and f. We have assumed also the Lipschitz condition for these functions with the same coefficient.*

Remark 4.14 *If Assumption $\tilde{H}\,[\,A\,]$ is satisfied then the matrix $A^{-1}(x, y, w)$: $\Omega_0 \to M[k, k]$ exists and $A^{-1} \in C(\Omega_0, M[k, k])$. Moreover, there are $\tilde{\alpha}, \tilde{\beta} \in \Theta$ such that*

$$\|A^{-1}(x, y, w)\| \leq \tilde{\alpha}(\kappa) \;\text{ on }\; [0, a] \times R^n \times C(B, R^k; \kappa), \tag{4.31}$$

$$\|A^{-1}(x, y, w) - A^{-1}(\bar{x}, \bar{y}, \bar{w})\| \leq \tilde{\beta}(\kappa)\, [\, |x - \bar{x}| + \|y - \bar{y}\| + \|w - \bar{w}\|_0\,] \tag{4.32}$$

on $[0, a] \times R^n \times C_L(B, R^k; \kappa)$.

Remark 4.15 *If Assumption $\tilde{H}\,[\,\varrho\,]$, $\tilde{H}\,[\,A\,]$ are satisfied and $\phi \in J[\lambda]$, $z \in K_{c.\phi}[d]$ then the derivative*

$$\frac{d}{d\tau} A[g, z](\tau, x, y)$$

exists almost everywhere on $[0, c]$ for $(x, y) \in [0, c] \times R^n$ and the function

$$\frac{d}{d\tau} A[g, z](\,\cdot\,, x, y)$$

is integrable on $[0, c]$.

Now we formulate the main theorem in this Section.

Theorem 4.16 *Suppose that Assumptions $\tilde{H}\,[\,\varrho\,]$, $\tilde{H}\,[\,A\,]$, $\tilde{H}\,[\,f\,]$ hold and $\phi \in J[\lambda]$.*

Then there exist $c \in (0, a]$ and $d = (d_0, d_1, d_2) \in R^3_+$ such that problem (4.23) - (4.25) has exactly one solution $(\tilde{g}, v) \in K_c[p, d_0] \times K_{c.\phi}[d]$. Furthermore, if $\bar{\phi} \in J[\lambda]$ and $(\bar{g}, \bar{v}) \in K_c[p, d_0] \times K_{c.\bar{\phi}}[d]$ is a solution of system (4.23), (4.24) with initial condition

$$z(x, y) = \bar{\phi}(x, y) \;\text{ on }\; E_0, \tag{4.33}$$

then there exists $M_c \in R_+$ such that $\|v - \bar{v}\|_c \leq M_c \|\phi - \bar{\phi}\|_0$. The function v is a solution of problem (4.21), (4.22) on $[-r_0, c] \times R^n$.

The proof will be divided into 3 lemmas. For $z \in K_{c.\phi}[d]$ we define $U = U_\phi z$ by

$$U(x, y) = A^{-1}(x, y, z_{(x,y)})\, \{A[g, z](0, x, y) * \phi[g](0, x, y)\} \tag{4.34}$$

$$+ A^{-1}(x, y, z_{(x,y)}) \int_0^x \left\{ \frac{d}{d\tau} A[g, z](\tau, x, y) * Z[g, z](\tau, x, y) + f[g, z](\tau, x, y) \right\} d\tau,$$

$$U(x, y) = \phi(x, y) \;\text{ on }\; E_0, \tag{4.35}$$

where the function $g = g[z]$ is the solution of (4.23), (4.25).

Lemma 4.17 *Suppose that*

1) Assumptions $\tilde{H}[\varrho]$, $\tilde{H}[A]$, $\tilde{H}[f]$ *are satisfied,* $c \in (0, a]$, $\phi \in J[\lambda]$ *and* $z \in K_{c.\phi}[d]$,

2) condition (4.26) holds true and

$$d_0 > \lambda_0,$$

$$d_1 > \max\{\tilde{\alpha}(d_0)[\alpha(d_0) + \lambda_2 \alpha(\lambda_0)\alpha(d_0)], \ \lambda_1\},$$

$$d_2 > \lambda_2[1 + \tilde{\alpha}(d_0)\alpha(\lambda_0)(2 + p)].$$

Then for $c \in (0, a]$ *sufficiently small we have:* $U_\phi : K_{c.\phi}[d] \to K_{c.\phi}[d]$.

The proof of the lemma is based on theorems on integral inequalities. Details can be found in [71] . Suppose that ϕ, $\bar{\phi} \in J[\lambda]$ and $z \in K_{c.\phi}[d]$, $\bar{z} \in K_{c.\bar{\phi}}[d]$. Denote by $\bar{g} = g[\bar{z}]$ the solution of functional integral problem (4.23), (4.25) with $\bar{z}$ and $\bar{\phi}$ instead of z and ϕ respectively. Now we consider two operators: U_ϕ and $U_{\bar{\phi}}$.

Lemma 4.18 *Suppose that*

1) the assumptions of Lemma 4.17 are satisfied,

2) $c \in (0, a]$ *is so small that we have*

$$U_\phi : K_{c.\phi}[d] \to K_{c.\phi}[d] \quad \text{and} \quad U_{\bar{\phi}} : K_{c.\bar{\phi}}[d] \to K_{c.\bar{\phi}}[d].$$

Under these assumptions there exist $M^* > 0$, $M_c > 0$ *such that for*

$$z \in K_{c.\phi}[d], \ \bar{z} \in K_{c.\bar{\phi}}[d]$$

we have

$$||U_\phi z - U_{\bar{\phi}}\bar{z}||_c \le M_c ||z - \bar{z}||_c + M^* ||\phi - \bar{\phi}||_0 \tag{4.36}$$

and $\lim_{c \to 0} M_c = 0$.

The proof of the lemma is based on theorems on integral inequalities. Details can be found in [71].

Lemma 4.19 *Suppose that Assumptions* $\tilde{H}[\varrho]$, $\tilde{H}[A]$, $\tilde{H}[f]$ *are satisfied and condition (4.26) holds true. Then for each* $\phi \in J[\lambda]$ *there exists* $c \in (0, a]$ *such that system (4.23) - (4.25) has exactly one solution* $(\tilde{g}, v) \in K_c[p, d_0] \times K_{c.\phi}[d]$. *If* $\bar{\phi} \in J[\lambda]$ *and* $(\bar{g}, \bar{v}) \in K_c[p, d_0] \times K_{c.\bar{\phi}}[d]$ *is a solution of system (4.23), (4.24), (4.33) then*

$$||v - \bar{v}||_c \le M^*(1 - \bar{q})^{-1} ||\phi - \bar{\phi}||_0 \tag{4.37}$$

where $\bar{q} \in (0, 1)$. *Moreover, the functions* v *and* $\bar{v}$ *are the solutions of (4.21), (4.22) and (4.21), (4.33) respectively.*

PROOF. It follows from Lemmas 4.11 and 4.17 that for sufficiently small $c \in (0, a]$ we have $U_\phi : K_{c.\phi}[d] \to K_{c.\phi}[d]$. Let $\bar{q} \in (0, 1)$. Hence, by Lemma 4.18 there is $c \in (0, a]$ such that for z, $\tilde{z} \in K_{c.\phi}[d]$ we have $||U_\phi z - U_\phi \tilde{z}||_c \le \bar{q}\, ||z - \tilde{z}||_c$. Thus U_ϕ

is a contraction. Denote by v the fixed point of U_ϕ on $K_{c.\phi}$. Then v is the solution of (4.23) - (4.25). We prove that v is the solution of (4.21). We have the identity

$$v(x,y) = A^{-1}(x,y,v_{(x,y)})\,\{A[\tilde{g},v](0,x,y) * \phi[\tilde{g}](0,x,y)\}$$

$$+A^{-1}(x,y,v_{(x,y)})\int_0^x \left\{\frac{d}{d\tau}A[\tilde{g},v](\tau,x,y) * Z[\tilde{g},v](\tau,x,y) + f[\tilde{g},v](\tau,x,y)\right\} d\tau$$

on $[0,c]\times R^n$, where $\tilde{g} = g[v]$. Hence, by integration by parts we obtain

$$A(x,y,v_{(x,y)})v(x,y) = A[\tilde{g},v](0,x,y) * \phi[\tilde{g}](0,x,y)$$

$$+A[\tilde{g},v](x,x,y) * Z[\tilde{g},v](x,x,y) - A[\tilde{g},v](0,x,y) * Z[\tilde{g},v](0,x,y)$$

$$+\int_0^x \left\{-A[\tilde{g},v](\tau,x,y) * \frac{d}{d\tau}Z[\tilde{g},v](\tau,x,y) + f[\tilde{g},v](\tau,x,y)\right\} d\tau$$

on $[0,c]\times R^n$, and consequently

$$\int_0^x \left\{-A[\tilde{g},v](\tau,x,y) * \frac{d}{d\tau}Z[\tilde{g},v](\tau,x,y) + f[\tilde{g},v](\tau,x,y)\right\} d\tau = 0 \qquad (4.38)$$

on $[0,c]\times R^n$. For $1\le i\le k$ the function $\tilde{g}_i = g_i[v]$ is the unique solution of Cauchy problem (4.29) with v instead of z. Therefore using the group property we obtain $y = \tilde{g}_i(x,0,\eta)$ if and only if $\eta = \tilde{g}_i(0,x,y)$. For a fixed $x\in[0,c]$ and $1\le i\le k$ the above relation represents a one - to - one transformation of the space R^n into itself. Furthermore, it is easy to check that this transformation together with its inverse is Lipschitz continuous. If $1\le i\le k$, then writing (4.38) by coordinates, using the transformation $y = \tilde{g}_i(x,0,\eta)$ and differentiating the resulting relation with respect to x we derive

$$\sum_{j=1}^{k} A_{ij}\left(x,\tilde{g}_i(x,0,\eta),v_{(x,\tilde{g}_i(x,0,\eta))}\right)$$

$$\times\left[D_x v_j(x,\tilde{g}_i(x,0,\eta)) + \sum_{l=1}^{n}\varrho_{il}\left(x,\tilde{g}_i(x,0,\eta),v_{(x,\tilde{g}_i(x,0,\eta))}\right)D_{y_l}v_j(x,\tilde{g}_i(x,0,\eta))\right]$$

$$= f_i\left(x,\tilde{g}_i(x,0,\eta),v_{(x,\tilde{g}_i(x,0,\eta))}\right),$$

for almost all $(x,\eta)\in[0,c]$ and for $1\le i\le k$. By using the transformation $\eta = \tilde{g}_i(0,x,y)$, which preserves the sets of measure zero, we obtain that system (4.21) is satisfied for almost all $(x,y)\in[0,c]\times R^n$. Initial problem (4.21), (4.22) is equivalent to the functional integral system $z = U_\phi z$. It follows then the solution v is unique. Finally, estimation (4.36) with $z = v$, $U_\phi z = v$, $\bar{z} = \bar{v}$, $U_{\bar{\phi}}\bar{z} = \bar{v}$ implies (4.37), which completes the proof of the Lemma.

Note that we have actually proved that there exists a solution of problem (4.21), (4.22) under additional assumption (4.26). If all the assumptions of Theorem 4.16 are satisfied then inequality (4.26) holds true for sufficiently small $c\in(0,a]$. Thus Theorem 4.16 is proved.

It is easy to see that quasilinear systems with a deviated argument and integral differential problems are particular cases of (4.21). Such systems were considered in [58, 71, 124, 125]. Suppose that

$$B_l : R^n \to M[k,k],\ B_l = \left[b_{ij}^{(l)}\right]_{i,j=1,\ldots,k},\quad l = 1,\ldots,\bar{k},\ \bar{k} \geq k,$$

and $\psi : R^n \to R^k$ are given functions. Consider system (4.21) with the boundary data on $\bar{k}$ arbitrary (not necessarily distinct) hyperplanes $x = a_l$, $l = 1,\ldots,\bar{k}$,

$$\sum_{l=1}^{\bar{k}} B_l(y)z(a_l, y) = \psi(y),\ y \in R^n. \tag{4.39}$$

The above boundary condition includes as a particular case the boundary condition of the Cesari type ([45 – 47]). If $b_{ij}^l = \delta_{li}\,\delta_{ij}$, then (4.39) reduces to the Nicolletti condition ([193, 200]). If furthermore, all $a_l = 0$, $1 \leq l \leq \bar{k}$, then we get usual Cauchy condition. Boundary condition arise, in a suitable version, from a laser problem in nonlinear optics ([15, 17, 19]). The existence result for the Cauchy problem can be extended for system (4.21) with boundary condition (4.39), see [57, 60].

Remark 4.20 *We have considered hyperbolic systems in the second canonical form and their solutions which satisfy the Lipschitz condition with respect to (x, y). If the matrix A depends on z in a classical sense i.e.*

$$A(x, y, z_{(x,y)}) = A(x, y, z(x, y)),$$

then we can consider solutions of an adequate systems which satisfy the generalized Lipschitz condition:

$$\|z(x,y) - z(\bar{x},\bar{y})\| \leq \left|\int_x^{\bar{x}} \omega(t)dt\right| + d_2\|y - \bar{y}\|$$

with $\omega \in L([-r_0, c], R_+)$ and $d_2 \in R_+$. Details can be found in [57]

4.7 Uniqueness of solutions

We start with functional differential inequalities in the Carathéodory sense. Let $a > 0$ and $r_0 \in R_+ = [0, +\infty)$ be fixed. For $\eta : [-r_0, a] \to R$ and $x \in [0, a]$ we define $\eta_{(x)} : [-r_0, 0] \to R$ by $\eta_{(x)}(t) = \eta(x + t)$, $t \in [-r_0, 0]$. Given the functions $\sigma : [0, a] \times C([-r_0, 0], R_+) \to R_+$ and $\chi : [-r_0, 0] \to R_+$, we consider the Cauchy problem

$$\eta'(x) = \sigma(x, \eta_{(x)}),\quad x \in [0, a], \tag{4.40}$$

$$\eta = \chi \quad \text{on } [-r_0, 0]. \tag{4.41}$$

Function σ is said to satisfy Carathéodory conditions if

(i) $\sigma(x, \cdot)$ is continuous for almost all $x \in [0, a]$,

(ii) $\sigma(\cdot, \omega)$ is measurable for every $\omega \in C([-r_0, 0], R_+)$ and there exists $m_\sigma \in L([0,a], R_+)$ such that $\sigma(x, \omega) \leq m_\sigma(x)$ for $\omega \in C([-r_0, 0], R_+)$ and for a.e. $x \in [0, a]$.

In this Section we consider Carathéodory solutions of problem (4.40), (4.41) or Carathéodory solutions of functional differential inequalities generated by (4.40), (4.41).

Lemma 4.21 *Assume that $\chi \in C([-r_0, 0], R_+)$. If σ satisfies Carathéodory condition and it is nondecreasing with respect to the functional variable, then problem (4.40), (4.41) admits a local maximum solution.*

PROOF. Let $0 < a_0 < a$ and let $\psi : [-r_0, a_0] \to R_+$ be a subsolution of problem (4.40), (4.41) i.e.

(i) ψ is continuous on $[-r_0, 0]$ and it is absolutely continuous on $[0, a_0]$,

(ii) $\psi'(x) \leq \sigma(x, \psi_{(x)})$ for almost all $x \in [0, a_0]$ and $\psi \leq \chi$ on $[-r_0, 0]$.

Let us consider the problem

$$\eta'(x) = \sigma(x, \eta_{(x)}) + 2, \quad x \in [0, a_0], \tag{4.42}$$

$$\eta(x) = \chi(x) + 2 \quad \text{for } x \in [-r_0, 0]. \tag{4.43}$$

Let $\varphi_0 : [-r_0, \tilde{a}] \to R_+$, $0 < \tilde{a} \leq a$, be a local solution of problem (4.42), (4.43). The existence of φ_0 follows from [93, 94]. Then there is $0 < c \leq \tilde{a}$ such that $\psi(x) \leq \varphi_0(x)$ for $x \in [0, c]$. Let us define the sequence $\{\varphi_m\}$, $\varphi_m : [-r_0, c] \to R_+$, as follows. Function φ_0 is given above and

$$\varphi_m(x) = \chi(x) + \frac{1}{m} \quad \text{for } x \in [-r_0, 0],$$

$$\varphi_m(x) = \chi(0) + \frac{1}{m} + \int_0^x \left[\sigma(t, (\varphi_{m-1})_{(t)}) + \frac{1}{m}\right] dt$$

for $x \in (0, c]$, $m \geq 1$. It is easy to see that $\psi(x) \leq \varphi_m(x) \leq \varphi_{m-1}(x)$, for $m \geq 1$, $x \in [-r_0, c]$. The functions $\{\varphi_m\}$ are equicontinuous and uniformly bounded. In force of Ascoli - Arzelà theorem, it follows that a suitable subsequence converges to a continuous function φ uniformly on $[-r_0, c]$. Of course $\varphi = \chi$ on $[-r_0, 0]$ and φ is absolutely continuous in $[0, c]$, moreover $\psi \leq \varphi$ on $[-r_0, c]$. It follows that φ is the maximum solution of (4.40), (4.41).

Assumption $\tilde{H}[\sigma]$. Suppose that

1) σ satisfies Carathéodory conditions and is nondecreasing with respect to the functional variable,

2) $\chi \in C([-r_0, 0], R_+)$ and $\varphi : [-r_0, c) \to R_+$ is the maximum solution of problem (4.40), (4.41).

Lemma 4.22 *Suppose that Assumption $\tilde{H}[\sigma]$ holds and*

1) $\psi : [-r_0, c) \to R_+$ is continuous on $[-r_0, 0]$ and absolutely continuous on $[0, c)$,

2) $\psi(x) \leq \chi(x)$ for $x \in [-r_0, 0]$ and $\psi'(x) \leq \sigma(x, \psi_{(x)})$ for almost all $x \in [0, c)$.

Then $\psi \leq \varphi$ on $[0, c)$.

PROOF. Let $0 < c_0 < c$ be fixed and let us consider the problem

$$\eta'(x) = \sigma(x, \eta_{(x)}) + \varepsilon \quad \text{for } x \in [0, c_0], \qquad \eta = \chi + \varepsilon \text{ on } [-r_0, 0].$$

There is $\varepsilon_0 > 0$ such that the solution $\eta(\cdot, \varepsilon)$ of the above problem exists on $[-r_0, c_0]$ for $0 < \varepsilon < \varepsilon_0$, ([93]). It is easy to see that $\psi(x) \leq \eta(x, \varepsilon)$ for $x \in [0, c_0]$, $0 < \varepsilon < \varepsilon_0$. Let us define the sequence $\{\varphi_m\}$, $\varphi_m : [-r_0, c_0] \to R_+$, as follows : $\varphi_0(x) = \eta(x, \varepsilon)$ and

$$\varphi_m(x) = \chi(x) + \frac{1}{m} \quad \text{for } x \in [-r_0, 0],$$

$$\varphi_m(x) = \chi(0) + \frac{1}{m} + \int_0^x \left[\sigma(t, (\varphi_{m-1})_{(t)}) + \frac{1}{m} \right] dt$$

for $x \in [0, c_0]$, $m \geq 1$. Then $\psi(x) \leq \varphi_m(x) \leq \varphi_{m-1}(x)$ on $[-r_0, c_0]$ for $m > \varepsilon^{-1}$. In force of Ascoli - Arzelà theorem, it follows that a suitable subsequence converges to a function φ uniformly in $[-r_0, c_0]$. Consequently $\psi \leq \varphi$ on $[0, c_0]$ and the thesis follows from the arbitrariness of $0 < c_0 < c$.

Lemma 4.23 *Suppose that Assumption $\tilde{H}[\sigma]$ holds and that the function $\psi \in C([-r_0, c), R_+)$ satisfies the initial inequality*

$$\psi(x) \leq \chi(x) \quad on \quad [-r_0, 0].$$

Suppose that

$$\psi(x) \leq \chi(0) + \int_0^x \sigma(t, \psi_{(t)})\, dt, \quad x \in [0, c).$$

Then $\psi(x) \leq \varphi(x)$ on $[0, c)$.

PROOF. Let us consider the function $u : [-r_0, c) \to R_+$ defined by $u(x) = \chi(x)$ for $x \in [-r_0, 0]$ and

$$u(x) = \chi(0) + \int_0^x \sigma(t, \psi_{(t)}) dt \quad \text{for } x \in [0, c).$$

Then $u'(x) \leq \sigma(x, \psi_{(x)}) \leq \sigma(x, u_{(x)})$ for almost all $x \in [0, c)$. Lemma 4.22 assures the thesis.

Let $V : C(B, R) \to C([-r_0, 0], R_+)$ be the operator defined by

$$(Vw)(t) = \max\{|w(t, s)| : s \in [-r, r]\}.$$

Note that the continuity of Vw follows from Theorem 33.1 in [195].

Assumption H $[\sigma]$. Suppose that

1) $\sigma : [0, a] \times C([-r_0, 0], R_+) \to R_+$ satisfies Carathéodory conditions and is nondecreasing with respect to the functional variable,

2) the maximum solution of problem (4.40), (4.41) with $\chi(x) = 0$ on $[-r_0, 0]$, is the null function.

Now we can state and prove a theorem on the uniqueness of solutions of problem (4.2), (4.3).

Theorem 4.24 *Suppose that Assumption H [σ] is satisfied and*

1) $f(x,\cdot)$ is continuous for a.e. $x \in [0,a]$ and $f(\cdot,y,w,q)$ is measurable for every $(y,w,q) \in R^n \times C(B,R^k) \times R^n$,

2) the partial derivatives

$$D_q f = \left[D_{q_j} f_i\right]_{i=1,\dots,k,\, j=1,\dots,n}$$

exist on $R^n \times C(B,R^k) \times R^n$ and for a.e. $x \in [0,a]$,

3) $D_q f(x,\cdot)$ is continuous for a.e. $x \in [0,a]$ and $D_q f(\cdot,y,w,q)$ is measurable for every $(y,w,q) \in R^n \times C(B,R^k) \times R^n$,

4) there exists $m_q \in L([0,a],R_+)$ such that $\|D_q f(x,y,w,q)\| \le m_q(x)$ for $(y,w,q) \in R^n \times C(B,R^k) \times R^n$ and for a.e. $x \in [0,a]$,

5) the estimate

$$\|f(x,y,w,q) - f(x,y,\bar{w},q)\| \le \sigma(x, V(w-\bar{w}))$$

holds on $R^n \times C(B,R^k) \times R^n$ for a.e. $x \in [0,a]$.

Then problem (4.2), (4.3) admits one bounded and uniformly continuous solution at the most.

PROOF. Suppose that u and $\tilde{u}$ are two bounded and uniformly continuous solutions of the problem. Put

$$Q_j(x,y,t) = \left(x, y, \tilde{u}_{(x,y)}, D_y\tilde{u}_j(x,y) + t[D_y(u_j - \tilde{u}_j)(x,y)]\right),$$

in force of Hadamard mean value theorem we have

$$D_x(u-\tilde{u})_j(x,y) - \sum_{i=1}^{n}\int_0^1 D_{q_i} f_j(Q_j(x,y,t))dt\, D_{y_i}(u_j - \tilde{u}_j)(x,y) \tag{4.44}$$

$$= f_j(x,y,u_{(x,y)}, D_y u_j(x,y)) - f_j(x,y,\tilde{u}_{(x,y)}, D_y u_j(x,y)).$$

Denoted by $g_j(\cdot,x,y)$ the solution of the Cauchy problem

$$\eta'(t) = -\int_0^1 D_q f_j(Q_j(t,\eta(t),\xi))\,d\xi, \quad \eta(x) = y,$$

from (4.44) it follows that

$$(u_j - \tilde{u}_j)(x,y) = (u_j - \tilde{u}_j)(0, g_j(0,x,y))$$

$$+\int_0^x \left[f_j\left(t, g(t,x,y), u_{(t,g(t,x,y))}, D_y u_j(t,g(t,x,y))\right)\right.$$

$$\left. - f_j\left(t, g(t,x,y), \tilde{u}_{(t,g(t,x,y))}, D_y u_j(t,g(t,x,y))\right)\right] dt.$$

Let $\psi(x) = \sup\{\,|(u-\tilde{u})(x,y)| : y \in R^n\,\}$, $-r_0 \le x \le a$. Then ψ is continuous on $[-r_0, 0]$ and absolutely continuous on $[0,a]$, ([32]). Moreover from assumption 5) we get

$$\psi(x) \le \psi(0) + \int_0^x \sigma(t, \psi_{(t)})\,dt, \quad x \in [0,a].$$

Since $\psi(x) = 0$ on $[-r_0, 0]$, from Theorem 4.24 we deduce that $u = \tilde{u}$. This completes the proof.

Now we formulate a particular case of Assumption H $[\sigma]$ which allows to get a uniqueness result as a corollary of Theorem 3.1.

Assumption H' $[\sigma]$. Suppose that

1) the function $\sigma_0 : [0, a_0] \times R_+ \to R_+$ satisfies Carathéodory conditions and is nondecreasing with respect to the second variable,

2) the null function is the maximum solution in Carathéodory sense of the initial problem

$$\eta'(x) = \sigma_0(x, \eta(x)), \quad \eta(0) = 0.$$

Let $\|\cdot\|_0$ denote the supremum norm in the space $C([-r_0, 0], R)$.

Theorem 4.25 *Suppose that assumption H' $[\sigma]$ holds and function f satisfies assumptions 1) - 4) of Theorem 4.24. Assume that the following estimation*

$$\| f(x, y, w, q) - f(x, y, \bar{w}, q) \| \leq \sigma_0(x, \|w - \bar{w}\|_0)$$

holds on $R^n \times C(B, R^k) \times R^n$ for a.e. $x \in [0, a]$. Then problem (4.2), (4.3) admits one bounded and uniformly continuous solution at the most.

PROOF. The assertion follows from Theorem 4.24 with $\sigma(x, \omega) = \sigma_0(x, \|\omega\|_0)$.

4.8 Function spaces

For simplicity of formulations and proving existence theorems we will consider the Cauchy problem (4.2), (4.3) for $k = 1$. We formulate the problem. Given the functions

$$f : [0, a] \times R^n \times C(B, R) \times R^n \to R, \qquad \varphi : E_0 \to R,$$

we consider the equation

$$D_x z(x, y) = f(x, y, z_{(x,y)}, D_y z(x, y)), \tag{4.45}$$

with the initial condition

$$z(x, y) = \varphi(x, y) \text{ on } E_0, \tag{4.46}$$

where $D_y z = (D_{y_1} z, ..., D_{y_n} z)$ and $z_{(x,y)} : B \to R$ is defined by (4.1) with $k = 1$. We denote by $C_L(B, R)$ the class of all functions $w \in C(B, R)$ such that

$$\|w\|_L = \sup\left\{ \frac{|w(t, s) - w(t, \bar{s})|}{\|s - \bar{s}\|} : (t, s), (t, \bar{s}) \in B \right\} < +\infty.$$

We define $\|w\|_{0.L} = \|w\|_0 + \|w\|_L$ for $w \in C_L(B, R)$. Let $C^{0.1}(B, R)$ be the set of all continuous functions $w : B \to R$ such that the derivatives $(D_{s_1} w, ..., D_{s_n} w) = D_s w$ exist and $D_s w \in C(B, R^n)$. For $w \in C^{0.1}(B, R)$ we put

$$\|w\|_1 = \|w\|_0 + \max\{ \|D_s w(t, s)\| : (t, s) \in B \}.$$

We denote by $C_L^{0.1}(B,R)$ the set of all the functions $w \in C^{0.1}(B,R)$ such that $\|w\|_{1.L} < +\infty$ where

$$\|w\|_{1.L} = \|w\|_1 + \sup\left\{ \frac{\|D_s w(t,s) - D_s w(t,\bar{s})\|}{\|s-\bar{s}\|} : (t,s),(t,\bar{s}) \in B \right\}.$$

We will consider the spaces

$$\Omega^{(0)} = [0,a] \times R^n \times C(B,R) \times R^n, \quad \Omega^{(1)} = [0,a] \times C^{0.1}(B,R^n) \times R^n$$

and

$$\Omega^{(1.L)} = [0,a] \times R^n \times C_L^{0.1}(B,R^n) \times R^n.$$

We put

$$E[t] = [-r_0,t] \times R^n \quad \text{and} \quad E^*[t] = [0,t] \times R^n$$

where $0 \le t \le a$ and we denote by

$$\|\cdot\|_{(t,R)}, \quad \|\cdot\|_{(t,R^n)} \quad \text{and} \quad \|\cdot\|_{[t,R^n]}$$

the supremum norms in the spaces $C(E[t],R)$, $C(E[t],R^n)$ and $C(E^*[t],R^n)$ respectively. Let Θ be the set defined in Section 4.2.

Assumption H $[D_q f]$. Suppose that

1) for every $(x,y,w,q) \in \Omega^{(1)}$ there exist the derivatives

$$(D_{q_1} f(x,y,w,q), \ldots, D_{q_n} f(x,y,w,q)) = D_q f(x,y,w,q)$$

and $D_q f(\cdot,y,w,q) \in L([0,a],R^n)$ for every $(y,w,q) \in R^n \times C^{0.1}(B,R) \times R^n$,

2) there is a function $\alpha \in \Theta$ such that

$$\|D_q f(x,y,w,q)\| \le \alpha(x, \|w\|_1) \quad \text{on} \quad \Omega^{(1)},$$

3) there is $\gamma \in \Theta$ such that for $(x,y,w,q) \in \Omega^{(1.L)}$, $\bar{y}, \bar{q} \in R^n$ and $h \in C^{0.1}(B,R)$ we have

$$\|D_q f(x,y,w,q) - D_q f(x,\bar{y},w+h,\bar{q})\| \le \gamma(x, \|w\|_{1.L})\,[\,\|y-\bar{y}\| + \|h\|_1 + \|q-\bar{q}\|\,].$$

Now we define some further functions spaces. Given $s = (s_0,s_1,s_2) \in R_+^3$, we denote by $C^{1.L}[E_0;s]$ the set of all the functions $\varphi : E_0 \to R$ such that

(i) $\varphi \in C(E_0,R)$, there exists $D_y\varphi(x,y)$ for $(x,y) \in E_0$ and $D_y\varphi \in C(E_0,R^n)$,

(ii) $|\varphi(x,y)| \le s_0$, $\|D_y\varphi(x,y)\| \le s_1$ on E_0, and

$$\|D_y\varphi(x,y) - D_y\varphi(x,\bar{y})\| \le s_2\|y-\bar{y}\| \quad \text{on} \quad E_0.$$

Let $\varphi \in C^{1.L}[E_0;s]$ be given and let $0 < c \le a$, $d = (d_0,d_1,d_2)$, $\lambda = (\lambda_0,\lambda_1)$ where $d_i \in R_+$, $d_i \ge s_i$ for i = 0,1,2, and $\lambda_i \in L([0,a],R_+)$, $i = 0,1$. We denote by $C_{\varphi.c}^{1.L}[d;\lambda]$ the set of all the functions $z : [-r_0,c] \times R^n \to R$ such that

$$z \in C([-r_0,c] \times R^n, R), \quad z(x,y) = \varphi(x,y) \quad \text{on} \quad E_0$$

and

(i) there exists $D_y z(x,y)$ on $[0,c] \times R^n$,
(ii) $|z(x,y)| \le d_0$ and $\|D_y z(x,y)\| \le d_1$ on $[-r_0, c] \times R^n$,
(iii) for $x, \bar{x} \in [0,c]$, $y, \bar{y} \in R^n$ we have

$$|z(x,y) - z(\bar{x},y)| \le |\int_x^{\bar{x}} \lambda_0(t)\,dt|,$$

$$\|D_y z(x,y) - D_y z(\bar{x},\bar{y})\| \le |\int_x^{\bar{x}} \lambda_1(t)\,dt| + d_2\|y - \bar{y}\|.$$

Let $p = (p_0, p_1) \in R_+^2$, and $\mu \in L([0,c], R_+)$. We denote by $C_c^{0.L}[p;\mu]$ the class of all the functions $v : [0,c] \times R^n \to R^n$ such that $v \in C([0,c] \times R^n, R^n)$ and
(i) $\|v(x,y)\| \le p_0$ on $[0,c] \times R^n$,
(ii) for $(x,y), (\bar{x},\bar{y}) \in [0,c] \times R^n$ we have

$$\|v(x,y) - v(\bar{x},\bar{y})\| \le |\int_x^{\bar{x}} \mu(t)\,dt| + p_1\|y - \bar{y}\|.$$

We will prove that under suitable assumptions on f and φ, and for sufficiently small c with $0 < c \le a$, there exists a solution $\bar{z}$ of problem (4.45), (4.46) such that $\bar{z} \in C_{\varphi.c}^{1.L}[d;\lambda]$ and $\bar{v} \in C_c^{0.L}[p;\mu]$ where $\bar{v} = D_y \bar{z}\,|_{[0,c]\times R^n}$.

4.9 Bicharacteristics of nonlinear functional differential equations

Suppose that $\varphi \in C^{1.L}[E_0; s]$ and $z \in C_{\varphi.c}^{1.L}[d;\lambda]$, $u \in C_c^{0,L}[p;\mu]$. We consider the Cauchy problem

$$\eta'(t) = -D_q f(\,t, \eta(t), z_{(t,\eta(t))}, u(t,\eta(t))\,), \qquad \eta(x) = y, \tag{4.47}$$

and denote by $g[z,u](\,\cdot\,, x, y)$ its solution.

Lemma 4.26 *Suppose that Assumption H $[D_q f]$ is satisfied and let*

$$\varphi,\ \bar{\varphi} \in C^{1.L}[E_0; s], \quad z \in C_{\varphi.c}^{1.L}[d;\lambda], \quad \bar{z} \in C_{\bar{\varphi}.c}^{1.L}[d;\lambda], \quad u,\ \bar{u} \in C_c^{0,L}[p;\mu]$$

be given.

Then the solutions $g[z,u](\,\cdot\,,x,y)$ and $g[\bar{z},\bar{u}](\,\cdot\,,x,y)$ exist on $[0,c]$, they are unique and we have the estimates

$$\|g[z,u](t,x,y) - g[z,u](t,\bar{x},\bar{y})\|$$

$$\le \left[\left| \int_x^{\bar{x}} \alpha(\tau,\tilde{d})d\tau \right| + \|y - \bar{y}\| \right] \exp\left[\bar{d} \left| \int_x^t \gamma(\tau,|d|)d\tau \right| \right] \tag{4.48}$$

where

$$|d| = d_0 + d_1 + d_2, \quad \tilde{d} = d_0 + d_1, \quad \bar{d} = 1 + d_1 + d_2 + p_1$$

and

$$\|g[z,u](t,x,y) - g[\bar{z},\bar{u}](t,x,y)\|$$

$$\leq \left| \int_x^t \gamma(\tau, |d|) \left[\|z - \bar{z}\|_{(\tau,R)} + \|D_y z - D_y \bar{z}\|_{(\tau,R^n)} \right.\right. \tag{4.49}$$

$$\left.\left. + \|u - \bar{u}\|_{[\tau,R^n]} \right] d\tau \right| \exp\left[\bar{d} \left| \int_x^t \gamma(\tau, |d|)\, d\tau \right| \right].$$

PROOF. The existence and uniqueness of the solution of (4.47) follows from classical theorems. On this purpose, note that the right-hand side of the differential system satisfies the Carathéodory assumption and the following Lipschitz condition holds:

$$\|D_q f(t, \xi, z_{(t,\xi)}, u(t,\xi)) - D_q f(t, \bar{\xi}, z_{(t,\bar{\xi})}, u(t,\bar{\xi}))\| \leq \bar{d}\gamma(t, |d|)\, \|\xi - \bar{\xi}\|.$$

It follows from Assumption H $[D_q f]$ and (4.47) that for $(t,x,y), (t,\bar{x},\bar{y}) \in [0,c] \times [0,c] \times R^n$ we have

$$\|g[z,u](t,x,y) - g[z,u](t,\bar{x},\bar{y})\| \leq \left| \int_x^{\bar{x}} \alpha(\tau, \tilde{d})\, d\tau \right| + \|y - \bar{y}\|$$

$$+ \left| \int_x^t \gamma(\tau, |d|) \left[\, \|g[z,u](\tau,x,y) - g[z,u](\tau,\bar{x},\bar{y})\| \right.\right.$$

$$+ \|z_{(\tau, g[z,u](\tau,x,y))} - z_{(\tau, g[z,u](\tau,\bar{x},\bar{y}))}\|_1$$

$$\left.\left. + \|u(\tau, g[z,u](\tau,x,y)) - u(\tau, g[z,u](\tau,\bar{x},\bar{y}))\| \,\right] d\tau \right|.$$

Since

$$\|z_{(\tau, g[z,u](\tau,x,y))} - z_{(\tau, g[z,u](\tau,\bar{x},\bar{y}))}\|_1 \leq (d_1 + d_2)\|g[z,u](\tau,x,y) - g[z,u](\tau,\bar{x},\bar{y})\|$$

and

$$\|u(\tau, g[z,u](\tau,x,y)) - u(\tau, g[z,u](\tau,\bar{x},\bar{y}))\| \leq p_1 \|g[z,u](\tau,x,y) - g[z,u](\tau,\bar{x},\bar{y})\|$$

then we have the integral inequality

$$\|g[z,u](t,x,y) - g[z,u](t,\bar{x},\bar{y})\| \leq \left| \int_x^{\bar{x}} \alpha(\tau, \tilde{d}) d\tau \right| + \|y - \bar{y}\|$$

$$+ \bar{d} \left| \int_x^t \gamma(\tau, |d|) \|g[z,u](\tau,x,y) - g[z,u](\tau,\bar{x},\bar{y})\| d\tau \right|.$$

It follows from Gronwall inequality that estimate (4.48) is satisfied. Now we prove inequality (4.49). For

$$z \in C^{1.L}_{\varphi.c}[d;\lambda], \quad \bar{z} \in C^{1.L}_{\bar{\varphi}.c}[d;\lambda] \text{ and } u, \bar{u} \in C^{0,L}_{c}[p;\mu]$$

we have

$$\| g[z,u](t,x,y) - g[\bar{z},\bar{u}](t,x,y) \|$$

$$\leq \left| \int_x^t \gamma(\tau,|d|) \left[\| g[z,u](\tau,x,y) - g[\bar{z},\bar{u}](\tau,x,y) \| \right.\right.$$

$$\left.\left. + \| z_{(\tau,g[z,u](\tau,x,y))} - \bar{z}_{(\tau,g[\bar{z},\bar{u}](\tau,x,y))} \|_1 \right] d\tau \right|$$

$$+ \left| \int_x^t \gamma(\tau,|d|) \, \| u(\tau,g[z,u](\tau,x,y)) - \bar{u}(\tau,g[\bar{z},\bar{u}](\tau,x,y)) \| d\tau \right|.$$

Since

$$\| z_{(\tau,g[z,u](\tau,x,y))} - \bar{z}_{(\tau,g[z,u](\tau,x,y))} \|_0 \leq \|z - \bar{z}\|_{(\tau,R)}$$

and

$$\max_{(t,s)\in B} \{ \| (D_y z)_{(\tau,g[z,u](\tau,x,y))}(t,s) - (D_y \bar{z})_{(\tau,g[z,u](\tau,x,y))}(t,s) \| \}$$

$$\leq \|D_y z - D_y \bar{z}\|_{(\tau,R^n)}, \quad (t,s) \in B,$$

we have

$$\|z_{(\tau,g[z,u](\tau,x,y))} - \bar{z}_{(\tau,g[\bar{z},\bar{u}](\tau,x,y))}\|_1$$

$$\leq (d_1 + d_2) \, \| g[z,u](\tau,x,y) - g[\bar{z},\bar{u}](\tau,x,y) \| + \| z - \bar{z} \|_{(\tau,R)}$$

$$+ \| D_y z - D_y \bar{z} \|_{(\tau,R^n)}.$$

In similar way we obtain

$$\| u(\tau,g[z,u](\tau,x,y)) - \bar{u}(\tau,g[\bar{z},\bar{u}](\tau,x,y)) \|$$

$$\leq p_1 \| g[z,u](\tau,x,y) - g[\bar{z},\bar{u}](\tau,x,y) \| + \| u - \bar{u} \|_{[\tau,R^n]}.$$

Then we get the integral inequality

$$\| g[z,u](t,x,y) - g[\bar{z},\bar{u}](t,x,y) \|$$

$$\leq \left| \int_x^t \gamma(\tau,|d|) \left[\|z - \bar{z}\|_{(\tau,R)} + \|D_y z - D_y \bar{z}\|_{(\tau,R^n)} + \|u - \bar{u}\|_{[\tau,R^n]} \right] d\tau \right|$$

$$+ \left| \bar{d} \int_x^t \gamma(\tau,|d|) \, \| g[z,u](\tau,x,y) - g[\bar{z},\bar{u}](\tau,x,y) \| d\tau \right|.$$

From Gronwall inequality we deduce (4.49). This proves Lemma 4.26.

4.10 Integral functional equations

We denote by $CL(B,R)$ the set of all linear and continuous real functions defined on $C(B,R)$ and by $\|\cdot\|_*$ the norm in $CL(B,R)$. We formulate now further assumptions on f.

Assumption H $[f, D_y f, D_w f]$. Suppose that

1) there is $\tilde{\gamma} \in \Theta$ such that we have

$$|f(x,y,w,q)| \le \tilde{\gamma}(x, \|w\|_0) \text{ on } \Omega^{(0)},$$

2) for every $(x,y,w,q) \in \Omega^{(1)}$ there exist

$$D_y f(x,y,w,q) \in R^n \quad \text{and} \quad D_w f(x,y,w,q) \in CL(B,R),$$

3) for $(x,y,w,q) \in \Omega^{(1)}$ we have the estimates on $\Omega^{(1)}$

$$\|D_y f(x,y,w,q)\| \le \alpha(x, \|w\|_1), \quad \|D_w f(x,y,w,q)\|_* \le \alpha(x, \|w\|_1),$$

4) for $(x,y,w,q) \in \Omega^{(1.L)}$ and for $\bar{y}, \bar{q} \in R^n, h \in C^{0.1}(B,R)$ we have the estimates

$$\|D_y f(x,y,w,q) - D_y f(x,\bar{y},w+h,\bar{q})\| \le \gamma(x, \|w\|_{1.L})[\,\|y-\bar{y}\| + \|h\|_1 + \|q-\bar{q}\|\,],$$

and

$$\|D_w f(x,y,w,q) - D_w f(x,\bar{y},w+h,\bar{q})\|_* \le \gamma(x, \|w\|_{1.L})[\,\|y-\bar{y}\| + \|h\|_1 + \|q-\bar{q}\|\,].$$

Remark 4.27 *We prove a theorem on the existence of solutions and we give an estimate of the domain of solutions. For simplicity of notation, we have assumed the same estimation for derivatives $D_y f$, $D_w f$, $D_q f$. We have assumed also the Lipschitz condition for these derivatives with the same coefficient.*

Suppose that $\varphi \in C^{1.L}[E_0; s]$. Let $C^{0.L}_{\varphi.c}[p;\mu]$, $0 < c \le a$, be the class of all functions $u : [-r_0, c] \times R^n \to R^n$ such that $u(x,y) = D_u\varphi(x,y)$ on E_0 and $u\,|_{E^*[c]} \in C^{0.L}_c[p;\mu]$. Now we formulate a system of integral functional equations which are generated by (4.45), (4.46). Write

$$P[z,u](\tau,x,y) = \big(\tau, g[z,u](\tau,x,y), z_{(\tau,g[z,u](\tau,x,y))}, u(\tau, g[z,u](\tau,x,y))\big). \tag{4.50}$$

Given $\varphi \in C^{1.L}[E_0;s]$, $z \in C^{1.L}_{\varphi.c}[d;\lambda]$, and $u \in C^{0.L}_{\varphi.c}[p;\mu]$, where $0 < c \le a$. We define

$$F[z,u](x,y) = \varphi(0, g(0,x,y))$$
$$+ \int_0^x \left[f(P[z,u](\tau,x,y)) - \sum_{k=1}^n D_{q_k} f(P[z,u](\tau,x,y))\, u_k(\tau, g(\tau,x,y)) \right] d\tau,$$

and

$$G_i[z,u](x,y) = D_{y_i}\varphi(0, g(0,x,y))$$
$$+ \int_0^x \left[D_{y_i} f(P[z,u](\tau,x,y)) + D_w f(P[z,u](\tau,x,y))(u_i)_{(\tau,g(\tau,x,y))} \right] d\tau,$$

where $i = 1, \ldots, n$. Moreover we put $G[z,u] = (G_1[z,u], ..., G_n[z,u])$. We will consider the following system of functional integral equations

$$z = F[z,u], \quad u = G[z,u], \quad z = \phi \text{ on } E_0, \quad u = D_y\varphi \text{ on } E_0, \tag{4.51}$$

$$g[z,u](t,x,y) = y + \int_t^x D_q f(P[z,u](\tau,x,y))d\tau. \tag{4.52}$$

Remark 4.28 *Integral functional system (4.51), (4.52) is obtained in the following way. We introduce first an additional unknown function $u = D_y z$ in (4.45). Then we consider the linearization of (4.45) with respect to u, i.e.*

$$D_x z(x,y) = f(U) + \sum_{j=1}^{n} D_{q_j} f(U)\left(D_{y_j} z(x,y) - u_j(x,y)\right) \tag{4.53}$$

where $U = \left(x, y, z_{(x,y)}, u(x,y)\right)$. By virtue of equation (4.45) we get the differential system for the unknown function u :

$$D_x u_i(x,y) = D_{y_i} f(U) + D_w f(U)(D_{y_i} z)_{(x,y)} + \sum_{j=1}^{n} D_{q_j} f(U) D_{y_j} u_i(x,y), \tag{4.54}$$

where $i = 1, \ldots, n$. Finally we put $D_y z = u$ in (4.54). Note that system (4.53), (4.54) has the following property: the differential equations of bicharacteristics for (4.53) and for (4.54) are the same and they have the form (4.47). If we consider (4.53), (4.54) along the bicharacteristic $g[z,u](\cdot, x, y)$, we obtain

$$\frac{d}{dt} z(t, g[z,u](t,x,y)) = f(P[z,u](t,x,y))$$

$$-\sum_{j=1}^{n} D_{q_j} f(P[z,u](t,x,y)) u_j(t, g[z,u](t,x,y)),$$

and

$$\frac{d}{dt} u_i(t, g[z,u](t,x,y)) = D_{y_i} f(P[z,u](t,x,y))$$

$$+D_w f(P[z,u](t,x,y))(u_i)_{(t,g[z,u](t,x,y))}$$

where $i = 1, \ldots, n$. By integrating on $[0,x]$ with respect to t, we get (4.51), (4.52).

The proof of the existence of the solution of (4.51), (4.52) will be based on the following method of successive approximations. Suppose that $\varphi \in C^{1.L}[E_0; s]$ and that Assumptions H $[D_q f]$, H $[f, D_y f, D_w f]$ are satisfied. We define the sequence $\{z^{(m)}, u^{(m)}\}$ in the following way. We put first

$$z^{(0)}(x,y) = \varphi(x,y) \text{ on } E_0, \quad z^{(0)}(x,y) = \varphi(0,y) \text{ on } [0,c] \times R^n, \tag{4.55}$$

$$u^{(0)}(x,y) = D_y\varphi(x,y) \text{ on } E_0, \quad u^{(0)}(x,y) = D_y\varphi(0,y) \text{ on } [0,c] \times R^n. \tag{4.56}$$

Then $z^{(0)} \in C^{1.L}_{\varphi.c}[d;\lambda]$ and $u^{(0)} \in C^{0.L}_{\varphi.c}[p;\mu]$. Now, if $(z^{(m)}, u^{(m)}) \in C^{1.L}_{\varphi.c}[d;\lambda] \times C^{0.L}_{\varphi.c}[p;\mu]$ are known functions, then

(i) $u^{(m+1)} = D_y\varphi(x,y)$ on E_0 and $u^{(m+1)}$ is a solution of the equation

$$u \;=\; G^{(m)}[\,u\,] \tag{4.57}$$

on $[0,c] \times R^n$, where $G^{(m)} = (\,G_1^{(m)}, \ldots, G_n^{(m)}\,)$ is defined by

$$G_i^{(m)}[\,u\,](x,y) = D_{y_i}\varphi(0, g(0,x,y)) + \int_0^x \Big[D_{y_i} f(P[z^{(m)},u](\tau,x,y))$$

$$+ D_w f(P[z^{(m)},u](\tau,x,y))(u_i^{(m)})_{(\tau, g(\tau,x,y))} \Big]\, d\tau, \quad i = 1, ..., n,$$

(ii) $z^{(m+1)}(x,y) = \varphi(x,y)$ on E_0 and

$$z^{(m+1)}(x,y) = F[\,z^{(m)}, u^{(m+1)}\,](x,y) \text{ on } [0,c] \times R^n. \tag{4.58}$$

We have denoted by g the function satisfying the equation

$$g[z^{(m)},u](t,x,y) = y + \int_t^x D_q f(P[z^{(m)},u](\tau,x,y))d\tau \tag{4.59}$$

and for simplicity we ignore the dependence of g on $z^{(m)}$ and u.

Remark 4.29 *Observe that the equations $u = G[\,z^{(m)}, u\,]$ and (4.57) are not identical. System (4.57) is obtained in the following way. Supposed that*

$$(\,z^{(m)}, u^{(m)}\,) \in C^{1.L}_{\varphi.c}[d;\lambda] \times C^{0.L}_{\varphi.c}[p;\mu]$$

are known functions, we consider system (4.54) with $z^{(m)}$ instead of z, i.e.

$$D_x u_i(x,y) = D_{y_i} f(\,x, y, z^{(m)}_{(x,y)}, u(x,y)\,)$$

$$+ D_w f(\,x, y, z^{(m)}_{(x,y)}, u(x,y)\,)(D_{y_i} z^{(m)})_{(x,y)}$$

$$+ \sum_{j=1}^{n} D_{q_j} f(\,x, y, z^{(m)}_{(x,y)}, u(x,y)\,) D_{y_j} u_i(x,y), \qquad i = 1, \ldots, n.$$

If we assume that $D_y z^{(m)} = u^{(m)}$ (see Theorem 4.30), then by integrating the above system along the bicharacteristic $g[z^{(m)},u](\,\cdot\,,x,y)$, we obtain (4.57).

We wish to emphasize that the main difficulty in carrying out of this construction is the problem of the existence of the sequence $\{\,z^{(m)}, u^{(m)}\,\}$.

4.11 The existence of the sequence of successive approximations

In this Section we prove that the sequence $\{ z^{(m)}, u^{(m)} \}$ exists and that

$$D_y z^{(m)}(x,y) = u^{(m)}(x,y) \quad \text{on} \quad [0,c] \times R^n$$

provides $c \in (0,a]$ is sufficiently small. Now we introduce some technical notations and assumptions which enable us to give the estimate of the constant c. We define the functions $\tilde{\Gamma}, \ \Gamma, \ \Gamma_0 : [0,a] \to R_+$ by

$$\tilde{\Gamma}(x) = s_1 + (\bar{d} + p_1) \int_0^x \alpha(\tau, \tilde{d}) d\tau + p_0 \, \bar{d} \int_0^x \gamma(\tau, |d|) \, d\tau,$$

$$\Gamma(x) = \left[s_2 + \int_0^x \left(\bar{d}(1 + p_0)\gamma(\tau, |d|) + p_1 \alpha(\tau, \tilde{d}) \right) d\tau \right] \exp\left[\bar{d} \int_0^x \gamma(\tau, |d|) d\tau \right],$$

$$\Gamma_0(x) = [\, \Gamma(x) + 1 + p_0 \,] \int_0^x \gamma(\tau, |d|) d\tau.$$

Write

$$\lambda_0(x) = \tilde{\gamma}(x, d_0) + p_0 \alpha(x, \tilde{d}) + \alpha(x, \tilde{d}) \, \tilde{\Gamma}(c) \exp\left[\bar{d} \int_0^c \gamma(\tau, |d|) d\tau \right],$$

$$\lambda_1(x) = \mu(x) = \alpha(x, \tilde{d}) \, [\, \Gamma(c) + 1 + p_0 \,].$$

Assumption H $[p, d, c]$. Suppose that $\varphi \in C^{1.L}[E_0; s]$ and
1) $d_i > s_i$ for $i = 1, 2, 3,$ and $p_0 = d_1, \ p_1 = d_2$,
2) the constant $c \in (0, a]$ is small enough to satisfy

$$\int_0^c \alpha(\tau, \tilde{d}) d\tau < 1, \qquad \Gamma_0(c) < 1, \qquad \Gamma(c) \leq p_1,$$

$$\left(1 - \int_0^c \alpha(\tau, \tilde{d}) d\tau \right)^{-1} \left[s_1 + \int_0^c \alpha(\tau, \tilde{d}) d\tau \right] \leq p_0,$$

$$s_0 + \int_0^c \left[\tilde{\gamma}(\tau, d_0) + p_0 \alpha(\tau, \tilde{d}) \right] d\tau \leq d_0.$$

Now we prove the main theorem in this Section.

Theorem 4.30 *If Assumptions $H\,[D_q f\,]$, $H\,[f, D_y f, D_w f\,]$, $H\,[p, d, c\,]$ are satisfied, then for any $m \geq 0$ we have*

(I_m) $z^{(m)}$ *and* $u^{(m)}$ *are defined on* $[-r_0, c] \times R^n$, *and*

$$z^{(m)} \in C^{1.L}_{\varphi.c}[d; \lambda], \quad u^{(m)} \in C^{0.L}_{\varphi.c}[p; \mu],$$

(II_m) $D_y z^{(m)}(x,y) = u^{(m)}(x,y)$ *on* $[0,c] \times R^n$.

PROOF. We will prove (I_m) and (II_m) by induction. It follows from (4.55), (4.56) that conditions (I_0) and (II_0) are satisfied. Supposed now that conditions (I_m) and (II_m) hold for a given $m \geq 0$, we will prove that there exist $u^{(m+1)}$: $[-r_0, c] \times R^n \to R^n$ and $u^{(m+1)} \in C^{0.L}_{\varphi.c}[p;\mu]$. We claim that

$$G^{(m)} : C^{0.L}_{\varphi.c}[p;\mu] \to C^{0.L}_{\varphi.c}[p;\mu]. \tag{4.60}$$

Indeed, it follows from Assumption H $[\,f, D_y f, D_w f\,]$ that for $u \in C^{0.L}_{\varphi.c}[p;\mu]$ we have

$$\|G^{(m)}[u](x,y)\| \leq s_1 + (1+p_0)\int_0^x \alpha(\tau,\tilde{d})\,d\tau \leq p_0$$

and

$$\|\,G^{(m)}[u](x,y) - G^{(m)}[u](\bar{x},\bar{y})\,\| \leq \Big|\int_x^{\bar{x}} \mu(t)dt\,\Big| + p_1\|y-\bar{y}\|$$

on $[0,c] \times R^n$. By the above inequalities we obtain (4.60). Now we prove that for $u,\ \tilde{u} \in C^{0.L}_{\varphi.c}[p;\mu]$ we have

$$\|\,G^{(m)}[u] - G^{(m)}[\tilde{u}]\,\|_{[c,R^n]} \leq \Gamma_0(c)\|u-\tilde{u}\|_{[c,R^n]}. \tag{4.61}$$

It follows from Assumptions H $[\,D_q f\,]$, H $[\,f, D_y f, D_w f\,]$ that

$$\|\,G^{(m)}[u](x,y) - G^{(m)}[\tilde{u}](x,y)\,\| \leq \|D_y\varphi(\,0, g(0,x,y)\,) - D_y\varphi(\,0,\tilde{g}(0,x,y)\,)\|$$

$$+\int_0^x \|\,D_y f(\,P[z^{(m)},u](\tau,x,y)\,) - D_y f(\,P[z^{(m)},\tilde{u}](\tau,x,y)\,)\,\|\,d\tau$$

$$+\sum_{j=1}^n \int_0^x \Big|\,D_w(\,P[z^{(m)},u](\tau,x,y)\,)\,(u_j^{(m)})_{(\tau,g(\tau,x,y))}$$

$$-D_w f(\,P[z^{(m)},\tilde{u}](\tau,x,y)\,)\,(u_j^{(m)})_{(\tau,\tilde{g}(\tau,x,y))}\,\Big|\,d\tau,$$

where $\tilde{g}$ is the solution of equation (4.59) with $\tilde{u}$ instead of u in the right-hand side. Then we have

$$\|\,G^{(m)}[\,u\,](x,y) - G^{(m)}[\,\tilde{u}\,](x,y)\,\| \leq s_2\,\|\,g(0,x,y) - \tilde{g}(0,x,y)\,\| \tag{4.62}$$

$$+\int_0^x \|\,g(\tau,x,y) - \tilde{g}(\tau,x,y)\,\|\,\Big[\bar{d}\gamma(\tau,|d|)\,(1+p_0) + p_1\alpha(\tau,\tilde{d})\Big]\,d\tau$$

$$+(1+p_0)\int_0^x \|u-\tilde{u}\|_{[\tau,R^n]}\;\gamma(\tau,|d|)\,d\tau.$$

It follows from Lemma 4.26 that for $0 \leq \tau \leq x \leq a,\ y \in R^n$ we have

$$\|\,g(\tau,x,y) - \tilde{g}(\tau,x,y)\,\|$$

$$\leq \int_\tau^x \gamma(\xi,|d|)\,\|u-\tilde{u}\|_{[\xi,R^n]}\,d\xi\,\exp\Big[\,\bar{d}\int_\tau^x \gamma(\xi,|d|)\,d\xi\,\Big],$$

this gives (4.61) when substituted in (4.62). From the Banach fixed point theorem it follows that there is exactly one $u^{(m+1)} \in C^{0.L}_{\varphi.c}[p;\mu]$ satisfying (4.57) on $[0,c]\times R^n$. Now we prove that the function $z^{(m+1)}$ given by (4.58) satisfies (II_{m+1}). Put

$$\Delta(x,y,\bar{y}) = z^{(m+1)}(x,\bar{y}) - z^{(m+1)}(x,y) - \sum_{j=1}^{n} u_j^{(m+1)}(x,y)(\bar{y}_j - y_j), \tag{4.63}$$

where $x \in [0,c]$, $y,\bar{y} \in R^n$. We prove that there exists $\tilde{C} \in R_+$ such that

$$|\Delta(x,y,\bar{y})| \leq \tilde{C}\, \|y-\bar{y}\|^2 \text{ for } (x,y),\ (x,\bar{y}) \in [0,c]\times R^n. \tag{4.64}$$

It follows from (4.57), (4.58) that

$$\Delta(x,y,\bar{y}) = F[\,z^{(m)},u^{(m+1)}\,](x,\bar{y}) - F[\,z^{(m)},u^{(m+1)}\,](x,y)$$

$$-\sum_{j=1}^{n} G_j^{(m)}[\,u^{(m+1)}\,](x,y)\,(\bar{y}_j - y_j).$$

Let

$$Q^{(m)}(s,\tau,x,y,\bar{y}) =$$

$$\Big(\tau, g(\tau,x,y) + s[g(\tau,x,\bar{y}) - g(\tau,x,y)], z^{(m)}_{(\tau,g(\tau,x,y))} + s[z^{(m)}_{(\tau,g(\tau,x,\bar{y}))} - z^{(m)}_{(\tau,g(\tau,x,y))}],$$

$$u^{(m+1)}(\tau,g(\tau,x,y)) + s[u^{(m+1)}(\tau,g(\tau,x,\bar{y})) - u^{(m+1)}(\tau,g(\tau,x,y))]\Big).$$

For simplicity of formulation next properties of the function $\Delta(x,y,\bar{y})$ we define

$$\bar{\Delta}(x,y,\bar{y}) = \varphi(0,g(0,x,\bar{y})) - \varphi(0,g(0,x,y))$$

$$-\sum_{j=1}^{n} D_{y_j}\varphi(0,g(0,x,y))\,[\,g_j(0,x,\bar{y}) - g_j(0,x,y)\,]$$

$$+\int_0^x \sum_{j=1}^{n} \int_0^1 \Big[D_{y_j} f\Big(Q^{(m)}(s,\tau,x,y,\bar{y})\Big)$$

$$-D_{y_j} f\Big(P[z^{(m)},u^{(m+1)}](\tau,x,y)\Big)\Big]\, ds\,[\,g_j(\tau,x,\bar{y}) - g_j(\tau,x,y)\,]\, d\tau$$

$$+\int_0^x \int_0^1 \Big[D_w f\Big(Q^{(m)}(s,\tau,x,y,\bar{y})\Big)$$

$$-D_w f\Big(P[z^{(m)},u^{(m+1)}](\tau,x,y)\Big)\Big]\, ds\, \Big[z^{(m)}_{(\tau,g(\tau,x,\bar{y}))} - z^{(m)}_{(\tau,g(\tau,x,y))}\Big]\, d\tau$$

$$+\int_0^x \sum_{j=1}^{n} \int_0^1 \Big[D_{q_j} f\Big(Q^{(m)}(s,\tau,x,y,\bar{y})\Big)$$

$$-D_{q_j} f\Big(P[z^{(m)},u^{(m+1)}](\tau,x,\bar{y})\Big)\Big]\, ds$$

$$\times \left[u_j^{(m+1)}(\tau, g(\tau, x, \bar{y})) - u_j^{(m+1)}(\tau, g(\tau, x, y)) \right] d\tau$$

$$+ \int_0^x D_w f \left(P[z^{(m)}, u^{(m+1)}](\tau, x, y) \right)$$

$$\times \left[z^{(m)}_{(\tau, g(\tau, x, \bar{y}))} - z^{(m)}_{(\tau, g(\tau, x, y))} - \sum_{j=1}^{n} (u_j^{(m)})_{(\tau, g(\tau, x, y))} \left(g_j(\tau, x, \bar{y}) - g_j(\tau, x, y) \right) \right] d\tau.$$

Moreover we put

$$\tilde{\Delta}(x, y, \bar{y}) = \sum_{j=1}^{n} D_{y_j} \varphi(0, g(0, x, y)) \left[g_j(0, x, \bar{y}) - g_j(0, x, y) - (\bar{y}_j - y_j) \right]$$

$$+ \int_0^x \sum_{j=1}^{n} D_{y_j} f \left(P[z^{(m)}, u^{(m+1)}](\tau, x, y) \right) \left[g_j(\tau, x, \bar{y}) - g_j(\tau, x, y) - (\bar{y}_j - y_j) \right] d\tau$$

$$+ \int_0^x D_w f \left(P[z^{(m)}, u^{(m+1)}](\tau, x, y) \right)$$

$$\times \left[\sum_{j=1}^{n} (u_j^{(m)})_{(\tau, g(\tau, x, y))} \left[g_j(\tau, x, \bar{y}) - g_j(\tau, x, y) - (\bar{y}_j - y_j) \right] \right] d\tau$$

$$- \int_0^x \sum_{j=1}^{n} \left[D_{q_j} f \left(P[z^{(m)}, u^{(m+1)}](\tau, x, \bar{y}) \right) \right.$$

$$\left. - D_{q_j} f \left(P[z^{(m)}, u^{(m+1)}](\tau, x, y) \right) \right] u_j^{(m+1)}(\tau, g(\tau, x, y))\, d\tau.$$

By using the Hadamard mean value theorem to the difference

$$F[\, z^{(m)}, u^{(m+1)} \,](x, \bar{y}) - F[\, z^{(m)}, u^{(m+1)} \,](x, y)$$

we get

$$\Delta(x, y, \bar{y}) = \bar{\Delta}(x, y, \bar{y}) \; + \; \tilde{\Delta}(x, y, \bar{y}). \tag{4.65}$$

The next claim is

$$\tilde{\Delta}(x, y, \bar{y}) = 0, \quad (x, y),\, (x, \bar{y}) \in [0, c] \times R^n. \tag{4.66}$$

It follows from (4.52) that for $j = 1, \ldots, n$ we have

$$g_j(\tau, x, \bar{y}) - g_j(\tau, x, y) - (\bar{y}_j - y_j) =$$

$$= \int_\tau^x \left[D_{q_j} f \left(P[z^{(m)}, u^{(m+1)}](\xi, x, \bar{y}) \right) - D_{q_j} f \left(P[z^{(m)}, u^{(m+1)}](\xi, x, y) \right) \right] d\xi.$$

Let us check that

$$\tilde{\Delta}(x, y, \bar{y}) = \tag{4.67}$$

$$= \sum_{j=1}^{n} \Big\{ \int_0^x \Big[D_{q_j} f\left(P[z^{(m)}, u^{(m+1)}](\tau, x, \bar{y}) \right) - D_{q_j} f\left(P[z^{(m)}, u^{(m+1)}](\tau, x, y) \right) \Big]$$

$$\times \Big[D_{y_j}\varphi(0, g(0, x, y)) + \int_0^\tau D_{y_j} f\left(P[z^{(m)}, u^{(m+1)}](s, x, y) \right) ds$$

$$+ \int_0^\tau D_w f\left(P[z^{(m)}, u^{(m+1)}](s, x, y) \right) (u_j^{(m)})_{(s, g(s,x,y))} ds$$

$$- u_j^{(m+1)}(\tau, g(\tau, x, y)) \Big] \, d\tau \Big\}.$$

We have

$$g(s, \tau, g(\tau, x, y)) = g(s, x, y), \; for \; (x, y) \in [0, c] \times R^n, \; \tau, \, s \in [0, x],$$

therefore we get that for $(x, y) \in [0, c] \times R^n$, $\tau \in [0, x]$

$$\begin{aligned} u_j^{(m+1)}(\tau, g(\tau, x, y)) &= D_{y_j}\varphi(0, g(0, x, y)) \\ &+ \int_0^\tau D_{y_j} f\left(P[z^{(m)}, u^{(m+1)}](s, x, y) \right) ds \\ &+ \int_0^\tau D_w f\left(P[z^{(m)}, u^{(m+1)}](s, x, y) \right) (u_j^{(m)})_{(s, g(s,x,y))} \, ds \end{aligned}$$

and we have $\tilde{\Delta}(x, y, \bar{y}) = 0$ by (4.57). Then we have proved that $\Delta(x, y, \bar{y}) = \bar{\Delta}(x, y, \bar{y})$. It follows from Assumptions H $[D_q f]$, H $[f, D_y f, D_w f]$ that the terms

$$\| D_y f(Q^{(m)}(s, \tau, x, y, \bar{y})) - D_y f(P[z^{(m)}, u^{(m+1)}](\tau, x, y)) \|,$$

$$\| D_q f(Q^{(m)}(s, \tau, x, y, \bar{y})) - D_q f(P[z^{(m)}, u^{(m+1)}](\tau, x, y)) \|$$

$$\| D_w f(Q^{(m)}(s, \tau, x, y, \bar{y})) - D_w f(P[z^{(m)}, u^{(m+1)}](\tau, x, y)) \|_*$$

are bounded from above by

$$\bar{d} \, \gamma_1(\tau, |d|) \, \| \, g(\tau, x, y) - g(\tau, x, \bar{y}) \, \|.$$

An easy computation shows that

$$\| \varphi(0, g(0, x, \bar{y})) - \varphi(0, g(0, x, y)) - \sum_{j=1}^{n} D_{y_j}\varphi(0, g(0, x, y)) \, [g_j(0, x, \bar{y}) - g_j(0, x, y)] \, \|$$

$$\leq s_2 \, \| \, g(0, x, y) - g(0, x, \bar{y}) \, \|^2,$$

and

$$\| z^{(m)}_{(\tau, g(\tau, x, y))} - z^{(m)}_{(\tau, g(\tau, x, \bar{y}))} \|_0 \leq d_1 \| \, g(\tau, x, y) - g(\tau, x, \bar{y}) \, \|,$$

$$\| u^{(m+1)}(\tau, g(\tau, x, y)) - u^{(m+1)}(\tau, g(\tau, x, \bar{y})) \| \leq p_1 \| \, g(\tau, x, y) - g(\tau, x, \bar{y}) \, \|.$$

Since

$$D_y z^{(m)}(x, y) = u^{(m)}(x, y) \;\; \text{on} \;\; [0, c] \times R^n,$$

we have

$$\|z^{(m)}_{(\tau,g(\tau,x,y))} - z^{(m)}_{(\tau,g(\tau,x,\bar{y}))} - \sum_{j=1}^{n}(u_j^{(m)})_{(\tau,g(\tau,x,y))}\,[\,g_j(\tau,x,y) - g_j(\tau,x,\bar{y})\,]\,\|_0$$

$$\leq d_2\|\,g(\tau,x,y) - g(\tau,x,\bar{y})\,\|^2.$$

The above estimates and the definition of $\bar{\Delta}$ imply

$$|\,\Delta(x,y,\bar{y})\,| \leq s_2\|\,g(0,x,y) - g(x,x,\bar{y})\,\|^2$$

$$+\int_0^x \|\,g(\tau,x,y) - g(\tau,x,\bar{y})\,\|^2 \left[d_2\,\alpha(\tau,\tilde{d}) + \bar{d}(1+d_1+p_1)\,\gamma(\tau,|d|)p_1\right] d\tau.$$

It follows from Lemma 4.26 that

$$\|\,g(\tau,x,y) - g(\tau,x,\bar{y})\,\| \leq \|y-\bar{y}\|\exp\left\{\bar{d}\int_0^c \gamma(\xi,|d|)\,d\xi\right\}.$$

Then there is $\tilde{C} > 0$ such that the estimate (4.64) holds, and consequently

$$D_y z^{(m+1}(x,y) = u^{(m+1)}(x,y) \;\text{ on }\; [0,c]\times R^n.$$

Now we prove that

$$z^{(m+1)} \in C^{1.L}_{\varphi.c}[d;\lambda]. \tag{4.68}$$

Of course $z^{(m+1)}$ is continuous on $[-r_0,c]\times R^n$ and $z^{(m+1)} = \varphi$ on E_0. Moreover from (II_{m+1}) it follows that

$$\|D_y z^{(m+1)}(x,y)\| \leq d_1,$$

$$\|D_y z^{(m+1)}(x,y) - D_y z^{(m+1)}(\bar{x},\bar{y})\| \leq \left|\int_x^{\bar{x}} \lambda_1(t)\,dt\right| + d_2\|y-\bar{y}\|.$$

on $[0,c]\times R^n$. Our assumptions imply the following estimates on $[0,c]\times R^n$:

$$|z^{(m+1)}(x,y)| \leq d_0 \qquad |z^{(m+1)}(x,y) - z^{(m+1)}(\bar{x},y)| \leq \left|\int_x^{\bar{x}} \lambda_0(t)\,dt\right|.$$

This completes the proof of Theorem 4.30.

4.12 Convergence of the sequence $\{z^{(m)}, u^{(m)}\}$.

We prove that the sequence $\{z^{(m)}, u^{(m)}\}$ is convergent if the constant c is sufficiently small. Now we introduce some notations and assumptions which enable us to give the estimate of the constant c. Write

$$\bar{\Gamma}(x) = 1 + p_0 + s_2 + \int_0^x \left[\bar{d}(1+p_0)\gamma(\tau,|d|) + p_1\alpha(\tau,\tilde{d})\right] d\tau.$$

We define the functions $\theta_0,\ \theta_1,\ \tilde{\theta}_0,\ \tilde{\theta}_1\ :\ [0,\ c]\ \to\ R_+$ by

$$\theta_0(x) = \bar{\Gamma}(c)\,\gamma(x, |d|),\quad \theta_1(x) = \bar{\Gamma}(c)\,\gamma(x, |d|) + \alpha(x, \tilde{d}),$$

$$\tilde{\theta}_i(x) = \Big[\tilde{\Gamma}(c) + p_0\Big]\,\gamma(x, |d|) + \alpha(x, \tilde{d}) + G(c)\theta_i(x),\quad i = 0, 1,$$

where

$$G(x) = \exp\Big[\int_0^x \theta_0(\tau)d\tau\Big]\int_0^x \Big[\Big(\tilde{\Gamma}(c) + p_0\Big)\,\gamma(\tau, |d|) + 2\alpha(\tau, \tilde{d})\Big]\,d\tau.$$

Assumption H $[c]$. Suppose that $c \in [0, a]$ is such a small constant that

$$\int_0^c \theta_1(\tau)d\tau \exp\Big[\int_0^c \theta_0(\tau)d\tau\Big] + \int_0^c \tilde{\theta}_1(\tau)d\tau < 1.$$

We can now formulate the result on the convergence of the sequence of successive approximations.

Theorem 4.31 *If Assumptions* $H\ [D_q f\,]$, $H\ [f, D_y f, D_w f\,]$, $H\ [p, d, c\,]$, $H\ [c]$ *are satisfied then the sequences* $\{z^{(m)}\}$ *and* $\{u^{(m)}\}$ *are uniformly convergent on* $[0, c] \times R^n$.

PROOF. For $t \in [0, c]$ and $m \geq 1$ we put

$$Z^{(m)}(t) = \sup\{\,|z^{(m)}(x, y) - z^{(m-1)}(x, y)|\,:\ (x, y) \in [0, t] \times R^n\,\},$$

$$U^{(m)}(t) = \sup\{\,\|u^{(m)}(x, y) - u^{(m-1)}(x, y)\|\,:\ (x, y) \in [0, t] \times R^n\,\}.$$

Let us prove that for $x \in [0, c]$ we have

$$U^{(m+1)}(x) \leq \int_0^x \Big[\theta_0(\tau)Z^{(m)}(\tau) + \theta_1(\tau)U^{(m)}(\tau)\Big]\,d\tau\ \exp\Big[\int_0^x \theta_0(\tau)d\tau\Big] \tag{4.69}$$

and

$$Z^{(m+1)}(x) \leq \int_0^x \Big[\tilde{\theta}_0(\tau)Z^{(m)}(\tau) + \tilde{\theta}_1(\tau)U^{(m)}(\tau)\Big]\,d\tau. \tag{4.70}$$

Write $g^{(k)}(\tau, x, y) = g[z^{(k-1)}, u^{(k)}](\tau, x, y)$. It follows from Lemma 4.26 that we have the estimate

$$\|\,g^{(m+1)}(\tau, x, y) - g^{(m)}(\tau, x, y)\,\| \tag{4.71}$$

$$\leq \int_0^x \gamma(\xi, |d|)\,\Big[\,Z^{(m)}(\xi) + U^{(m)}(\xi) + U^{(m+1)}(\xi)\Big]\,d\xi$$

where $(x, y) \in [0, c] \times R^n$, $0 \leq \tau \leq x$. Then we obtain the integral inequality on $[0, c]$

$$U^{(m+1)}(x) \leq \int_0^x \theta_0(\tau)U^{(m+1)}(\tau)\,d\tau + \int_0^x \Big[\theta_0(\tau)Z^{(m)}(\tau) + \theta_1(\tau)U^{(m)}(\tau)\Big]\,d\tau.$$

The above estimate and Gronwall inequality imply inequality (4.69) for $x \in [0, c]$. According to (4.58) and assumptions H $[D_q f]$, H $[f, D_y f, D_q f]$ we have on $[0, c] \times R^n$

$$|z^{(m+1)}(x,y) - z^{(m)}(x,y)| \le s_1 \| g^{(m+1)}(0,x,y) - g^{(m)}(0,x,y) \|$$

$$+ \int_0^x \left[\alpha(\tau, \tilde{d}) + p_0 \gamma(\tau, |d|) \right] \left[\bar{d} \, \| g^{(m+1)}(\tau,x,y) - g^{(m)}(\tau,x,y) \| \right.$$

$$\left. + Z^{(m)}(\tau) + U^{(m)}(\tau) + U^{(m+1)}(\tau) \right] d\tau$$

$$+ \int_0^x \alpha(\tau, \tilde{d}) \left[p_1 \| g^{(m+1)}(\tau,x,y) - g^{(m)}(\tau,x,y) \| + U^{(m+1)}(\tau) \right] d\tau.$$

It follows from the above estimates and from (4.69) (4.71) that inequality (4.70) holds. By Assumption H $[c]$ and by (4.69), (4.70) we conclude that there exists $\delta \in (0, 1)$ such that

$$U^{(m+1)}(x) + Z^{(m+1)}(x) \le \delta \left[U^{(m)}(x) + Z^{(m)}(x) \right], \qquad x \in [0, c], \quad m \ge 0. \quad (4.72)$$

From (4.55) and from Assumptions H $[D_q f]$, H $[f, D_y f, D_w f]$ it follows that there is $\bar{C} \in R_+$ such that

$$Z^{(1)}(x) \le \bar{C} + \int_0^x \left[\tilde{\gamma}(\tau, d_0) + \alpha(\tau, \tilde{d}) p_0 \right] d\tau,$$

$$U^{(1)} \le \bar{C} + (1 + p_0) \int_0^x \alpha(\tau, \tilde{d}) d\tau,$$

where $x \in [0, c]$. Finally the convergence of the sequence $\{z^{(m)}, u^{(m)}\}$ follows from (4.72). This completes the proof of Theorem 4.31.

4.13 The main theorem

We are able now to state the main result on the existence of weak solutions.

Theorem 4.32 *Under Assumptions H $[D_q f]$, H $[f, D_y f, D_w f]$, H $[p, d, c]$, H $[c]$ for every $\varphi \in C^{1.L}[E_0; s]$ a solution $\bar{z} : [-r_0, c] \times R^n \to R$ of problem (4.45), (4.46) exists. Moreover $\bar{z} \in C_{\varphi.c}[d; \lambda]$ and $D_y \bar{z} \in C^{0.L}_{\varphi.c}[p; \mu]$.*

Proof. From Theorem 4.31 we deduce that two functions $\bar{z} \in C^{1.L}_{\varphi.c}[d, \lambda]$ and $\bar{u} \in C^{0.L}_{\varphi.c}[p; \mu]$ exist such that $\{ z^{(m)} \}$ converges to $\bar{z}$ and $\{ u^{(m)} \}$ converges to $\bar{u}$ uniformly on $[0, c] \times R^n$. Furthermore, we have that $D_y \bar{z}$ exists on $[0, c] \times R^n$ and $D_y \bar{z} = \bar{u}$. Thus we get that

$$\bar{z}(x,y) = \varphi(0, \bar{g}(0,x,y))$$

$$+ \int_0^x \left[f\,(P[\bar{z}, D_y \bar{z}](\tau,x,y)) - \sum_{j=1}^n D_{q_j} f\,(P[\bar{z}, D_y \bar{z}](\tau,x,y))\, D_{y_j} \bar{z}(\tau,x,y) \right] d\tau$$

and $\bar{g}$ satisfies

$$\bar{g}(t,x,y) = y + \int_t^x D_q f(\,P[\bar{z}, D_y \bar{z}](\tau,x,y)\,)\,d\tau.$$

Now we prove that $\bar{z}$ is a weak solution of (4.45). For a given $y \in R^n$, let us put $s = \bar{g}(0,x,y)$. It follows from Lemma 4.26 that $\bar{g}(t,x,y) = \bar{g}(t,0,s)$ for $t,\ x \in [0,c]$ and $y = \bar{g}(x,0,s)$. Moreover, from (4.51) we get

$$\bar{z}(x, g(x,0,s)) = \varphi(0,s) + \tag{4.73}$$

$$\int_0^x f\left(\tau, \bar{g}(\tau,0,s), \bar{z}_{(\tau,\bar{g}(\tau,0,s))}, D_y\bar{z}(\tau,\bar{g}(\tau,0,s))\right)\,d\tau$$

$$-\sum_{j=1}^n \int_0^x D_{q_j} f\left(\tau, \bar{g}(\tau,0,s), \bar{z}_{(\tau,\bar{g}(\tau,0,s))}, D_y\bar{z}(\tau,\bar{g}(\tau,0,s))\right) D_{y_j}\bar{z}(\tau,\bar{g}(\tau,0,s))\,d\tau.$$

The relations $s = \bar{g}(0,x,y)$ and $y = \bar{g}(x,0,s)$ are equivalent for $y, s \in R^n$. By differentiating (4.73) with respect to x and by putting again $y = \bar{g}(x,0,s)$, we obtain that $\bar{z}$ satisfies (4.45) for almost all $x \in [0,c]$ with fixed $y \in R^n$. Since $\bar{z}$ satisfies initial condition (4.46), the proof is complete.

Remark 4.33 *If in Theorem 4.32 we assume that f is continuous, then we get classical solutions for our Cauchy problem.*

The next theorem gives sufficient conditions for the local existence of a solution of the Cauchy problem (4.45), (4.46) without the estimate of the existence domain.

Theorem 4.34 *If Assumptions H $[D_q f]$, H $[f, D_y f, D_w f]$ are satisfied then there exists $c > 0$ such that problem (4.45), (4.46) admits the solution $u : [-r_0, c] \times R^n \to R$ such that $u \in C^{1.L}_{\varphi.c}[d;\lambda]$ and $D_y u \in C^{0.L}_{\varphi.c}[p;\mu]$ for some $d \in R^3$, $p \in R^2$ and $\lambda_i \in L([0,c], R_+)$, $i = 0,1$, $\mu \in L([0,c], R_+)$.*

Assumptions H $[p,d,c]$ and H $[c]$ give the estimate of the existence domain of the solution of problem (4.45), (4.46).

4.14 Some noteworthy particular cases

We wish to emphasize that our hereditary setting contains in particular some well known delay structures. Given the functions $\varphi : E_0 \to R$, $\tilde{f} : [0,a] \times R^n \times R \times R^n \to R$, and $\alpha : [0,a] \to R$, $\beta = (\beta_1, \ldots, \beta_n) : [0,a] \times R^n \to R^n$, we consider the operator f defined by

$$f(x,y,w,q) = \tilde{f}\left(x, y, w(\alpha(x) - x, \beta(x,y) - y), q\right), \quad (x,y,w,q) \in \Omega^{(0)}. \tag{4.74}$$

In this case (4.45) is equivalent to the differential equation with a deviated argument

$$D_x z(x,y) = \tilde{f}(\,x, y, z(\alpha(x), \beta(x,y)), D_y z(x,y)\,) \tag{4.75}$$

with the initial condition (4.46). Now we formulate our existence result for problem (4.46), (4.75).

Assumption H $[\tilde{f}]$. Suppose that the function $\tilde{f}$ satisfies the conditions:

1) there is $\tilde{\gamma} \in L([0,a], R_+)$ such that for $(x, \zeta, q) \in [0,a] \times R^n \times R \times R^n$, we have $|\tilde{f}(x,y,\zeta,q)| \leq \tilde{\gamma}(x)$,

2) the derivatives $D_y\tilde{f}$, $D_\zeta\tilde{f}$, $D_q\tilde{f}$ exist on $[0,a] \times R^n \times R \times R^n$ and there is $\tilde{\alpha} \in L([0,a] \times R_+)$ such that for $(x,y,\zeta,q) \in [0,a] \times R^n \times R \times R^n$ we have

$$\|D_y\tilde{f}(x,y,\zeta,q)\| \leq \tilde{\alpha}(x), \quad |D_\zeta\tilde{f}(x,y,\zeta,q)| \leq \tilde{\alpha}(x), \quad \|D_q\tilde{f}(x,y,\zeta,q)\| \leq \tilde{\alpha}(x),$$

3) there is $\tilde{\beta} \in L([0,a], R_+)$ such the terms

$$\|D_y\tilde{f}(x,y,\zeta,q) - D_y\tilde{f}(x,\bar{y},\bar{\zeta},\bar{q})\|, \ \|D_\zeta\tilde{f}(x,y,\zeta,q) - D_\zeta\tilde{f}(x,\bar{y},\bar{\zeta},\bar{q})\|,$$

$$\|D_q\tilde{f}(x,y,\zeta,q) - D_q\tilde{f}(x,\bar{y},\bar{\zeta},\bar{q})\|$$

are bounded from above by

$$\beta(x)\left[\|y-\bar{y}\| + |\zeta - \bar{\zeta}| + \|q - \bar{q}\|\right]$$

where (x,y,ζ,q), $(\bar{x},\bar{y},\bar{\zeta},\bar{q}) \in [0,a] \times R^n \times R \times R^n$.

Assumption H $[\alpha, \beta]$ Suppose that the functions α and β satisfy the conditions:

1) $\alpha \in C([0,a], R)$, $\beta \in C([0,a] \times R^n, R^n)$ and $-r_0 \leq \alpha(x) - x \leq 0$, $-r \leq \beta(x,y) - y \leq r$ on $[0,a] \times R^n$,

2) the derivative $D_y\beta = [D_{y_i}\beta_j]_{i,j=1,\ldots,n}$ exists on $[0,a] \times R^n$ and

$$\|D_y\beta(x,y)\| \leq L_0, \quad \|D_y\beta(x,y) - D_y\beta(x,\bar{y})\| \leq L_1\|y - \bar{y}\|.$$

Then the main theorem reduces to the following one.

Theorem 4.35 *Suppose that Assumptions H* $[\tilde{f}]$, *H* $[\alpha, \beta]$ *are satisfied and let* $\varphi \in C^{1.L}[E_0; s]$, *and* $d_i > s_i$ for $i = 1,2,3$. *Then there exist* $c \in (0,a]$, $\lambda = (\lambda_0, \lambda_1) \in L([0,c], R_+) \times L([0,c], R_+)$ *and a function* $\bar{z} \in C^{1.L}_{\varphi.c}[d;\lambda]$ *such that* $\bar{z}$ *is a solution of (4.75) on* $[-r_0, c] \times R^n$.

PROOF. In order to apply Theorem 4.32 to the function f given by (4.74), let us show that the Assumptions H $[D_qf]$, H $[f, D_yf, D_wf]$ are fulfilled. Observe that in $\Omega^{(1)}$ we have for $i = 1,\ldots,n$

$$D_{y_i}f(x,y,w,q) = D_{y_i}\tilde{f}(x,y,w(\alpha(x)-x,\beta(x,y)-y),q)$$

$$+D_p\tilde{f}(x,y,w(\alpha(x)-x,\beta(x,y)-y),q)$$

$$\times\left[\sum_{j=1}^{n} D_{s_j}w(\alpha(x)-x,\beta(x,y)-y)\ (D_{y_i}\beta_j(x,y) - \delta_{ij})\right],$$

where δ_{ij} is the Kronecker symbol and

$$D_wf(x,y,w,q)h = D_p\tilde{f}(x,y,w(\alpha(x)-x,\beta(x,y)-y),q)\,h(\alpha(x)-x,\beta(x,y)-y),$$

$$D_{q_i} f(x,y,w,q) = D_{q_i} \tilde{f}(\,x,y,w(\alpha(x)-x,\beta(x,y)-y),q\,).$$

Thus it is easy to see that we have $|f(x,y,w,q)| \le \tilde{\gamma}(x)$ on $\Omega^{(0)}$, and

$$\|D_y f(x,y,w,q)\| \le \tilde{\alpha}_0(x)\,[1+(1+L_0)\|w\|_1]\,,$$

$$\|D_w f(x,y,w,q)\|_* \le \tilde{\alpha}(x) \quad \text{on } \Omega^{(1)}.$$

Moreover, the following Lipschitz condition is satisfied:

$$\|D_y f(x,y,w,q) - D_y f(x,\bar{y},w+h,\bar{q})\| \le \Big\{\, \tilde{\beta}(x)\ [1+(1+L_0)\|w\|_1]^2$$

$$+\tilde{\alpha}(x)\left[(1+L_0)^2\|w\|_{1.L} + L_1\|w\|_1\right] \Big\}\,[\,\|y-\bar{y}\| + \|h\|_1 + \|q-\bar{q}\|\,]\,.$$

It is easy to see that the derivatives $D_q f$ and $D_w f$ satisfy the Lipschitz condition established in Assumptions H $[\,D_q f\,]$, H $[f, D_y f, D_w f]$. For a sufficiently small $c \in (0,a]$, condition 2) of Assumption H $[\,p,d,c\,]$ and Assumption H $[\,c\,]$ are satisfied. Hence the assertion follows as an immediate application of Theorem 4.32.

Remark 4.36 *If we consider the function f defined by*

$$f(\,x,y,w,q\,) = \tilde{f}(\,x,y,\int_B w(t,s)\,dt\,ds, q\,)$$

then equation (4.45) reduces to the differential integral equation

$$D_x z(x,y) = \tilde{f}(\,x,y,\int_{-r_0}^{0}\int_{-r}^{r} z(x+t,y+s)\,dt\,ds, D_y z(x,y)\,).$$

It is easy to formulate sufficient conditions for the existence of the solutions of the above equation with the initial condition (4.46) as an application of Theorem 4.32.

Remark 4.37 *It is easy to see that Theorem 4.32 can be extended for systems of functional differential equations.*

Remark 4.38 *It is important in Assumptions H $[D_q f\,]$, H $[f, D_y, D_w f\,]$ that we have assumed the local Lipschitz condition for the derivatives $D_y f$, $D_q f$, $D_w f$ on some special function spaces. Let us consider the simplest assumptions. Suppose that there is $\bar{L}$ such that*

$$\|\,D_y f(x,y,w,q) - D_y f(x,\bar{y},w+h,\bar{q})\,\| \le \bar{L}\ [\,\|y-\bar{y}\| + \|h\|_0 + \|q-\bar{q}\|\,] \quad (4.76)$$

and that suitable inequalities for the derivatives $D_q f$, $D_w f$ are satisfied. Of course, our results are true under the above stronger assumptions. Now we show that our formulation of Lipschitz conditions is important. More precisely, that there is a class of nonlinear equations satisfying Assumptions H $[D_q f\,]$, H $[f, D_y f, D_w f\,]$ and do not satisfying conditions of type (4.76). It is easily seen that equation with a deviated argument (4.75) does not satisfy conditions of type (4.76).

Chapter 5

Mixed Problems for Nonlinear Equations

5.1 Introduction

We formulate the functional differential problem. Let $a > 0$, $r_0 \in R_+$, $b = (b_1, \ldots, b_n) \in R^n$ and $r = (r_1, \ldots, r_n) \in R^n_+$ be given where $b_i > 0$ for $1 \le i \le n$. Suppose that $\kappa \in \mathbf{Z}$, $0 \le \kappa \le n$, is fixed. For each $y = (y_1, \ldots, y_n) \in R^n$ we write $y = (y', y'')$ where $y' = (y_1, \ldots, y_\kappa)$, $y'' = (y_{\kappa+1}, \ldots, y_n)$. We have $y' = y$ if $\kappa = n$ and $y'' = y$ if $\kappa = 0$. We define the sets

$$E = (0, a) \times [-b', b') \times (-b'', b''], \quad B = [-r_0, 0] \times [0, r'] \times [-r'', 0].$$

Let $c = (c_1, \ldots, c_n) = b + r$ and

$$E_0 = [-r_0, 0] \times [-b', c'] \times [-c'', b''],$$

$$\partial_0 E = (\,(0, a) \times [-b', c'] \times [-c'', b''\,]) \setminus E, \quad E^* = E_0 \cup E \cup \partial_0 E.$$

Suppose that $z : E^* \to R$ and $(x, y) \in [0, a] \times [-b, b]$ are fixed. We define the function $z_{(x,y)} : B \to R$ as follows

$$z_{(x,y)}(t, s) = z(x + t, y + s), \quad (t, s) \in B.$$

The function $z_{(x,y)}$ is the restriction of z to the set $[x - r_0, x] \times [y', y' + r'] \times [y'' - r'', y'']$ and this restriction is shifted to the set B. Elements of the space $C(B, R)$ will be denoted be w, $\bar{w}$ and so on. We denote by $\| \cdot \|_0$ the supremum norm in the space $C(B, R)$. The symbol $\| \cdot \|_0$ will also denote the supremum norm in the space $C([-r_0, 0], R)$. Put $\Omega = E \times C(B, R) \times R^n$ and let $f : \Omega \to R$, $\phi : E_0 \cup \partial_0 E \to R$ be given functions. We will deal with the following mixed problem

$$D_x z(x, y) = f(\,x, y, z_{(x,y)}, D_y z(x, y)\,), \tag{5.1}$$

$$z(x, y) = \phi(x, y) \text{ on } E_0 \cup \partial_0 E. \tag{5.2}$$

A function $\tilde{z} : E^* \to R$ is a classical solution of problem (5.1), (5.2) if

(i) $\tilde{z} \in C(E^*, R)$ and $\tilde{z}$ admits the derivatives $D_x\tilde{z}$ and $D_y\tilde{z}$ on E,

(ii) $\tilde{z}$ satisfies the functional differential equation (5.1) on E and condition (5.2) holds.

We will consider generalized solutions of problem (5.1), (5.2) also. We denote by F the Niemycki operator corresponding to equation (5.1) i. e.

$$F[z](x,y) = f(x, y, z_{(x,y)}, D_y z(x,y)).$$

We consider the following questions related to mixed problems: differential inequalities, comparison theorems and applications, Chaplygin method for almost linear problems, difference methods for nonlinear equations, existence of generalized and classical solutions. Results of this Chapter are based on the papers [28, 29, 63, 67, 111, 149].

5.2 FUNCTIONAL DIFFERENTIAL INEQUALITIES

We start with the following theorem on strong inequalities.

Theorem 5.1 *Suppose that*

1) the function $f : \Omega \to R$ of the variables (x, y, w, q) satisfies the following monotonicity condition : if $w, \bar{w} \in C(B, R)$, $w(t,s) \le \bar{w}(t,s)$ on B and $w(0,0) = \bar{w}(0,0)$ then $f(x,y,w,q) \le f(x,y,\bar{w},q)$,

2) the derivatives $(D_{q_1}f, \dots, D_{q_n}f) = D_q f$ exist on Ω and

$$D_{q_i} f(x,y,w,q) \ge 0 \text{ on } \Omega \text{ for } 1 \le i \le \kappa, \tag{5.3}$$

$$D_{q_i} f(x,y,w,q) \le 0 \text{ on } \Omega \text{ for } \kappa+1 \le i \le n, \tag{5.4}$$

3) $u, v \in C(E^*, R)$ are two functions such that the derivatives $D_x u$, $D_y u$, $D_x v$, $D_y v$ exist on E,

4) the initial-boundary inequality $u(x,y) < v(x,y)$ holds on $E_0 \cup \partial_0 E$,

5) the functional differential inequality

$$D_x u(x,y) - F[u](x,y) < D_x v(x,y) - F[v](x,y). \tag{5.5}$$

is satisfied for $(x,y) \in E$.

Under these assumptions we have $u(x,y) < v(x,y)$ on E.

PROOF. Suppose, by contradiction, that the assertion fails to be true; then the set

$$J_+ = \{x \in (0,a) : (u-v)(x,y) \ge 0 \text{ for some } y \in [-b,b]\}$$

is not empty. If we put $\bar{x} = \inf J_+$, from the initial-boundary inequality it is clear that $\bar{x} > 0$ and there exists $\bar{y} = (\bar{y}_1, \dots, \bar{y}_n) \in [-b', b') \times (-b'', b'']$ such that

$$u(x,y) < v(x,y) \text{ for } (x,y) \in E \cap ((0,\bar{x}) \times R^n) \text{ and } u(\bar{x},\bar{y}) = v(\bar{x},\bar{y}).$$

Then

$$D_x(u-v)(\bar{x},\bar{y}) \ge 0. \tag{5.6}$$

Note that three sets (possibly empty) of integers I^+, I^-, I^0 exist such that $I^+ \cup I^- \cup I^0 = \{1, \ldots, n\}$ and $\bar{y}_i = b_i$ for $i \in I^+$, $\bar{y}_i = -b_i$ for $i \in I^-$, $-b_i < \bar{y}_i < b_i$ for $i \in I^0$. Since $(\bar{x}, \bar{y}) \notin \partial_0 E$ then $\{1, \ldots, \kappa\} \cap I^+ = \emptyset$ and $\{\kappa + 1, \ldots, n\} \cap I^- = \emptyset$. From the last relation we obtain

$$D_{y_i}(u - v)(\bar{x}, \bar{y}) \geq 0 \text{ for } i \in I^+, \;\; D_{y_i}(u - v)(\bar{x}, \bar{y}) \leq 0 \text{ for } i \in I^-, \tag{5.7}$$

and

$$D_{y_i}(u - v)(\bar{x}, \bar{y}) = 0 \text{ for } i \in I^0. \tag{5.8}$$

Thus, from (5.3) - (5.5) and (5.7), (5.8) we deduce that

$$D_x(u - v)(\bar{x}, \bar{y}) < F[u](\bar{x}, \bar{y}) - F[v](\bar{x}, \bar{y})$$

$$\leq f(\bar{x}, \bar{y}, v_{(\bar{x},\bar{y})}, D_y u(\bar{x}, \bar{y})) - f(\bar{x}, \bar{y}, v_{(\bar{x},\bar{y})}, D_y v(\bar{x}, \bar{y}))$$

$$= \sum_{i=1}^{n} D_{q_i} f(P_i) D_{y_i}(u - v)(\bar{x}, \bar{y}) \leq 0,$$

where $P_i \in \Omega$, $1 \leq i \leq n$, are intermediate points. The last inequality contradicts (5.6) and thus the assertion follows.

Remark 5.2 *The assumption 5) of Theorem 5.1 can be replaced by the following condition:*

5') denoted by

$$\Sigma = \{(x, y) \in E : \; u(t, s) < v(t, s) \text{ on } E \cap ((0, x) \times R^n) \text{ and } u(x, y) = v(x, y)\},$$

then for $(x, y) \in \Sigma$ *we have*

$$D_x u(x, y) - F[u](x, y) < D_x v(x, y) - F[v](x, y).$$

Remark 5.3 *Suppose that the function*

$$\tilde{f} : E \times R \times C(B, R) \times R^n \to R$$

is nondecreasing with respect to the functional variable and

$$f(x, y, w, q) = \tilde{f}(x, y, w(0, 0), w, q).$$

Then f *satisfies the monotonicity condition 1) from Theorem 5.1.*

Now we prove a theorem on weak functional differential inequalities. In this Chapter we denote by $V : C(B, R) \to F([-r_0, 0], R_+)$ the operator given by

$$(Vw)(t) = \max\{ |w(t, s)| : \; s = (s', s'') \in [0, r'] \times [-r'', 0] \}, \tag{5.9}$$

where $-r_0 \leq t \leq 0$. It is easy to see that V has the following property: if $w \in C(B, R)$ then $Vw \in C([-r_0, 0], R_+)$.

Assumption H $[\sigma]$. Suppose that the function $\sigma : [0,a) \times C([-r_0,0], R_+) \to R_+$ satisfies the conditions:

1) if $\zeta, \bar{\zeta} \in C([-r_0,0], R_+)$, $\zeta(t) \le \bar{\zeta}(t)$ on $[-r_0,0]$ and $\zeta(0) = \bar{\zeta}(0)$ then $\sigma(x,\zeta) \le \sigma(x,\bar{\zeta})$,

2) $\sigma(x,\theta) = 0$ for $x \in [0,a)$, where $\theta : [-r_0,0] \to R$ is given by $\theta(t) = 0$ on $[-r_0,0]$,

2) σ is continuous and the function $\bar{\omega}(x) = 0$ for $x \in [-r_0,a)$ is the unique solution of the problem

$$\omega'(x) = \sigma(x, \omega_{(x)}), \quad \omega(x) = 0 \text{ for } x \in [-r_0,0]. \tag{5.10}$$

Remark 5.4 *Suppose that* $\tilde{\sigma} : [0,a) \times R_+ \times C([-r_0,0], R_+) \to R_+$ *is nondecreasing with respect to the functional variable and* $\sigma(x,\zeta) = \tilde{\sigma}(x,\zeta(0),\zeta)$. *Then* σ *satisfies condition 1) of Assumption H* $[\sigma]$.

Theorem 5.5 *Suppose that*

1) the conditions 1) - 3) of Theorem 5.1 and Assumption H $[\sigma]$ *are satisfied,*

2) for every w, $\bar{w} \in C(B,R)$ *if* $\bar{w} \ge w$ *then*

$$f(x,y,\bar{w},q) - f(x,y,w,q) \le \sigma(x, V(\bar{w}-w)) \tag{5.11}$$

where $(x,y,q) \in E \times R^n$,

3) the initial-boundary inequality $u(x,y) \le v(x,y)$ *holds on* $E_0 \cup \partial_0 E$,

4) the functional differential inequality

$$D_x u(x,y) - F[u](x,y) \le D_x v(x,y) - F[v](x,y) \tag{5.12}$$

is satisfied on E.

Under these assumptions we have for $(x,y) \in E$

$$u(x,y) \le v(x,y). \tag{5.13}$$

PROOF. Let $0 < a_0 < a$ be fixed. For every $\varepsilon > 0$ we denote by $\omega(\cdot,\varepsilon)$ the right-hand maximum solution of the problem

$$\omega'(x) = \sigma(x, \omega_{(x)}) + \varepsilon, \quad \omega(x) = \varepsilon \text{ on } [-r_0,0]. \tag{5.14}$$

Then (see [93, 94]) there exists a number $\varepsilon_0 > 0$ such that for every $0 < \varepsilon < \varepsilon_0$ the solution $\omega(\cdot,\varepsilon)$ is defined on $[-r_0,a_0)$ and $\lim_{\varepsilon\to 0} \omega(x,\varepsilon) = 0$ uniformly with respect to $x \in [-r_0,a_0)$. Let us consider now the function $u_\varepsilon : [-r_0,a_0) \times [-b',c'] \times [-c'',b''] \to R$ defined by $u_\varepsilon(x,y) = u(x,y) - \omega(x,\varepsilon)$ where $0 < \varepsilon < \varepsilon_0$. From the initial-boundary inequality for u, v it follows that

$$u_\varepsilon(x,y) < v(x,y) \text{ on } (E_0 \cup \partial_0 E) \cap ([-r_0, a_0) \times R^n).$$

Let $(x,y) \in E \cap ((0,a_0) \times R^n)$ be fixed. We shall prove that

$$D_x u_\varepsilon(x,y) - F[u_\varepsilon](x,y) < D_x v(x,y) - F[v](x,y). \tag{5.15}$$

It follows from (5.11), (5.12) that

$$D_x u_\varepsilon(x,y) - F[u_\varepsilon](x,y) \le D_x v(x,y) - F[v](x,y)$$
$$-\sigma(x,(\omega(\cdot,\varepsilon))_{(x)}) - \varepsilon + \sigma(x, V(u-u_\varepsilon)_{(x,y)})$$
$$= D_x v(x,y) - F[v](x,y) - \varepsilon,$$

which proves inequality (5.15). By virtue of Theorem 5.1 we have that $u(x,y) - \omega(x,y) < v(x,y)$ on $E \cap ((0,a_0) \times R^n)$. Now, by passing to the limit for $\varepsilon \to 0$ we obtain by inequality (5.13) on $E \cap ((0,a_0) \times R^n)$. By the arbitrariness of $0 < a_0 < a$, the assertion follows.

Remark 5.6 *The assumption 4) of Theorem 5.5 can be replaced by the following condition*

4') the functional differential inequality (5.12) is satisfied on $\tilde{\Sigma}$ *where* $\tilde{\Sigma} = \{(x,y) \in E : u(x,y) > v(x,y)\}$. *In this case we use Theorem 5.1 with assumption 5') instead of 5), see Remark 5.2.*

Remark 5.7 *Suppose that in Theorem 5.5 the functional differential inequality (5.12) is replaced by the corresponding strong inequality. Then the strong estimation* $u(x,y) < v(x,y)$ *holds on* E. *This assertion can be proved by applying Theorem 5.5 and then repeating the argument used in the proof of Theorem 5.1.*

As an application of Theorem 5.5, the following result can be derived.

Remark 5.8 *Suppose that conditions 1) and 2) of Theorem 5.5 hold. Then mixed problem (5.1), (5.2) admits one solution at the most.*

Remark 5.9 *A particular case of the comparison problem in Assumption* $H[\sigma]$ *is the following one. Let* $\sigma_0 : [0,a) \times R_+ \to R_+$ *be a function such that*

(i) σ_0 *is continuous and* $\sigma_0(x,0) = 0$ *for* $x \in [0,a)$,

(ii) σ_0 *is nondecreasing with respect to the last variable,*

(iii) for every w, $\bar{w} \in C(B,R)$ *if* $\bar{w}(t,s) \ge w(t,s)$ *on* B *then*

$$f(x,y,\bar{w},q) - f(x,y,w,q) \le \sigma_0(x, \|\bar{w} - w\|_0)$$

where $(x,y,q) \in E \times R^n$.

(iv) the function $\bar{\omega}(x) = 0$, $x \in [0,a)$, *is the unique solution of the problem*

$$\omega'(x) = \sigma_0(x,\omega(x)), \quad \omega(0) = 0.$$

If f *satisfies conditions 1) and 2) of Theorem 5.1 then problem (5.1), (5.2) admits one solution at the most.*

Suppose that the function $\tilde{f} : E \times R \times C(B,R) \times R^n \to R$ is given. We prove a theorem on weak functional differential inequalities generated by the equation

$$D_x z(x,y) = \tilde{f}(x,y,z(x,y),z_{(x,y)},D_y z(x,y)).$$

Write $\tilde{F}[z](x,y) = \tilde{f}(x,y,z(x,y),z_{(x,y)},D_y z(x,y))$.

Theorem 5.10 *Suppose that*

1) the function $\tilde{f}$ is nondecreasing with respect to the functional variable,

2) the derivatives $(D_{q_1}\tilde{f},\dots,D_{q_n}\tilde{f})$ exist and $D_{q_i}\tilde{f}(P) \ge 0$ for $1 \le i \le \kappa$, $D_{q_i}\tilde{f}(P) \le 0$ for $\kappa+1 \le i \le n$, where $P \in E \times R \times C(B,R) \times R^n$,

3) there is a function $\tilde{\sigma} : [0,a) \times R_+ \times R_+ \to R_+$ such that

(i) $\tilde{\sigma}$ is continuous and $\tilde{\sigma}(x,0,0) = 0$ for $x \in [0,a)$,

(ii) $\tilde{\sigma}$ is nondecreasing with respect to the last variable and the function $\bar{\omega}(x) = 0$, $x \in [0,a)$ is the unique solution of the problem

$$\omega'(x) = \tilde{\sigma}(x,\omega(x),\omega(x)), \quad \omega(0) = 0,$$

(iii) for every $\bar{p} \ge p$ and $\bar{w} \ge w$ where $p, \bar{p} \in R$, $w, \bar{w} \in C(B,R)$ we have

$$\tilde{f}(x,y,\bar{p},\bar{w},q) - \tilde{f}(x,y,p,w,q) \le \tilde{\sigma}(x,\bar{p}-p,\|\bar{w}-w\|_0)$$

4) $u, v \in C(E^*,R)$ are two functions such that the derivatives $D_x u$, $D_y u$, $D_x v$, $D_y v$ exist on E and $u(x,y) \le v(x,y)$ on $E_0 \cup \partial_0 E$,

5) the functional differential inequality

$$D_x u(x,y) - \tilde{F}[u](x,y) \le D_x v(x,y) - \tilde{F}[v](x,y)$$

is satisfied on E.

Under these assumptions we have $u(x,y) \le v(x,y)$ on E.

This theorem follows from Theorem 5.5 for $f(x,y,w,q) = \tilde{f}(x,y,w(0,0),w,q)$ and $\sigma(x,\zeta) = \tilde{\sigma}(x,\zeta(0),\|\zeta\|_0)$.

Remark 5.11 *The results presented in this Section can be extended to weakly coupled systems of functional differential problems. Some quasi-monotonicity conditions are required in this case.*

5.3 Comparison theorems for mixed problems

We will estimate a function of several variables satisfying a functional differential inequality by a solution of ordinary functional differential problem. Let us consider the following set of assumptions.

Assumption H $[\sigma,\Lambda]$**.** Suppose that

1) $\Lambda = (\lambda_1,\dots,\lambda_n) : E \to R^n$ is a function such that

$$\lambda_i(x,y) \ge 0 \text{ for } 1 \le i \le \kappa, \quad \lambda_i(x,y) \le 0 \text{ for } \kappa+1 \le i \le n,$$

2) the function $\sigma : [0,a) \times C([-r_0,0],R_+) \to R_+$ satisfies the following condition of monotonicity: if $\zeta,\bar{\zeta} \in C([-r_0,0],R_+)$, $\zeta(t) \le \bar{\zeta}(t)$ on $[-r_0,0]$ and $\zeta(0) = \bar{\zeta}(0)$ then $\sigma(x,\zeta) \le \sigma(x,\bar{\zeta})$,

3) σ is continuous and for every $\eta \in C([-r_0,0],R_+)$ there exists on $[-r_0,a)$ the right-hand maximum solution of the problem

$$\omega'(x) = \sigma(x,\omega_{(x)}), \quad \omega(x) = \eta(x) \text{ on } [-r_0,0]. \tag{5.16}$$

Let us state now a comparison lemma which will be used in the following.

Lemma 5.12 *Suppose that conditions 2), 3) of Assumption $H[\sigma, \Lambda]$ are satisfied and moreover*

1) let $\tilde{\omega} \in C([-r_0, a), R_+)$ and $\eta \in C([-r_0, 0], R_+)$ be such that $\tilde{\omega}(x) \leq \eta(x)$ on $[-r_0, 0]$,

2) denoted by $\omega(\cdot, \eta)$ the right-hand maximum solution of (5.16) and by $J_+ = \{x \in (0, a) : \tilde{\omega}(x) > \omega(x, \eta)\}$ suppose that

$$D_- \tilde{\omega}(x) \leq \sigma(x, \tilde{\omega}_{(x)}) \quad for \quad x \in J_+.$$

Then we have $\tilde{\omega}(x) \leq \omega(x, \eta)$ for $x \in [-r_0, a)$.

PROOF. Let $0 < a_0 < a$ be fixed. Consider the Cauchy problem

$$\omega'(x) = \sigma(x, \omega_{(x)}) + \varepsilon, \quad \omega(x) = \eta(x) + \varepsilon \text{ on } [-r_0, 0].$$

There is $\varepsilon > 0$ such that for $0 < \varepsilon < \varepsilon_0$ the maximum solution $\omega(\cdot, \eta, \varepsilon)$ of the above problem is defined on $[-r_0, a_0)$ and $\lim_{\varepsilon \to 0} \omega(x, \eta, \varepsilon) = \omega(x, \eta)$ uniformly on $[-r_0, a_0)$. It is easy to prove that $\tilde{\omega}(x) < \omega(x, \eta, \varepsilon)$ on $[-r_0, a)$ for $0 < \varepsilon < \varepsilon_0$. Now, passing to the limit for $\varepsilon \to 0$ we obtain $\tilde{\omega}(x) \leq \omega(x, \eta)$ on $[0, a_0)$. By the arbitrariness of $a_0 \in (0, a)$, the assertion follows.

Theorem 5.13 *Suppose that Assumption $H[\sigma, \Lambda]$ is satisfied. Let $u \in C(E^*, R)$ be a function such that*

1) the derivatives $D_x u$, $D_y u$ exist on E,

2) $\eta \in C([-r_0, 0], R_+)$ is nondecreasing and

$$|u(x, y)| \leq \eta(x) \quad on \quad E_0, \qquad |u(x, y)| \leq \eta(0) \quad on \quad \partial_0 E, \tag{5.17}$$

3) the functional differential inequality is satisfied on E

$$\left| D_x u(x, y) - \sum_{i=1}^{n} \lambda_i(x, y) D_{y_i} u(x, y) \right| \leq \sigma(x, V u_{(x,y)}). \tag{5.18}$$

Under these assumptions we have

$$|u(x, y)| \leq \omega(x, \eta) \quad for \quad (x, y) \in E \tag{5.19}$$

where $\omega(\cdot, \eta)$ is the right-hand maximum solution of problem (5.16).

PROOF. Let us define

$$\tilde{\omega}(x) = \max\{|u(x, y)| : \; y \in [-b', c'] \times [-c'', b'']\}, \; x \in [-r_0, a). \tag{5.20}$$

Then $\tilde{\omega} \in C([-r_0, a), R_+)$ and the estimate (5.19) is equivalent to

$$\tilde{\omega}(x) \leq \omega(x, \eta), \quad x \in [-r_0, a). \tag{5.21}$$

We will prove relation (5.21) by using Lemma 5.12. Note that $\tilde{\omega}(x) \leq \eta(x)$ for $x \in [-r_0, 0]$, in force of estimate (5.17). Thus, put $J_+ = \{x \in (0, a) : \; \tilde{\omega}(x) > \omega(x, \eta)\}$, we have to prove that

$$D_- \tilde{\omega}(x) \leq \sigma(x, \tilde{\omega}_{(x)}) \text{ for } x \in J_+. \tag{5.22}$$

Let $\tilde{x} \in J_+$ be fixed; then $\tilde{\omega}(\tilde{x}) > 0$ and there is $\tilde{y} = (\tilde{y}_1, \dots, \tilde{y}_n) \in [-b', c'] \times [-c'', b'']$ such that $\tilde{\omega}(\tilde{x}) = |u(\tilde{x}, \tilde{y})|$. Let I^+, I^-, I^0 be the sets of integers (possibly empty) such that $I^+ \cup I^- \cup I^0 = \{1, \dots, n\}$ and

$$\tilde{y}_i \geq b_i \text{ for } i \in I^+, \quad \tilde{y}_i \leq -b_i \text{ for } i \in I^-,$$

$$-b_i < \tilde{y}_i < b_i \text{ for } i \in I_0.$$

Since $(\tilde{x}, \tilde{y}) \notin \partial_0 E$, then we have that $\{1, \dots, \kappa\} \cap I^+ = \emptyset$, and $\{\kappa+1, \dots, n\} \cap I^- = \emptyset$. It follows that, either (i) $\tilde{\omega}(\tilde{x}) = u(\tilde{x}, \tilde{y})$ or (ii) $\tilde{\omega}(\tilde{x}) = -u(\tilde{x}, \tilde{y})$. If the case (i) holds, we have

$$D_{y_i} u(\tilde{x}, \tilde{y}) \geq 0 \text{ for } i \in I^+, \quad D_{y_i} u(\tilde{x}, \tilde{y}) \leq 0 \text{ for } i \in I^-,$$

$$D_{y_i} u(\tilde{x}, \tilde{y}) = 0 \quad \text{for } i \in I_0.$$

From (5.18), (5.20) and observing that $u(x, y) \leq \tilde{\omega}(x)$ for $x \in [-r_0, \tilde{x}]$, $y \in [-b', c'] \times [-c'', b'']$, we have

$$D_- \tilde{\omega}(\tilde{x}) \leq \sigma(\, \tilde{x}, V u_{(\tilde{x}, \tilde{y})}\,) + \sum_{i=1}^{n} \lambda_i(\tilde{x}, \tilde{y})\, D_{y_i} u(\tilde{x}, \tilde{y}) \leq \sigma(\, \tilde{x}, \tilde{\omega}_{(\tilde{x})}\,),$$

which proves (5.22) in the case (i). If the condition $\tilde{\omega}(\tilde{x}) = -u(\tilde{x}, \tilde{y})$ is satisfied, then (5.22) can be proved again in an analogous way. Thus, an application of Lemma 5.12 completes the proof.

Let us consider now two problems: problem (5.1), (5.2) and the following one

$$D_x z(x, y) = \tilde{f}(\, x, y, z_{(x,y)}, D_y z(x, y)\,), \tag{5.23}$$

$$z(x, y) = \tilde{\varphi}(x, y) \text{ for } (x, y) \in E_0 \cup \partial_0 E, \tag{5.24}$$

where $\tilde{f} : \Omega \to R$ and $\tilde{\varphi} : E_0 \cup \partial_0 E \to R$ are given functions. In this situation the following result can be proved.

Theorem 5.14 *Suppose that conditions 2) and 3) of Assumption H $[\sigma, \Lambda]$ hold and moreover suppose that*

1) the following estimates are satisfied

$$|\, f(x, y, w, q) - \tilde{f}(x, y, \bar{w}, q)\,| \leq \sigma(\, x, V(w - \bar{w})\,)$$

on Ω and

$$|\,(\varphi - \tilde{\varphi})(x, y)\,| \leq \eta(x) \text{ on } E_0, \quad |\,(\varphi - \tilde{\varphi})(x, y)\,| \leq \eta(0) \text{ on } \partial_0 E,$$

where $\eta \in C([-r_0, 0], R_+)$ is a nondecreasing function,

2) the derivatives $(\, D_{q_1} f, \dots, D_{q_n} f\,) = D_q f$ exist on Ω, $D_q f \in C(\Omega, R^n)$ and conditions (5.3), (5.4) are satisfied,

3) u and v are solutions of (5.1), (5.2) and (5.23), (5.24) respectively.

Under these assumptions we have

$$|\,(u - v)(x, y)\,| \leq \omega(x, \eta) \text{ on } E \tag{5.25}$$

where $\omega(\,\cdot\,, \eta)$ is the right-hand maximum solution of (5.16).

PROOF. If we put $\tilde{z}(x,y) = (u-v)(x,y)$ on E^*, then $|\tilde{z}(x,y)| \leq \eta(x)$ on E_0 and $|\tilde{z}(x,y)| \leq \eta(0)$ on $\partial_0 E$. Moreover, the functional differential inequality is satisfied

$$\left| D_x\tilde{z}(x,y) - \sum_{i=1}^{n} \int_0^1 D_{q_i} f(\,P(x,y,t)\,)dt\; D_{y_i}\tilde{z}(x,y) \right| \leq \sigma(\,x, V\tilde{z}_{(x,y)}\,),$$

where $(x,y) \in E$ and $P(x,y,t) = (\,x,y,u_{(x,y)}, D_y v(x,y) + tD_y(u-v)(x,y)\,)$. Hence (5.25) follows by applying Theorem 5.13 to the function $\tilde{z}$.

Theorem 5.15 *Suppose that conditions 2) and 3) of Assumption $H\,[\sigma,\Lambda\,]$ hold and $\bar{\omega}(x) = 0,\ x \in [-r_0, a)$, is the right-hand maximum solution of problem (5.16) corresponding to $\eta(x) = 0,\ x \in [-r_0, 0]$. Moreover suppose that*

1) the function f satisfies condition 2) of Theorem 5.14 and the following estimation holds on Ω :

$$|\,f(x,y,w,q) - f(x,y,\bar{w},q)\,| \leq \sigma(\,x, V(w-\bar{w})\,).$$

Under these assumptions, mixed problem (5.1), (5.2) admits one solution on E at the most.

The above uniqueness result is a consequence of Theorem 5.14 for $f = \tilde{f}$. Now we formulate a result on the continuous dependence of solutions on given functions.

Theorem 5.16 *Suppose that all the assumptions of Theorem 5.14 are satisfied and that the function $\bar{\omega}(x) = 0,\ x \in [-r_0, a)$, is the right-hand maximum solution of problem (5.16) corresponding to $\eta(x) = 0$ for $x \in [-r_0, 0]$. Let $0 < a_0 < a$ be fixed.*

Under these assumptions, for every $\varepsilon > 0$ there exists $\delta > 0$ such that, if $|\,f(x,y,w,q) - \tilde{f}(x,y,\bar{w},q)\,| < \delta$ on $(\,E \cap ((0,a_0] \times R^n)\,) \times C(B,R) \times R^n$ and $|\,(\varphi - \tilde{\varphi})(x,y)\,| < \delta$ on $(\,E_0 \cup \partial_0 E\,) \cap (\,[-r_0, a_0] \times R^n\,)$, then $|\,(u-v)(x,y)\,| < \varepsilon$ for $(x,y) \in E \cap (\,[-r_0, a_0] \times R^n\,)$.

The above theorem follows from Theorem 5.13.

Remark 5.17 *The results presented in this Section can be extended to weakly coupled systems of functional differential problems. Some quasi-monotonicity conditions are required in this case.*

5.4 CHAPLYGIN METHOD FOR MIXED PROBLEMS

Suppose that $f = (\,f_1, \ldots, f_n\,) : E \to R$, $g : E \times C(B,R) \to R$, $\phi : E_0 \cup \partial_0 E \to R$ are given functions. Consider the almost linear equation

$$D_x z(x,y) = g(\,x,y,z_{(x,y)}\,) + \sum_{i=1}^{n} f_i(x,y)\, D_{y_i} z(x,y) \tag{5.26}$$

with the initial-boundary condition

$$z(x,y) = \phi(x,y) \quad \text{on } E_0 \cup \partial_0 E. \tag{5.27}$$

In this Section we give sufficient condition for the existence of two monotone sequences $\{u^{(m)}\}$, $\{v^{(m)}\}$ such that if $\bar{z}$ is a solution of problem (5.26), (5.27) then $u^{(m)} \leq \bar{z} \leq v^{(m)}$ on E and $\{u^{(m)}\}$, $\{v^{(m)}\}$ are uniformly convergent to $\bar{z}$. The convergence that we get is of the Newton type, which means that

$$0 \leq (\bar{z} - u^{(m)})(x,y) \leq \frac{2A}{2^{2^m}} \quad \text{and} \quad 0 \leq (v^{(m)} - \bar{z}) \leq \frac{2A}{2^{2^m}} \text{ on } E$$

where A is a constant not dependent on m. The functions $u^{(m)}$ and $v^{(m)}$ are defined as solutions of some linear functional differential equations obtained by the linearization of (5.26) with respect to the functional variable. This method of approximating of solutions of differential equations was introduced by Chaplygin in [49]. In the original Chaplygin method only one approximating sequence was defined (cf. [49, 158]). This method was applied by W. Mlak and E. Schechter [166] to the system of the first-order semilinear partial differential equations and was extended by M. Nowotarska [170] to the case of an infinite system.

Note that because the model of functional dependence in (5.26) is based on the use of the operator $z_{(x,y)}$, the given function g is a functional operator on $C(B,R)$ with respect to the last variable. The Chaplygin method for partial functional differential equations with another model of functional dependence was considered in [104] but linearization only with respect to the nonfunctional argument was allowed here. The ideas presented in [104] were generalized on parabolic functional differential problems by S. Brzychczy in [35, 37]. The Chaplygin method for parabolic functional differential problems with the linearization with respect to the functional variable was investigated in [152].

We prove in this Section that Chaplygin sequences are monotonic using theorems on functional differential inequalities. We also give the estimate of the difference between the exact solution of (5.26), (5.27) and its approximations using the comparison theorem for functional differential inequalities. This estimate is of the same order as in the nonfunctional case [166] but better that of [104]. This is due to the fact that in our method, unlike the method of [166], the linearization concerns the functional variable. The results of this Section are based on the paper [63]. We denote by $CL(B,R)$ the set of all linear and continuous real functions defined on $C(B,R)$ and by $\|\cdot\|_*$ the norm in the space $CL(B,R)$. Let $\bar{F}$ be the Niemycki operator corresponding to (5.26):

$$\bar{F}[z](x,y) = g(x,y,z_{(x,y)}) + \sum_{i=1}^{n} f_i(x,y)\, D_{y_i} z(x,y).$$

Assumption H $[f,g]$. Suppose that

1) the functions f, g are continuous and

$$f_i(x,y) \geq 0 \text{ for } 1 \leq i \leq \kappa, \quad f_i(x,y) \leq 0 \text{ for } \kappa+1 \leq i \leq n,$$

where $(x,y) \in E$,

2) the derivative $D_w g(x,y,w)$ exists on $E \times C(B,R)$ and

$$D_w g(x,y,w)\, h \geq 0 \text{ if } h \in C(B,R_+),$$

$$D_w g(x,y,w)\,h \geq D_w g(x,y,\bar{w})\,h \text{ if } \bar{w} \geq w \text{ and } h \in C(B,R_+).$$

Assumption H $[u,v]$. There are two functions $u,\ v \in C(E^*,R)$ such that

$$D_x u(x,y) \leq \bar{F}[\,u\,](x,y) \;\; D_x v(x,y) \geq \bar{F}[\,v\,](x,y) \text{ on } E$$

and $u(x,y) \leq \phi(x,y) \leq v(x,y)$ on $E_0 \cup \partial_0 E$.

For given functions $u,\ v \in C(E^*,R)$ define

$$G(\,\cdot\,;u) : E \times C(B,R) \to R, \quad H(\,\cdot\,;u,v) : E \times C(B,R) \to R$$

by

$$G(x,y,w,q;u) = g(x,y,u_{(x,y)}) + D_w g(x,y,u_{(x,y)})\,(w - u_{(x,y)}) + \sum_{i=1} f_i(x,y)\,q_i,$$

$$H(x,y,w,q;u,v) = g(x,y,v_{(x,y)}) + D_w g(x,y,u_{(x,y)})\,(w - v_{(x,y)}) + \sum_{i=1}^{n} f_i(x,y)\,q_i.$$

Now we prove a theorem which is essential in the proof of the monotonicity of the Chaplygin sequence.

Theorem 5.18 *Suppose that Assumptions* $H\,[\,f,g\,]$ *and* $H\,[\,u,v\,]$ *are satisfied and*

1) $\bar{z} : E^* \to R$ *is a solution of (5.26), (5.27),*

2) $\bar{u},\ \bar{v} : E^* \to R$ *are solutions of equations*

$$D_x z(x,y) = G(\,x,y,z_{(x,y)},D_y z(x,y);u\,) \tag{5.28}$$

and

$$D_x z(x,y) = H(\,x,y,z_{(x,y)},D_y z(x,y);u,v\,) \tag{5.29}$$

respectively,

3) the initial-boundary inequalities

$$u(x,y) \leq \bar{u}(x,y) \leq \bar{z}(x,y) \leq \bar{v}(x,y) \leq v(x,y) \text{ on } E_0 \cup \partial_0 E \tag{5.30}$$

are satisfied.

Then

$$u(x,y) \leq \bar{u}(x,y) \leq \bar{z}(x,y) \leq \bar{v}(x,y) \leq v(x,y) \text{ on } E \tag{5.31}$$

and functional differential inequalities

$$D_x \bar{u}(x,y) \leq \bar{F}[\,\bar{u}\,](x,y) \tag{5.32}$$

$$D_x \bar{v}(x,y) \geq \bar{F}[\,\bar{v}\,](x,y) \tag{5.33}$$

are satisfied for $(x,y) \in E$.

PROOF. We first prove that $u \le \bar{u}$ on E. Since u satisfies the functional differential inequality

$$D_x u(x,y) \le G(x,y,u_{(x,y)},D_y u(x,y);u), \quad (x,y) \in E,$$

and $\bar{u}$ is a solution of (5.28) we have our claim by (5.30) and Theorem 5.5. Our next goal is to prove that $\bar{u} \le \bar{z}$ on E. For any $(x,y) \in E$ we have

$$\bar{F}[\bar{u}](x,y) - D_x\bar{u}(x,y)$$

$$= g(x,y,\bar{u}_{(x,y)}) - g(x,y,u_{(x,y)}) - D_w g(x,y,u_{(x,y)})\,(\bar{u}_{(x,y)} - u_{(x,y)}) \ge 0.$$

This inequality is exactly (5.32). Since $\bar{z}$ is a solution of (5.26), (5.27) we get $\bar{u} \le \bar{z}$ on E. Analogously, since $\bar{v}$ is a solution of (5.29) and

$$D_x v(x,y) \ge H(x,y,v_{(x,y)},D_y v(x,y);u,v), \quad (x,y) \in E,$$

we have $\bar{v} \le v$ on E by (5.30 and Theorem 5.5. Finally in order to get $\bar{z} \le \bar{v}$ on E we need an auxiliary inequality $\bar{u} \le \bar{v}$ on E. The latter inequality follows from the fact that $\bar{v}$ is a solution of (5.29) and from the relation

$$H(x,y,\bar{u}_{(x,y)},D_y\bar{u}(x,y);u,v) - D_x\bar{u}(x,y)$$

$$= D_w g(x,y,y_{(x,y)} + \xi(v_{(x,y)} - u_{(x,y)}))\,(v_{(x,y)} - u_{(x,y)})$$
$$- D_w g(x,y,u_{(x,y)})\,(v_{(x,y)} - u_{(x,y)})$$

where $(x,y) \in E$, and $\xi \in (0,1)$. Since $u \le v$ on E^* we get from Assumption H $[f,g]$ that

$$D_x\bar{u}(x,y) \le H(x,y,\bar{u}_{(x,y)},D_y\bar{u}(x,y);u,v) \text{ on } E.$$

By initial estimate (5.30) and Theorem 5.5 we get $\bar{u} \le \bar{v}$ on E. With the help of this inequality we have

$$v_{(x,y)} + \xi(\bar{v}_{(x,y)} - v_{(x,y)}) \ge v_{(x,y)} \ge \bar{u}_{(x,y)} \ge u_{(x,y)}$$

for any $(x,y) \in E$, $\xi \in (0,1)$ and consequently, by the already proved relation $v \ge \bar{v}$ on E and Assumption H $[f,g]$ we get

$$\bar{F}[\bar{v}](x,y) - D_x\bar{v}(x,y) = D_w g(x,y,v_{(x,y)} + \xi(\bar{v}_{(x,y)} - v_{(x,y)}))\,(\bar{v}_{(x,y)} - v_{(x,y)})$$
$$- D_w g(x,y,u_{(x,y)} - v_{(x,y)}))\,(\bar{v}_{(x,y)} - v_{(x,y)}) \le 0 \text{ on } E,$$

which is (5.33). Since $\bar{z}$ is a solution of (5.26) we get the inequality $\bar{v} \ge \bar{z}$ on E using (5.30) and Theorem 5.5. This completes the proof of Theorem 5.18.

Now we formulate a monotone iterative method for problem (5.26), (5.27). Suppose that we have a sequence

$$\{u^{(m)},\ v^{(m)}\}, \quad u^{(m)},\ v^{(m)} : E^* \to R \tag{5.34}$$

such that

(i) the functions $u^{(0)}$, $v^{(0)}$ satisfy on E the functional differential inequalities

$$D_x u^{(0)}(x,y) \le \bar{F}[\,u^{(0)}\,](x,y),$$

$$D_x v^{(0)}(x,y) \ge \bar{F}[\,v^{(0)}\,](x,y),$$

(ii) for every $m \ge 1$ the function $u^{(m)}$ is a solution of the linear equation

$$D_x z(x,y) = G(\,x,y,z_{(x,y)},D_y z(x,y);u^{(m-1)}\,),$$

(iii) for every $m \ge 1$ the function $v^{(m)}$ is a solution of the linear equation

$$D_x z(x,y) = H(\,x,y,z_{(x,y)},D_y z(x,y);u^{(m-1)},v^{(m-1)}\,).$$

The sequence (5.34) satisfying the above conditions will be called the Chaplygin sequence for problem (5.26), (5.27). Now we give a theorem on the monotonicity of the Chaplygin sequence.

Theorem 5.19 *Suppose that Assumption H [f, g] is satisfied and the Chaplygin sequence (5.34) satisfies the initial-boundary estimates*

$$u^{(m-1)}(x,y) \le u^{(m)}(x,y) \le \bar{z}(x,y) \le v^{(m)}(x,y) \le v^{(m-1)}(x,y) \tag{5.35}$$

on $E_0 \cup \partial_0 E$ for $m \ge 1$ where $\bar{z} : E^* \to R$ is a solution of (5.26), (5.27).
Then we have for $m \ge 1$:

$$u^{(m-1)}(x,y) \le u^{(m)}(x,y) \le \bar{z}(x,y) \le v^{(m)}(x,y) \le v^{(m-1)}(x,y) \text{ on } E. \tag{5.36}$$

The above result follows from Theorem 5.18.

5.5 Approximate solutions

Now we give a theorem on the convergence of the Chaplygin sequence and on the estimate of the difference between the exact and approximate solutions of problem (5.26), (5.27).

Assumption H $[D_w g]$. Suppose that there are K_1, $K_2 \in R_+$ such that

$$\|D_w g(x,y,w)\|_* \le K_1$$

and

$$\|D_w g(x,y,w,) - D_w g(x,y,\bar{w})\|_* \le K_2 \|w - \bar{w}\|_0$$

on $E \times C(B,R)$.

Theorem 5.20 *Suppose that Assumptions H [f, g], H $[D_w g]$ are satisfied and*
1) $\bar{z} \to R$ is a solution of (5.26), (5.27) and the Chaplygin sequence (5.34) satisfies initial-boundary inequalities (5.35),

2) the functions $u^{(0)}$, $v^{(0)}$ *satisfy the condition* $|(u^{(0)}-v^{(0)})(x,y)| \le A$ *on* E^* *where* $A = \{4aK_2 \exp[K_1 a]\}^{-1}$,

3) the initial estimates

$$(v^{(m)} - u^{(m)})(x,y) \le A\left\{2^{2^m} \exp[K_1 a]\right\}^{-1}$$

hold true for $m \ge 1$.

Then we have

$$(v^{(m)} - u^{(m)})(x,y) \le \frac{2A}{2^{2^m}} \quad on \ E \tag{5.37}$$

for $m \ge 0$.

PROOF. We prove (5.37) by induction. It is immediately seen that (5.37) is satisfied for $m = 0$. Suppose that it holds for some $m \ge 1$. Then for all $(x,y) \in E$ we have

$$D_x[v^{(m+1)} - u^{(m+1)}](x,y) = \sum_{i=1}^{n} f_i(x,y)\, D_{y_i}(v^{(m+1)} - u^{(m+1)})(x,y)$$

$$+g(x,y,v^{(m)}_{(x,y)}) + D_w g(x,y,u^{(m)}_{(x,y)})(v^{(m+1)}_{(x,y)} - v^{(m)}_{(x,y)})$$

$$-g(x,y,u^{(m)}_{(x,y)}) - D_w g(x,y,u^{(m)}_{(x,y)})(u^{(m+1)}_{(x,y)} - u^{(m)}_{(x,y)}),$$

and consequently

$$D_x[v^{(m+1)} - u^{(m+1)}](x,y)$$

$$= \left[D_w g(x,y,u^{(m)}_{(x,y)} + \xi(v^{(m)}_{(x,y)} - u^{(m)}_{(x,y)})) - D_w g(x,y,u^{(m)}_{(x,y)})\right](v^{(m)}_{(x,y)} - u^{(m)}_{(x,y)})$$

$$+D_w g(x,y,u^{(m)}_{(x,y)})(v^{(m+1)}_{(x,y)} - u^{(m+1)}_{(x,y)})$$

$$+\sum_{i=1}^{n} f_i(x,y)\, D_{y_i}(v^{(m+1)} - u^{(m+1)})(x,y)$$

where $\xi \in (0,1)$. If for any $m \ge 0$ we write $\tilde{z}^{(m)} = v^{(m)} - u^{(m)}$ then by Assumption H $[D_w g]$ and by inductive hypothesis we have

$$\left| D_x \tilde{z}^{(m+1)}(x,y) - \sum_{i=1}^{n} f_i(x,y)\, D_{y_i} \tilde{z}^{(m+1)}(x,y) \right|$$

$$\le K_2 \| \tilde{z}^{(m)}_{(x,y)} \|_0^2 + K_1 \| \tilde{z}^{(m+1)}_{(x,y)} \|_0 \le K_2 \left(\frac{2A}{2^{2^m}}\right)^2 + K_1 \| \tilde{z}^{(m+1)}_{(x,y)} \|_0 .$$

The above estimate together with the initial estimates and Theorem 5.13 yield

$$\left| \tilde{z}^{(m+1)}(x,y) \right| \le A\left\{2^{2^m} \exp[K_1 a]\right\}^{-1} \exp[K_1 x]$$

$$+K_2 \left(\frac{2A}{2^{2^m}}\right)^2 \int_0^x \exp[K_1(x-t)]$$

$$\le \frac{A}{2^{2^{m+1}}} + \frac{4A^2 K_2 a \exp[K_1 a]}{2^{2^{m+1}}} = \frac{2A}{2^{2^{m+1}}}.$$

Thus Theorem 5.20 follows by induction.

Remark 5.21 *If all the assumptions of Theorem 5.20 are satisfied then we have the following error estimates:*

$$(\,\tilde{z}-u^{(m)}\,)(x,y)\le\frac{2A}{2^{2^m}}\quad\text{and}\quad(\,v^{(m)}-\tilde{z}\,)(x,y)\le\frac{2A}{2^{2^m}}$$

for $(x,y)\in E,\ m\ge 0$.

5.6 Difference methods for mixed problems

Let $E,\ E_0,\ \partial_0E,\ E^*,\ B$ and Ω be the sets defined in Section 5.1. Suppose that $f:\Omega\to R$ and $\varphi:E_0\cup\partial_0E\to R$ are given functions. Consider the mixed problem

$$D_xz(x,y)=f(\,x,y,z_{(x,y)},D_yz(x,y)\,)\tag{5.38}$$

$$z(x,y)=\varphi(x,y)\quad\text{on}\quad E_0\cup\partial_0E.\tag{5.39}$$

We formulate a difference problem corresponding to (5.38), (5.39). We define a mesh in E^* and B. Suppose that $h=(h_0,h')$ where $h'=(h_1,\ldots,h_n)$ stand for steps of the mesh. For $(i,m)\in\mathbf{Z}^{1+n}$ where $m=(m_1,\ldots,m_n)$ we define $y^{(m)}$ by $y^{(m)}=(\,y^{(m_1)},\ldots,y^{(m_n)}\,)=m*h$ and $x^{(i)}=ih_0$. Denote by Δ the set of all h such that there are $M=(M_1,\ldots,M_n)\in\mathbf{Z}^n$ and $M_0\in\mathbf{Z}$ such that $M*h'=r$ and $M_0h_0=r_0$. We assume that $\Delta\ne\emptyset$ and that there is a sequence $\{h^{(j)}\}$, $h^{(j)}\in\Delta$, such that $\lim_{j\to\infty}h^{(j)}=0$. Let

$$R_h^{1+n}=\{\,(x^{(i)},y^{(m)}):\ \ (i,m)\in\mathbf{Z}^{1+n}\,\}.$$

We define the sets

$$E_{0.h}=E\cap R_h^{1+n},\ \ B_h=B\cap R_h^{1+n},\ \ E_h=E\cap R_h^{1+n},$$

and

$$\partial_0E_h=\partial_0E\cap R_h^{1+n},\quad E_h^*=E_{0.h}\cup E_h\cup\partial_0E_h.$$

Elements of the set E_h^* will be denoted by $(x^{(i)},y^{(m)})$ or (x,y). For a function $z:E_h^*\to R$ and a point $(x^{(i)},y^{(m)})\in E_h^*$ we write $z^{(i,m)}=z(x^{(i)},y^{(m)})$ and

$$\|z\|_{i.h}=\max\{\,|z^{(j,m)}|:\ \ (x^{(j)},y^{(m)})\in E_h^*,\ \ j\le i\,\}.$$

For the above z and for a point $(x^{(i)},y^{(m)})\in E_h$ we define the function $z_{(i,m)}:B_h\to R$ by the formula

$$z_{(i,m)}(t,s)=z(x^{(i)}+t,y^{(m)}+s),\ \ (t,s)\in B_h.$$

Write $y^{(m')}=(y^{(m)})'$ and $y^{(m'')}=(y^{(m)})''$. The function $z_{(i,m)}$ is the restriction of z to the set $(\,[x^{(i)}-r_0,x^{(i)}]\times[y^{(m')},y^{(m')}+r']\times[y^{(m'')}-r'',y^{(m'')}]\,)\cap R_h^{1+n}$ and this restriction is shifted to the set B_h. Let $\delta_0,\ \delta=(\,\delta_1,\ldots,\delta_n\,)$ be the difference operators given by

$$\delta_0=\frac{1}{h_0}\left(z^{(i+1,m)}-z^{(i,m)}\right),$$

$$\delta_j z^{(i,m)} = \frac{1}{h_j}\left(z^{(i,m+e_j)} - z^{(i,m)}\right) \quad \text{for } 1 \le j \le \kappa,$$

$$\delta_j z^{(i,m)} = \frac{1}{h_j}\left(z^{(i,m)} - z^{(i,m-e_j)}\right) \quad \text{for } \kappa+1 \le j \le n.$$

Write $\Omega_h = E_h \times \mathbf{F}(B_h, R) \times R^n$ and suppose that $f_h : \Omega_h \to R$, $\varphi_h : E_{0.h} \cup \partial_0 E_{0.h} \to R$, are given functions. Consider the functional difference equation with the initial-boundary condition

$$\delta_0 z^{(i,m)} = f_h(\, x^{(i)}, y^{(m)}, z_{(i,m)}, \delta z^{i,m)}\,) \tag{5.40}$$

$$z^{(i,m)} = \varphi_h^{(i,m)} \quad \text{on } E_{0.h} \cup \partial_0 E_{0.h}. \tag{5.41}$$

We will approximate solutions of (5.38), (5.39) by solutions of (5.40), (5.41).

5.7 Functional difference equations with mixed conditions

We start with a theorem on the error estimate of approximate solutions for a general class of functional difference equations. Let

$$E_h' = \{\, (x^{(i)}, y^{(m)}) \in E_h : \ (x^{(i+1)}, y^{(m)}) \in E_h \,\}.$$

Suppose that $F_h : E_h' \times \mathbf{F}(B_h, R) \to R$ is a given operator. For $w \in \mathbf{F}(B_h, R)$ and $(x^{(i)}, y^{(m)})$ we write $F_h[\,(i,m), w\,] = F_h(x^{(i)}, y^{(m)}, w)$. For given functions $\varphi_h : E_{0.h} \cup \partial_0 E_{0.h} \to R$ we consider the functional difference equation

$$z^{(i+1,m)} = F_h[\,(i,m), z_{(i,m)}\,] \tag{5.42}$$

with initial-boundary condition (5.41). It is easy to see that there exists exactly one solution $u_h : E_h^* \to$ of problem (5.41), (5.42). There is $N_0 \in \mathbf{N}$ such that $N_0 h_0 < a \le (N_0 + 1) h_0$. Write

$$I_{0.h} = \{\, x^{(i)} : \ -M_0 \le i \le 0 \,\}, \ \ I_h = \{\, x^{(i)} : \ 0 \le i \le N_0 \,\},$$

and $I_h' = I_h \setminus \{\, x^{(N_0)} \,\}$. For a function $\eta : I_{0.h} \cup I_h \to R$ we put $\eta^{(i)} = \eta(x^{(i)})$, $x^{(i)} \in I_{0.h} \cup I_h$. For the above η and for a point $x^{(i)} \in I_h$ we define the function $\eta_{(i)} : I_{0.h} \to R$ by

$$\eta_{(i)}(t) = \eta(x^{(i)} + t), \ \ t \in I_{0.h}.$$

Suppose that we have a function $\sigma_h : I_h' \times \mathbf{F}(I_{0.h}, R_+) \to R_+$. If $(x^{(i)}, \omega) \in I_h' \times \mathbf{F}(I_{0.h}, R_+)$ then we write $\sigma_h[\,i, \omega\,] = \sigma(\, x^{(i)}, \omega\,)$. In the sequel we will need the following operator $V_h : \mathbf{F}(B_h, R) \to \mathbf{F}(I_{0.h}, R_+)$. If $w \in \mathbf{F}(B_h, R)$ and $x^{(i)} \in I_{0.h}$ then we put

$$(V_h w)\,(x^{(i)}) = \max\{\, |w^{(i,m)}| : \ (x^{(i)}, y^{(m)}) \in B_h \,\}.$$

Suppose that $v_h : E_h^* \to R$ and $\gamma_{0.h} : I_{0.h} \cup I_h \to R_+$, $\gamma_h : I_h' \to R_+$ are such functions that

$$\left| v_h^{(i+1,m)} - F_h[\,(i,m),(v_h)_{(i,m)}\,] \right| \le \gamma_h^{(i)} \quad \text{on } E_h' \tag{5.43}$$

and

$$\left| v_h^{(i,m)} - \varphi_h^{(i,m)} \right| \le \gamma_{0.h} \text{ on } E_{0.h} \cup \partial_0 E_{0.h}. \tag{5.44}$$

The function v_h satisfying the above relations is considered as an approximate solution of problem (5.41), (5.42). We give a theorem on the estimate of the difference between the exact and approximate solutions of (5.41), (5.42).

Theorem 5.22 *Suppose that*

1) the operators $F_h : E_h' \times \mathbf{F}(B_h, R) \to R$ and $\sigma_h : I_h' \times \mathbf{F}(I_{0.h}, R_+) \to R_+$ satisfy the conditions:

(i) σ_h is nondecreasing with respect to the functional variable,

(ii) the estimate

$$|\,F_h[\,(i,m),w\,] - F_h[\,(i,m),\bar{w}\,]\,| \le \sigma_h[\,i, V_h(w-\bar{w})\,] \tag{5.45}$$

holds on $E_h' \times \mathbf{F}(B_h, R)$,

2) $\varphi_h : E_{0.h} \cup \partial_0 E_{0.h} \to R$ is given and $u_h : E_h^* \to R$ is a solution of problem (5.42), (5.41),

3) the functions $v_h : E_h^* \to R_+$ and $\gamma_h : I_h' \to R_+$, $\gamma_{0.h} : I_{0.h} \cup I_h \to R_+$ are such that estimates (5.43), (5.44) hold,

4) the function $\beta_h : I_{0.h} \cup I_h \to R_+$ satisfies the recurrent inequality

$$\beta_h^{(i+1)} \ge \sigma_h[\,i,(\beta_h)_{(i)}\,] + \gamma_h^{(i)}, \quad 0 \le i \le N_0 - 1, \tag{5.46}$$

and

$$\beta_h^{(i)} \ge \gamma_{0.h}^{(i)} \quad \text{on } I_{0.h} \cup I_h. \tag{5.47}$$

Under these assumptions we have

$$\left| u_h^{(i,m)} - v_h^{(i,m)} \right| \le \beta_h^{(i)} \tag{5.48}$$

on E_h.

PROOF. We prove assertion (5.48) by induction on i. It follows from (5.44), (5.47) that inequality (5.48) holds true for $i = 0$. Suppose that estimate (5.48) is satisfied for

$$0 \le i \le j,\ (x^{(i)}, y^{(m)}) \in E_h.$$

Then using assumption (5.47) we obtain $V_h(u_h - v_h)_{(j)} \le (\beta_h)_{(j)}$. It follows from (5.45) that

$$\begin{aligned} &\left| u_h^{(j+1,m)} - v_h^{(j+1,m)} \right| \\ &\le \left|\,F_h[\,(j,m),(u_h)_{(j,m)}\,] - F_h[\,(j,m),(v_h)_{(j,m)}\,]\,\right| \\ &\quad + \left| F_h[\,(j,m),(v_h)_{j,m)}\,] - v_h^{(j+1,m)} \right| \\ &\le \sigma_h[\,j, V_h(u_h - v_h)_{(j)}\,] + \gamma_h^{(j)} \le \sigma_h[\,j,(\beta_h)_{(j)}\,] + \gamma_h^{(j)} \le \beta_h^{(j+1)}, \end{aligned}$$

where $(x^{(j+1)}, y^{(m)}) \in E_h$. Hence, the proof is completed by induction.

Remark 5.23 *If all the assumptions of Theorem 5.22 are satisfied and the function β_h is nondecreasing on $I_{0.h} \cup I_h$ then we have the estimate $\|u_h - v_h\|_{i.h} \le \beta_h^{(i)}$ for $0 \le i \le N_0$. In applications we will consider such functions β_h satisfying (5.46) that $\lim_{h\to 0} \beta_h^{(i)} = 0$ for $0 \le i \le N_0$. This implies $\lim_{h\to 0} \|u_h - v_h\|_{i.h} = 0$ for $0 \le i \le N_0$.*

5.8 Convergence of difference methods

We will need the following assumption on a comparison function σ_h.

Assumption H $[\sigma_h]$. Suppose that the function $\sigma_h : I_h' \times \mathbf{F}(I_{0.h}, R_+) \to R_+$ satisfies the conditions:

1) σ_h is nondecreasing with respect to the functional variable,

2) $\sigma_h[i, \theta_h] = 0$ for $0 \le i \le N_0 - 1$ where $\theta_h^{(j)} = 0$ on $I_{0.h}$ and the difference problem

$$\eta^{(i+1)} = \eta^{(i)} + h_0 \sigma_h[i, \eta_{(i)}] \text{ for } 0 \le i \le N_0 - 1, \tag{5.49}$$

$$\eta^{(i)} = 0 \quad \text{on} \quad I_{0.h} \tag{5.50}$$

is stable in the following sense: if $\gamma,\ \alpha_0 : \Delta \to R_+$ are such functions that $\lim_{h\to 0} \gamma(h) = 0,\ \lim_{h\to 0} \alpha_0(h) = 0$ and $\eta_h : I_{0.h} \cup I_h \to R_+$ is a solution of the problem

$$\eta^{(i+1)} = \eta^{(i)} + h_0 \sigma_h[i, \eta_{(i)}] + h_0 \gamma(h) \text{ for } 0 \le i \le N_0 - 1, \tag{5.51}$$

$$\eta^{(i)} = \alpha_0(h) \quad \text{on} \quad I_{0.h} \tag{5.52}$$

then there exists a function $\beta : \Delta \to R_+$ such that $\lim_{h\to 0} \beta(h) = 0$ and $\eta_h^{(i)} \le \beta(h)$ on I_h.

Assumption H $[f_h, \sigma_h]$. Suppose that the function $f_h : \Omega_h \to R_+$ satisfies the conditions:

1) the derivatives $(D_{q_1} f_h, \ldots, D_{q_n} f_h) = D_q f_h$ exist on Ω_h and $D_q f(x, y, w, \cdot) \in C(R^n, R^n)$,

2) for each $P = (x, y, w, q) \in \Omega_h$ we have

$$D_{q_i} f_h(P) \ge 0 \text{ for } 1 \le i \le \kappa, \qquad D_{q_i} f_h(P) \le 0 \text{ for } \kappa + 1 \le i \le n,$$

and

$$1 - h_0 \sum_{j=1}^{n} \frac{1}{h_j} \,|\, D_{q_j} f_h(P) \,| \ge 0,$$

3) the function σ_h satisfies Assumption H $[\sigma_h]$ and

$$\left| f_h(x^{(i)}, y^{(m)}, w, q) - f_h(x^{(i)}, y^{(m)}, \bar{w}, q) \right| \le \sigma_h[i, V_h(w - \bar{w})] \quad \text{on } \Omega_h.$$

Now we prove a theorem on the convergence of the method (5.40), (5.41).

Theorem 5.24 *Suppose that Assumption $H\,[f_h, \sigma_h]$ is satisfied and*

1) the function $u_h : E_h^* \to R$ is a solution of problem (5.40), (5.41) and there exists a function $\alpha_0 : \Delta \to R_+$ such that

$$\left| \varphi^{(i,m)} - \varphi_h^{(i,m)} \right| \leq \alpha_0(h) \text{ on } E_{0.h} \cup \partial_0 E_{0.h}, \tag{5.53}$$

and $\lim_{h\to 0} \alpha_0(h) = 0$,

2) the function $v : E^* \to R$ is a solution of problem (5.38), (5.39), the derivatives $D_x v$, $D_y v$ are continuous and bounded on E,

3) there exists a function $\tilde{\beta} : \Delta \to R_+$ such that

$$\left| f_h(x^{(i)}, y^{(m)}, (v_h)_{(i,m)}, \delta v_h^{(i,m)}) - f(x^{(i)}, y^{(m)}, v_{(x^{(i)}, y^{(m)})}, \delta v^{(i,m)}) \right| \leq \tilde{\beta}(h)$$

on E_h' and $\lim_{h\to 0} \tilde{\beta}(h) = 0$ where v_h is the restriction of v to the set E_h^*.

Then there exists a function $\beta : \Delta \to R_+$ such that

$$\left| u_h^{(i,m)} - v_h^{(i,m)} \right| \leq \beta(h) \quad \text{on } E_h \tag{5.54}$$

and $\lim_{h\to 0} \beta(h) = 0$.

PROOF. We apply Theorem 5.22 to prove (5.54). Let the function $\Gamma_h : E_h' \to R$ be defined by

$$\delta_0 v_h^{(i,m)} = f_h(x^{(i)}, y^{(m)}, (v_h)_{(i,m)}, \delta v_h^{(i,m)}) + \Gamma_h^{(i,m)}.$$

It follows from assumptions 2), 3) that there exists a function $\gamma : \Delta \to R_+$ such that $\left| \Gamma_h^{(i,m)} \right| \leq \gamma(h)$ on E_h' and $\lim_{h\to 0} \gamma(h) = 0$. We define the operator $F_h : E_h' \times \mathbf{F}(B_h, R) \to R$ as follows:

$$F_h[(i,m), w)] = w^{(0,\theta)} + h_0 f_h(x^{(i)}, y^{(m)}, w, \delta w^{(0,\theta)})$$

where $\theta = (0, \ldots, 0) \in R^n$. Then u_h satisfies the relation

$$u_h^{(i+1,m)} = F_h[(i,m), (u_h)_{(i,m)}] \text{ for } (x^{(i)}, y^{(m)}) \in E_h',$$

and

$$\left| v_h^{(i+1,m)} - F_h[(i,m), (v_h)_{(i,m)}] \right| \leq h_0 \gamma(h), \text{ for } (x^{(i)}, y^{(m)}) \in E_h'.$$

Now we estimate the difference $F_h[(i,m), w] - F_h[(i,m), \bar{w}]$ on $E_h' \times \mathbf{F}(B_h, R)$. It follows from conditions 2), 3) of Assumption H $[f_h, \sigma_h]$ that

$$| F_h[(i,m), w] - F_h[(i,m), \bar{w}] |$$

$$\leq \left| (w - \bar{w})^{(0,\theta)} + h_0 \left[f_h(x^{(i)}, y^{(m)}, w, \delta w^{(0,\theta)}) - f_h(x^{(i)}, y^{(m)}, w, \delta \bar{w}^{(0,\theta)}) \right] \right|$$

$$+ h_0 \left| \left[f_h(x^{(i)}, y^{(m)}, w, \delta \bar{w}^{(0,\theta)}) - f_h(x^{(i)}, y^{(m)}, \bar{w}, \delta \bar{w}^{(0,\theta)}) \right] \right|$$

$$\leq \Bigg| (w-\bar{w})^{(0,\theta)} + h_0 \sum_{j=1}^{\kappa} D_{q_j} f_h(Q) \frac{1}{h_j} \Big[(w-\bar{w})^{(0,\theta+e_j)} - (w-\bar{w})^{(0,\theta)} \Big]$$

$$+ h_0 \sum_{j=\kappa+1}^{n} D_{q_j} f_h(Q) \frac{1}{h_j} \Big[(w-\bar{w})^{(0,\theta)} - (w-\bar{w})^{(0,\theta-e_j)} \Big] \Bigg|$$

$$+ h_0\, \sigma_h(\, x^{(i)}, V_h(w-\bar{w})\,).$$

Then we have

$$| F_h[(i,m),w] - F_h[(i,m),\bar{w}] |$$

$$\leq \Bigg| (w-\bar{w})^{(0,\theta)} \Bigg[1 - h_0 \sum_{j=1}^{n} \frac{1}{h_j} | D_{q_j} f_h(Q) | \Bigg] \Bigg|$$

$$+ h_0 \sum_{j=1}^{\kappa} \frac{1}{h_j} D_{q_j} f_h(Q) \Big| (w-\bar{w})^{(0,\theta+e_j)} \Big| - h_0 \sum_{j=\kappa+1}^{n} \frac{1}{h_j} D_{q_j} f_h(Q) \Big| (w-\bar{w})^{(0,\theta-e_j)} \Big|$$

$$+ h_0\, \sigma_h(\, x^{(i)}, V_h(w-\bar{w})\,),$$

where Q is an intermediate point. We conclude that the operator f_h satisfies the condition

$$| F_h[(i,m),w] - F_h[(i,m),\bar{w}] | \leq V_h(w-\bar{w})^{(0)} + h_0 \sigma_h[\, i, V_h(w-\bar{w})\,]$$

on $E'_h \times \mathbf{F}(B_h, R)$. Consider the problem (5.51), (5.52) and its solution $\eta_h : I_{0.h} \cup I_h \to R_+$. It follows from Theorem 5.22 that

$$\Big| u_h^{(i,m)} - v_h^{(i,m)} \Big| \leq \eta_h^{(i)} \quad \text{on } E'_h. \tag{5.55}$$

Now we obtain the assertion of the Theorem 5.24 from the stability of problem (5.49), (5.50) (see Assumption H $[\sigma_h]$).

Remark 5.25 *Assumption 3) of the Theorem is called the compatibility condition of equation (5.40) with respect to (5.38).*

Remark 5.26 *If all the assumptions of Theorem 5.24 are satisfied then condition (5.55) gives an error estimate for the approximate solution u_h of (5.38), (5.39). In the next Section we give examples of functions σ_h for which problem (5.51), (5.52) can be solved and the function η_h can be estimated.*

5.9 Interpolating operators

Now we define the operator $T_h : \mathbf{F}(B_h, R) \to \mathbf{F}(B, R)$ as follows. Put

$$S_* = \{\, (j,s) : \; j \in \{0,1\}, \;\; s = (s_1, \ldots, s_n), \; s_i \in \{0,1\} \;\text{ for }\; 1 \leq i \leq n \}.$$

Let $w \in \mathbf{F}(B_h, R)$ and $(x,y) \in B$. There exists $(x^{(i)}, y^{(m)}) \in B_h$ such that $x^{(i)} \leq x \leq x^{(i+1)}$, $y^{(m)} \leq y \leq y^{(m+1)}$ and $(x^{(i+1)}, y^{(m+1)}) \in B_h$ where $m+1 = (m_1 + 1, \ldots, m_n + 1)$. We define

$$(T_h w)(x,y) = \sum_{(j,s)\in S_*} w^{(i+j,m+s)} \left(\frac{Y - Y^{(i,m)}}{h}\right)^{(j,s)} \left(1 - \frac{Y - Y^{(i,m)}}{h}\right)^{1-(j,s)}$$

where

$$\left(\frac{Y - Y^{(i,m)}}{h}\right)^{(j,s)} = \left(\frac{x - x^{(i)}}{h_0}\right)^{j} \prod_{k=1}^{n} \left(\frac{y_k - y_k^{(m_k)}}{h_k}\right)^{s_k}$$

and

$$\left(1 - \frac{Y - Y^{(i,m)}}{h}\right)^{1-(j,s)} = \left(1 - \frac{x - x^{(i)}}{h_0}\right)^{1-j} \prod_{k=1}^{n} \left(1 - \frac{y_k - y_k^{(m_k)}}{h_k}\right)^{1-s_k}$$

and we take $0^0 = 1$ in the above formulas. Then we have defined $T_h : B \to R$. It is easy to see that $T_h w$ is continuous on B. Let $L_{h_0} : \mathbf{F}(I_{0.h}, R) \to \mathbf{F}([-r_0, 0], R)$ be the operator defined by

$$(L_{h_0}\omega)(x) = \omega^{(i+1)} \frac{x - x^{(i)}}{h_0} + \omega^{(i)} \left(1 - \frac{x - x^{(i)}}{h_0}\right)$$

where $x^{(i)} \leq x \leq x^{(i+1)}$, and $\omega \in \mathbf{F}(I_{0.h}, R)$. It is easy to see that L_{h_0} is a particular case of T_h. Namely, if we put $n = 0$ in the definition of T_h then we obtain L_{h_0}.

Theorem 5.27 *Suppose that the function $w : B \to R$ is of class C^2 and denote by w_h the restriction of w to the set B_h. Let $\tilde{C}$ be such a constant that*

$$|D_{xx}w(x,y)|, \ |D_{xy_k}w(x,y)|, \ \left|D_{y_j y_k}w(x,y)\right| \leq \tilde{C} \ \text{on} \ B \tag{5.56}$$

where j, $k = 1, \ldots, n$. Then

$$\|T_h w_h - w\|_0 \leq \tilde{C} \left[h_0^2 + 2h_0 \sum_{k=1}^{n} h_k + \sum_{j,k=1}^{n} h_j h_k \right]. \tag{5.57}$$

PROOF. Let $(x,y) \in B$ and $x^{(i)} \leq x \leq x^{(i+1)}$, $y^{(m)} \leq y \leq y^{(m+1)}$ where $(x^{(i)}, y^{(m)})$, $(x^{(i+1)}, y^{(m+1)}) \in B_h$. It is easy to prove by induction with respect to n that

$$\sum_{(j,s)\in S_*} \left(\frac{Y - Y^{(i,m)}}{h}\right)^{(j,s)} \left(1 - \frac{Y - Y^{(i,m)}}{h}\right)^{1-(j,s)} = 1, \tag{5.58}$$

and

$$\sum_{(j,s)\in S_*} \left(\frac{Y - Y^{(i,m)}}{h}\right)^{(j,s)} \left(1 - \frac{Y - Y^{(i,m)}}{h}\right)^{1-(j,s)} jh_0 = x - x^{(j)}, \tag{5.59}$$

$$\sum_{(j,s)\in S_*}\left(\frac{Y-Y^{(i,m)}}{h}\right)^{(j,s)}\left(1-\frac{Y-Y^{(i,m)}}{h}\right)^{1-(j,s)} s_k h_k = y_k - y_k^{(m_k)}, \tag{5.60}$$

where $k = 1, \ldots, n$. Put

$$A[(j,s),P] = w^{(i,m)} + jh_0 D_x w^{(i,m)} + \sum_{k=1}^{n} s_k h_k D_{y_k} w^{(i,m)} + \frac{1}{2}(jh_0)^2 D_{xx} w(P)$$

$$+ jh_0 \sum_{k=1}^{n} s_k h_k D_{xy_k} w(P) + \frac{1}{2}\sum_{k,k'=1}^{n} s_k h_k s_{k'} h_{k'} D_{y_k y_{k'}} w(P)$$

and

$$B[(x,y),Q] = w^{(i,m)} + (x - x^{(i)}) D_x w^{(i,m)} + \sum_{k=1}^{n} (y_k - y_k^{(m_k)}) D_{y_k} w^{(i,m)}$$

$$+ \frac{1}{2}(x - x^{(i)})^2 D_{xx} w(Q) + (x - x^{(i)}) \sum_{k=1}^{n} (y_k - y_k^{(m_k)}) D_{xy_k} w(Q)$$

$$+ \frac{1}{2}\sum_{k,k'=1}^{n} (y_k - y_k^{(m_k)})(y_{k'} - y_{k'}^{(m_{k'})}) D_{y_k y_{k'}} w(Q).$$

There are P, $Q \in B$ such that

$$(T_h w_h)(x,y) - w(x,y)$$

$$= \sum_{(j,s)\in S_*}\left(\frac{Y-Y^{(i,m)}}{h}\right)^{(j,s)}\left(1-\frac{Y-Y^{(i,m)}}{h}\right)^{1-(j,s)} A[(j,s),P] - B[(x,y),Q].$$

The estimates (5.56) and relations (5.58) - (5.60) imply

$$|(T_h w_h)(x,y) - w(x,y)| \le \tilde{C}\left[h_0^2 + 2h_0 \sum_{k=1}^{n} h_k + \sum_{j,k=1}^{n} h_j h_k\right]$$

where $(x,y) \in B$. This implies assertion (5.57).

5.10 The Euler method for mixed problems

Now we consider functional differential problem (5.38), (5.39) and the difference equation

$$\delta_0 z^{(i,m)} = f(x^{(i)}, y^{(m)}, T_h z_{(i,m)}, \delta z^{(i,m)}) \tag{5.61}$$

with initial-boundary condition (5.41). We formulate sufficient conditions for the convergence of the method and we give an error estimate.

Assumption H $[f, \sigma]$. Suppose that the function $f : \Omega \to R$ is continuous and

1) there exists a function $\sigma : [0, a_0) \times R_+ \to R_+$, $a_0 > a$, such that

(i) σ is continuous and $\sigma(x, 0) = 0$ for $x \in [0, a_0)$,

(ii) σ is nondecreasing with respect to both variables and $\bar{\omega}(x) = 0$ on $[0, a_0)$ is the maximum solution of the problem

$$\omega'(x) = \sigma(x, \omega(x)), \quad \omega(0) = 0,$$

(iii) the estimate

$$| f(x, y, w, q) - f(x, y, \bar{w}, q) | \le \sigma(x, \|w - \bar{w}\|_0)$$

is satisfied on Ω,

2) the derivatives $(D_{q_1} f, \ldots, D_{q_n} f) = D_q f$ exist on Ω and $D_q f(x, y, w, \cdot) \in C(R^n, R^n)$,

3) the estimates $D_{q_i} f(P) \ge 0$ for $1 \le i \le \kappa$, $D_{q_i} f(P) \le 0$ for $\kappa + 1 \le n$, and

$$1 - h_0 \sum_{j=1}^{n} \frac{1}{h_j} | D_{q_j} f(P) | \ge 0$$

are satisfied for $P \in \Omega$.

Theorem 5.28 *Suppose that Assumption $H\,[\,f, \sigma\,]$ is satisfied and*

1) the function $u_h : E_h^* \to R$ is a solution of problem (5.61), (5.41),

2) the function $v : E^* \to R$ is a solution of (5.38), (5.39), v is of class C^2 on E^* and the derivatives of the second order of v are bounded,

3) there is a function $\alpha_0 : \Delta \to R_+$ such that estimate (5.53) holds and $\lim_{h \to 0} \alpha_0(h) = 0$.

Then there exists a number $\varepsilon_0 > 0$ and a function $\gamma : \Delta \to R_+$ such that we have for $\|h\| < \varepsilon_0$

$$\|u_h - v_h\|_{i.h} \le \gamma(h), \quad 0 \le i \le N_0, \tag{5.62}$$

and $\lim_{h \to 0} \gamma(h) = 0$.

PROOF. We prove that the function

$$f_h(x, y, w, q) = f(x, y, T_h w, q), \quad (x, y, w, q) \in \Omega_h,$$

satisfies all the assumptions of Theorem 5.22. It follows from Theorem 5.27 that there is a function $\tilde{\beta} : \Delta \to R_+$ such that

$$\left| f(x^{(i)}, y^{(m)}, T_h(v_h)_{(i,m)}, \delta v_h^{(i,m)}) - f(x^{(i)}, y^{(m)}, v_{(x^{(i)}, y^{(m)})}, \delta v^{(i,m)}) \right| \le \tilde{\beta}(h)$$

on E_h' and $\lim_{h \to 0} \tilde{\beta}(h) = 0$. Then functional difference equation (5.61) satisfies the compatibility condition with respect to equation (5.38). Consider the difference problem

$$\eta^{(i+1)} = \eta^{(i)} + h_0 \sigma(x^{(i)}, \eta^{(i)}), \quad 0 \le i \le N_0 - 1, \quad \eta^{(0)} = 0. \tag{5.63}$$

We prove that the above problem is stable in the sense of Assumption H $[\sigma_h]$. Let $\eta_h : I_h \to R_+$ be a solution of the problem

$$\eta^{(i+1)} = \eta^{(i)} + h_0\sigma(x^{(i)}, \eta^{(i)}) + h_0\gamma(h), \quad 0 \le i \le N_0 - 1, \quad \eta^{(0)} = \alpha_0(h), \tag{5.64}$$

and the functions α_0, $\gamma : \Delta \to R_+$ satisfy $\lim_{h\to 0}\gamma(h) = 0$, $\lim_{h\to 0}\alpha_0(h) = 0$. Consider also the Cauchy problem

$$\omega'(x) = \sigma(x, \omega(x)) + \gamma(h), \quad \gamma(0) = \alpha_0(h). \tag{5.65}$$

There exists $\varepsilon_0 > 0$ such that the maximum solution ω_h of (5.65) is defined on $[0, a)$ for $\|h\| < \varepsilon_0$. Moreover we have $\lim_{h\to 0}\omega_h(x) = 0$ uniformly on $[0, a)$. The function ω_h is convex on $[0, a)$, therefore we have

$$\omega_h^{(i+1)} \ge \omega_h^{(i)} + h_0\sigma(x^{(i)}, \omega_h^{(i)}) + h_0\gamma(h) \quad \text{for } 0 \le i \le N_0 - 1.$$

Since η_h satisfies (5.64) and $\eta_h^{(0)} = \omega_h^{(0)}$ then we have $\eta_h^{(i)} \le \omega_h^{(i)}$ for $0 \le i \le N_0$ which completes the proof of the stability of problem (5.63). The estimate (5.62) is satisfied with $\gamma(h) = \omega_h(a)$. This completes the proof of the theorem.

Now we give an error estimate for method (5.61), (5.41).

Theorem 5.29 *Suppose that*

1) the assumptions of Theorem 5.28 are satisfied with $\sigma(x, p) = Lp$ (i. e. the function f satisfies the Lipschitz condition with respect to the functional variable),

2) the constant $\tilde{C} \in R_+$ is such that

$$|D_{xx}v(x,y)| \le \tilde{C}, \quad |D_{xy_j}v(x,y)| \le \tilde{C}, \quad |D_{y_jy_k}v(x,y)| \le \tilde{C}$$

where $j, k = 1, \ldots n$ and $(x, y) \in E^*$,

3) there exists $\tilde{M} = (\tilde{M}_1, \ldots, \tilde{M}_n) \in R_+$ such that

$$|D_{q_j}f(P)| \le \tilde{M}_j \quad \text{for } P \in \Omega, \quad j = 1, \ldots, n.$$

Then

$$\|u_h - v_h\|_{i.h} \le \tilde{\eta}_h^{(i)}, \quad i = 0, 1, \ldots, N_0, \tag{5.66}$$

where

$$\tilde{\eta}_h^{(i)} = \alpha_0(h)\,(1 + Lh_0)^i + h_0\tilde{\gamma}(h)\left[1 + (1 + Lh_0) + \ldots + (1 + Lh_0)^{i-1}\right]$$

and

$$\tilde{\gamma}(h) = \frac{h_0}{2}\tilde{C} + L\tilde{C}\left[h_0^2 + 2h_0\sum_{k=1}^{n} h_k + \sum_{j,k=1}^{n} h_j h_k\right] + \frac{\tilde{C}}{2}\sum_{j=1}^{n}\tilde{M}_j h_j.$$

PROOF. The difference operators δ_0 and δ satisfy the conditions

$$|\delta_0 v_h^{(i,m)} - D_x v^{(i,m)}| \le \frac{h_0}{2}\tilde{C},$$

$$|\,\delta_j v_h^{(i,m)} - D_{y_j} v^{(i,m)}\,| \le \frac{h_j}{2}\tilde{C}, \;\; j = 1,\dots,n.$$

It follows from Theorem 5.27 and from the above estimates that

$$\left|\, \delta_0 v_h^{(i,m)} - f(\,x^{(i)}, y^{(m)}, T_h(v_h)_{(i,m)}, \delta v_h^{(i,m)}\,)\,\right| \le \tilde{\gamma}(h) \;\text{ on }\; E'_h.$$

The function $\tilde{\eta}_h$ is a solution of the problem

$$\eta^{(i+1)} = (1 + Lh_0)\,\eta^{(i)} + h_0\tilde{\gamma}(h), \;\; 0 \le i \le N_0 - 1, \;\;\; \eta^{(0)} = \alpha_0(h),$$

which is equivalent to (5.64) for $\sigma(\,x,p\,) = Lp$. Then we obtain assertion (5.66) from Theorem 5.22.

Now we formulate a difference method for differential integral equations. Suppose that $\kappa = n$. Then we have

$$B = [-r_0, 0] \times [0, r], \;\; E = (0, a) \times [-b, b)$$

and

$$E_0 = [-r_0, 0] \times [-b, c], \; \partial_0 E = (\,(0, a) \times [-b, c]\,) \setminus E$$

where $c = b + r$. Suppose that $\tilde{f} : E \times R \times R^n \to R$ is a given function and $f : \Omega \to R$ is defined by

$$f(x, y, w, q) = \tilde{f}(\,x, y, \int_B w(t, s)dt\,ds, q\,).$$

Then equation (5.38) reduces to the following differential integral equation

$$D_x z(x, y) = \tilde{f}(\,x, y, \int_D z(x + t, y + s)dtds, \, D_y z(x, y)\,). \tag{5.67}$$

Consider the initial-boundary value problem consisting of (5.67), (5.39) and the difference method (5.61), (5.41) with the above given f. It is easy to see that for $(j, s) \in S_*$ we have

$$\int_{x^{(i)}}^{x^{(i+1)}} \int_{y^{(m)}}^{y^{(m+1)}} \left(\frac{Y - Y^{(i,m)}}{h}\right)^{(j,s)} \left(1 - \frac{Y - Y^{(i,m)}}{h}\right)^{1-(j,s)} dx\,dy = \frac{1}{2^{n+1}} \prod_{k=0}^{n} h_k.$$

For the vector $M = (M_1, \dots, M_n) \in \mathbf{Z}^n$ satisfying the condition $M * h' = r$ we put $M - 1 = (M_1 - 1, \dots, M_n - 1)$. Then for a function $w : B_h \to R$ we have

$$\int_B (T_h w)\,(x, y)\,dx\,dy = \frac{1}{2^{n+1}} \left[\prod_{k=0}^{n} h_k\right] \sum_{\xi=-M_0}^{-1} \sum_{\tau=0}^{M-1} \sum_{(j,s)\in S_*} w^{(\xi+j, \tau+s)}.$$

The difference method for problem (5.67), (5.39) has the form

$$\delta_0 z^{(i,m)} = \tilde{f}\left(\,x^{(i)}, y^{(m)}, I[\,z_{(i,m)}\,], \,\delta z^{(i,m)}\,\right),$$

$$z^{(i,m)} = \varphi_h^{(i,m)} \quad \text{on} \quad E_{0.h} \cup \partial_0 E_h,$$

where

$$I[\, z_{(i,m)} \,] = \frac{1}{2^{n+1}} \left[\prod_{k=0}^{n} h_k \right] \sum_{\xi=-M_0}^{-1} \sum_{\tau=0}^{M-1} \sum_{(j,s)\in S_*} z^{(i+\xi+j,m+\tau+s)}.$$

It is easy to formulate sufficient conditions for the convergence and for the error estimate of the above difference method.

5.11 Bicharacteristics for mixed problems

Put $\tilde{\Omega} = \bar{E} \times C(B,R)$ and assume that $\varrho = (\,\varrho_1,\ldots,\varrho_n\,) : \tilde{\Omega} \to R^n$, $f : \tilde{\Omega} \to R$, $\phi : E_0 \cup \partial_0 E \to R$ are given functions. We will deal with the quasilinear equation

$$D_x z(x,y) + \sum_{i=1}^{n} \varrho_i(\, x, y, z_{(x,y)}\,) D_{y_i} z(x,y) = f(\, x, y, z_{(x,y)}\,) \tag{5.68}$$

with the initial-boundary condition

$$z(x,y) = \phi(x,y) \tag{5.69}$$

on $E_0 \cup \partial_0 E$. Write $E_x^* = E^* \cap (\,[-r_0, x] \times R^n\,)$ and $E_x = [0,x] \times [-b,b]$ where $0 \le x \le a$. Let $\|\cdot\|_x$ be the supremum norm in the space $C(E_x^*, R)$. We will denote by $\|\cdot\|_{(x)}$ the supremum norm in the space $C(E_x, R)$. A function $u : E_c^* \to R$, $0 < c \le a$, is the Carathéodory solution of problem (5.68), (5.69) if

(i) u is continuous and the derivatives $D_x u$, $D_y u$ exist almost everywhere on E_c,

(ii) u satisfies equation (5.68) almost everywhere on E_c and condition (5.69) holds on $(\,E \cup \partial_0 E\,) \cap (\,[-r_0, c] \times R^n\,)$.

We define now some function spaces. Let $C_L(B,R)$ be the class of all functions $w \in C(B,R)$ such that

$$\|w\|_L = \sup \left\{ \frac{|w(t,s) - w(\bar{t},\bar{s})|}{|t-\bar{t}| + \|s-\bar{s}\|} \;:\; (t,s),\ (\bar{t},\bar{s}) \in B \right\} < +\infty.$$

Let us denote by $\|\cdot\|_0$ the supremum norm in the space $C(B,R)$ and

$$\|w\|_{0.L} = \|w\|_0 + \|w\|_L \quad \text{for} \quad w \in C_L(B,R).$$

Write

$$C(B,R;\xi) = \{\, w \in C(B,R) : \ \|w\|_0 \le \xi \,\},$$

$$C_L(B,R;\xi) = \{\, w \in C_L(B,R) : \ \|w\|_{0.L} \le \xi \,\},$$

where $\xi \in R_+$. Let $p = (p_0, p_1, p_2) \in R_+^3$. Denote by $J[p]$ the class of all functions $\phi \in C(E_0 \cup \partial_0 E, R)$ such that

$$|\phi(x,y)| \le p_0, \ |\phi(x,y) - \phi(\bar{x},\bar{y})| \le p_1|x-\bar{x}| + p_2\|y-\bar{y}\| \quad \text{on } E_0 \cup \partial_0 E.$$

Suppose that $c \in (0, a]$, $q = (q_0, q_1, q_2) \in R^3_+$, $q_i \geq p_i$, $i = 0, 1, 2$. Let $K_{c.\phi}[q]$ be the class of all functions $z \in C(E^*_c, R)$ such that

(i) $z(x, y) = \phi(x, y)$ on $(E_0 \cup \partial_0 E) \cap ([-r_0, c] \times R^n)$,

(ii) $|z(x, y)| \leq q_0$ and $|z(x, y) - z(\bar{x}, \bar{y})| \leq q_1|x - \bar{x}| + q_2\|y - \bar{y}\|$ on E_c.

Put $|q| = q_0 + q_1 + q_2$. Let Θ be the class of all functions $\gamma : R_+ \to (0, \infty)$ which are nondecreasing on R_+ and $\lim_{t \to 0+} \gamma(t) = 0$.

Assumption H $[\varrho]$. Suppose that

1) the function $\varrho(\cdot, y, w) : [0, a] \to R^n$ is measurable for every fixed $(y, w) \in [-b, b] \times C(B, R)$, and $\varrho(x, \cdot) : [-b, b] \times C(B, R) \to R^n$ is continuous for almost all $x \in [0, a]$,

2) $\varrho_i(x, y, w) \leq 0$ for $1 \leq i \leq \kappa$ and $\varrho_i(x, y, w) \geq 0$ for $\kappa + 1 \leq i \leq n$ on $\tilde{\Omega}$,

3) there exist functions $\alpha_1, \beta_1 \in \Theta$ such that $\|\varrho(x, y, w)\| \leq \alpha_1(\xi)$ for $(y, w) \in [-b, b] \times C(B, R; \xi)$ almost everywhere on $[0, a]$ and

$$\| \varrho(x, y, w) - \varrho(x, \bar{y}, \bar{w}) \| \leq \beta_1(\xi) [\|y - \bar{y}\| + \|w - \bar{w}\|_0]$$

for (y, w), $(\bar{y}, \bar{w}) \in C_L(B, R; \xi)$ almost everywhere on $[0, a]$.

We define bicharacteristics of equation (5.68) as follows. Suppose that $\phi \in J[p]$, $c \in (0, a]$ and $z \in K_{c.\phi}[q]$. Consider the Cauchy problem

$$\eta'(t) = \varrho(t, \eta(t), z_{(t, \eta(t))}), \quad \eta(x) = y, \tag{5.70}$$

where $(x, y) \in \bar{E}$. Denote by $g[z](\cdot, x, y)$ the solution of the above problem. The function $g[z]$ is the bicharacteristic of equation (5.68) corresponding to z. Let $\delta[z](x, y)$ be the left end of the maximal interval on which the solution $g[z](\cdot, x, y)$ is defined. Write

$$\Gamma_+ = \{ (x, y) \in \partial E : \text{ there exists } i, \ 1 \leq i \leq \kappa, \text{ such that } y_i = b_i \},$$

$$\Gamma_- = \{ (x, y) \in \partial E : \text{ there exists } i, \ \kappa + 1 \leq i \leq n, \text{ such that } y_i = -b_i \},$$

and $\Gamma_0 = \{0\} \times [-b, b]$, $\Gamma = \Gamma_0 \cup \Gamma_+ \cup \Gamma_-$.

Lemma 5.30 *Suppose that Assumption H* $[\varrho]$ *is satisfied and* $c \in (0, a]$, ϕ, $\bar{\phi} \in J[p]$, $z \in K_{c.\phi}[q]$, $\bar{z} \in K_{c.\bar{\phi}}[q]$. *Then the solutions* $g[z](\cdot, x, y)$ *and* $g[\bar{z}](\cdot, x, y)$ *exist on the intervals* $I_{(x,y)}$ *and* $\bar{I}_{(x,y)}$ *such that for* $\zeta = \delta[z](x, y)$ *and* $\bar{\zeta} = \delta[\bar{z}](x, y)$ *we have*

$$(\zeta, g[z](\zeta, x, y)) \in \Gamma \text{ and } (\bar{\zeta}, g[\bar{z}](\bar{\zeta}, x, y)) \in \Gamma.$$

Solutions of problem (5.70) are unique. Moreover we have the estimates

$$\| g[z](t, x, y) - g[z](t, \bar{x}, \bar{y}) \| \leq C [\|y - \bar{y}\| + \alpha_1(q_0)|x - \bar{x}|] \tag{5.71}$$

for $t \in I_{(x,y)} \cap I_{(\bar{x},\bar{y})}$, (x, y), $(\bar{x}, \bar{y}) \in \bar{E}$, *and*

$$\| g[z](t, x, y) - g[\bar{z}](t, x, y) \| \leq C \beta_1(|q|) c \|z - \bar{z}\|_x, \tag{5.72}$$

for $t \in I_{(x,y)} \cap \bar{I}_{(x,y)}$, $(x, y) \in E$, *where* $\bar{I}_{(x,y)}$ *is the domain of the bicharacteristic* $g[\bar{z}](\cdot, x, y)$ *and* $C = \exp[c(1 + q_2)\beta_1(|q|)]$.

PROOF. The existence and uniqueness of solutions of (5.70) follows from classical theorems on Carathéodory solutions of initial problems. The function $g[z](\cdot, x, y)$ satisfies the integral equation

$$g[z](t,x,y) = y + \int_x^t \varrho\left(\tau, g[z](\tau,x,y), z_{(\tau, g[z](\tau,x,y))}\right) d\tau$$

and

$$\|z_{(\tau,g[z](\tau,x,y))}\|_0 \le q_0,$$
$$\|z_{(\tau,g[z](\tau,x,y))} - z_{(\tau,g[z](\tau,\bar{x},\bar{y}))}\|_0 \le q_2 \,\|g[z](\tau,x,y) - g[z](\tau,\bar{x},\bar{y})\,\|.$$

It follows from Assumption H $[\varrho]$ that the function $g[z](\cdot,x,y) - g[z](\cdot,\bar{x},\bar{y})$ satisfies the integral inequality

$$\|g[z](t,x,y) - g[z](t,\bar{x},\bar{y})\| \le \|y-\bar{y}\| + \alpha_1(q_0)|x-\bar{x}|$$
$$+(1+q_2)\beta_1(|q|)\left|\int_{x'}^t \|g[z](\tau,x,y) - g[z](\tau,\bar{x},\bar{y})\|\,d\tau\right|, \quad t \in I_{(x,y)} \cap I_{(\bar{x},\bar{y})},$$

for some $x' \in (0,c)$. Now we obtain (5.71) by the Gronwall inequality. For $z \in K_{c.\phi}[q]$ and $\bar{z} \in K_{c.\bar{\phi}}[q]$ we have the estimate

$$\|\, z_{(t,g[z](t,x,y))} - \bar{z}_{(t,g[\bar{z}](t,x,y))}\,\|_0 \le q_2 \,\|\, g[z](t,x,y) - g[\bar{z}](t,x,y)\,\| + \|\, z-\bar{z}\|_x.$$

It follows that the function $g[z](\cdot,x,y) - g[\bar{z}](\cdot,x,y)$ satisfies the integral inequality

$$\|\, g[z](t,x,y) - g[\bar{z}](t,x,y)\,\| \le c\beta_1(|q|)\,\|\, z-\bar{z}\,\|_x$$
$$+(1+q_2)\,\beta_1(|q|)\left|\int_x^t \|\, g[z](\tau,x,y) - g[\bar{z}](\tau,x,y)\,\|\,d\tau\right|, \quad t \in I_{(x,y)} \cap \bar{I}_{(x,y)}.$$

We get (5.72) by the Gronwall inequality. This proves Lemma 5.30.

Assumption $\tilde{H}$ $[\varrho]$. Suppose that there is a function $\beta : R_+ \to (0,\infty)$ such that

$$\varrho_i(x,y,w) \le -\beta(\xi) \text{ on } C(B,R;\xi) \text{ for } 1 \le i \le \kappa,$$

and

$$\varrho_i(x,y,w) \ge \beta(\xi) \text{ on } C(B,R;\xi) \text{ for } \kappa+1 \le i \le n.$$

Now we prove a lemma on the regularity of the function $\delta[z]$.

Lemma 5.31 *Suppose that Assumptions H $[\varrho]$, $\tilde{H}$ $[\varrho]$ are satisfied and $c \in (0,a]$, ϕ, $\bar{\phi} \in J[p]$, $z \in K_{c.\phi}[q]$, $\bar{z} \in K_{c.\bar{\phi}}[q]$. Then the functions $\delta[z]$ and $\delta[\bar{z}]$ are continuous on $\bar{E}$. Moreover we have the estimates*

$$|\,\delta[z](x,y) - \delta[z](\bar{x},\bar{y})\,| \le \frac{C}{\beta(q_0)}\,[\,\|y-\bar{y}\| + \alpha_1(q_0)|x-\bar{x}|\,] \tag{5.73}$$

and

$$|\,\delta[z](x,y) - \delta[\bar{z}](x,y)\,| \le \frac{C}{\beta(q_0)}\beta_1(|q|)c\,\|z-\bar{z}\|_x \tag{5.74}$$

on $\bar{E}$.

PROOF. The continuity of $\delta[z]$ and $\delta[\bar{z}]$ on $\bar{E}$ follows from classical theorems on continuous dependence on initial conditions for Carathéodory solutions of initial problems. Now we prove (5.73). This estimate is obvious in the case $\delta[z](x,y) = \delta[z](\bar{x},\bar{y}) = 0$ (i.e. in the case when solutions of problem (5.70) are defined on $[0,x]$ and $[0,\bar{x}]$). Suppose now that $0 \leq \delta[z](x,y) < \delta[z](\bar{x},\bar{y})$. Then for $\bar{\zeta} = \delta[z](\bar{x},\bar{y})$ we have $(\bar{\zeta}, g[z](\bar{\zeta},\bar{x},\bar{y})) \in \Gamma_+ \cup \Gamma_-$ and there exists i, $1 \leq i \leq n$, such that

$$g_i[z](\bar{\zeta},\bar{x},\bar{y}) = b_i \ \text{ if } \ 1 \leq i \leq \kappa, \tag{5.75}$$

or

$$g_i[z](\bar{\zeta},\bar{x},\bar{y}) = -b_i \ \text{ if } \ \kappa+1 \leq i \leq n. \tag{5.76}$$

Consider the case (5.75). Then

$$\begin{aligned}
&-\beta(q_0)\ [\,\delta[z](\bar{x},\bar{y}) - \delta[z](x,y)\,] \\
&\geq \int_{\delta[z](x,y)}^{\delta[z](\bar{x},\bar{y})} \varrho_i\left(\tau, g[z](\tau,x,y), z_{(\tau,g[z](\tau,x,y))}\right)\, d\tau \\
&= g_i[z](\,\delta[z](\bar{x},\bar{y}), x, y\,) - g_i[z](\,\delta[z](x,y), x, y\,) \\
&\geq g_i[z](\,\delta[z](\bar{x},\bar{y}), x, y\,) - g_i[z](\,\delta[z](\bar{x},\bar{y}), \bar{x}, \bar{y}\,) \\
&\geq -C\ [\,\|y-\bar{y}\| + \alpha_1(q_0)|x-\bar{x}|\,].
\end{aligned}$$

Thus we see that the proof of (5.73) is completed in the case (5.75). In a similar way we prove (5.73) in the case (5.76). Now we consider estimate (5.74). The inequality is obvious if $\delta[z](x,y) = \delta[\bar{z}](x,y) = 0$. Suppose now that $0 \leq \delta[z](x,y) < \delta[\bar{z}](x,y)$. Then for $\zeta = \delta[\bar{z}](x,y)$ we have $(\zeta, g[\bar{z}](\zeta,x,y)) \in \Gamma_+ \cup \Gamma_-$ and there is i, $1 \leq i \leq n$, such that

$$g_i[\bar{z}]\zeta, x, y) = b_i \text{ if } 1 \leq i \leq \kappa, \tag{5.77}$$

or

$$g_i[\bar{z}](\zeta, x, y) = -b_i \text{ if } \kappa+1 \leq i \leq n. \tag{5.78}$$

Consider the case (5.77). Then

$$\begin{aligned}
&-\beta(q_0)\ [\,\delta[\bar{z}](x,y) - \delta[z](x,y)\,] \\
&\geq \int_{\delta[z](x,y)}^{\delta[\bar{z}](x,y)} \varrho_i\left(\tau, g[z](\tau,x,y), z_{(\tau,g[z](\tau,x,y))}\right)\, d\tau \\
&\geq g_i[z](\,\delta[\bar{z}](x,y), x, y\,) - g_i[\bar{z}](\,\delta[\bar{z}](x,y), x, y\,) \\
&\geq -C\,\beta_1(|q|)\, c\, \|z-\bar{z}\|_x.
\end{aligned}$$

The above inequality implies (5.74) in the case (5.77). In the same way we prove (5.74) in the case (5.78). This completes the proof of Lemma 5.31.

5.12 Functional integral equations

Now we define the integral operator corresponding to mixed problem (5.68), (5.69). Suppose that $\phi \in J[p]$, $c \in (0, a]$, $z \in K_{c.\phi}[q]$ and $g[z](\cdot, x, y)$ is the bicharacteristic of equation (5.68) corresponding to z. Let $I_{(x,y)}$ be the domain of the function $g[z](\cdot, x, y)$. Consider the operator U_ϕ defined on $K_{c.\phi}[q]$ by

$$(U_\phi z)(x, y) = \phi(\delta[z](x, y), g[z](\delta[z](x, y), x, y)$$

$$+ \int_{\delta[z](x,y)}^{x} f\left(\tau, g[z](\tau, x, y), z_{(\tau, g[z](\tau,x,y))}\right) d\tau \text{ for } (x, y) \in E_c,$$

and

$$(U_\phi z)(x, y) = \phi(x, y) \text{ on } (E_0 \cup \partial_0 E) \cap ([-r_0, c] \times R^n),$$

where $\delta[z](x, y)$ is the left end of the interval $I_{(x,y)}$.

Remark 5.32 *The operator U_ϕ is obtained in the following way. Consider equation (5.68) along bicharacteristic $g[z](\cdot, x, y)$. Then we get*

$$\frac{d}{dt} z(t, g[z](t, x, y)) = f(t, g[z](t, x, y), z_{(t, g[z](t,x,y))}).$$

By integrating the above equation with respect to t on the interval $[\delta[z](x, y), x]$ we get the integral equation $z = U_\phi z$ with the above given U_ϕ.

Assumption H $[f]$. Suppose that

1) the function $f(\cdot, y, w) : [0, a] \to R$ is measurable for $(y, w) \in [-b, b] \times C(B, R)$ and $f(x, \cdot) : [-b, b] \times C(B, R) \to R$ is continuous for almost all $x \in [0, a]$,

2) there exist functions α_2, $\beta_2 \in \Theta$ such that $|f(x, y, w)| \leq \alpha_2(\xi)$ on $[-b, b] \times C(B, R; \xi)$ almost everywhere on $[0, a]$ and

$$|f(x, y, w) - f(x, \bar{y}, \bar{w})| \leq \beta_2(\xi) [\|y - \bar{y}\| + \|w - \bar{w}\|_0]$$

for (y, w), $(\bar{y}, \bar{w}) \in [-b, b] \times C_L(B, R; \xi)$ almost everywhere on $[0, a]$.

Assumption H $[c, q]$. Suppose that the constants c, $q = (q_0, q_1, q_2)$ satisfy the conditions

1) $q_0 \geq p_0 + c\,\alpha_2(q_0)$,

2) $q_1 \geq \max\{p_1, \alpha_2(q_0) + A\,\alpha_1(q_0)\}$ and $q_2 \geq A$, where

$$A = C\left[\frac{1}{\beta(q_0)}(p_1 + p_2\alpha_1(q_0) + \alpha_2(q_0)) + p_2 + c(1 + q_2)\beta_2(|q|)\right].$$

Lemma 5.33 *If $\phi \in J[p]$ and Assumptions $H[\varrho]$, $\tilde{H}[\varrho]$, $H[f]$, $H[c, q]$ are satisfied then $U_\phi : K_{c.\phi}[q] \to K_{c.\phi}[q]$.*

Proof. Suppose that $z \in K_{c.\phi}[q]$. Then

$$|(U_\phi z)(x, y)| \leq p_0 + c\,\alpha_2(q_0) \leq q_0 \text{ on } E_c^*. \tag{5.79}$$

Now we prove the function $U_\phi z$ satisfies the Lipschitz condition with respect to (x,y) with the constants (q_1,q_2). If (x,y), $(\bar{x},\bar{y}) \in E_c$ then

$$|(U_\phi z)(x,y) - (U_\phi z)(\bar{x},\bar{y})| \le A_\phi + A_f$$

where

$$A_\phi = |\phi(\delta[z](x,y), g[z](\delta[z](x,y),x,y) - \phi(\delta[z](\bar{x},\bar{y}), g[z](\delta[z](\bar{x},\bar{y}),\bar{x},\bar{y})|,$$

$$A_f = \Big| \int_{\delta[z](x,y)}^{x} f\left(\tau, g[z](\tau,x,y), z_{(\tau,g[z](\tau,x,y))}\right) d\tau$$
$$- \int_{\delta[z](\bar{x},\bar{y})}^{\bar{x}} f\left(\tau, g[z](\tau,\bar{x},\bar{y}), z_{(\tau,g[z](\tau,\bar{x},\bar{y}))}\right) d\tau \Big|.$$

It follows from Lemmas 5.30 and 5.31 that

$$A_\phi \le C\left[p_2 + \frac{1}{\beta(q_0)}(p_1 + p_2\,\alpha_1(q_0))\right][\,\|y-\bar{y}\| + \alpha_1(q_0)|x-\bar{x}|\,]$$

and

$$A_f \le C\left[c\,\beta_2(|q|)\,(1+q_2) + \frac{1}{\beta(q_0)}\alpha_2(q_0)\right]$$
$$\times[\,\|y-\bar{y}\| + \alpha_1(q_0)|x-\bar{x}|\,] + \alpha_2(q_0)\,|x-\bar{x}|.$$

Then using Assumption H $[c,q]$ we get

$$|(U_\phi z)(x,y) - (U_\phi z)(\bar{x},\bar{y})| \le q_1|x-\bar{x}| + q_2\|y-\bar{y}\| \text{ on } E_c.$$

The above inequality and (5.79) imply Lemma 5.33.

Now we prove that for sufficiently small $c \in (0,a]$ the operator U_ϕ is a contraction. Write

$$\Lambda_c = c\,\beta_1(|q|)\,A + c\,\beta_2(|q|).$$

Lemma 5.34 *Suppose that*

1) $\phi \in J[p]$ and Assumption H $[\varrho]$, $\tilde{H}$ $[\varrho]$, H $[f]$, H $[c,q]$ are satisfied,

2) $\Lambda_c < 1$.

Then there exists exactly one function $u \in K_{c.\phi}[q]$ satisfying the equation $z = U_\phi z$.

PROOF. Lemma 5.33 shows that $U_\phi : K_{c.\phi}[q] \to K_{c.\phi}[q]$. Now we prove that

$$\|U_\phi z - U_\phi \bar{z}\|_c \le \Lambda_c \|z-\bar{z}\|_c \tag{5.80}$$

for z, $\bar{z} \in K_{c.\phi}[q]$. For $(x,y) \in E_c$ we have

$$|(U_\phi z)(x,y) - (U_\phi \bar{z})(x,y)| \le B_\phi + B_f$$

where

$$B_\phi = |\phi(\delta[z](x,y), g[z](\delta[z](x,y),x,y) - \phi(\delta[\bar{z}](x,y), g[\bar{z}](\delta[\bar{z}](x,y),x,y)|$$

and

$$B_f = \left| \int_{\delta[z](x,y)}^{x} f\left(\tau, g[z](\tau,x,y), z_{(\tau,g[z](\tau,x,y))}\right) d\tau \right.$$

$$\left. - \int_{\delta[\bar{z}](x,y)}^{x} f\left(\tau, g[\bar{z}](\tau,x,y), \bar{z}_{(\tau,g[\bar{z}](\tau,x,y))}\right) d\tau \right| .$$

The following estimates are consequences of Assumptions H $[\varrho]$, H $[f]$ and Lemmas 5.30, 5.31 :

$$B_\phi \le C\,\beta_1(|q|)\,c\left[p_2 + \frac{1}{\beta(q_0)}\,(p_1 + p_2\,\alpha_1(q_0))\right] \|z - \bar{z}\|_x ,$$

$$B_f \le c\,\beta_2(|q|)\,\{\,c\,\beta_2(|q|)\,(1+q_2)\,C\,\|z-\bar{z}\|_x$$

$$+\|z-\bar{z}\|_x\,\} + \frac{C}{\beta(q_0)}\alpha_2(q_0)\,\|z-\bar{z}\|_x .$$

The above inequalities imply (5.80). By the Banach fixed point theorem there exists exactly one $u \in K_{c.\phi}[q]$ satisfying the equation $z = U_\phi z$. This completes the proof of Lemma 5.34.

Theorem 5.35 *Suppose that Assumptions H $[\varrho]$, $\tilde{H}$ $[\varrho]$, H $[f]$, H $[c,q]$ are satisfied and $\phi \in J[p]$. Let $c \in (0,a]$ be such a constant the $\Lambda_c < 1$.*

Then there exists exactly one solution u of problem (5.68), (5.69) on E_c^*.

PROOF. Lemma 5.34 shows that there exists a unique solution $u \in K_{c.\phi}[q]$ of equation $z = U_\phi z$. Then u satisfies the initial-boundary condition (5.69) on $(E_0 \cup \partial_0 E) \cap ([-r_0, c] \times R^n)$. Now we prove that u is the Carathéodory solution of (5.68) on E_c. We write $\delta(x,y)$ and $g(\cdot,x,y) = (g_1(\cdot,x,y), \ldots g(\cdot,x,y))$ instead of $\delta[u](x,y)$ and $g[u](\cdot,x,y)$. Put

$$E^{(0)} = \{\,(x,y) \in E:\ 0 < x < c,\ \delta(x,y) = 0\,\},$$

$$E^{(i)} = \{\,(x,y) \in E:\ 0 < x < c,\ g_i(\delta(x,y),x,y) = b_i\,\},\quad 1 \le i \le \kappa,$$

$$E^{(i)} = \{\,(x,y) \in E:\ 0 < x < c,\ g_i(\delta(x,y),x,y) = -b_i\,\},\quad \kappa+1 \le i \le n.$$

At first we prove that u satisfies (5.68) almost everywhere on $E^{(0)}$. For a fixed x, $(x,y) \in E^{(0)}$, we put $\eta = g(0,x,y)$. Let $I_{(x,y)}$ be the domain of the bicharacteristic $g(\cdot,x,y)$. It follows that $g(t,x,y) = g(t,0,\eta)$ for $t \in I_{(x,y)}$ and $y = g(x,0,\eta)$. The relations $\eta = g(0,x,y)$ and $y = g(0,x,\eta)$, where $(x,y) \in E^{(0)}$, $\eta \in (-b,b)$, are equivalent. Then the relations

$$u(x,y) = (U_\phi)(x,y),\ \ (x,y) \in E^{(0)},$$

and

$$u(x,g(x,0,\eta)) = (U_\phi u)(x,g(x,0,\eta)),\ \ x \in I_{(0,\eta)},\ \eta \in (-b,b),$$

are equivalent. It follows that

$$u(x,g(x,0,\eta)) = \phi(0,\eta) + \int_0^x f(\,\tau, g(\tau,0,\eta), u_{(\tau,g(\tau,0,\eta))}\,)\,d\tau, \quad \eta \in (-b,b). \quad (5.81)$$

By differentiating (5.81) with respect to x we get

$$D_x u(x,g(x,0,\eta)) + \sum_{j=1}^{n} D_{y_j} u(x,g(x,0,\eta)) \frac{d}{dx} g_j(x,0,\eta)$$

$$= f(\,x, g(x,0,\eta), u_{(x,g(x,0,\eta))}\,)$$

for almost all $x \in I_{(0,\eta)}$. Making use of the transformation $y = g(x,0,\eta)$ and (5.70) we get (5.68) almost everywhere on $E^{(0)}$. Now we prove the u satisfies equation (5.68) on $E^{(n)}$. Let $(x,y) \in E^{(n)}$ and

$$\lambda = \delta(x,y), \quad \eta' = (\eta_1, \ldots, \eta_{n-1}) = g'(\delta(x,y), x, y), \quad (5.82)$$

where

$$g'(\,\cdot\,, x, y) = (\,g_1(\,\cdot\,, x, y), \ldots, g_{n-1}(\,\cdot\,, x, y)\,)\,.$$

For $(x,y) \in E^{(n)}$ we have $-b_n = g_n(\delta(x,y), x, y)$. Consider the family of bicharacteristics $g(\,\cdot\,, \lambda, (\eta', -b_n))$ with

$$\lambda \in (0,c), \quad \eta' \in (-b_1, b_1) \times \ldots \times (-b_{n-1}, b_{n-1}).$$

The relations (5.82) and $y = g(x, \lambda, (\eta', -b_n))$ are equivalent. Then the relations

$$u(x,y) = (\,U_\phi u\,)(x,y), \quad (x,y) \in E^{(n)},$$

and

$$u(x, g(x,\lambda,(\eta',-b_n))) = (\,U_\phi u\,)(x, g(x,\lambda,(\eta',-b_n))), \quad (5.83)$$

where $\lambda \in (0,c)$, $x \in I_{(\lambda,(\eta',-b_n))}$, and $\eta' \in (-b_1,b_1) \times \ldots \times (-b_{n-1}, b_{n-1})$, are equivalent. It follows from (5.83) that

$$u(x, g(x,\lambda,(\eta',-b_n))) = \phi(\,\lambda, (\eta',-b_n)\,) +$$

$$\int_\lambda^x f\left(\tau, g(\tau,\lambda,(\eta',-b_n)), u_{(\tau,g(\tau,\lambda,(\eta',-b_n)))}\right) d\tau.$$

By differentiating the above identity with respect to x we get

$$D_x u(x, g(x,\lambda,(\eta',-b_n))) + \sum_{j=1}^{n} D_{y_j} u(x, g(x,\lambda,(\eta',-b_n))) \frac{d}{dx} g_j(x,\lambda,(\eta',-b_n))$$

$$= f\left(x, g(x,\lambda,(\eta',-b_n)), u_{(x,g(x,\lambda,(\eta',-b_n)))}\right),$$

for almost all $x \in I_{(\lambda,(\eta',-b_n))}$. Making use of the transformation $y = g(x, \lambda, (\eta' - b_n))$ and (5.70) we get (5.68) almost everywhere on $E^{(n)}$. In a similar way we prove that u satisfies (5.68) almost everywhere on $E^{(i)}$ for $1 \le i \le n-1$. Since

$$E \cap (\,(0,c) \times R^n\,) = E^{(0)} \cup E^{(1)} \cup \ldots \cup E^{(n)}$$

it follows that u is the Carathéodory solution of (5.68), (5.69) on E_c^*. This completes the proof of Theorem 5.35.

Remark 5.36 *Suppose that all the assumptions on Theorem 5.35 are satisfied and* $\tilde{u} : E_c^* \to R$ *is a solution of (5.68) with the initial-boundary condition* $z(x,y) = \tilde{\phi}(x,y)$ *on* $E_0 \cup \partial_0 E$ *with* $\tilde{\phi} \in J[p]$. *Then for each* $x \in (0,c)$ *there is* $\tilde{\Lambda}_x \in R_+$ *such that*

$$\|u - \tilde{u}\|_x \leq \tilde{\Lambda}_x \|\phi - \tilde{\phi}\|_{(0.x)}$$

for $x \in (0,c)$, *where* $\| \cdot \|_{(0.x)}$ *is the supremum norm in the space*

$$C\left((E_0 \cup \partial_0 E) \cap ([-r_0, x] \times R^n), R \right).$$

The proof of the above property of mixed problems is based on Lemmas 5.30 and 5.31. We omit details.

Remark 5.37 *Suppose that*

1) the functions ϱ, f *and* ϕ *are continuous and the derivatives* $D_y\varrho$, $D_w\varrho$, $D_y f$, $D_w f$ *exist on* $\tilde{\Omega}$,

2) the functions $D_y\varrho$, $D_w\varrho$, $D_y f$, $D_w f$ *satisfy the Lipschitz condition with respect to* (y,z) *on* $\tilde{\Omega}$,

3) the function ϕ *is continuous and the derivative* $D_u\phi$ *exists on* $E_0 \cup \partial_0 E$,

4) the function $D_y\phi$ *satisfies the Lipschitz condition with respect to* y *on* $E_0 \cup \partial_0 E$,

5) the compatibility condition

$$D_x\phi(x,y) + \sum_{i=1}^{n} \varrho(\, x, y, \phi_{(x,y)}\,) D_{y_i}\phi(x,y) = f(x,y,\phi_{(x,y)})$$

where $(x,y) \in \Gamma$, *is satisfied.*

Then there exists $c \in (0,a]$ *such that the classical solution* $\bar{u} : E_c^* \to R$ *of problem (5.68), (5.69) exists and is unique.*

The proof of the existence of classical solution of (5.68), (5.69) is based on the method of bicharacteristics. Details can be found in [64].

5.13 Bicharacteristics of Nonlinear Mixed Problems

In Sections 5.13 - 5.15 we develop the existence theory of weak solutions of mixed problem (5.38), (5.39). Let $c \in (0,a]$. A function $u : E_c^* \to R$ is a weak solution of (5.38), (5.39) provides

(i) $u \in C(E_c^*, R)$ and $D_y u(x,y)$ exists on E_c,

(ii) the function $u(\cdot, y) : [-r_0, c] \to R^k$ is absolutely continuous on $[0,c]$ for each $y \in [-b,b]$,

(iii) for each $y \in [-b', b') \times (-b'', b'']$ equation (5.38) is satisfied for almost all $x \in [0,c]$ and condition (5.39) holds on $(\, E_0 \cup \partial_0 E\,) \cap (\, [-r_0, c] \times R^n]\,)$.

We will discuss the question of the existence and uniqueness of solutions of problem (5.38), (5.39). We will denote by $\| \cdot \|_x$ the supremum norms in the spaces

$C(E_x^*, R)$ and $C(E_x^*, R^n)$. Analogously, we will use the symbol $\|\cdot\|_{(x)}$ to denote supremum norm in $C(E_x, R)$ and $C(E_x, R^n)$. Let $C_L(B, R)$ be the space defined in Section 5.11. We denote by $C^{0.1}(B, R)$ be the set of all $w \in C(B, R)$ such that the derivatives $(D_{s_1} w(t,s), \ldots, D_{s_n} w(t,s)) = D_s w(t,s)$ exist and $D_s w \in C(B, R^n)$. For $w \in C^{0.1}(B, R)$ we put

$$\|w\|_1 = \|w\|_0 + \max\{\, \|D_s w(t,s)\| : (t,s) \in B \,\}.$$

We denote by $C_L^{0.1}(B, R)$ the set of all the functions $w \in C^{0.1}(B, R)$ such that $\|w\|_{1.L} < +\infty$ where

$$\|w\|_{1.L} = \|w\|_1 + \sup\left\{ \frac{\|D_s w(t,s) - D_s w(\bar t, \bar s)\|}{|t - \bar t| + \|s - \bar s\|} : (t,s), (\bar t, \bar s) \in B \right\}.$$

We will consider the spaces

$$\Omega^{(0)} = \bar E \times C(B, R) \times R^n, \quad \Omega^{(1)} = \bar E \times C^{0.1}(B, R) \times R^n$$

and

$$\Omega^{(1.L)} = \bar E \times C_L^{0.1}(B, R) \times R^n.$$

Now we define some further functions spaces. Given $s = (s_0, s_1, s_2) \in R_+^3$, we denote by $C^{1.L}[s]$ the set of all the functions $\phi : E_0 \cup \partial_0 E \to R$ such that

(i) $\phi \in C(E_0 \cup \partial_0 E, R)$ and there exists $D_y \phi(x,y)$ for $(x,y) \in E_0 \cup \partial_0 E$,

(ii) the estimates $|\phi(x,y)| \le s_0$, $\|D_y \phi(x,y)\| \le s_1$ and

$$|\phi(x,y) - \phi(\bar x, y)| \le s_1 |x - \bar x),$$

$$\|D_y \phi(x,y) - D_y \phi(\bar x, \bar y)\| \le s_2 \left[\, |x - \bar x| + \|y - \bar y\| \,\right],$$

are satisfied on $E_0 \cup \partial_0 E$.

Let $\phi \in C^{1.L}[s]$ be given and let $0 < c \le a$, $d = (d_0, d_1, d_2) \in R_+^3$, $d_i \ge s_i$ for i = 0,1,2. We denote by $C_{c.\phi}^{1.L}[d]$ the set of all functions $z : E_c^* \to R$ such that $z \in C(E_c^*, R)$, $z(x,y) = \phi(x,y)$ on $(E_0 \cup \partial_0 E) \cap ([-r_0, c] \times R^n)$ and

(i) there exists $D_y z(x,y)$ on E_c,

(ii) the estimates $|z(x,y)| \le d_0$, $\|D_y z(x,y)\| \le d_1$ and $|z(x,y) - z(\bar x, y)| \le d_1$, $\|D_y z(x,y) - D_y z(\bar x, \bar y)\| \le d_2 \left[\, |x - \bar x| + \|y - \bar y\| \,\right]$ are satisfied on E_c.

Let $p = (p_0, p_1) \in R_+^2$, and $p_0 \ge s_1$, $p_1 \ge s_2$. We denote by $C_c^{0.L}[p]$ the class of all functions $v : E_c \to R^n$ such that

(i) $v \in C(E_c, R^n)$ and $\|v(x,y)\| \le p_0$ on E_c,

(ii) for $(x,y), (\bar x, \bar y) \in E_c$ we have $\|v(x,y) - v(\bar x, \bar y)\| \le p_1 \left[\, |x - \bar x| + \|y - \bar y\| \,\right]$.

We will prove that under suitable assumptions on f and ϕ, and for sufficiently small c with $0 < c \le a$, there exists a solution $\bar z$ of problem (5.38), (5.39) such that $\bar z \in C_{c.\phi}^{1.L}[d]$ and $D_y \bar z \in C_c^{0.L}[p]$.

Assumption H $[\,D_q f\,]$. Suppose that

1) the derivatives $(\,D_{q_1} f, \ldots, D_{q_n} f\,) = D_q f$ exist on $\Omega^{(1)}$ and $D_q f(\cdot, y, w, q) \in L([0,a], R^n)$ for every $(y, w, q) \in [-b, b] \times C^{0.1}(B, R) \times R^n$,

2) there is a function $\alpha \in \Theta$ such that

$$\|\, D_q f(x, y, w, q) \,\| \le \alpha(\|w\|_1) \quad \text{on} \quad \Omega^{(1)},$$

3) there is $\gamma \in \Theta$ such that for $(x,y,w,q) \in \Omega^{(1.L)}$, $\bar{y} \in [-b,b]$, $\bar{q} \in R^n$ and $h \in C^{0.1}(B,R)$ we have

$$\|D_q f(x,y,w,q) - D_q f(x,\bar{y},w+h,\bar{q})\| \leq \gamma(\|w\|_{1.L})\,[\,\|y-\bar{y}\| + \|h\|_1 + \|q-\bar{q}\|\,],$$

4) $D_{q_i} f(x,y,w,q) \geq 0$ for $1 \leq i \leq \kappa$ and $D_{q_i} f(x,y,w,q) \leq 0$ for $\kappa+1 \leq i \leq n$ on $\Omega^{(1)}$.

Suppose that $\phi \in C^{1.L}[E_0;s]$ and $z \in C^{1.L}_{c.\phi}[d]$, $u \in C^{0,L}_c[p]$. We consider the Cauchy problem

$$\eta'(t) = -D_q f(\,t,\eta(t), z_{(t,\eta(t))}, u(t,\eta(t))\,), \qquad \eta(x) = y, \tag{5.84}$$

and denote by $g[z,u](\,\cdot\,,x,y)$ its solution. The function $g[z,u](\,\cdot\,,x,y)$ is the bicharacteristic of equation (5.38) corresponding to (z,u). Let $I_{(x,y)}$ be the domain of $g[z,u](\,\cdot\,,x,y)$ and $\delta[z,u](x,y)$ is the left end of the maximal interval on which the bicharacteristic $g[z,u](\,\cdot\,,x,y)$ is defined.

Lemma 5.38 *Suppose that Assumption $H\,[D_q f\,]$ is satisfied and let*

$$\phi,\ \bar{\phi} \in C^{1.L}[s], \quad z \in C^{1.L}_{c.\phi}[d], \quad \bar{z} \in C^{1.L}_{c.\bar{\phi}}[d], \quad u,\ \bar{u} \in C^{0,L}_c[p]$$

be given. Then the solutions $g[z,u](\,\cdot\,,x,y)$ and $g[\bar{z},\bar{u}](\,\cdot\,,x,y)$ exist on the intervals $I_{(x,y)}$ and $\bar{I}_{(x,y)}$ such that for $\zeta = \delta[z,u](x,y)$ and $\bar{\zeta} = \delta[\bar{z},\bar{u}](x,y)$ we have

$$(\,\zeta, g[z,u](\zeta,x,y)\,) \in \Gamma \quad \text{and} \quad (\,\bar{\zeta}, g[\bar{z},\bar{u}](\bar{\zeta},x,y)\,) \in \Gamma.$$

The bicharacteristics are unique on $I_{(x,y)}$ and $\bar{I}_{(x,y)}$. Moreover we have the estimates

$$\|g[z,u](t,x,y) - g[z,u](t,\bar{x},\bar{y})\| \leq \exp\left[\,c\,\bar{d}\,\gamma(|d|)\,\right]\left[\,\|y-\bar{y}\| + \alpha(\tilde{d})|x-\bar{x}|\,\right]$$

for $t \in I_{(x,y)} \cap I_{(\bar{x},\bar{y})}$, (x,y), $(\bar{x},\bar{y}) \in E_c$, where $|d| = d_0 + d_1 + d_2$, $\tilde{d} = d_0 + d_1$, $\bar{d} = 1 + d_1 + d_2 + p_1$ and

$$\|g[z,u](t,x,y) - g[\bar{z},\bar{u}](t,x,y)\|$$

$$\leq c\,\gamma(|d|)\,\exp\left[\,c\,\gamma(|d|)\,\bar{d}\,\right]\,[\,\|z-\bar{z}\|_x + \|D_y z - D_y \bar{z}\|_x + \|u-\bar{u}\|_{(x)}\,]$$

for $(x,y) \in E_c$, $t \in I_{(x,y)} \cap \bar{I}_{(x,y)}$, $t \leq x$.

The proof of the above Lemma is similar to the proof of Lemma 5.30. Details are omitted.

Assumption $\tilde{H}\ [D_q f\,]$. Suppose that there is a function $\beta : R_+ \to (0,+\infty)$ such that

$$D_{q_i} f(x,y,w,q) \geq \beta(\|w\|_1) \ \text{ for } 1 \leq i \leq \kappa,$$

$$D_{q_i} f(x,y,w,q) \leq -\beta(\|w\|_1) \ \text{ for } \kappa+1 \leq i \leq n,$$

on $\Omega^{(1)}$.

Now we give a lemma on regularity of function $\delta[z,u]$.

Lemma 5.39 *Suppose that Assumptions $H\,[D_q f\,]$ and $\tilde{H}\,[D_q f\,]$ are satisfied and*

$$c \in (0,a],\ \phi,\ \bar{\phi} \in C^{1.L}[s],\ z \in C^{1.L}_{c.\phi}[d],\ \bar{z} \in C^{1.L}_{c.\bar{\phi}}[d],\ u,\ \bar{u} \in C^{0.L}_{c}[p].$$

Then the function $\delta[z,u]$ and $\delta[\bar{z},\bar{u}]$ are continuous on $\bar{E}$ and

$$\|\,\delta[z,u](x,y) - \delta[z,u](\bar{x},\bar{y})\,\| \le \frac{1}{\beta(\tilde{d})}\ \exp[c\bar{d}\gamma(|d|)]\ \Big[\ \|y-\bar{y}\| + \alpha(\tilde{d})|x-\bar{x}\ \Big]$$

and

$$\|\,\delta[z,u](x,y) - \delta[\bar{z},\bar{u}](x,y)\,\|$$

$$\le \frac{1}{\beta(\tilde{d})} c\,\gamma(|d|)\ \exp\big[\,c\,\gamma(|d|)\,\bar{d}\,\big]\ \big[\,\|z-\bar{z}\|_x + \|D_y z - D_y\bar{z}\|_x + \|u-\bar{u}\|_{(x)}\,\big]$$

on E_c.

The proof of the above Lemma is based on Lemma 5.38, see the proof of Lemma 5.31.

5.14 Integral functional equations

We denote by $CL(B,R)$ the set of all linear and continuous real functions defined on $C(B,R)$ and by $\|\cdot\|_*$ the norm in $CL(B,R)$. We formulate now next assumptions on f.

Assumption H $[\,f, D_y f, D_w f\,]$. Suppose that

1) there is $\tilde{\gamma} \in \Theta$ such that

$$|\,f(x,y,w,q)\,| \le \tilde{\gamma}(\|w\|_0)\ \text{ on }\ \Omega^{(0)},$$

2) for every $P = (x,y,w,q) \in \Omega^{(1)}$ there exist $D_y f(P) \in R^n$, $D_w f(P) \in CL(B,R)$ and the estimates

$$\|D_y f(x,y,w,q)\| \le \alpha(\|w\|_1), \quad \|D_w f(x,y,w,q)\|_* \le \alpha(\|w\|_1),$$

are satisfied on $\Omega^{(1)}$,

3) for $(x,y,w,q) \in \Omega^{(1.L)}$ and for $\bar{y} \in [-b,b]$, $\bar{q} \in R^n$, $h \in C^{0.1}(B,R)$ we have the estimates

$$\|D_y f(x,y,w,q) - D_y f(x,\bar{y},w+h,\bar{q})\| \le \gamma(\|w\|_{1.L})[\,\|y-\bar{y}\| + \|h\|_1 + \|q-\bar{q}\|\,]$$

and

$$\|D_w f(x,y,w,q) - D_w f(x,\bar{y},w+h,\bar{q})\|_* \le \gamma(\|w\|_{1.L})[\,\|y-\bar{y}\| + \|h\|_1 + \|q-\bar{q}\|\,].$$

Remark 5.40 *We give a theorem on the existence of solutions of problem (5.38, (5.39). For simplicity of the formulation of the result we have assumed the same estimation for the derivatives $D_y f$, $D_w f$, $D_q f$. We also have assumed the Lipschitz condition for these derivatives with the same coefficient.*

Suppose that $\varphi \in C^{1.L}[s]$ and $0 < c \leq a$. Let $C^{(0.L}_{c.\varphi}[p]$ be the class of all functions $u : E_c^* \to R^n$ such that $u(x,y) = D_y\varphi$ on $(E_0 \cup \partial_0 E) \cap ([-r_0, c] \times R^n)$ and $u\,|_{E_c} \in C^{0.L}_c[p]$. Now we formulate a system of integral functional equations which are generated by (5.38), (5.39). Write

$$P[z,u](\tau,x,y) = (\tau, g[z,u](\tau,x,y), z_{(\tau,g[z,u](\tau,x,y))}, u(\tau, g[z,u](\tau,x,y))),$$

$$\Phi[z,u](x,y) = \phi(\delta[z,u](x,y), g[z,u](\delta[z,u](x,y),x,y)),$$

$$\psi_i[z,u](x,y) = D_{y_i}\phi(\delta[z,u](x,y), g[z,u](\delta[z,u](x,y),x,y)),\ i = 1,\ldots,n,$$

and

$$\Psi[z,u](x,y) = (\psi_1[z,u](x,y),\ldots,\psi_n[z,u](x,y)).$$

Given $\phi \in C^{1.L}[s]$, $z \in C^{1.L}_{c.\phi}[d]$, and $u \in C^{0.L}_{c.\varphi}[p]$, where $0 < c \leq a$. We define

$$F[z,u](x,y) = \Phi[z,u](x,y)$$

$$+ \int_{\delta[z,u](x,y)}^{x} \left[f(P[z,u](\tau,x,y)) - \sum_{k=1}^{n} D_{q_k} f(P[z,u](\tau,x,y))\, u_k(\tau, g(\tau,x,y)) \right] d\tau,$$

and

$$G_i[z,u](x,y) = \psi_i[z,u](x,y)$$

$$+ \int_{\delta[z,u](x,y)}^{x} [D_{y_i} f(P[z,u](\tau,x,y)) + D_w f(P[z,u](\tau,x,y))(u_i)_{(\tau,g(\tau,x,y))}]\, d\tau$$

where $i = 1,\ldots,n$. Moreover we put $G[z,u] = (G_1[z,u],...,G_n[z,u])$. We will consider the following system of functional integral equations

$$z = F[z,u],\quad u = G[z,u], \tag{5.85}$$

$$g[z,u](t,x,y) = y + \int_t^x D_q f(P[z,u](\tau,x,y))d\tau \tag{5.86}$$

with initial-boundary conditions

$$z = \phi \text{ on } E_0 \cup \partial_0 E,\quad u = D_y\varphi \text{ on } (E_0 \cup \partial_0 E) \cap ([-r_0, c] \times R^n). \tag{5.87}$$

Remark 5.41 *Integral functional system (5.85) - (5.87) is obtained in the following way. We introduce first an additional unknown function $u = D_y z$ in (5.38). Then we consider the linearization of (5.38) with respect to u, i.e.*

$$D_x z(x,y) = f(U) + \sum_{j=1}^{n} D_{q_j} f(U)\,(D_{y_j} z(x,y) - u_j(x,y)) \tag{5.88}$$

where $U = (x, y, z_{(x,y)}, u(x,y))$. By virtue of equation (5.38) we get the differential system for the unknown function u :

$$D_x u_i(x,y) = D_{y_i} f(U) + D_w f(U)(D_{y_i} z)_{(x,y)} + \sum_{j=1}^{n} D_{q_j} f(U) D_{y_j} u_i(x,y), \tag{5.89}$$

where $i = 1, \ldots, n$. Finally we put $D_y z = u$ in (5.89). System (5.88), (5.89) has the following property: the differential equations of bicharacteristics for (5.88) and for (5.89) are the same and they have the form (5.84).

If we consider (5.88), (5.89) along the bicharacteristic $g[z, u](\,\cdot\,, x, y)$, we obtain

$$\frac{d}{dt}\, z(t, g[z,u](t,x,y)) = f(P[z,u](t,x,y))$$

$$-\sum_{j=1}^{n} D_{q_j} f(P[z,u](t,x,y)) u_j(t, g[z,u](t,x,y)),$$

and

$$\frac{d}{dt}\, u_i(t, g[z,u](t,x,y)) = D_{y_i} f(P[z,u](t,x,y))$$

$$+ D_w f(P[z,u](t,x,y))\,(u_i)_{(t, g[z,u](t,x,y))}\,,$$

where $i = 1, \ldots, n$. By integrating the above equations on $[\,\delta[z,u](x,y), x\,]$ with respect to t, we get (5.85), (5.86).

The existence result for (5.85) - (5.87) is be based on the following method of successive approximations. Suppose that $\phi \in C^{1.L}[s]$ and that Assumptions H $[\,D_q f\,]$, $\tilde{H}$ $[\,D_q f\,]$, H $[\,f, D_y f, D_w f\,]$ are satisfied. We define the sequence

$$\{z^{(m)}, u^{(m)}\}, \; z^{(m)} : E_c^* \to R, \;\; u^{(m)} : E_c^* \to R^n,$$

in the following way. Let $z^{(0)} : E_c^* \to R$, $u^{(0)} : E_c^* \to R^n$ be arbitrary functions such that $z^{(0)} \in C_{c.\phi}^{1.L}[d]$ and $u^{(0)} \in C_{c.\phi}^{0.L}[p]$. Now, if $(z^{(m)}, u^{(m)}) \in C_{c.\phi}^{1.L}[d] \times C_{c.\phi}^{0.L}[p]$ are known functions, then $u^{(m+1)}$ is a solution of the equation $u = G^{(m)}[\,u\,]$ with the initial-boundary condition

$$u(x,y) = D_y \phi(x,y) \;\text{ on }\; (\,E_0 \cup \partial_0 E\,) \cap (\,[-r_0, c] \times R^n\,)$$

and

$$z^{(m+1)}(x,y) = F[\,z^{(m)}, u^{(m+1)}\,](x,y) \;\text{ on }\; E_c,$$

$$z^{(m+1)}(x,y) = \phi(x,y) \;\text{ on }\; (\,E_0 \cup \partial_0 E\,) \cap (\,[-r_0, c] \times R^n\,)$$

where $G^{(m)} = \left(G_1^{(m)}, \ldots, G_n^{(m)} \right)$ is defined by

$$G_i^{(m)}[\,u\,](x,y) = \psi_i[z^{(m)}, u](x,y) + \int_{\delta[z^{(m)},u](x,y)}^{x} \Big[D_{y_i} f(P[z^{(m)}, u](\tau, x, y))$$

$$+ D_w f(P[z^{(m)}, u](\tau,x,y))(u_i^{(m)})_{(\tau, g(\tau,x,y))} \Big]\, d\tau, \quad i = 1, \ldots, n.$$

We wish to emphasize that the main difficulty in carrying out of this construction is the problem of the existence of the sequence $\{z^{(m)}, u^{(m)}\}$.

5.15 The existence of solutions of nonlinear mixed problems

Now we give the main existence result for mixed problem (5.38), (5.39)

Theorem 5.42 *If Assumptions $H\,[D_q f\,]$, $\tilde{H}\,[D_q f\,]$, $H\,[f, D_y f, D_w f]$, are satisfied and $\phi \in C^{1.L}[s]$ then there are $q = (q_0, q_1, q_2) \in R_+^3$, $p = (p_0, p_1) \in R_+^2$ and $c \in (0, a]$ such that*

(i) then for any $m \geq 0$ we have

(I_m) $z^{(m)}$ and $u^{(m)}$ are defined on E_c^* and $z^{(m)} \in C_{c.\phi}^{1.L}[q]$, $u^{(m)} \in C_{c.\phi}^{0.L}[p]$,

(II_m) $D_y z^{(m)}(x,y) = u^{(m)}(x,y)$ on E_c,

(ii) there are the limits

$$\bar{z}(x,y) = \lim_{m\to\infty} z^{(m)}(x,y), \quad \bar{u}(x,y) = \lim_{m\to\infty} u^{(m)}(x,y) \quad uniformly\ on\ E_c,$$

and $D_y\bar{z} = \bar{u}$,

(iii) the function $\bar{z}$ is the solution of problem (5.38) (5.39) on E_c^*.

The proof of this theorem can be found in [67], see also the proof of Theorem (5.35).

Remark 5.43 *If in Theorem 5.42 we assume that f is continuous, then we get classical solutions for our mixed problem.*

We wish to emphasize that our hereditary setting contains in particular some well known delay structures. Given the functions $\phi : E_0 \cup \partial_0 E \to R$, $\tilde{f} : \bar{E} \times R \times R^n \to R$, and $\alpha : [0,a] \to R$, $\beta = (\beta_1, \ldots, \beta_n) : \bar{E} \to R^n$, we consider the operator f defined by

$$f(x,y,w,q) = \tilde{f}\,(\,x, y, w(\alpha(x) - x, \beta(x,y) - y), q\,)\,, \quad (x,y,w,q) \in \Omega^{(0)}.$$

In this case (5.38) is equivalent to the differential equation with a deviated argument

$$D_x z(x,y) = \tilde{f}(\,x, y, z(\alpha(x), \beta(x,y)), D_y z(x,y)\,) \tag{5.90}$$

with the initial-boundary condition (5.39).
The existence result for problem (5.90), (5.39) can be easy deduced from the main Theorem 5.42. If we consider the function defined by

$$f(x,y,w,q) = \tilde{f}(\,x, y, \int_B w(t,s)\,dt\,ds, q\,)$$

then problem (5.38) reduces to the differential integral equation

$$D_x z(x,y) = \tilde{f}(\,x, y, \int_B z(x+t, y+s)\,dt\,ds, D_y z(x,y)\,)$$

with the initial-boundary condition (5.39. The existence result for the above equation with initial-boundary condition (5.39) can be easy deduced from Theorem 5.42.

Remark 5.44 *It is easy to see that Theorem 5.42 can be extended for systems of functional differential equations.*

5.16 Uniqueness of weak solutions of mixed problems

We give a theorem on uniqueness of weak solutions of problem (5.38), (5.39) under the assumptions that f satisfies the Perron type conditions with respect to the functional variable.

Assumption H $[\sigma]$**.** Suppose that

1) $\sigma : [0,a] \times C([-r_0,0],R_+) \to R_+$ satisfies Carathéodory conditions and is nondecreasing with respect to the functional variable,

2) the maximum solution of Cauchy problem

$$\omega'(x) = \sigma(x, \omega_{(x)}), \; x \in [0,a], \quad \omega(x) = 0 \text{ on } [0,a],$$

is $\bar{\omega}(x) = 0$ on $[-r_0, a]$.

Theorem 5.45 *Suppose that Assumption H* $[\sigma]$ *is satisfied and*

1) the function $f(x,\,\cdot\,) : [-b,b] \times C(B,R) \times R^n \to R$ *is continuous for a.e.* $x \in [0,a]$ *and* $f(\,\cdot\,, y, w, q) : [0,a] \to R$ *is measurable for every* $(y,w,q) \in [-b,b] \times C(B,R) \times R^n$,

2) the partial derivatives $D_q f = (D_{q_1} f, \ldots, D_{q_n} f)$ *exist on* $[-b,b] \times C(B,R) \times R^n$ *and for a.e.* $x \in [0,a]$,

3) $D_q f(x,\,\cdot\,)$ *is continuous for a.e.* $x \in [0,a]$, $D_q f(\,\cdot\,, y, w, q)$ *is measurable for every* $(y,w,q) \in [-b,b] \times C(B,R) \times R^n$ *and* $D_{q_i} f(x,y,w,q) \geq 0$ *for* $1 \leq i \leq \kappa$, $D_{q_i} f(x,y,w,q) \leq 0$ *for* $\kappa + 1 \leq i \leq n$,

4) there exists $m_q \in L([0,a],R_+)$ *such that* $\| D_q f(x,y,w,q) \| \leq m_q(x)$ *for* $(y,w,q) \in [-b,b] \times C(B,R) \times R^n$ *and for a.e.* $x \in [0,a]$,

5) the estimate

$$\| f(x,y,w,q) - f(x,y,\bar{w},q) \| \leq \sigma(x, V(w - \bar{w}))$$

holds on $[-b,b] \times C(B,R) \times R^n$ *for a.e.* $x \in [0,a]$.

Then problem (5.38), (5.39) admits one weak solution solution at the most.

Proof. Suppose that u and $\tilde{u}$ are two solutions of the problem. Put

$$Q(x,y,t) = \big(x, y, \tilde{u}_{(x,y)}, D_y\tilde{u}(x,y) + t[D_y u(x,y) - D_y \tilde{u}(x,y)]\big),$$

in force of Hadamard mean value theorem we have

$$D_x(u - \tilde{u})(x,y) - \sum_{i=1}^{n} \int_0^1 D_{q_i} f(Q(x,y,t))dt\, D_{y_i}(u - \tilde{u})(x,y) \tag{5.91}$$

$$= f(x, y, u_{(x,y)}, D_y u(x,y)) - f(x, y, \tilde{u}_{(x,y)}, D_y u(x,y)).$$

Denoted by $g(\,\cdot\,, x, y)$ the solution of the Cauchy problem

$$\eta'(t) = -\int_0^1 D_q f(Q(t, \eta(t), \xi))d\xi, \quad \eta(x) = y,$$

from (5.91) it follows that

$$(u-\tilde{u})(x,y) = (u-\tilde{u})(\,0, g(0,x,y)\,)$$

$$+\int_0^x [f\,(\,t, g(t,x,y), u_{(t,g(t,x,y))}, D_y u(t, g(t,x,y)\,)$$

$$-f\,(\,t, g(t,x,y), \tilde{u}_{(t,g(t,x,y))}, D_y u(t, g(t,x,y)\,)\,]\,dt.$$

Let $\psi(x) = \max\{\,|(u-\tilde{u})(x,y)| : y \in [-b,b]\,\}$, $-r_0 \le x \le a$. Then ψ is continuous on $[-r_0, a)$ and absolutely continuous on $[0,a)$ ([32]), moreover from assumption 5) we get

$$\psi(x) \le \psi(0) + \int_0^x \sigma(\,t, \psi_{(t)}\,)dt, \quad x \in [0,a].$$

Since $\psi(x) = 0$ on $[-r_0, 0]$, from Lemma 4.23 we deduce that $u = \tilde{u}$. This completes the proof.

Now we formulate a particular case of assumption H $[\sigma]$ which allows to get an uniqueness result as a corollary of Theorem 5.45.

Assumption H' $[\,\sigma\,]$. Suppose that

1) the function $\sigma_0 : [0,a_0] \times R_+ \to R_+$ satisfies Carathéodory conditions and is nondecreasing with respect to the second variable,

2) the null function is the maximum solution in Carathéodory sense of problem

$$\eta'(x) = \sigma_0(\,x, \eta(x)\,), \quad \eta(0) = 0.$$

Theorem 5.46 *Suppose that assumption H' $[\sigma\,]$ holds and function f satisfies assumptions 1) - 4) of Theorem 5.45. Assume that the following estimation*

$$\|\,f(x,y,w,q) - f(x,y,\bar{w},q)\,\| \le \sigma_0(\,x, \|w-\bar{w}\|_0\,)$$

holds on $[-b,b') \times (-b'',b''] \times C(B,R) \times R^n$ for a.e. $x \in [0,a]$.

Then problem (5.38), (5.39) admits one weak solution at the most.

PROOF. The assertion follows from Theorem 5.45 with

$$\sigma(\,x,\omega\,) = \sigma_0(\,x, \|\omega\|_0\,).$$

Remark 5.47 *The results presented in this Section can be extended to weakly coupled systems of functional differential problems.*

Chapter 6

Numerical Method of Lines

6.1 Introduction

The method of lines for partial differential equations consists in replacing spatial derivatives by difference expressions. Then the partial equation is transformed into a system of ordinary differential equations. The method is used for approximation of solutions of nonlinear differential problems of parabolic type by solutions of ordinary equations ([91, 153, 219, 220, 222, 225, 238]). The method is also treated as a tool for proving of existence theorems for differential problems corresponding to parabolic equations [223, 224, 227] or first-order hyperbolic systems [101, 157]. Simple examples of the method of lines for nonlinear functional differential equations were considered in [29, 108, 128]. The method for equations of higher orders is studied in [91]. The book [189] demonstrates lots of examples of the use of the numerical method of lines. Convergence analysis of one step difference methods generated by the numerical method of lines was investigated in [186].

Our results are based on [28, 29] and [237, 239, 240].

Let $B = [-r_0, 0] \times [-r, r]$, $E_0 = [-r_0, 0] \times R^n$, $E = [0, a] \times R^n$ where $r_0 \in R_+$, $r = (r_1, \ldots, r_n) \in R_+^n$ and $a > 0$. Let $\Omega = E \times C(B, R) \times R^n$ and suppose that $f : \Omega \to R$ and $\phi : E_0 \to R$ are given functions. In this Chapter we consider the Cauchy problem

$$D_x z(x, y) = f(\, x, y, z_{(x,y)}, D_y z(x, y)\,) \tag{6.1}$$

$$z = \phi \text{ on } E_0. \tag{6.2}$$

We present a theory of the numerical method of lines for functional differential problems. In Sections 6.2 - 6.5 we consider the Cauchy problem in an unbounded domain. Sections 6.6 - 6.8 deal with mixed problems.

Now we define a mesh in R^n. For $y, \bar{y} \in R^n$ we write $y * \bar{y} = (y_1\bar{y}_1, \ldots, y_n\bar{y}_n)$. Suppose that for $h = (h_1, \ldots, h_n)$, where $h_i > 0$, there exists $N = (N_1, \ldots, N_n) \in \mathbf{N}^n$ such that $N * h = r$. We denote by Δ the set of all h having the above property. We assume that $\Delta \neq \emptyset$ and that there is a sequence $\{h^{(j)}\}$, $h^{(j)} = (h_1^{(j)}, \ldots, h_n^{(j)}) \in$

Δ, such that $\lim_{j\to\infty} h^{(j)} = 0$. Let

$$y^{(m)} = m * h, \quad y^{(m)} = (\, y_1^{(m_1)}, \ldots, y_n^{(m_n)}\,) \text{ where } m = (m_1, \ldots, m_n) \in \mathbf{Z}^n.$$

Write $R_h^n = \{y^{(m)} : \ m \in \mathbf{Z}^n\}$ and $E_h = [0,a] \times R_h^n$, $E_{0.h} = [-r_0, 0] \times R_h^n$, $B_h = B \cap E_{0.h}$. Elements of the set $E_{0.h} \cup E_h$ will be denoted by $(x, y^{(m)})$ or (x,y). For a function $z : E_{0.h} \cup E_h \to R^n$ and a point $(x, y^{(m)}) \in E_{0.h} \cup E_h$ we write $z^{(m)}(x) = z(x, y^{(m)})$. Let $\delta = (\, \delta_1, \ldots, \delta_n\,)$ be a difference operator defined in the following way. Put $S = \{\, s = (s_1, \ldots, s_n) : \ s_i \in \{-1, 0, 1\}$ for $1 \le i \le n\,\}$ and

$$\delta_i z^{(m)}(x) = \frac{1}{h_i} \sum_{s\in S} c_{s.m}^{(i)}\, z^{(m+s)}(x), \quad 1 \le i \le n,$$

where $z : E_{0.h} \cup E_h \to R$ and $c_{s.m}^{(i)}$ are given numbers. Write $\delta z = (\, \delta_1 z, \ldots, \delta_n z\,)$. We will approximate the spatial derivatives $D_y z^{(m)}(x)$ by means of $\delta z^{(m)}(x)$. In the next part of the Chapter we adopt additional assumptions on $c_{s.m}^{(i)}$. Since the coefficients $c_{s.m}^{(i)}$ depend on $m \in \mathbf{Z}^n$, the approximation of the spatial derivative may be different in different points on the spatial mesh. For $z : E_{0.h} \cup E_h \to R$, $x \in [0,a]$, $m \in \mathbf{Z}^n$, we define the function $z_{(x,m)} : B_h \to R$ by

$$z_{(x,m)}(t,y) = z(x+t, y^{(m)} + y), \quad (t,y) \in B_h.$$

Let $\Omega_h = E_h \times C(B_h, R) \times R^n$ and suppose that $f_h : \Omega_h \to R$, $\phi_h : E_{0.h} \to R$ are given functions. We will approximate solutions of (6.1), (6.2) by means of solutions of the infinite system of ordinary functional differential equations

$$D_x z^{(m)}(x) = f_h(\, x, y^{(m)}, z_{(x,m)}, \delta z^{(m)}(x)\,), \quad (x, y^{(m)}) \in E_h, \tag{6.3}$$

with the initial condition

$$z^{(m)}(x) = \phi_h^{(m)}(x) \ \text{ on } E_{0.h}. \tag{6.4}$$

Let F_h be the Niemycki operator corresponding to (6.3) i. e.

$$F_h[\,z\,]^{(m)}(x) = f_h(\, x, y^{(m)}, z_{(x,m)}, \delta z^{(m)}(x)\,).$$

The initial value problem (6.3), (6.4) will be called the numerical method of lines for (6.1), (6.2). The function f_h and the difference operator δ characterize a general class of differential difference schemes which can be applied to (6.1), (6.2). We give sufficient conditions for the solvability of problem (6.3), (6.4) and for the convergence of the sequence $\{u_j\}$ of solutions of (6.3), (6.4) to a solution of (6.1), (6.2). We will consider unbounded solutions of (6.1), (6.2) and (6.3), (6.4).

Theorems concerning the numerical method of lines will be based on comparison results where a function satisfying some differential difference inequalities on an unbounded domain is estimated by a solution of an adequate ordinary functional differential problem. The first result of this type was given by S. Łojasiewicz ([157], Lemma 1) for a function of two independent variables satisfying linear differential difference inequalities with constant coefficients. It is essential fact in

[157] that the differential difference inequalities are periodic with respect to the spatial variable. The comparison result from [157] is extended in [28] (see also [237]) on differential difference inequalities with a functional variable and on nonlinear comparison problems. It is an important fact in [28, 237] that the finite systems of differential difference equations have been considered. In this Chapter we consider infinite systems of equations (6.3) or infinite systems of differential difference inequalities. The comparison theorem given in Section 6.2 is new also in the nonfunctional case.

6.2 Comparison theorem

In this Chapter we denote by $\|\cdot\|$ the Euclidean norm in R^n and by $\|\cdot\|_0$ the supremum norm in the space $C([-r_0,0],R)$ and in the space $C(B,R)$. We use the symbol $\|\cdot\|_h$ to denote the supremum norm in the space $C(B_h,R)$. We will prove that problem (6.3), (6.4) has exactly one solution in the following class of functions. A function $z:[-r_0,c]\times R^n_h\to R$ where $0\le c\le a$, will be called the function of class Σ if the following conditions are satisfied:

(i) $z(\cdot,y^{(m)})\in C([-r_0,c],R)$ for all $m\in\mathbf{Z}^n$,

(ii) there exists a function $\Psi:R_+\to R_+$ such that Ψ has a continuous derivative on R_+, the function Ψ' is bounded and nonnegative,

(iii) for $t\in R_+$ we have $\max\{\,|z^{(m)}(x)|:\ x\in[-r_0,c],\ \|y^{(m)}\|\le t\}\le\Psi(t)$ and $\lim_{t\to+\infty}\Psi(t)=+\infty$.

In a comparison theorem we will estimate a function of several variables by means of a function of one variable. Therefore we will need the following operator $V_h:C(B_h,R)\to C([-r_0,0],R_+)$. If $w\in C(B_h,R)$ then

$$(V_hw)(t)=\max\Big\{\,|w(t,y^{(m)})|:\ -N\le m\le N\Big\},\quad t\in[-r_0,0].$$

Suppose that $\sigma:[0,a]\times C([-r_0,0],R_+)\to R_+$ and $\lambda=(\lambda_1,\ldots,\lambda_n):E_h\times C(B_h,R)\to R^n$ are given functions. In this Section we consider the differential difference inequalities

$$\Big|\,D_xz^{(m)}(x)-\sum_{i=1}^n\lambda_i(\,x,y^{(m)},z_{(x,m)}\,)\,\delta_iz^{(m)}(x)\,\Big|\le\sigma(\,x,V_hz_{(x,m)}\,),\tag{6.5}$$

where $(x,y^{(m)})\in(0,a]\times R^n_h$. We prove that a function satisfying the above inequality can be estimated by a solution of a suitable Cauchy problem for ordinary functional differential equation. This estimation will be used in an existence theorem for problem (6.3), (6.4) and in the investigation of the stability of the method of lines.

Theorem 6.1 *Suppose that*

1) the function $\sigma:[0,a]\times C([-r_0,0],R_+)\to R_+$ is continuous and there is $K\in R_+$ such that

$$|\,\sigma(x,w)-\sigma(x,\bar w)\,|\le K\,\|w-\bar w\|_0\ \ on\ \ [0,a]\times C([-r_0,0],R_+),\tag{6.6}$$

2) $u : E_{0.h} \cup E_h \to R$ is of class Σ, there exists the derivative $D_x u(x,y)$ for $(x,y) \in (0,a] \times R^n_h$ and u satisfies differential difference inequality (6.5) on $(0,a] \times R^n_h$,

3) the function $\sigma(x,\cdot)$ is nondecreasing and there are $L = (L_1,\dots,L_n) \in R^n_+$ and $\eta \in C([-r_0,0],R_+)$ such that $|\lambda_i(x,y,w)| \le L_i$ on $E_h \times C(B_h,R)$ for $i = 1,\dots,n$, and

$$|u(x,y)| \le \eta(x) \quad \text{on } E_{0.h}, \tag{6.7}$$

4) the operator δ and the function λ satisfy the condition:

$$c^{(i)}_{s.m}\, \lambda_i(x,y^{(m)},w) \ge 0 \quad \text{for } s \in S\setminus\{0\}, \text{ on } (0,a] \times R^n_h, \quad i = 1,\dots,n,$$

5) there exists $C_1 > 0$ such that

$$\sum_{s\in S} |c^{(i)}_{s.m}| \le C_1 \quad \text{for } i = 1,\dots,n, \ m \in \mathbf{Z}^n.$$

Under these assumptions we have $|u(x,y)| \le \omega(x,\eta)$ on E_h, where $\omega(\cdot,\eta)$ is the solution of the problem

$$\omega'(x) = \sigma(x,\omega_{(x)}), \qquad \omega(x) = \eta(x) \text{ on } [-r_0,0].$$

PROOF. We first show that $u(x,y) \le \omega(x,\eta)$ on $[0,a] \times R^n_h$. Let us define $v(x,y) = u(x,y) - \omega(x,\eta)$ on $E_{0.h} \cup E_h$ and $\Psi_0(t) = \max\{ v(x,y) : (x,y) \in E_h, \|y\| \le t\}$, $t \in R_+$. It follows from assumption 2) that there exists a function $\Psi : R_+ \to R_+$ such that conditions (i), (ii) of the definition of the class Σ are satisfied and $\Psi_0(t) \le \Psi(t)$ for $t \in R_+$. Let $C > 0$ be the constant such that $\Psi'(t) \le C$ for $t \in R_+$. We define the function $H : E_0 \cup E \to R$ in the following way:

$$H(x,y) = \exp\left[\Psi\left(\sqrt{1+\|y\|^2}\right) + x\,(1 + K\exp(C\|r\|) + C_1\exp(C\|h\|)\Gamma)\right]$$

for $(x,y) \in E$, where

$$\Gamma = \sum_{i=1}^n \frac{L_i}{h_i},$$

and

$$H(x,y) = \exp\left[\Psi\left(\sqrt{1+\|y\|^2}\right)\right] \quad \text{for } (x,y) \in E_0.$$

It can be shown that H is a solution of the differential difference inequality

$$D_x z^{(m)}(x) \ge z^{(m)}(x) + K\|z_{(x,m)}\|_h + \sum_{i=1}^n L_i \left| \delta_i z^{(m)}(x) \right| \tag{6.8}$$

on $(0,a] \times R^n_h$. Let $\varepsilon > 0$ be fixed. It follows that there exists $A > 0$ such that

$$\frac{v(x,y)}{H(x,y)} < \varepsilon \quad \text{for } (x,y) \in (0,a] \times R^n_h \text{ and } \|y\| \ge A. \tag{6.9}$$

Write

$$U(x,y) = \frac{u(x,y)}{H(x,y)}, \qquad V(x,y) = \frac{v(x,y)}{H(x,y)}.$$

where $(x,y) \in E_{0.h} \cup E_h$. We prove that

$$V(x,y) < \varepsilon \text{ for } (x,y) \in E_{0.h} \cup E_h, \ \ \|y\| < A. \tag{6.10}$$

Initial inequality (6.7) implies $V(x,y) \leq 0$ for $(x,y) \in E_{0.h}$. Suppose that assertion (6.10) is false. Then there exists $\tilde{x} \in (0,a]$ and $\tilde{m} \in \mathbf{Z}^n$ such that $\|y^{(\tilde{m})}\| < A$ and

$$V^{(m)}(x) < \varepsilon \text{ for } x \in [0,\tilde{x}), \ \ m \in \mathbf{Z}^n, \tag{6.11}$$

$$V^{(\tilde{m})}(\tilde{x}) = \max\{ V^{(m)}(\tilde{x}) : \ m \in \mathbf{Z}^n \} = \varepsilon. \tag{6.12}$$

Then we have

$$\|V_{(\tilde{x},\tilde{m})}\|_h = V^{(\tilde{m})}(\tilde{x}). \tag{6.13}$$

Using the definition of the function v and (6.5) we get

$$D_x v^{(\tilde{m})}(\tilde{x}) + \sigma(\tilde{x}, \omega_{(\tilde{x})}) \leq \sigma(\tilde{x}, V_h u_{(\tilde{x},\tilde{m})})$$

$$+ \sum_{i=1}^{n} \lambda_i(\tilde{x}, y^{(\tilde{m})}, u_{(\tilde{x},\tilde{m})})\, \delta_i u^{(\tilde{m})}(\tilde{x}).$$

We have $\delta v^{(\tilde{m})}(\tilde{x}) = \delta u^{(\tilde{m})}(\tilde{x})$, therefore from (6.11), (6.12) and from assumptions 3), 4) it follows that

$$\sum_{i=1}^{n} \lambda_i(\tilde{x}, y^{(\tilde{m})}, u_{(\tilde{x},\tilde{m})})\, \delta_i v^{(\tilde{m})}(\tilde{x})$$

$$= \sum_{i=1}^{n} \lambda_i(\tilde{x}, y^{(\tilde{m})}, u_{(\tilde{x},\tilde{m})}) \frac{1}{h_i} \sum_{s\in S} c^{(i)}_{s.\tilde{m}}\, v^{(\tilde{m}+s)}(\tilde{x})$$

$$= \sum_{i=1}^{n} \sum_{s\in S} \frac{1}{h_i}\, c^{(i)}_{s.\tilde{m}}\, \lambda(\tilde{x}, y^{(\tilde{m})}, u_{(\tilde{x},\tilde{m})})\, V^{(\tilde{m}+s)}(\tilde{x})\, H^{(\tilde{m}+s)}(\tilde{x})$$

$$\leq V^{(\tilde{m})}(\tilde{x}) \sum_{i=1}^{n} |\, \lambda_i(\tilde{x}, y^{(\tilde{m})}, u_{(\tilde{x},\tilde{m})})\, |\; |\, \delta_i H^{(\tilde{m})}(\tilde{x})\, |$$

$$\leq V(\tilde{x}, y^{(\tilde{m})}) \sum_{i=1}^{n} L_i\, |\, \delta_i H^{(\tilde{m})}(\tilde{x})\, |.$$

Hence

$$D_x v^{(\tilde{m})}(\tilde{x}) + \sigma(\tilde{x}, \omega_{(\tilde{x})}) \leq \sigma(\tilde{x}, V_h u_{(\tilde{x},\tilde{m})})$$

$$+ V^{(\tilde{m})}(\tilde{x}) \sum_{i=1}^{n} L_i |\, \delta_i H^{(\tilde{m})}(\tilde{x})\, |. \tag{6.14}$$

Since $V^{(\tilde m)}(\tilde x) > 0$, $D_x V^{(\tilde m)}(\tilde x) \geq 0$, $H^{(\tilde m)}(\tilde x) > 0$, $D_x H^{(\tilde m)}(\tilde x) > 0$, it follows from (6.6), (6) that

$$0 < D_x v^{(\tilde m)}(\tilde x) = D_x V^{(\tilde m)}(\tilde x)\, H^{(\tilde m)}(\tilde x) + V^{(\tilde m)}(\tilde x)\, D_x H^{(\tilde m)}(\tilde x)$$

$$\leq K \,\|\, V_h u_{(\tilde x, \tilde m)} - \omega_{(\tilde x)} \|_0 + V^{(\tilde m)}(\tilde x) \sum_{i=1}^{n} L_i \,|\, \delta_i H^{(\tilde m)}(\tilde x)\,|.$$

Therefore we have

$$0 < D_x V^{(\tilde m)}(\tilde x)\, H^{(\tilde m)}(\tilde x) \tag{6.15}$$

$$\leq K \,\|\, V_h u_{(\tilde x, \tilde m)} - \omega_{(\tilde x)} \,\|_0 + V(\tilde x, y^{(\tilde m)}) \left[\sum_{i=1}^{n} L_i \,|\, \delta_i H^{(\tilde m)}(\tilde x)\,| - D_x H^{(\tilde m)}(\tilde x) \right].$$

It is easy to verify that condition (6.13) implies the inequality

$$\|\, V_h u_{(\tilde x, \tilde m)} - \omega_{(\tilde x)} \,\|_0 \leq V^{(\tilde m)}(\tilde x) \,\|\, H_{(\tilde x, y^{(\tilde m)})} \,\|_0,$$

which together with (6.15) and (6.8) gives a contradiction. Then (6.10) is proved. It follows from (6.9), (6.10) that $V(x,y) < \varepsilon$ on E_h where $\varepsilon > 0$ is arbitrary. Then we have $u(x,y) \leq \omega(x,\eta)$ on E_h. The proof of the inequality $-u(x,y) \leq \omega(x,\eta)$ on E_h is analogous. This completes the proof of the theorem.

6.3 Existence theorem and stability

In this Section we shall show the solvability of problem (6.3), (6.4) and the stability of the method of lines. Roughly speaking, method (6.3), (6.4) is called stable if the values of functions v_h nearly satisfying (6.3), (6.4) tend to the values of the solutions u_h of problem (6.3), (6.4) if h tends to zero. We consider solutions of (6.3), (6.4) of the class Σ.

Theorem 6.2 *Suppose that*

1) the function f_h characterizing the method of lines satisfies the Lipschitz condition

$$|\, f_h(x,y,w,q) - f_h(x,y,\bar w,q)\,| \leq K \|w - \bar w\|_h \quad on \;\; \Omega_h \tag{6.16}$$

with $K > 0$,

2) there exist partial derivatives $D_q F_h = (\, D_{q_1} F_h, \ldots, D_{q_n} F_h\,)$ on Ω_h, and there are $(L_1, \ldots, L_n) \in R^n_+$ such that

$$|\, D_{q_i} f_h(x,y,w,q)\,| \leq L_i, \; i = 1, \ldots, n, \;\; on \;\; \Omega_h, \tag{6.17}$$

3) for $(x, y^{(m)}, w, q) \in \Omega_h$, $s \in S \setminus \{0\}$ the inequalities are satisfied

$$D_{q_i} f_h(x, y^{(m)}, w, q)\, c^{(i)}_{s.m} \geq 0, \; i = 1, \ldots, n, \tag{6.18}$$

4) there exists $C_1 \in R_+$ such that

$$\sum_{s \in S} |c^{(i)}_{s.m}| \leq C_1 \;\; for \;\; i = 1, \ldots, n, \; m \in \mathbf{Z}^n,$$

5) the function $\phi_h : E_{0.h} \to R$, $h \in \Delta$, *is of class* Σ *and there exists a function* $v_h : E_{0.h} \cup E_h \to R$, $h \in \Delta$, *belonging to class* Σ, *having the derivative* $D_x v_h(x, y^{(m)})$ *for* $x \in (0, a]$, $m \in \mathbf{Z}^n$ *and satisfying the estimates*

$$|\,(v_h - \phi_h)^{(m)}(x)\,| \leq \gamma_0(h) \quad on \quad E_{0.h}$$

and

$$\Big| D_x v_h(x, y^{(m)}) - F_h[\,v_h\,]^{(m)}(x) \Big| \leq \gamma_1(h) \quad on \quad E_h$$

where $\gamma_i : \Delta \to R_+$, $i = 0, 1$, *are such functions that* $\lim_{h\to 0} \gamma_i(h) = 0$.
Then there exists a solution $u_h : E_{0.h} \cup E_h \to R$ *of problem (6.3), (6.4) belonging to class* Σ *and a function* $\omega : [0, a] \times \Delta \to R_+$ *such that* $\big|(\,u_h - v_h\,)^{(m)}(x)\big| \leq \omega(x, h)$ *on* E_h *and* $\lim_{h\to 0} \omega(x, h) = 0$ *uniformly with respect to* $x \in [0, a]$.

PROOF. Let $h \in \Delta$ be fixed. We shall show the existence and uniqueness of the solution of problem (6.3), (6.4). Let X be the set of all sequences

$$\zeta = \{\,\zeta^{(m)}\,\}_{m\in\mathbf{Z}^n}, \quad \zeta^{(m)} \in R,$$

such that $\sup\{\,|\zeta^{(m)}| : \, m \in \mathbf{Z}^n\,\} < \infty$. Then X is the Banach space if we define the norm $\|\zeta\|_X = \sup\{\,|\zeta^{(m)}| : m \in \mathbf{Z}^n\,\}$. Then $C([-r_0, 0], X)$ is also the Banach space with the norm $\|w\|_{CX} = \max\{\,\|w(t)\|_X : \, t \in [-r_0, 0]\,\}$ where $w \in C(\,[-r_0, 0], X\,)$. Define the family of operators $A_m : C(\,[-r_0, 0], X\,) \to C(B_h, R)$, $m \in \mathbf{Z}^n$, in the following way: $(A_m w)(t, y) = w(t, y^{(m)} + y)$, $(t, y) \in B_h$. Let the function $\tilde{\phi}_h : E_{0.h} \cup E_h \to R$ be defined by

$$\tilde{\phi}_h(x, y) = \phi_h(x, y) \text{ on } E_{0.h} \text{ and } \tilde{\phi}(x, y) = \phi(0, y) \text{ on } E_h.$$

We define the function $g_h : [0, a] \times C([-r_0, 0], X) \to X$, $g_h = \{\,g_h^{(m)}\,\}_{m\in\mathbf{Z}^n}$, by

$$g_h^{(m)}(x, w) = f_h(\,x, y^{(m)}, A_m w + (\tilde{\phi})_{(x,m)}, \delta w^{(m)}(0) + \delta\tilde{\phi}^{(m)}(x)\,),$$

where $x \in [0, a]$, $w \in C([-r_0, 0], X)$, $m \in \mathbf{Z}^n$. Consider the Cauchy problem

$$\zeta'(x) = g_h(\,x, \zeta_{(x)}\,) \text{ for } x \in [0, a], \quad \zeta(x) = 0 \text{ on } [-r_0, 0], \tag{6.19}$$

in the Banach space X. Let us show that the function g_h satisfies the Lipschitz condition with respect to functional variable. We have

$$|\,g_h^{(m)}(x, w) - g_h^{(m)}(x, \bar{w})\,|$$

$$\leq \Big| \, f_h(\,x, y^{(m)}, A_m w + (\tilde{\phi})_{(x,m)}, \delta w^{(m)}(0) + \delta\tilde{\phi}^{(m)}(0)\,)$$

$$- f_h(\,x, y^{(m)}, A_m \bar{w} + (\tilde{\phi})_{(x,m)}, \delta \bar{w}^{(m)}(0) + \delta\tilde{\phi}^{(m)}(0)\,) \Big|$$

$$\leq K\,\|A_m w - A_m \bar{w}\|_h + \sum_{i=1}^{n} |\,\delta_i(w - \bar{w})^{(m)}(0)\,|,$$

and consequently

$$| g_h^{(m)}(x,w) - g_h^{(m)}(x,\bar{w}) |$$

$$\leq K \max\Big\{ | (w-\bar{w})^{(m+s)}(t) | : \ t \in [-r_0,0], \ -N \leq s \leq N \Big\}$$

$$+C_1 \sum_{i=1}^{n} \frac{L_i}{h_i} \max\Big\{ | (w-\bar{w})^{(m+s)}(0) | : \ s \in S \Big\}$$

$$\leq \left[K + C_1 \sum_{i=1}^{n} \frac{L_i}{h_i} \right] \|w - \bar{w}\|_{CX},$$

where $x \in [0,a]$, w, $\bar{w} \in C([-r_0,0],X)$ and $m \in \mathbf{Z}^n$. Therefore we have

$$\|g_h(x,w) - g_h(x,\bar{w})\|_X \leq L[h] \, \|w-\bar{w}\|_{CX}$$

where $L[h]$ is a constant depending on h. System (6.19) is the Cauchy problem for functional differential equations in the Banach space. The right-hand side of the equation satisfies the global Lipschitz condition with respect to the functional variable. Hence problem (6.19) has exactly one solution $\zeta_h : [-r_0,a] \to X$. The function u_h defined by $u_h^{(m)}(x) = \zeta_h^{(m)} - \tilde{\phi}_h^{(m)}(x)$, $x \in [-r_0,a]$, $m \in \mathbf{Z}^n$, is the solution of (6.3), (6.4) belonging to class Σ. Now we shall prove the stability of problem (6.3), (6.4). Let us define $\omega : [-r_0,a] \times \Delta \to R_+$ in the following way:

$$\omega(x,h) = \left[\frac{1}{K}\gamma_1(h) + \gamma_0(h) \right] \exp(Kx) - \frac{1}{K}\gamma_1(h), \quad x \in [0,a],$$

$$\omega(x,h) = \gamma_0(h) \ \text{ for } \ x \in [-r_0,0].$$

The function $\omega(\,\cdot\,,h)$ is the solution of the problem

$$\omega'(x) = K\|\omega_{(x)}\|_0 + \gamma_1(h), \ x \in [0,a], \qquad \omega(x) = \gamma_0(h) \text{ on } [-r_0,0]$$

and $\lim_{h\to 0} \omega(x,h) = 0$ uniformly with respect to $x \in [-r_0,a]$. Since

$$D_x\,(\,u_h - v_h\,)^{(m)}\,(x)$$

$$= [\, f_h(\,x, y^{(m)}, (u_h)_{(x,m)}, \delta u_h^{(m)}(x)\,) - f_h(\,x, y^{(m)}, (u_h)_{(x,m)}, \delta v_h^{(m)}(x)\,)\,]$$

$$+f_h(\,x, y^{(m)}, (u_h)_{(x,m)}, \delta v_h^{(m)}(x)\,) - f_h(\,x, y^{(m)}, (v_h)_{(x,m)}, \delta v_h^{(m)}(x)\,)$$

$$+F_h[\,v_h\,]^{(m)}(x) - D_x v_h^{(m)}(x), \quad x \in [0,a], \ m \in \mathbf{Z}^n,$$

it follows that

$$\left| D_x(u_h - v_h)^{(m)}(x) - \sum_{i=1}^{n} \int_0^1 D_{q_i} f_h\,(\,Q_m(x,\tau)\,)\, d\tau \, \delta_i(u_h - v_h)^{(m)}(x) \right|$$

$$\leq K\|(u_h)_{(x,m)} - (v_h)_{(x,m)}\|_h + \gamma_1(h)$$

$$= K\|V_h(u_h - v_h)_{(x,m)}\|_0 + \gamma_1(h)$$

for $x \in [0,a]$, $m \in \mathbf{Z}^n$, where

$$Q_m(x,\tau) = (\, x, y^{(m)}, (u_h)_{(x,m)}, \delta v_h^{(m)}(x) + \tau\, \delta(u_h - v_h)^{(m)}(x)\,).$$

The function $u_h - v_h$ is of class Σ. Then all the assumptions of of Theorem 6.1 are satisfied and we have $|\,(\,u_h - v_h\,)^{(m)}(x)\,| \le \omega(x,h)$ on E_h, which completes the proof of the Theorem 6.2.

6.4 Convergence of the method of lines

Now we prove the main theorem for problems (6.1), (6.2) and (6.3), (6.4).

Theorem 6.3 *Suppose that the assumptions 1) - 4) of Theorem 6.2 are satisfied and*

1) the operator δ satisfies the conditions

$$\sum_{s\in S} c_{s.m}^{(i)} = 0,\ \sum_{s\in S} s_j c_{s.m}^{(i)} = \delta_{ij} \ \text{ for } \ i,j = 1,\dots,n,\ m \in \mathbf{Z}^n,$$

where δ_{ij} is the Kronecker symbol,

2) there exists $C_2 > 0$ such that $h_i h_j^{-1} \le C_2$, $i,j = 1,\dots,n$,

3) $v \in C(E_0 \cup E, R)$ is a solution of problem (6.1), (6.2), the function $v\,|_E$ is of class C^2 and $|\,D_{y_i y_j} v(x,y)\,| \le C_3$ on E, $i,j = 1,\dots,n$, for a certain $C_3 \ge 0$, and the function $v_h = v\,|_{E_{0.h}\cup E_h}$ is of class Σ,

4) $f \in C(\Omega, R)$ and there are $(L_1,\dots,L_n) \in R_+^n$ such that $|D_{q_i} f(x,y,w,q)| \le L_i$, $1 \le i \le n$, on Ω,

5) there exist functions $\beta_i : \Delta \to R_+$, $i = 0,1$, such that $\lim_{h\to 0}\beta_i(h) = 0$ and

$$|(\,\phi - \phi_h\,)^{(m)}(x)| \le \beta_0(h) \ \text{on } E_{0.h} \tag{6.20}$$

$$\Big|\, f(\,x, y^{(m)}, v_{(x,y^{(m)})}, \delta v^{(m)}(x)\,) - F_h[\,v_h\,]^{(m)}(x)\,\Big| \le \beta_1(h) \ \text{on } E_h. \tag{6.21}$$

Then there exist $\bar a > 0$ and a function $\bar\beta : \Delta \to R_+$ such that $|\,(\,u_h - v_h\,)^{(m)}(x)\,| \le \bar\beta(h)$ and $\lim_{h\to 0}\bar\beta(h) = 0$, where $h \in \Delta$, $\|h\| \in (0,\bar a]$, $x \in [0,a]$, $m \in \mathbf{Z}^n$.

PROOF. We will apply Theorem 6.2 in the proof of the above statement. From (6.20) it follows that $|\,(v_h - \phi_h)^{(m)}(x)\,| \le \beta_0(h)$ on $E_{0.h}$. By an elementary calculation we obtain from assumptions 1) - 3) that

$$\Big|\, D_{y_i} v^{(m)}(x) - \delta_i v_h^{(m)}(x)\,\Big| \le \frac{1}{2} n^2 C_1 C_2 C_3\, \|h\|,\ 1 \le i \le n, \ \text{ on } E_h. \tag{6.22}$$

Thus from assumptions 4) and 5) there exists a function $\beta : \Delta \to R_+$ such that $\lim_{h\to 0}\beta(h) = 0$ and

$$\Big|\, D_x v_h^{(m)}(x) - F_h[\,v_h\,]^{(m)}(x)\,\Big|$$

$$\le \Big|\, f(\,x, y^{(m)}, v_{(x,y^{(m)})}, D_y v^{(m)}(x)\,) - f(\,x, y^{(m)}, v_{(x,y^{(m)})}, \delta v^{(m)}(x)\,)\,\Big|$$

$$+\left| f(x, y^{(m)}, v_{(x,y^{(m)})}, \delta v^{(m)}(x)) - F_h[v_h]^{(m)}(x) \right|$$
$$\leq \beta(h) + \beta_1(h) \text{ on } E_h.$$

The assumptions of the Theorem 6.2 are satisfied with $\gamma_0 = \beta_0$, $\gamma_1 = \beta + \beta_1$. Taking $\bar{\beta} = \omega(a, h)$, $h \in \Delta$, we have the assertion of our theorem.

6.5 Examples of the numerical methods of lines

We will denote by $T_h : C(B_h, R) \to \mathbf{F}(B, R)$ the interpolating operator defined in the following way. Put $S_+ = \{s = (s_1, \ldots, s_n) : \ s_i \in \{0, 1\} \text{ for } 1 \leq i \leq n\ \}$. If $(x, y) \in B$ then there exists $m \in \mathbf{Z}^n$ such that $y^{(m)} \leq y \leq y^{(m+1)}$, $(x, y^{(m)})$, $(x, y^{(m+1)}) \in B_h$ where $m+1 = (m_1+1, \ldots, m_n+1)$. For $w \in \mathbf{F}_c(B_h, R)$ we set

$$(T_h w)(x, y) = \sum_{s \in S_+} w^{(m+s)}(x) \left(\frac{y - y^{(m)}}{h}\right)^s \left(1 - \frac{y - y^{(m)}}{h}\right)^{1-s}, \tag{6.23}$$

where

$$\left(\frac{y - y^{(m)}}{h}\right)^s = \prod_{i=1}^{n} \left(\frac{y_i - y_i^{(m_i)}}{h_i}\right)^{s_i}, \tag{6.24}$$

and

$$\left(1 - \frac{y - y^{(m)}}{h}\right)^{1-s} = \prod_{i=1}^{n} \left(1 - \frac{y_i - y_i^{(m_i)}}{h_i}\right)^{1-s_i}, \tag{6.25}$$

and we put $0^0 = 1$ in the above definitions. It is easy to check that $T_h w \in C(B, R)$ if $w \in C(B_h, R)$.

Example 1. Let us consider problem (6.1), (6.2) and the method of lines (6.3), (6.4) where the function $f_h : \Omega_h \to R$, $h \in \Delta$, is defined by

$$f_h(x, y^{(m)}, z_{(x,m)}, q) = f(x, y^{(m)}, T_h z_{(x,m)}, q)$$

on $E_h \times \mathbf{F}(E_{0.h} \cup E_h, R) \times R^n$. We denote $e_i = (0, \ldots, 0, 1, 0, \ldots, 0) \in R^n$ and 1 standing on the $i-th$ place, $1 \leq i \leq n$. We assume that the function $D_q f$ satisfies the conditions

$$D_{q_i} f(x, y, w, q) \geq 0 \text{ on } \Omega \text{ for } 1 \leq i \leq \kappa, \tag{6.26}$$

$$D_{q_i} f(x, y, w, q) \leq 0 \text{ on } \Omega \text{ for } \kappa + 1 \leq i \leq n, \tag{6.27}$$

where $0 \leq \kappa \leq n$ is fixed. We define the difference operator δ by

$$\delta_i z^{(m)}(x) = \frac{1}{h_i}\left[z^{(m+e_i)}(x) - z^{(m)}(x)\right], \ i = 1, \ldots, \kappa, \tag{6.28}$$

$$\delta_i z^{(m)}(x) = \frac{1}{h_i}\left[z^{(m)}(x) - z^{(m-e_i)}(x)\right], \ i = \kappa + 1, \ldots, n. \tag{6.29}$$

Then condition (6.18) is satisfied. If we define

$$\delta_i z^{(m)}(x) = \frac{1}{h_i}\left[2z^{(m+e_i)}(x) - 3z^{(m)}(x) + z^{(m-e_i)}(x)\right], \; i = 1,\dots,n, \tag{6.30}$$

where $x \in [0,a]$, $m \in \mathbf{Z}^n$, and assume that $D_{q_i} f(x,y,w,q) \geq 0$ on Ω, $i = 1,\dots,n$ then condition (6.18) is satisfied.

Let us consider now the difference operator defined by

$$\delta_i z^{(m)}(x)$$

$$= \frac{1}{h_i}\left[\sum_{j=1,\, j\neq i}^{n} \left(z^{(m+e_j)}(x) + z^{(m-e_j)}(x)\right) + z^{(m+e_i)}(x) - (2n-1)z^{(m)}(x)\right]$$

where $1 \leq i \leq \kappa$ and

$$\delta_i z^{(m)}(x)$$

$$= \frac{1}{h_i}\left[(2n-1)z^{(m)}(x) - \sum_{j=1, j\neq i}^{n} \left(z^{(m+e_j)}(x) + z^{(m-e_j)}(x)\right) - z^{(m-e_i)}(x)\right]$$

where $\kappa + 1 \leq i \leq n$. If the function $D_q f$ satisfies conditions (6.26), (6.27) then assumption (6.18) is satisfied.

Example 2. Let $B' = [-r,r]$. Consider the differential integral equation

$$D_x z(x,y) = f(\,x,y, \int_{B'} z_{(x,y)}(s)\,ds, D_y z(x,y)\,) \tag{6.31}$$

with the initial condition

$$z(0,y) = \phi(y) \quad \text{for } y \in R^n, \tag{6.32}$$

where the function $z_{(x,y)} : B' \to R$ is given by $z_{(x,y)}(s) = z(x,y+s)$, $s \in B'$. One of the method of lines for the Cauchy problem (6.31), (6.32) is the following

$$D_x z^{(m)}(x) = f(\,x, y^{(m)}, \int_{B'} T_h z_{(x,m)}(s)\,ds,\; \delta z^{(m)}(x)\,),$$

$$z(0,y^{(m)}) = \phi(y^{(m)}) \quad \text{for } m \in \mathbf{Z}^n.$$

We have obtained the Cauchy problem for the infinite system of ordinary differential equations where

$$\int_{B'} T_h z_{(x,m)}(s)ds = \frac{1}{2^n}\left[\prod_{i=1}^{n} h_i\right] \sum_{\xi=-N}^{N-1} \sum_{s\in S_+} z^{(m+\xi+s)}(x) \tag{6.33}$$

and $N - 1 = (N_1 - 1,\dots,N_n - 1)$.

Example 3. Let $\tilde{f} : E \times R \times R^n \to R$, and $\alpha : E \to [0,a]$, $\beta : E \to R^n$ be given functions. Define $f(x,y,w,q) = \tilde{f}(\,x,y,w(\alpha(x,y) - x, \beta(x,y) - y), q\,)$. Equation (6.1) reduces to the differential equation with a deviated argument

$$D_x z(x,y) = \tilde{f}(\,x,y,z(\alpha(x,y),\beta(x,y)), D_y z(x,y)\,).$$

If we set

$$f_h(x, y^{(m)}, z_{(x,m)}, q) = \tilde{f}(\,x, y^{(m)}, T_h z_{(x,m)}(\alpha(x, y^{(m)}) - x, \beta(x, y^{(m)}) - y^{(m)}), q\,)$$

then we obtain method (6.3), (6.4).

Now we give a theorem on the error estimate for the numerical method of lines. Consider problem (6.1), (6.2) and the method of lines

$$D_x z^{(m)}(x) = f(\,x, y^{(m)}, T_h z_{(x,m)}, \delta z^{m)}(x)\,) \tag{6.34}$$

$$z^{(m)}(x) = \phi_h^{(m)}(x) \text{ on } E_{0.h}, \tag{6.35}$$

where δ is defined by (6.28), (6.29) and $\phi_h : E_{0.h} \to R$ is a given function.

Theorem 6.4 *Suppose that*

1) the function $f : \Omega \to R$ is continuous and satisfies the Lipschitz condition

$$|\,f(x, y, w, q) - f(x, y, \bar{w}, q)\,| \le K\, \|w - \bar{w}\|_0$$

with $K > 0$,

2) there exist partial derivatives

$$(\,D_{q_1} f, \ldots, D_{q_n} f\,) = D_q f$$

on Ω, conditions (6.26), (6.27) are satisfied and there are $(L_1, \ldots, L_n) \in R_+^n$ such that $|\,D_{q_i} f(x, y, w, q)\,| \le L_i$ on Ω for $1 \le i \le n$,

3) the function $\phi_h : E_{0.h} \to R$, $h \in \Delta$, is of class Σ,

4) $v \in C(E_0 \cup E, R)$ is a solution of problem (6.1), (6.2), the function $v\,|_E$ is of class C^2 and there is $\tilde{C} \in R_+$ such that $|\,D_{y_i y_j} v(x, y)\,| \le \tilde{C}$ on E, $i, j = 1, \ldots, n$,

5) the function $v_h = v\,|_{E_{0.h} \cup E_h}$ is of class Σ and there is a function $\beta_0 : \Delta \to R_+$ such that $|\,(\phi - \phi_h)(x, y)\,| \le \beta_0(h)$ on $E_{0.h}$ and $\lim_{h \to 0} \beta_0(h) = 0$.

Then there exists a solution $u_h : E_{0.h} \cup E_h \to R$ of problem (6.34), (6.35) belonging to class Σ and we have the following error estimate

$$\left|(u_h - v_h)^{(m)}(x)\right| \le \left[\beta_0(h) + \frac{\beta(h)}{K}\right] \exp[Kx] - \frac{\beta(h)}{K} \quad \textit{on } E_h, \tag{6.36}$$

where

$$\beta(h) = \frac{1}{2} \sum_{i=1}^{n} L_i h_i + K\tilde{C} \sum_{i,j=1}^{n} h_i\, h_j.$$

PROOF. The existence of the solution $u_h : E_{0.h} \cup E_h \to R$ of problem (6.34), (6.35) follows from Theorem 6.2. Now we prove estimation (6.36). Let the function $\Gamma : E_h \times \Delta \to R$ be defined by

$$D_x v_h^{(m)}(x) = f(\,x, y^{(m)}, T_h(v_h)_{(x,m)}, \delta v_h^{(m)}(x)\,) + \Gamma^{(m)}(x, h),$$

where $(x, y^{(m)}) \in E_h$. We prove that

$$\left|\Gamma^{(m)}(x, h)\right| \le \beta(h) \text{ for } (x, y^{(m)}, h) \in E_h \times \Delta. \tag{6.37}$$

The assumptions 1) and 2) imply

$$\left| \Gamma^{(m)}(x,h) \right| \tag{6.38}$$

$$= \left| f(x, y^{(m)}, v_{(x,y^{(m)})}, D_y v^{(m)}(x)) - f(x, y^{(m)}, T_h(v_h)_{(x,m)}, \delta v_h^{(m)}(x)) \right|$$

$$\leq K \, \| v_{(x,y^{(m)})} - T_h(v_h)_{(x,m)} \|_0 + \sum_{i=1}^{n} L_i \left| D_{y_i} v^{(m)}(x) - \delta_i v_h^{(m)}(x) \right| .$$

We prove that

$$\| v_{(x,y^{(m)})} - T_h(v_h)_{(x,m)} \|_0 \leq \tilde{C} \sum_{i,j=1}^{n} h_i \, h_j . \tag{6.39}$$

Let $(t,y) \in B$. There is $\bar{m} = (\bar{m}_1, \ldots, \bar{m}_n) \in \mathbf{Z}^n$ such that $y^{(\bar{m})} \leq y^{(m)} + y \leq y^{(\bar{m}+1)}$, $(x, y^{(\bar{m})})$, $(x, y^{(\bar{m})}) \in B_h$ where $\bar{m} + 1 = (\bar{m}_1 + 1, \ldots, \bar{m}_n + 1)$. Then we have

$$v_{(x,y^{(m)})}(t,y) - T_h(v_h)_{(x,m)}(t,y) = v(x+t, y^{(m)} + y)$$

$$- \sum_{s \in S_+} v_h^{(\bar{m}+s)}(x+t) \left(\frac{y^{(m)} + y - y^{(\bar{m})}}{h} \right)^s \left(1 - \frac{y^{(m)} + y - y^{(\bar{m})}}{h} \right)^{1-s}$$

The operator T_h has the following properties

$$\sum_{s \in S_+} \left(\frac{y^{(m)} + y - y^{(\bar{m})}}{h} \right)^s \left(1 - \frac{y^{(m)} + y - y^{(\bar{m})}}{h} \right)^{1-s} = 1$$

and

$$\sum_{s \in S_+} \left(\frac{y^{(m)} + y - y^{(\bar{m})}}{h} \right)^s \left(1 - \frac{y^{(m)} + y - y^{(\bar{m})}}{h} \right)^{1-s} h_j s_j$$

$$= y_j^{(m_j)} + y_j - y_j^{(\bar{m}_j)}, \quad 1 \leq j \leq n.$$

Now, using the Taylor formula of the second order for v and the above relations we get

$$v_{(x,y^{(m)})}(t,y) - T_h(v_h)_{(x,m)}(t,y)$$

$$= \frac{1}{2} \sum_{i,j=1}^{n} D_{y_j y_j} v(Q) \left(y_i^{(m_i)} + y_i - y_i^{(\bar{m}_i)} \right) \left(y_j^{(m_j)} + y_j - y_j^{(\bar{m}_j)} \right)$$

$$- \frac{1}{2} \sum_{s \in S_+} \left(\frac{y^{(m)} + y - y^{(\bar{m})}}{h} \right)^s \left(1 - \frac{y^{(m)} + y - y^{(\bar{m})}}{h} \right)^{1-s}$$

$$\times \sum_{i,j=1}^{n} D_{y_i y_j} v(\tilde{Q}) \, s_i h_i \, s_j h_j$$

where $Q,\, \tilde{Q} \in [0,a] \times R^n$, and consequently

$$\left| v_{(x,y^{(m)})}(t,y) - T_h(v_h)_{(x,m)}(t,y) \right| \le \tilde{C} \sum_{i,j=1}^{n} h_i h_j, \quad (t,y) \in B,$$

which completes the proof of (6.39). It follows from assumption 4) that

$$\left| D_{y_i} v^{(m)}(x) - \delta_i v_h^{(m)}(x) \right| \le \frac{h_i}{2} \tilde{C} \ \text{ on } E_h \ \ 1 \le i \le n. \tag{6.40}$$

Estimates (6.38) - (6.40) imply (6.37). It is easy to prove that the function $u_h - v_h$ satisfies the differential difference inequality

$$\left| D_x (u_h - v_h)^{(m)}(x) - \sum_{i=1}^{n} \int_0^1 D_{q_i} f(\, Q(x,m,t)\,)\, dt\ \delta_i (u_h - v_h)^{(m)}(x) \right|$$

$$\le K \| u_h - v_h \|_h + \beta(h) \ \text{ on } E_h$$

where $Q(x,m,t) = (\, x, y^{(m)}, T_h(v_h)_{(x,m)}, \delta u_h^{(m)}(x) + t\,\delta(v_h - u_h)^{(m)}(x)\,)$. Denote by $\bar{\omega}(\,\cdot\,,h)$ the solution of the problem

$$\omega'(x) = K\,\omega(x) + \beta(h), \quad \omega(0) = \beta_0(h).$$

Since $\left| (u_h - v_h)^{(m)}(x) \right| \le \beta_0(h)$ on E_h, it follows from Theorem 6.1 that

$$\left| (u_h - v_h)^{(m)}(x) \right| \le \bar{\omega}(x,h) = \left[\beta_0(h) + \frac{\beta(h)}{K} \right] \exp{(Kx)} - \frac{\beta(h)|}{K} \ \text{ on } E_h.$$

This completes the proof of the theorem.

Remark 6.5 *It is easy to see that Theorem 6.4 can be applied for differential equations with a deviated argument and for differential integral problems.*

6.6 Differential difference inequalities for mixed problems

Let $a > 0$, $r_0 \in R_+$, $b = (b_1, \ldots, b_n) \in R^n$ and $r = (r_1, \ldots, r_n) \in R^n_+$ be given where $b_i > 0$ for $1 \le i \le n$. Suppose that $\kappa \in \mathbf{Z}$, $0 \le \kappa \le n$, is fixed. Let B, E, $E_0, \partial_0 E$, E^* be the sets defined in Section 5.1. Put $\Omega = E \times C(B,R) \times R^n$ and suppose that $f : \Omega \to R$ and $\varphi : E_0 \cup \partial_0 E \to R$ are given functions. Consider the mixed problem

$$D_x z(x,y) = f(\, x, y, z_{(x,y)}, D_y z(x,y)\,) \tag{6.41}$$

$$z(x,y) = \varphi(x,y) \ \text{ on } \ E_0 \cup \partial_0 E. \tag{6.42}$$

We are interested now in the discretization with respect to the spatial variable of the results given in Chapter V. In particular, we will present here a discrete version of the comparison Theorem 5.13. First of all we define a mesh in E^* and B with respect to the spatial variable. Let Δ be the set defined in Section 6.1. For $h = (h_1, \dots, h_n) \in \Delta$ we put $y^{(m)} = m * h$. Let

$$R^{1+n}_{x.h} = \{ (x, y^{(m)}) : x \in R, \ m \in \mathbf{Z}^n \}.$$

We define the sets

$$B_h = B \cap R^{1+n}_{x.h}, \quad E_{0.h} = E_0 \cap R^{1+n}_{x.h}, \quad E_h = E \cap R^{1+n}_{x.h}$$

and

$$\partial_0 E_h = \partial_0 E \cap R^{1+n}_{x.h}, \quad E^*_h = E^* \cap R^{1+n}_{x.h}.$$

Elements of E^*_h will be denoted by $(x, y^{(m)})$ or (x, y). Let $\mathbf{F}_c(B_h, R)$ be the set of all functions $w : B_h \to R$ such that $w(\,\cdot\,, y) \in C([-r_0, 0], R)$ for each fixed y. In the same way we define the sets $\mathbf{F}_c(E^*_h, R)$, $\mathbf{F}_c(E_h, R)$ and $\mathbf{F}_c(E_{0.h} \cup \partial_0 E_h, R)$. For a function $z \in \mathbf{F}_c(E^*_h, R)$ and a point $(x, y^{(m)}) \in E^*_h$ we write $z^{(m)}(x) = z(x, y^{(m)})$. For the above z and for a point $(x, y^{(m)}) \in E_h$ we define the function $z_{(x,m)} : B_h \to R$ by the formula

$$z_{(x,m)}(t, y^{(s)}) = z^{(m+s)}(x + t), \quad (t, y^{(s)}) \in B_h.$$

It is easy to see that $z_{(x,m)} \in \mathbf{F}_c(B_h, R)$ for the above z and $(x, y^{(m)})$. Given $z : E^*_h \to R$, let $\delta = (\delta_1, \dots, \delta_n)$ be the difference operator defined by

$$\delta_j z^{(m)}(x) = \frac{1}{h_j} \left[z^{(m+e_j)}(x) - z^{(m)}(x) \right] \quad \text{for } 1 \le j \le \kappa, \tag{6.43}$$

$$\delta_j z^{(m)}(x) = \frac{1}{h_j} \left[z^{(m)}(x) - z^{(m-e_j)}(x) \right] \quad \text{for } \kappa + 1 \le i \le n \tag{6.44}$$

and $\delta z = (\delta_1 z, \dots, \delta_n z)$.

For $w \in \mathbf{F}_c(B_h, R)$ we define the function $V_h w : [-r_0, 0] \to R_+$ by

$$(V_h w)(x) = \max\{ |w^{(m)}(x)| : \ y^{(m)} \in [0, r'] \times [-r'', 0] \}, \quad x \in [-r_0, 0].$$

Let us consider the following set of assumptions.

Assumption H $[\Lambda, \sigma]$. Suppose that the functions

$$\Lambda = (\lambda_1, \dots, \lambda_n) : E_h \times \mathbf{F}_c(B_h, R) \to R^n \ \text{ and } \ \sigma : [0, a) \times C([-r_0, 0], R_+) \to R_+$$

satisfy the conditions:

1) $\lambda_i(x, y, w) \ge 0$ on $E_h \times \mathbf{F}_c(B_h, R)$ for $1 \le i \le \kappa$ and $\lambda_i(x, y, w) \le 0$ on $E_h \times \mathbf{F}_c(B_h, R)$ for $\kappa + 1 \le i \le n$,

2) σ is continuous and is nondecreasing with respect to the functional variable,

3) for each $\eta \in C([-r_0, 0], R_+)$ there is on $[-r_0, a)$ the maximum solution $\omega(\,\cdot\,, \eta)$ of the Cauchy problem

$$\omega'(x) = \sigma(\, x, \omega_{(x)}\,), \quad \omega(x) = \eta(x) \ \text{ on } \ [-r_0, 0]. \tag{6.45}$$

Theorem 6.6 *Suppose that Assumption $H\,[\Lambda,\sigma\,]$ is satisfied and*
1) the function $u \in \mathbf{F}_c(E_h^*, R)$ satisfies the initial inequality

$$|\,u^{(m)}(x)\,| \le \eta(x), \quad (x, y^{(m)}) \in E_{0.h}, \tag{6.46}$$

where $\eta \in C([-r_0, 0], R_+)$ and the boundary inequality holds

$$|u^{(m)}(x)| \le \omega(x, \eta), \quad (x, y^{(m)}) \in \partial_0 E_h, \tag{6.47}$$

2) the functional differential difference inequality

$$\left| D_- u^{(m)}(x) - \sum_{i=1}^{n} \lambda_i(x, y^{(m)}, u_{(x,m)})\, \delta_i u^{(m)}(x) \right| \le \sigma(\,x, V_h u_{(x,m)}\,) \tag{6.48}$$

is satisfied on E_h.

Under these assumptions we have

$$|\,u^{(m)}(x)\,| \le \omega(x, \eta) \text{ on } E_h. \tag{6.49}$$

Proof. Consider the function $\tilde{\omega}(x) = \max\{\, |u^{(m)}(x)| \, : \, (x, y^{(m)}) \in E_h^* \,\}$ where $x \in [-r_0, a)$. Then $\tilde{\omega} \in C([-r_0, a), R_+)$ and estimation (6.49) is equivalent to

$$\tilde{\omega}(x) \le \omega(x, \eta), \quad x \in [-r_0, a). \tag{6.50}$$

We shall prove (6.50) by using Lemma 5.12. In order to do this, let $J_+ = \{x \in (0, a) : \tilde{\omega}(x) > \omega(x, \eta)\,\}$. We only need to show that

$$D_- \tilde{\omega}(x) \le \sigma(\,x, \tilde{\omega}_{(x)}\,) \text{ for } x \in J_+ \tag{6.51}$$

and

$$\tilde{\omega}(x) \le \eta(x) \text{ on } [-r_0, 0]. \tag{6.52}$$

Let $x \in J_+$ be fixed. Then $\tilde{\omega}(x) > 0$ and there exists $m \in \mathbf{Z}^n$ such that $\tilde{\omega}(x) = |\,u^{(m)}(x)\,|$. From (6.47) it follows that $(x, y^{(m)}) \in E_h$. Thus two possibilities can happen, either (i) $\tilde{\omega}(x) = u^{(m)}(x)$ or (ii) $\tilde{\omega}(x) = -u^{(m)}(x)$. In the case that (i) holds, we have

$$D_- u^{(m)}(x) \le \sigma(\,x, V_h u_{(x,m)}\,)$$
$$+ \sum_{j=1}^{\kappa} \lambda_j(\,x, y^{(m)} u_{(x,m)}\,) \frac{1}{h_j} \left[u^{(m+e_j)}(x) - u^{(m)}(x) \right]$$
$$+ \sum_{j=\kappa+1}^{n} \lambda_j(\,x, y^{(m)} u_{(x,m)}\,) \frac{1}{h_j} \left[u^{(m)}(x) - u^{(m-e_j)}(x) \right] \le \sigma(\,x, \tilde{\omega}_{(x)}\,).$$

Inequality (6.51) is proved in the case that (i) holds. Analogously (6.51) can be proved if (ii) is satisfied. From (6.46) we have that initial estimation (6.52) holds, thus an application of Lemma 5.12 completes the proof of (6.49).

6.7 Method of lines for mixed problem

Write $\Omega_h = E_h \times \mathbf{F}_c(B_h, R) \times R^n$ and suppose that $f_h : \Omega_h \to R$ and $\varphi_h \in \mathbf{F}_c(E_{0.h} \cup \partial_0 E_h, R)$ are given functions. We will approximate solutions of problem (6.41), (6.42) by solutions of the system of ordinary functional differential equations

$$D_x z^{(m)}(x) = f_h(x, y^{(m)}, z_{(x,m)}, \delta z^{(m)}(x)) \tag{6.53}$$

with the initial-boundary condition

$$z^{(m)}(x) = \varphi_h^{(m)}(x) \text{ on } E_{0.h} \cup \partial_0 E_h. \tag{6.54}$$

Let $\mathbf{F}_h$ be the Niemycki operator corresponding to (6.53) i. e. $F_h[z]^{(m)}(x) = f_h(x, y^{(m)}, z_{(x,m)}, \delta z^{(m)}(x))$. Consider the following set of assumptions.

Assumption H $[\sigma, f_h]$. Suppose that

1) the function $\sigma : [0, a_0) \times C([-r_0, 0], R_+) \to R_+$ where $a_0 > a$, satisfies conditions 2), 3) of Assumption H $[\Lambda, \sigma]$ and $\bar{\omega}(x) = 0$ for $x \in [-r_0, a_0)$ is the unique solution of problem (6.45) with $\eta = 0$ on $[-r_0, 0]$,

2) the function f_h of the variables (x, y, w, q) is continuous with respect to x and it has the derivatives $(D_{q_1} f_h, \ldots, D_{q_n} f_h) = D_q f_h$ on Ω_h,

3) $D_q f(x, y, w, \cdot) \in C(R^n, R^n)$ and for $P = (x, y, w, q) \in \Omega_h$ we have

$$D_{q_i} f_h(P) \geq 0 \text{ for } 1 \leq i \leq \kappa \text{ and } D_{q_i} f_h(P) \leq 0 \text{ for } \kappa + 1 \leq i \leq n,$$

4) the estimate

$$|f_h(x, y, w, q) - f_h(x, y, \bar{w}, q)| \leq \sigma(x, V_h(w - \bar{w}))$$

is satisfied on Ω_h.

Assumption H $[z_0]$. There is $z_0 \in \mathbf{F}_c(E_h^*, R)$ such that

1) z_0 satisfies initial-boundary condition (6.54) and the derivative $D_x z_0^{(m)}(x)$ exists on E_h,

2) there is $\gamma \in C([0, a), R_+)$ such that $|D_x z_0^{(m)}(x) - F_h[z_0]^{(m)}(x)| \leq \gamma(x)$ on E_h and the maximum solution ω_0 of the problem

$$\omega'(x) = \sigma(x, \omega_{(x)}) + \gamma(x), \quad \omega(x) = 0 \text{ on } [-r_0, 0] \tag{6.55}$$

is defined on $[-r_0, a)$.

Let us consider the sequence of functions $\{\omega_j\}$, $\omega_j : [-r_0, a) \to R_+$, $j \geq 0$, given by

(i) ω_0 is the maximum solution of problem (6.55),

(ii) if ω_j is given then $\omega_{j+1}(x) = 0$ on $[-r_0, 0]$ and

$$\omega_{j+1}(x) = \int_0^x \sigma(t, (\omega_j)_{(t)})\, dt \quad \text{for } x \in [0, a).$$

Lemma 6.7 *Suppose that σ satisfies condition 1) of Assumption H $[\sigma, f_h]$ and Assumption H $[z_0]$ holds. Then $\omega_{j+1}(x) \leq \omega_j(x)$ for $x \in [0, a)$, $j \geq 0$, and $\lim_{j\to\infty} \omega_j(x) = 0$ uniformly on $[0, a)$.*

We omit the simple proof of the above Lemma.

Theorem 6.8 *If Assumptions H* $[\sigma, f_h]$ *and H* $[z_0]$ *are satisfied then there exists exactly one solution* $u_h : E_h^* \to R$ *of problem (6.53), (6.54).*

PROOF. Let the sequence functions $\{z_j\}$, $z_j : E_h^* \to R$, $j \geq 0$, be defined by :
(i) z_0 is given in Assumption H $[z_0]$,
(ii) if z_j is the known function then z_{j+1} is the solution of the equation

$$D_x z^{(m)}(x) = f_h(x, y^{(m)}, (z_j)_{(x,m)}, \delta z^{(m)}(x))$$

with initial-boundary condition (6.54).
Now we prove that

$$\left| (z_{j+i} - z_j)^{(m)}(x) \right| \leq \omega_j(x) \text{ on } E_h, \tag{6.56}$$

for $i, j \geq 0$. First we prove (6.56) for $j = 0$ and $i \geq 0$. It follows that estimate (6.56) is satisfied for $j = 0$, $i = 0$. If we assume that $|(z_j - z_0)^{(m)}(x)| \leq \omega_0(x)$ on E_h, then using Hadamard mean value theorem and Assumption H $[\sigma, f_h]$ we conclude that the function $z_{i+1} - z_0$ satisfies the differential difference inequality

$$\cdot \left| D_x (z_{i+1} - z_0)^{(m)}(x) - \sum_{l=1}^{n} \int_0^1 D_{q_l} f_h(P^{(m)}(x,t)) dt \, \delta_l (z_{i+1} - z_0)^{(m)}(x) \right|$$

$$\leq \sigma(x, (\omega_0)_{(x)}) \text{ on } E_h,$$

where $P^{(m)}(x,t) = (x, y^{(m)}, (z_i)_{(x,m)}, \delta_0 z_0^{(m)}(x) + t\,\delta(z_{i+1} - z_0)^{(m)}(x))$. It follows from Theorem 6.6 that $|(z_{j+1} - z_0)^{(m)}(x)| \leq \omega_0(x)$ on E_h. Now suppose that we have the estimate $|(z_{j+i} - z_j)^{(m)}(x)| \leq \omega_j(x)$ on E_h for $i \geq 0$, with fixed $j \in \mathbf{N}$. Then, using the Hadamard theorem and Assumption H $[\sigma, f_h]$ we conclude that the function $z_{j+1+i} - z_{j+1}$ satisfies the differential difference inequality

$$\left| D_x (z_{j+1+i} - z_{j+1})^{(m)}(x) - \sum_{l=1}^{n} \int_0^1 D_{q_l} f_h(Q^{(m)}(x,t)) dt \delta_l (z_{j+1+i} - z_{j+1})^{(m)}(x) \right|$$

$$\leq \sigma(x, (\omega_j)_{(x)}) \text{ on } E_h,$$

where $Q^{(m)}(x,t) = (x, y^{(m)}, (z_{j+i})_{(x,m)}, \delta z_{j+1}^{(m)}(x) + t\,\delta(z_{j+1+i} - z_{j+1})^{(m)}(x))$. It follows from Theorem 6.6 that $|(z_{j+1+i} - z_{j+1})^{(m)}(x)| \leq \omega_*(x)$ on E_h, where $\omega_* : [-r_0, a) \to R_+$ is the maximum solution of the problem

$$\omega'(x) = \sigma(x, (\omega_j)_{(x)}), \quad \omega(x) = 0 \text{ on } [-r_0, 0].$$

Since $\omega_* = \omega_{j+1}$ then the proof of (6.56) is completed by induction. It follows from Lemma 6.7 and from (6.56) that the sequence $\{z_j\}$ is uniformly convergent on E_h. Let $u_h = \lim_{j\to\infty} z_j$. It follows that

$$z_{j+1}^{(m)}(x) = \varphi_h^{(m)}(0) + \int_0^x f_h(t, y^{(m)}, (z_j)_{(t,m)}, \delta z_{j+1}^{(m)}(t)) \, dt \text{ on } E_h,$$

where $j \geq 0$. Now, by passing to the limit for $j \to \infty$, we obtain that u_h is the unique solution of (6.53), (6.54).

Now we prove a theorem on the stability of the numerical method of lines.

Theorem 6.9 *Suppose that Assumptions $H[\sigma, f_h]$ and $H[z_0]$ are satisfied and the function $v_h : E_h^* \to R$, $\gamma_0, \gamma_1 : \Delta \to R_+$ are such that*

$$\left| D_x v_h^{(m)}(x) - F_h[v_h]^{(m)}(x) \right| \leq \gamma_1(h) \text{ on } E_h \tag{6.57}$$

and

$$\left| (v_h - \varphi_h)^{(m)}(x) \right| \leq \gamma_0(h) \text{ on } E_{0.h} \cup \partial_0 E_h$$

where $\lim_{h\to 0} \gamma_0(h) = 0$, $\lim_{h\to 0} \gamma_1(h) = 0$.

Then there is $\tilde{a} > 0$ and a function $\tilde{\omega} : [-r_0, a) \times \Delta \to R_+$ such that for $h \in \Delta$, $\|h\| \in (0, \tilde{a})$ we have

$$|(u_h - v_h)^{(m)}(x)| \leq \tilde{\omega}(x, h) \text{ on } E_h \tag{6.58}$$

and

$$\lim_{h'\to 0} \tilde{\omega}(x, h) = 0 \text{ } uniformly \text{ on } [0, a). \tag{6.59}$$

PROOF. There is $\tilde{a} > 0$ such that for $\|h\| \in (0, \tilde{a})$ the maximum solution $\tilde{\omega}(\cdot, h)$ of the Cauchy problem

$$\omega'(x) = \sigma(x, \omega_{(x)}) + \gamma_1(h), \qquad \omega(x) = \gamma_0(h) \text{ on } [-r_0, 0],$$

is defined on $[-r_0, a)$ and condition (6.59) holds. Put

$$\tilde{Q}^{(m)}(x, t) = (x, y^{(m)}, (u_h)_{(x,m}, \delta v_h^{(m)}(x) + t\,\delta(u_h - v_h)^{(m)}(x)).$$

It follows from Assumption H $[\sigma, f_h]$ and from Hadamard mean value theorem that the function $u_h - v_h$ satisfies the functional difference inequality

$$\left| D_x (u_h - v_h)^{(m)}(x) - \sum_{i=1}^{n} \int_0^1 D_{q_i} f_h(\tilde{Q}(x, t))\, dt\, \delta_i (u_h - v_h)^{(m)}(x) \right|$$

$$\leq \sigma(x, V_h(u_h - v_h)) + \gamma_1(h) \text{ on } E_h$$

and $|(u_h - v_h)^{(m)}(x)| \leq \gamma_0(h)$ on $E_{0.h} \cup \partial_0 E_h$. Now we obtain estimation (6.58) from Theorem 6.6.

The next theorem deals with the convergence of the numerical method of lines.

Theorem 6.10 *Suppose that Assumptions $H[\sigma, f_h]$, $H[z_0]$ are satisfied and*

1) $f \in C(\Omega, R)$ and $v : E^* \to R$ is a solution of problem (6.41), (6.42), the function $v|_E$ is of class C^1 and the derivatives $D_x v$, $D_y v$ are bounded on E,

2) there is a function $\bar{\gamma} : \Delta \to R$ such that

$$\left| F_h[v_h]^{(m)}(x) - f(x, y^{(m)}, v_{(x,y^{(m)})}, \delta v^{(m)}(x)) \right| \leq \bar{\gamma}(h) \text{ on } E_h$$

and $\lim_{h\to 0}\bar{\gamma}(h) = 0$, *where* v_h *is the restriction of* v *to the set* E_h^*,

3) there is a function $\gamma_0 : \Delta \to R_+$ *such that* $|(\varphi - \varphi_h)^{(m)}(x)| \le \gamma_0(h)$ *on* $E_{0.h} \cup \partial_0 E_h$ *and* $\lim_{h\to 0}\gamma_0(h) = 0$.

Then there is $\tilde{a} > 0$ *and a function* $\tilde{\omega} : [-r_0, a) \times \Delta \to R_+$ *such that condition (6.58) is satisfied for* $\|h\| \in (0, \tilde{a})$ *and (6.59) holds.*

PROOF. It follows from the assumption 2) that there is $\gamma_1 : \Delta \to R_+$ such that condition (6.57) is satisfied on E_h. Hence (6.58), (6.59) follow by applying Theorem 6.9.

Remark 6.11 *Let* $T_h : \mathbf{F}(B_h, R) \to \mathbf{F}(B, R)$ *be the operator defined in the following way. Let* $w \in \mathbf{F}(B_h, R)$ *and* $(x, y) \in B$. *There exists* $m \in \mathbf{Z}^n$ *such that* $y^{(m)} \le y \le y^{(m+1)}$ *and* $(x, y^{(m)})$, $(x, y^{(m+1)}) \in B_h$. *Then we define* $(T_h w)(x, y)$ *by formulas (6.23) - (6.25). The simplest example of the numerical method of lines for problem (6.41), (6.42) is the system*

$$D_x z^{(m)}(x) = f(\,x, y^{(m)}, T_h z_{(x,y^{(m)})}, \delta z^{(m)}(x)\,)$$

with initial-boundary condition (6.54). Note that the above system satisfies the assumption 2) of Theorem 6.10. This assumption is called the compatibility condition of the method with respect to (6.41).

Now we give a theorem on the error estimate for the numerical method of lines. Let $T_h : \mathbf{F}(B_h, R) \to \mathbf{F}(B, R)$ be the operator defined in Remark 6.11. Consider problem (6.41), (6.42) and the method of lines

$$D_x z^{(m)}(x) = f(\,x, y^{(m)}, T_h z_{(x,y^{(m)})}, \delta z^{(m)}(x)\,) \tag{6.60}$$

with initial-boundary condition (6.54).

Theorem 6.12 *Suppose that*

1) the function $f : \Omega \to R$ *is continuous and satisfies the Lipschitz condition*

$$|\,f(x, y, w, q) - f(x, y, \bar{w}, q)\,| \le K\,\|w - \bar{w}\|_0$$

with $K > 0$,

2) there exist partial derivatives $(\,D_{q_1} f, \ldots, D_{q_n} f\,) = D_q f$ *and*

$$D_{q_i} f(x, y, w, q) \ge 0 \;\text{ on }\; \Omega \quad \text{for } 1 \le i \le \kappa, \tag{6.61}$$

$$D_{q_i} f(x, y, w, q) \le 0 \;\text{ on }\; \Omega \quad \text{for } \kappa + 1 \le i \le n, \tag{6.62}$$

3) $D_q f(x, y, w, \cdot\,) \in C(R^n, R^n)$ *and there are* $(\,L_1, \ldots, L_n\,) \in R_+^n$ *such that*

$$|\,D_{q_i} f(x, y, w, q)\,| \le L_i \;\text{ on }\; \Omega, \;\; i = 1, \ldots, n, \tag{6.63}$$

4) the function $v \in C(E^*, R)$ *is a solution of problem (6.41), (6.42),* $v\,|_E$ *is of class* C^2 *and there is* $\tilde{C}$ *such that* $\big|\,D_{y_i y_j} v(x, y)\,\big| \le \tilde{C}$ *on* E, $i, j = 1, \ldots, n$,

5) there is a function $\beta_0 : \Delta \to R_+$ *such that* $|\,(\varphi_h - \varphi)(x, y)\,| \le \beta_0(h)$ *on* $E_{0.h} \cup \partial E_{0.h}$.

Then there exists a solution $u_h : E_h^* \to R$ of problem (6.60), (6.54) and we have the following error estimate

$$|(u_h - v)(x,y)| \le \left[\beta_0(h) + \frac{\beta(h)}{K}\right] \exp[Kx] - \frac{\beta(h)}{K} \quad \textit{on } E_h \tag{6.64}$$

with

$$\beta(h) = \frac{1}{2}\sum_{i=1}^{n} L_i h_i + K\tilde{C}\sum_{i,j=1}^{n} h_i h_j.$$

PROOF. The proof is based on the following idea. The function $u_h - v_h$ where $v_h = v\,|_{E_h^*}$ satisfies the differential difference inequality

$$\left| D_x\,(u_h - v_h)\,(x, y^{(m)}) - \sum_{i=1}^{n}\int_0^1 D_{q_i} f(\,Q^{(m)}(x,t)\,)\,dt\,\delta_i\,(u_h - v_h\,)^{(m)}\,(x)\right|$$

$$\le K\|\,u_h - v_h\,\|_h + \beta(h) \;\text{ on }\; E_h,$$

with $Q^{(m)}(x,t) \in \Omega$. The initial-boundary estimate $|(u_h - v_h)(x,y)| \le \beta_0(h)$ on $E_{0.h} \cup \partial E_{0.h}$ and Theorem 6.6 imply the statement (6.64). This completes the proof of the Theorem.

Remark 6.13 *It is easy to see that the above theorem can be applied to differential equations with a deviated argument and differential integral problems.*

6.8 Modified Method of Lines

The aim of this Section is to give a modification of the numerical method of lines (6.60), (6.54) corresponding to the mixed problem (6.41), (6.42). We will denote by T_h the interpolating operator $T_h : \mathbf{F}_c(E_h^*, R) \to \mathbf{F}(E^*, R)$ defined in the following way. Suppose that $w \in \mathbf{F}_c(E_h^*, R)$ and $(x,y) \in E^*$. Then there is $m \in \mathbf{Z}^n$ such that $y^{(m)} \le y \le y^{(m+1)}$ and $(x, y^{(m)})$, $(x, y^{(m+1)}) \in E_h^*$ and we define $(T_h w)(x,y)$ by formulas (6.23) - (6.25). We see at once that $T_h w \in C(E^*, R)$ if $w \in \mathbf{F}_c(E_h^*, R)$. Let us consider now the sequence of steps $\{\,h^{(k)}\,\}$, $h^{(k)} = (\,h_1^{(k)}, \dots, h_n^{(k)}\,) \in \Delta$, such that $\lim_{k\to\infty} h^{(k)} = 0$. Let

$$B_{h^{(k)}},\;\; E_{0,h^{(k)}},\;\; E_{h^{(k)}},\;\; \partial_0 E_{h^{(k)}},\;\; E_{h^{(k)}}^*$$

be the sets defined in Section 6.7. For simplicity of notation, we write

$$B_k = B_{h^{(k)}},\;\; E_{0.k} = E_{0,h^{(k)}},\;\; E_k = E_{h^{(k)}},\;\; \partial_0 E_k = \partial_0 E_{h^{(k)}},\;\; E_k^* = E_{h^{(k)}}^*$$

and $T_k = T_{h^{(k)}}$. Suppose that $\varphi_k : E_{0.k} \cup \partial_0 E_k \to R$, $k \ge 0$, are given functions. We define the sequence $\{\,u_k\,\}$, $u_k : E_k^* \to R$, as follows. The function u_0 is given by

$$u_0(x,y) = \varphi_0(x,y) \text{ for } (x,y) \in E_{0.k} \cup \partial_0 E_0,$$

$$u_0(x,y) = \varphi_0(0,y) \text{ for } (x,y) \in E_0.$$

If $u_k : E_k^* \to R$ is a given function then $u_{k+1} E_k^* \to R$ is a solution of the problem

$$D_x z^{(m)}(x) = f(\,x, y^{(m)}, (T_k u_k)_{(x,y^{(m)})}, \delta z^{(m)}(x)\,), \tag{6.65}$$

where $(x, y^{(m)}) \in E_{k+1}$ and

$$z^{(m)}(x) = \varphi_k^{(m)}(x) \quad \text{on} \quad E_{0.k+1} \cup \partial_0 E_{k+1}, \tag{6.66}$$

and the operator $\delta = (\delta_1, \ldots, \delta_n)$ is given by (6.43), (6.44). In Section 6.7 the solution v of problem (6.41), (6.42) was approximated by solutions of functional differential problems, whereas (6.65), (6.66) is the initial problem for an ordinary differential system. Let us consider now the following set of assumptions.

Assumption H $[\,f, \sigma\,]$. Suppose that $f \in C(\Omega, R)$ and

1) the function $\sigma : [0, a_0) \times R_+ \to R,\ a_0 > a$, is continuous and is nondecreasing with respect to the second variable,

2) the function $\bar{\omega}(x) = 0,\ x \in [0, a_0)$, is the unique solution of the Cauchy problem

$$\omega'(x) = \sigma(\,x, \omega(x)\,), \quad \omega(0) = 0,$$

and the estimate

$$|\, f(x, y, w, q) - f(x, y, \bar{w}, q)\,| \le \sigma(\,x, \|w - \bar{w}\|_0\,)$$

is satisfied on Ω,

3) there exist partial derivatives $(\,D_{q_1} f, \ldots, D_{q_n} f\,) = D_q f$ on Ω and conditions (6.61), (6.62) are satisfied,

4) $D_q f(x, y, w, \cdot\,) \in C(R^n, R^n)$ and there are $(\,L_1, \ldots, L_n\,) \in R_+^n$ such that estimates (6.63) are satisfied.

Now we prove a theorem on the convergence of the sequence $\{\,u_k\,\}$.

Theorem 6.14 *Suppose that Assumption H $[\,f, \sigma\,]$ is satisfied and*

1) $v \in C(E^*, R)$ is a solution of problem (6.41), (6.42), the function $v\,|_E$ is of class C^2 and there is $\tilde{C} \in R_+$ such that $\left|\, D_{y_i y_j} v(x,y) \,\right| \le \tilde{C}$ on Ω for $i, j = 1, \ldots, n$,

2) $\varphi_k \in \mathbf{F}_c(E_{0.k} \cup \partial_0 E_k, R)$ for $k \ge 0$ and there are functions $\beta,\ \beta_0 : \Delta \to R_+$ such that

$$|\,(v - u_0)(x,y)\,| \le \beta(h^{(0)}) \quad on \quad E_0^*,$$

$$|\,(\varphi - \varphi_k)(x,y)\,| \le \beta_0(h^{(k)}) \quad on \quad E_{0.k} \cup \partial_0 E_k, \quad k \ge 0,$$

and $\lim_{h \to 0} \beta(h) = 0,\ \lim_{h \to 0} \beta_0(h) = 0$.

Under these assumptions we have the following statements:

(i) for each $k \in \mathbf{N}$ there exists a solutions $u_k : E_k^* \to R$ of problem (6.65), (6.66),

(ii) there is a function $\gamma : \Delta \to R_+$ such that $|\,(v - u_k)(x,y)\,| \le \gamma(h^{(k)})$, $(x,y) \in E_k$, $k \ge 0$, and $\lim_{h \to 0} \gamma(h) = 0$.

Proof. The existence of the sequence $\{\,u_k\,\}$ follows from Assumption H $[\,f, \sigma\,]$. Write

$$\alpha_k(x) = \max\ \{\,|\,v(t,y) - u_k(t,y)\,| : \ (t,y) \in E_k^*, \ \ t \le x\,\},$$

where $x \in [0, a)$, $k \geq 0$, and

$$\beta_1(h) = \tilde{C} \sum_{i,j=1}^{n} h_i h_j, \quad \beta_2(h) = \frac{1}{2} \sum_{i=1}^{n} L_i h_i, \quad h \in \Delta.$$

Let us denote by $\{ \omega_k \}$, $\omega_k : [0, a] \to R_+$, the sequence of functions given by $\omega_0(x) = \beta(h^{(0)})$, $x \in [0, a]$, if ω_k is known function that

$$\omega_{k+1}(x) = \beta_0(h^{(k+1)}) + \int_0^x \Big[\sigma(t, \beta_1(h^{(k)}) + \omega_k(t)) + \beta_2(h^{(k+1)}) \Big] \, dt, \quad x \in [0, a].$$

We prove that for $k \geq 0$ we have

$$\alpha_k(x) \leq \omega_k(x), \quad x \in [0, a). \tag{6.67}$$

It follows from assumption 2) that estimate (6.67) is satisfied for $k = 0$. Suppose now that we have $\alpha_k(x) \leq \omega_k(x)$, $x \in [0, a)$, for a fixed $k \in \mathbf{N}$. Let us denote by $\Gamma_{k+1} : E_{k+1} \to R$ the function given by

$$D_x v^{(m)}(x) = f(x, y^{(m)}, v_{(x,y^{(m)})}, \delta v^{(m)}(x)) + \Gamma_{k+1}^{(m)}(x), \quad (x, y^{(m)}) \in E_{k+1}.$$

It follows that $\left| \Gamma_{k+1}^{(m)}(x) \right| \leq \beta_2(h^{(k+1)})$ on E_{k+1}. Let $v_k = v \mid_{E_k^*}$. As in the proof of Theorem 6.4 we have

$$\begin{aligned} &\| v_{(x,y^{(m)})} - (T_k u_k)_{(x,y^{(m)})} \|_0 \\ &\leq \| v_{(x,y^{(m)})} - (T_k v_k)_{(x,y^{(m)})} \|_0 + \| (T_k v_k)_{(x,y^{(m)})} - (T_k u_k)_{(x,y^{(m)})} \|_0 \\ &\leq \beta_1(h^{(k)}) + \alpha_k(x), \quad (x, y^{(m)}) \in E_{k+1}, \end{aligned}$$

and therefore

$$\left| f(x, y^{(m)}, v_{(x,y^{(m)})}, q) - f(x, y^{(m)}, T_k(u_k)_{(x,y^{(m)})}, q) \right| \leq \sigma(x, \beta_1(h^{(k)}) + \alpha_k(x)).$$

According to the Hadamard mean value theorem it follows that the function $v - u_{k+1}$ satisfies the differential difference inequality

$$\left| D_x(v - u_{k+1})^{(m)}(x) - \sum_{i=1}^{n} \int_0^1 D_{q_i} f(Q^{(k,m)}(x,t)) \, dt \, \delta_i (v - u_{k+1})^{(m)}(x) \right|$$

$$\leq \sigma(x, \beta(h^{(k)}) + \alpha_k(x)) + \beta_2(h^{(k+1)}), \quad (x, y^{(m)} \in E_{k+1},$$

where

$$Q^{(k,m)}(x, t) = (x, y^{(m)}, v_{(x,y^{(m)})}, \delta v^{(m)}(x) + t \, \delta(u_{k+1} - v)^{(m)}(x))$$

and the initial-boundary estimate

$$| (v - u_{k+1})^{(m)}(x) | \leq \beta_0(h^{(k+1)}), \quad (x, y^{(m)} \in E_{0,k+1} \cup \partial_0 E_{k+1},$$

is satisfied. It follows from Theorem 6.6 that

$$\left| (v - u_{k+1})^{(m)}(x) \right| \leq \omega_{k+1}(x), \tag{6.68}$$

where $(x, y^{(m)}) \in E_{k+1}$. Since the function ω_{k+1} is nondecreasing on $[0, a)$, estimate (6.68) shows that $\alpha_{k+1}(x) \leq \omega_{k+1}(x)$ on $[0, a)$. Now we obtain the statement (6.67) by induction. An easy computation shows that $\lim_{k\to\infty} \omega_k(x) = 0$ uniformly on $[0, a]$. This proves the statement (ii) and the theorem follows.

Remark 6.15 *If f satisfies the Lipschitz condition with respect to the functional variable then we can obtain a result on the error estimate for the modified method of lines (6.65), (6.66). Details are omitted.*

Remark 6.16 *In this Chapter we approximate solutions of partial functional differential equations by solution of ordinary functional differential problems. Difference methods for such equations were considered by various authors. We refer here the survey paper [236].*

Chapter 7

Generalized Solutions

7.1 Introduction

The following classes of weak solutions of hyperbolic functional differential problems are considered in literature. Mixed problems for almost linear systems in two independent variables were considered in [167], see also [102]. A continuous function is a solution of a mixed problem if it satisfies integral functional system arising from functional differential system by integrating along bicharacteristics. The paper [167] initiated investigations of first order functional differential equations. Existence theorems and differential inequalities generated by almost linear functional differential problems can be found in [42, 107, 123]. Distributional solutions of almost linear problems were considered in [204]. The method used in this paper is constructive, it is based on a difference scheme. Barbashin type functional differential problems are discussed in [102].

The class of Carathéodory solutions consists of all functions which are continuous, have their derivatives almost everywhere in some domain, and the set of all points where the differential equation is not fulfilled is of Lebesgue measure zero. The main existence and uniqueness results for quasilinear systems and initial or mixed problems in the class of almost everywhere solutions can be found in [56 - 58] and [60, 64, 72, 124, 125, 210, 211, 214, 215]. Existence results are based on the following idea: functional differential problems are equivalent to integral functional equations arising from original problems by integrating along bicharacteristics. Under suitable assumptions, solutions of functional integral equations are Carathéodory solutions of differential problems. Carathéodory solutions for nonlinear equations were studied in [59, 127, 212].

Cinquini Cibrario solutions to differential equations were first treated in [53 - 55]. This class of solutions is placed between classical solutions and solutions in the Carathéodory sense. What is more both inclusions are strict. This class of generalized solutions seems to be important if we look for classical solutions because the assumptions that the right-hand side of an equation is continuous is sufficient to prove that every Cinquini Cibrario solution to it is a classical solution. Generalized solutions in the Cinquini Cibrario sense are investigated in the case that assumptions for given functions are extended. Theorems on the uniqueness of Cinquini Cibrario solutions of functional differential problems with nonlinear comparison functions are proved by means of integral functional inequalities of the Volterra type [122]. Existence results for initial or mixed problems are obtained in [67, 146, 149] by a linearization procedure and by a construction of functional integral systems for unknown functions and for their derivatives with respect to the spatial variables. Under natural assumptions on given functions, solutions of integral equations generate weak solutions of original problems.

An essential extension of some ideas concerning distributional solutions of Hamilton-Jacobi equations is given in [12, 74, 76, 77, 155] where viscosity solutions are considered. Uniqueness and existence theorems have been developed for various types of equations and boundary value problems, and also some applications to problems of differential games have been studied. An extensive survey of developments in the area of second order equations and viscosity solutions up to year 1992 was given in [75]. Functional differential problems and viscosity solutions were considered in [187, 188] and [205 – 207]. Uniqueness results are based on a differential inequalities methods. Existence theorems were obtained by the vanishing viscosity method. The monograph [192] develops another approach that can be considered as a nonclassical method of characteristics for Hamilton - Jacobi equations. Generalized solutions (called the minimax solutions) is assumed to be flow invariant with respect to the so-called characteristic inclusions. This monograph includes existence and uniqueness results, examples of numerical modelling, applications to the theory of control and differential games. For further bibliography see the references in the papers cited above, especially in [67, 122, 152, 192].

The following classes of generalized solutions are considered in this Chapter. Solutions in the Friedrichs sense of initial problems for semilinear systems in two independent variables are developed in Sections 7.2 - 7.6. Next we consider nonlinear initial problems and Carathéodory solutions satisfying entropy conditions. Functional differential problems with unbounded delay are considered in Section 7.8. The last part of this Chapter is connected with viscosity solutions of functional differential equations and inequalities.

7.2 Quasi - equicontinuous operators for semilinear systems

Let $\bar{\Xi} \subset R^2$ be a compact set and $\lambda = (\lambda_1, \ldots, \lambda_m) \in C(\bar{\Xi}, R^m)$. Let J_0 be the segment $J_0 = \{(x,y) : x = 0, \ y \in [b_0, b_1]\}$, $b_0 < b_1$, and assume that $J_0 \subset \bar{\Xi}$. For $(x,y) \in \bar{\Xi}$ and $x \geq 0$ we define $\lambda_*(x,y) = \max\{\lambda_i(x,y) : 1 \leq i \leq m\}$, $\lambda^*(x,y) = \min\{\lambda_i(x,y) : 1 \leq i \leq m\}$. Let ω_* be the right-hand maximum solution of the problem

$$\omega'(x) = \lambda_*(x, \omega(x)), \quad \omega(0) = b_0,$$

and ω^* be the right-hand minimum solution of

$$\omega'(x) = \lambda^*(x, \omega(x)), \quad \omega(0) = b_1,$$

We assume that ω_* and ω^* exist on $[0,\, a]$, $a > 0$. Let

$$E = \{(x,y):\ x \in [0,\, a],\ \omega_*(x) \le y \le \omega^*(x)\},\ \ E_0 = [-r_0,\, 0] \times [b_0, b_1] \qquad (7.1)$$

where $r_0 \in R_+$. Assume that $E \subset \bar{\Xi}$. Write $\Omega = E \times R^m \times C(E_0 \cup E, R^m)$ and suppose that $\phi : E_0 \to R^m$, $f : \Omega \to R^m$ are given functions. The present Section develops the semilinear functional differential system

$$D_x z(x,y) + \Lambda(x,y)\, D_y z(x,y) = f(\,x, y, z(x,y), z\,) \qquad (7.2)$$

with the initial condition

$$z(x,y) = \phi(x,y) \ \text{ for } \ (x,y) \in E_0, \qquad (7.3)$$

where $z(x,y) = \text{colon}\,[\,z_1(x,y), \dots, z_m(x,y)\,]$, $\Lambda(x,y)$ is a diagonal matrix

$$\Lambda(x,y) = \text{diag}\ [\,\lambda_1(x,y), \dots, \lambda_m(x,y)\,]$$

and

$$f(x,y,p,z) = \text{colon}\ [\,f_1(x,y,p,z), \dots, f_m(x,y,p,z)\,],$$

$$\phi(x,y) = \text{colon}\ [\,\phi_1(x,y), \dots, \phi_m(x,y)\,],\ \ p = (\,p_1, \dots, p_m\,).$$

Let $E_x = (\,E_0 \cup E\,) \cap (\,[-r_0, x] \times R\,)$, $x \in [0, a]$. We assume that the function f satisfies the following Volterra condition: for each $(x,y) \in E$ there is a set $E[x,y] \subset E_x$ such that if z, $\bar{z} \in C(E_0 \cup E, R^m)$ and $z(t,s) = \bar{z}(t,s)$ for $(t,s) \in E[x,y]$ then $f(x,y,p,z) = f(x,y,p,\bar{z})$. For $x \in [0,a]$ and $z \in C(E_0 \cup E, R^m)$ define

$$\|\, z\, \|_x = \max\{\, \|z(t,s)\| :\ (t,s) \in E_x\,\}$$

where $\|\cdot\|$ is the Euclidean norm in R^m. In particular, $\|\cdot\|_a$ is the norm in the space $C(E_0 \cup E, R^m)$. The solution of the ordinary differential equation $\omega'(t) = \lambda_i(\,t, \omega(t)\,)$ through the point $(x,y) \in E$ is denoted by $g_i(\,\cdot\,, x, y)$ for $1 \le i \le m$. For each $(x,y) \in E$ the function $g_i(\,\cdot\,, x, y)$ is the i - th characteristic of (7.2). We assume that the characteristic $g_i(\,\cdot\,, x, y)$ is uniquely defined for every (x,y) in E and $1 \le i \le m$. Furthermore, it is assumed that condition on the matrix Λ are such that the solution g_i is a continuous function of (t,x,y) for every $1 \le i \le m$. When the above conditions on Λ are satisfied, E is said to be a domain of determinacy for system (7.2). For a function $z \in C(E_0 \cup E, R^m)$ we put

$$F_i[\,z\,](x,y) = f_i(\,x, y, z(x,y), z\,)$$

where $i = 1, \dots, m$ and $F = (\,F_1, \dots,\ F_m\,)$. For the set of characteristics

$$g(\,\cdot\,, x, y) = (\,g_1(\,\cdot\,, x, y), \dots, g_m(\,\cdot\,, x, y)\,)$$

we denote

$$F[\,z\,](t, g(t,x,y)) = (\,F_1[\,z\,](t, g_1(t,x,y)), \dots, F_m[\,z\,](t, g_m(t,x,y))\,)\,.$$

A solution of (7.2), (7.3) is taken in the wider sense of Friedrichs; namely, a continuous vector - valued function $\bar{z}$ is a solution in the wider sense if $\bar{z}$ satisfies the integral equation

$$z(x,y) = \Phi(x,y) + \int_0^x F[\,z\,](\,t, g(t,x,y))\,dt \tag{7.4}$$

for $(x,y) \in E$ with

$$z(x,y) = \phi(x,y) \text{ on } E_0 \tag{7.5}$$

and

$$\Phi(x,y) = (\,\Phi_1(x,y), \ldots, \Phi_m(x,y)\,) = (\,\phi_1(0, g_1(0,x,y)), \ldots, \phi_m(0, g_m(0,x,y))\,)$$

where $g(\,\cdot\,, x, y)$ represents the set of characteristics through (x,y). Unless otherwise noted, a solution shall refer to a solution of this sense. If the solution in the wider sense is continuously differentiable, it is then a classical solution of (7.2), (7.3). Consider Cauchy problem (7.2), (7.3). Let F^* be the operator defined by

$$F^*[\,z\,](x,y) = \int_0^x F[\,z\,](\,\tau, g(\tau,x,y)\,)\,d\tau,\ (x,y) \in E, \tag{7.6}$$

where $z \in C(E_0 \cup E, R^m)$. Let $C_\phi(E_0 \cup E, R^m)$ be the set of all vector - valued functions continuous on $E_0 \cup E$ which are identical with ϕ on E_0. We assume that the function f is continuous on Ω. Since the function $\Phi : E_0 \cup E \to R^m$ given by

$$\Phi(x,y) = \phi(x,y) \text{ for } (x,y) \in E_0, \quad \Phi(x,y) = \phi(\,0, g(0,x,y)\,) \text{ for } (x,y) \in E$$

is continuous and $\Phi \in C_\phi(E_0 \cup E, R^m)$ then there exists $\tilde{M} \geq 0$ such that

$$\|\, f(\,x, y, \Phi(x,y), \Phi\,)\,\| \leq \tilde{M} \text{ on } E.$$

It follows from the continuity of f that there is $\varrho > 0$ such that

$$\|\, f(\,x, y, z(x,y), z\,) - f(\,x, y, \Phi(x,y), \Phi\,)\,\| \leq 1 \text{ on } E$$

whenever $z \in C_\phi(E_0 \cup E, R^m)$ and $\|z - \Phi\|_a \leq \varrho$. The class of all elements $z \in C_\phi(E_0 \cup E, R^m)$ with $\|z - \Phi\|_a \leq \varrho$ we denote by $C_{\phi.\varrho}(E_0 \cup E, R^m)$. For $z \in C_{\phi.\varrho}(E_0 \cup E, R^m)$ we have

$$\|\, f(\,x, y, z(x,y), z\,)\,\| \leq M \text{ on } E \tag{7.7}$$

where $M = \tilde{M} + 1$. For any sequence $\{\,u^{(k)}\,\}$, $u^{(k)} \in C_{\phi.\varrho}(E_0 \cup E, R^m)$, define

$$\varepsilon^{(k)}(x,y) = u^{(k)}(x,y) - F^*[\,u^{(k)}\,](x,y),\ (x,y) \in E. \tag{7.8}$$

If $\{\,\varepsilon^{(k)}\,\}$ is a sequence converging uniformly on E to the function Φ and if F^* is an equicontinuous and equibounded operator on $C_{\phi.\varrho}(E_0 \cup E, R^m)$, then there is a subsequence of $\{\,u^{(k)}\,\}$ converging uniformly on E to a solution of (7.4). A continuity of F^* is not sufficient for the existence of a solution; hence, continuity of F^* is not sufficient for F^* to be equicontinuous. It follows from (7.7) that continuity of f is sufficient for F^* to be equibounded on $C_{\phi.\varrho}(E_0 \cup E, R^m)$. It is a sufficient condition for $\{ u^{(k)} \}$ to be equibounded. In the application of Ascoli-Arzelá theorem, $\{ u^{(k)} \}$ must be a sequence of equicontinuous functions on E. A condition on F^* which implies that the sequence $\{ u^{(k)} \}$, $u^{(k)} \in C_{\phi.\varrho}(E_0 \cup E, R^m)$, is a sequence of equicontinuous functions on E, is a sufficient condition for the existence of a subsequence of $\{ u^{(k)} \}$ which is uniformly convergent on E to a solution of (7.4). We introduce such a condition. In the present Chapter a condition on f more restrictive than continuity but not so restrictive as to imply uniqueness was obtained by applying the method of Langan [142 – 144]. If we assume in theorems of this Chapter that the right-hand sides of the systems with partial derivatives do not include the functional variable, then we obtain essentially more general theorems then those results in [142 – 144].

The operator F is said to be quasi-equicontinuous if for every sequence $\{ u^{(k)} \}$, $u^{(k)} \in C_{\phi.\varrho}(E_0 \cup E, R^m)$, the uniform convergence of the sequence of continuous functions $\{ u^{(k)} - F^*[u^{(k)}] \}$ implies $\{ u^{(k)} \}$ is a sequence of equicontinuous functions. Now we give examples of quasi-equicontinuous operators. We will consider the property of equicontinuity along a given characteristic family. More precisely, a sequence $\{ w^{(k)} \}$, $w^{(k)} \in C(E_0 \cup R, R)$, is said to be equicontinuous on E along given characteristic family g_i if $| w^{(k)}(x_1, y_1) - w^{(k)}(x_2, y_2) | < \varepsilon$, $k \geq 1$, whenever $|x_1 - x_2| + |y_1 - y_2| < \delta(\varepsilon)$, (x_1, y_1), $(x_2, y_2) \in E$, and $y_2 = g_i(x_2, x_1, y_1)$. The following lemmas are important in our considerations ([107, 143]).

Lemma 7.1 *A sequence of functions $\{ w^{(k)} \}$, $w^{(k)} \in C(E_0 \cup E, R)$, equicontinuous on E along a given characteristic family g_i and also along a second distinct characteristic family g_j, (i.e. $g_i(\tau, x, y) \neq g_j(\tau_1, x_1, y_1)$ for $(\tau, x, y) \neq (\tau_1, x_1, y_1)$) is a sequence of equicontinuous functions on E if the sequence $\{ w^{(k)} |_{J_0} \}$ converges to a continuous function on J_0.*

Lemma 7.2 *Suppose that*

1) for $k \geq 1$ we have $u^{(k)} = (u_1^{(k)}, \ldots, u_m^{(k)}) \in C_{\phi.\varrho}(E_0 \cup E, R^m)$ and $\{ \varepsilon^{(k)} \} = \{ \varepsilon_1^{(k)}, \ldots, \varepsilon_m^{(k)} \}$ is defined by (7.8),

2) f is continuous on Ω and $\{ \varepsilon^{(k)} \}$ is uniformly convergent on E.

Under these assumptions, for each i, $1 \leq i \leq m$, the sequence $\{ u_i^{(k)} \}$ is equicontinuous on E along the characteristic family g_i.

Lemma 7.3 *Suppose that*

1) the function $G : [0, a] \times C([-r_0, a], R^m) \to R^m$ is continuous, satisfies the Volterra condition and is nondecreasing with respect to the functional variable,

2) $\tilde{z} \in C([-r_0, a], R^m)$ and for some $\eta \in R^m$ we have

$$\tilde{z}(x) \leq \eta + \int_0^x G(s, \tilde{z})\, ds, \quad x \in [0, a],$$

3) $\omega(\cdot, \eta)$ is the right-hand maximum solution of the problem

$$\omega'(x) = G(x, \omega) \quad \omega(x) = \eta \text{ on } [-r_0, 0],$$

defined on $[-r_0, a]$.

Under these assumptions we have $\tilde{z}(x) \le \omega(x, \eta)$ *on* $[0, a]$.

PROOF. Put

$$\beta(x) = \eta \quad \text{for } x \in [-r_0, 0] \text{ and } \beta(x) = \eta + \int_0^x G(s, \tilde{z})\, ds \text{ for } x \in [0, a].$$

Then, by assumption 2) and by the monotonicity condition for G we have

$$\beta'(x) \le G(x, \beta) \text{ for } x \in [0, a].$$

It follows from a theorem on functional differential inequalities that $\beta(x) \le \omega(x, \eta)$ on $[0, a]$. In view of $\tilde{z}(x) \le \beta(x)$ on $[0, a]$, we obtain the assertion of Lemma 7.3.

Now we formulate differential integral problems. Let

$$E = \{ (x, y) : x \in [0, a],\ b_0 + x \le y \le b_1 - x \} \text{ and } E_0 = \{0\} \times [b_0, b_1],$$

where $2a \le b_1 - b_0$. Let

$$S_\tau = \{ (x, y) \in E : x = \tau \} \text{ and } T_s = \{ (x, y) \in E : y = s \}.$$

For $(x, y) \in E$ we define

$$u(x, y) = (v(x, y), w(x, y)) = (v_1(x, y), \ldots, v_\kappa(x, y), w_1(x, y), \ldots, w_{m-\kappa}(x, y)),$$

where $0 < \kappa < m$ is fixed. The object of our investigations is the system of Volterra differential integral equations

$$D_x v(x, y) + D_y v(x, y) = \tilde{G}(x, y, u(x, y), \int_{H(x,y)} u(\xi, \eta)\, d\xi\, d\eta), \tag{7.9}$$

$$D_x w(x, y) - D_y w(x, y) = \tilde{H}(x, y, u(x, y), \int_{H(x,y)} u(\xi, \eta)\, d\xi\, d\eta) \tag{7.10}$$

with the initial condition

$$u(0, y) = \phi(y) \text{ for } y \in [b_0, b_1], \tag{7.11}$$

where $\tilde{G} = (g_1, \ldots, g_\kappa)$, $\tilde{H} = (h_1, \ldots, h_{m-\kappa})$ and

$$\int_{H(x,y)} u(\xi, \eta)\, d\xi\, d\eta$$

$$= \left(\int_{H_0(x,y)} u(\xi, \eta)\, d\xi\, d\eta,\ \int_{H_1(x,y)} u(\xi, \eta)\, d\eta,\ \int_{H_2(x,y)} u(\xi, \eta)\, d\xi \right).$$

Assumption $\mathbf{H_0}$. Suppose that the sets H_0, H_1, H_2 satisfy the conditions:

1) if $(x, y) \in E$ then $H_0(x, y)$ is the Lebesgue - measurable set, $H_0(x, y) \subset E_x$ and

$$\lim_{(t,s)\to(0,0)} L_2 [H_0(x + t, y + s) \div H_0(x, y)] = 0$$

uniformly with respect to $(x,y) \in E$, where L_2 is the 2 - dimensional Lebesgue measure and the sign $\div$ denotes the symmetric difference of two sets,

2) if $(x,y) \in E$ then $H_1(x,y) \subset E_x$ is the 1 - dimensional Lebesgue - measurable set, $H_1(x,y) \subset E_x$ and

$$\lim_{(t,s)\to(0,0)} L_1 \{ [-(t,s) + H_1(x+t, y+s)] \div H_1(x,y) \} = 0$$

uniformly in $(x,y) \in E$, where L_1 is the 1 - dimensional Lebesgue measure and $-(t,s) + Y$ denotes the set of all elements $-(t,s) + \zeta$, where $\zeta \in Y$,

3) if $(x,y) \in E$ then $H_2(x,y) \subset E_x$ is the 1 - dimensional Lebesgue - measurable set, $H_1(x,y) \subset T_y$ and

$$\lim_{(t,s)\to(0,0)} L_1 \{ [-(t,s) + H_2(x+t, y+s)] \div H_2(x,y) \} = 0$$

uniformly in $(x,y) \in E$.

Remark 7.4 *If $(x,y) \in E$ then*

$$\int_{H_1(x,y)} u(\xi,\eta)\, d\eta$$

denotes the Lebesgue integral with respect to η and $\xi = x$. In a similar way we define the integral

$$\int_{H_2(x,y)} u(\xi,\eta)\, d\xi.$$

We adopt the following notations: If $p = (p_1, \ldots, p_m) \in R^m$ then

$$p' = (p_1, \ldots, p_\kappa),\ p'' = (p_{\kappa+1}, \ldots, p_m) \text{ and } p = (p', p'').$$

For $p \in R^m$ we define

$$\bar{p}^{(\kappa)} = (\bar{p}_1, \ldots, \bar{p}_\kappa, p_{\kappa+1}, \ldots, p_m)$$

and

$$\bar{p}_{(m-\kappa)} = (p_1, \ldots, p_\kappa, \bar{p}_{\kappa+1}, \ldots, \bar{p}_m).$$

We denote by $\|\cdot\|_k$ the Euclidean norm in R^k and we omit the index k if $k = m$. Let

$$\Sigma = \{ (x,y,\mu,p,q,r) \in E \times R^{4m} : \quad (x,y) \in E,$$
$$\|\mu\| \le M,\ \|p\| \le M\, L_2(E),\ \|q\| \le M\, b,\ \|r\| \le M\, a \}$$

where $b = b_1 - b_0$.

Assumption H $[\tilde{G}, \tilde{H}]$. Suppose that

1) $\tilde{G} \in C(\Sigma, R^\kappa)$, $\tilde{H} \in C(\Sigma, R^{m-\kappa})$ and $\phi \in C([b_0, b_1], R^m)$, $\|\phi(y)\| \le M$ for $y \in [b_0, b_1]$,

2) there exists a function $\sigma : [0,a] \times R_+^3 \to R_+$ such that

(i) σ is continuous, $\sigma(x,0,0,0) = 0$ on $[0,a]$ and it is nondecreasing with respect to all variables except x,

(ii) the estimates

$$\| \tilde{G}(x,y,\mu,p,q,r) - \tilde{G}(x,y,\bar{\mu}^{(\kappa)},p,\bar{q}^{(\kappa)},\bar{r}^{(\kappa)}) \|_\kappa$$
$$\leq \sigma(x, \|\mu' - \bar{\mu}'\|_\kappa, \|q' - \bar{q}'\|_\kappa, \|r - \bar{r}'\|_\kappa)$$

and

$$\| \tilde{H}(x,y,\mu,p,q,r) - \tilde{H}(x,y,\bar{\mu}_{(m-\kappa)},p,\bar{q}_{(m-\kappa)},\bar{r}_{(m-\kappa)}) \|_{m-\kappa}$$
$$\leq \sigma(x, \|\mu'' - \bar{\mu}''\|_{m-\kappa}, \|q'' - \bar{q}''\|_{m-\kappa}, \|r'' - \bar{r}''\|_{m-\kappa})$$

are satisfied on Σ,

3) there exists $\varepsilon_0 > 0$ such that for $0 < \varepsilon < \varepsilon_0$ the right-hand maximum solution $\omega(\cdot,\varepsilon)$ of the differential integral problem

$$\omega'(x) = \sigma(x,\omega(x),\varepsilon + b\omega(x),\varepsilon + \int_0^x \omega(s)\,ds), \quad \omega(0) = \varepsilon,$$

is defined on $[0,a]$ and $\lim_{\varepsilon\to 0}\omega(x,\varepsilon) = 0$ uniformly on $[0,a]$.

Theorem 7.5 *If Assumptions H_0, $H[\tilde{G},\tilde{H}\,]$ are satisfied then the operator F given by*

$$F[z](x,y)$$
$$= \left(\tilde{G}(x,y,z(x,y), \int_{H(x,y)} z(\xi,\eta)\,d\xi\,d\eta),\ \tilde{H}(x,y,z(x,y), \int_{H(x,y)} z(\xi,\eta)\,d\xi\,d\eta) \right)$$

is quasi-equicontinuous.

The proof of the theorem is based on the following idea. Let $\{u^{(k)}\} = \{v^{(k)}, w^{(k)}\}$ be a sequence of vector - valued functions such that $u^{(k)} \in C_{\phi.\varrho}(E_0 \cup R, R^m)$ and $\| u^{(k)}(x,y) \| \leq M$ for $k \geq 1$, $(x,y) \in E$. Let

$$\varepsilon^{(k)}(x,y) = (\varepsilon'_k(x,y), \varepsilon''_k(x,y)) = \left(v^{(k)}(x,y), w^{(k)}(x,y) \right)$$
$$- \left(\int_0^x \tilde{G}(s, s-x+y, u^{(k)}(s,s-x+y), \int_{H(s,s-x+y)} u^{(k)}(\xi,\eta)\,d\xi\,d\eta)\,ds, \right.$$
$$\left. \int_0^x \tilde{H}(s,-s+x+y,u^{(k)}(s,-s+x+y), \int_{H(s,-s+x+y)} u^{(k)}(\xi,\eta)\,d\xi\,d\eta)\,ds \right),$$

where $(x,y) \in E$. The functions $\varepsilon^{(k)}$, $k \geq 1$, are continuous on E ([107]). By definition, F is quasi-equicontinuous only if $\{u^{(k)}\}$ is a sequence of equicontinuous functions. It follows from Lemma 7.2 that $\{v^{(k)}\}$ is the sequence of equicontinuous functions on E along the characteristics $g_i(\tau,x,y) = \tau - x + y$, $1 \leq i \leq \kappa$; whereas, $\{w^{(k)}\}$ is a sequence of equicontinuous functions on E along the characteristics $g_i(\tau,x,y) = -\tau + x + y$, $\kappa + 1 \leq i \leq m$.

The sequences $\{v^{(k)}\}$ is equicontinuous on E along the characteristic family $g_i(\tau,x,y)=-\tau+x+y$, $\kappa+1\le i\le m$, and the sequence $\{w^{(k)}\}$ is equicontinuous along the characteristic family $g_i(\tau,x,y)=\tau-x+y$, where $1\le i\le\kappa$. We omit details concerning the above properties of the sequence $\{u^{(k)}\}$. Now Theorem 7.5 follows from Lemma 7.2 .

Consider a differential integral system with a constant matrix

$$D_x u(x,y)+\Lambda D_y u(x,y)=f(x,y,u(x,y),\int_{H(x,y)}u(\xi,\eta)\,d\xi\,d\eta) \tag{7.12}$$

and the initial condition

$$u(0,y)=\phi(y)\ \text{ for }\ y\in[b_0,b_1] \tag{7.13}$$

where

$$\Lambda=\operatorname{diag}[\lambda_1,\dots,\lambda_m].$$

Let $\lambda_*=\max\{\lambda_i:\ 1\le i\le m\}$, $\lambda^*=\min\{\lambda_i:\ 1\le i\le m\}$ and

$$E=\{(x,y):\ x\in[0,a],\ b_0+\lambda_* x\le y\le b_1+\lambda^* x\},\quad E_0=\{0\}\times[b_0,b_1]. \tag{7.14}$$

For $p=(p_1,\dots,p_k)\in R^k$, $1\le k\le m$, we write $[|p|]_k=\sum_{i=1}^k|p_i|$.

Assumption H $[f]$. Suppose that

1) $f\in C(\Sigma,R^m)$, $\phi\in C([b_0,b_1],R^m)$ and $\|\phi(y)\|\le M$ on $[b_0,b_1]$,

2) there exists a function $\sigma=(\sigma_1,\dots,\sigma_m)\in C([0,a]\times R^3_+,R^m_+)$ such that

$$\left|f_i(x,y,\mu,p,q,r)-f_i(x,y,\bar\mu^{(m-1)},p,\bar q^{(m-1)},\bar r^{(m-1)})\right|$$
$$\le\sigma_i(x,[|\mu'-\bar\mu'|]_{m-1},[|q'-\bar q'|]_{m-1},[|r'-\bar r'|]_{m-1})$$

where $i=1,2,\dots,m-1$, and

$$|f_m(x,y,\mu,p,q,r)-f_m(x,y,\bar\mu_{(1)},p,\bar q_{(1)},\bar r_{(1)})|$$
$$\le\sigma_m(x,|\mu_m-\bar\mu_m|,|q_m-\bar\mu_m|,|r_m-\bar r_m|)$$

on Σ,

3) there exists $\varepsilon_0>0$ such that for $0\le\varepsilon\le\varepsilon_0$ the right-hand maximum solution $\omega(\cdot,\varepsilon)=(\omega_1(\cdot,\varepsilon),\dots,\omega_m(\cdot,\varepsilon))$ of the Cauchy problem

$$\omega_i'(x)=\sigma_i(x,[|\omega(x)|]_{m-1},\varepsilon+b[|\omega(x)|]_{m-1},\varepsilon+\int_0^x[|\omega(s)|]_{m-1}\,ds),$$

for $i=1,\dots,m-1$ and

$$\omega_m'(x)=\sigma_m(x,\omega_m(x),\varepsilon+b\,\omega_m(x),\varepsilon+\int_0^x\omega_m(s)\,ds)$$

$$\omega(0)=(\varepsilon,\dots,\varepsilon),$$

is defined on $[0,a]$ and $\lim_{\varepsilon\to0}\omega(x,\varepsilon)=0$ uniformly on $[0,a]$,

4) $\lambda_m\ne\lambda_i$ for $i=1,\dots,m-1$.

Theorem 7.6 *Suppose that Assumptions H, H [f] are satisfied. Then the operator F defined by*

$$F[u](x,y) = f(x,y,u(x,y), \int_{H(x,y)} u(\xi,\eta)\,d\xi\,d\eta),\ (x,y) \in E,$$

is quasi-equicontinuous.

The proof of the theorem is based on the following idea. Consider the sequence $\{u^{(k)}\}$ where $u^{(k)} \in C_{\phi.\varrho}(E_0 \cup E, R^m)$ and $\|u^{(k)}(x,y)\| \leq M$ for $k \geq 1$ and $(x,y) \in E$. Write

$$\varepsilon^{(k)}(x,y) = u^{(k)}(x,y)$$

$$- \int_0^x f(s, g(s,x,y), u^{(k)}(s, g(s,x,y)), \int_{H(s,g(s,x,y))} u^{(k)}(\xi,\eta)\,d\xi\,d\eta)\,ds,$$

where $(x,y) \in E$ and $g = (g_1, \ldots, g_m)$ is the set of characteristics: $g(s,x,y) = \lambda_i(s-x)+y$, $i = 1, \ldots, m$. We assume that $\{\varepsilon^{(k)}\}$ is uniformly convergent on E. It follows from Lemma 7.2 that $\{u_i^{(k)}\}$ is a sequence of equicontinuous functions on E along the characteristic $g_i(\cdot, x, y)$. It follows that the the sequences $\{u_i^{(k)}\}$, $1 \leq i \leq m-1$, are equicontinuous on E along the characteristic $g_m(\cdot, x, y)$ and the sequence $\{u_m^{(k)}\}$ is equicontinuous on the characteristic family $g_i(\cdot, x, y)$, $1 \leq i \leq m-1$. Details can be found in [107]. Now Theorem 7.6 follows from Lemma 7.1.

7.3 Existence of solutions

In the present Section an existence theorem is given for quasi-equicontinuous functional differential problem (7.2), (7.3). The existence theorem is not restricted to systems with unique solutions.

Theorem 7.7 *Suppose that*

1) the matrix $\Lambda = diag\,[\lambda_1, \ldots, \lambda_m]$ is continuous on the compact set $\bar{\Xi} \subset R^2$,

2) the function $f : \Omega \to R^m$ is continuous and satisfies the Volterra condition,

3) $\phi \in C(E_0, R^m)$ and the operator F corresponding to the function f is quasi-equicontinuous on

$$C_{\phi.\varrho}(E_0 \cup E, R^m),$$

where $Ma \leq \varrho$ and ϱ, M are defined in Section 7.2.

Under these assumptions there exists a solution $\tilde{u} \in C_{\phi.\varrho}(E_0 \cup E, R^m)$ of the initial problem (7.2), (7.3).

Proof. Let the sequence $\{u^{(k)}\}$ be defined by

$$u^{(k)}(x,y) = \phi(x,y) \ \text{ on } \ E_0 \cup \left[E \cap ([0,k^{-1}] \times R)\right],$$

$$u^{(k)}(x,y) = \int_0^{x-k^{-1}} F[u^{(k)}](s, \varphi(s,x,y))\,ds \ \text{ on } \ E \cap ([k^{-1}, a] \times R).$$

We prove that

$$\|u^{(k)} - \phi\|_x \le \varrho \text{ for } x \in [0,a]. \tag{7.15}$$

For $x \in [0, k^{-1}]$ we have $\|u^{(k)} - \phi\|_x = 0$. Suppose that $\|u^{(k)} - \phi\|_x \le \varrho$ where $0 \le x \le n\,k^{-1}$ for some positive integer n. If $n\,k^{-1} \le x \le (n+1)k^{-1} \le a$ then

$$u^{(k)}\,|_{E_{x-k^{-1}}} \in C_{\phi.\varrho}(E_0 \cup E_{x-k^{-1}}, R^m)$$

and $\|u^{(k)} - \phi\|_x \le M(x - k^{-1}) \le M\,a \le \varrho$. By induction we obtain (7.15). It follows from (7.15) that $u^{(k)} \in C_{\phi.\varrho}(E_0 \cup E, R^m)$. Moreover, $u^{(k)}$ satisfies the integral equation

$$u^{(k)}(x,y) = \varepsilon^{(k)}(x,y) + \int_0^x F[u^{(k)}](\,s, g(s,x,y)\,)\,ds,\ (x,y) \in E,$$

where

$$\varepsilon^{(k)}(x,y) = \phi(x,y) - \int_0^x F[u^{(k)}](\,s, g(s,x,y)\,)\,ds \text{ if } 0 \le x \le k^{-1},$$

$$\varepsilon^{(k)}(x,y) = \phi(x,y) - \int_{x-k^{-1}}^x F[u^{(k)}](\,s, g(s,x,y)\,)\,ds \text{ if } k^{-1} \le x \le a.$$

Since $u^{(k)} \in C_{\phi.\varrho}(E_0 \cup E, R^m)$, it follows that $\lim_{k\to\infty} \varepsilon^{(k)}(x,y) = \phi(x,y)$ uniformly on E. The operator F is quasi-equicontinuous on $C_{\phi.\varrho}(E_0 \cup E, R^m)$, therefore there exists a subsequence of $\{\,u^{(k)}\,\}$ which is convergent uniformly to a solution of (7.2), (7.3). This completes the proof of Theorem 7.7.

We next state a uniqueness theorem of the Perron type. Let $S_x = [b_0, b_1]$ for $x \in [-r_0, 0]$ and $S_x = [\omega_*(x),\ \omega^*(x)]$ for $x \in (0,a]$. For a function $z \in C(E_0 \cup E, R^m)$ we define $Vz : [-r_0, a] \to R_+$ by

$$(Vz)(x) = \max\{\,\|z(x,y)\| : \ y \in S_x\,\},\ \ x \in [-r_0, a].$$

Note that Vz is a continuous function on $[-r_0, a]$.

Assumption H $[\,\sigma\,]$. Suppose that

1) the function $\sigma : [0,a] \times R_+ \times C([-r_0,a], R_+) \to R_+$ is continuous and satisfies the Volterra condition,

2) σ is nondecreasing with respect to the second and third variables,

3) the maximum solution of the initial problem

$$\omega'(x) = \sigma(\,x, \omega(x), \omega\,),\ \ \omega(x) = 0 \text{ on } [-r_0, 0], \tag{7.16}$$

is $\bar\omega(x) = 0$ for $x \in [-r_0, a]$.

Theorem 7.8 *Suppose that Assumption H $[\sigma\,]$ is satisfied and*

1) the function $f : \Omega \to R^m$ is continuous and satisfies the Volterra condition,

2) the estimate

$$\|\,f(x,y,p,u) - f(x,y,\bar p,v)\,\| \le \sigma(\,x, \|p - \bar p\|, V(u - v)\,)$$

is satisfied on Ω.

Under these assumptions the Cauchy problem (7.2), (7.3) admits at most one solution $\bar u \in C(E_0 \cup E, R^m)$.

PROOF. If $\bar{u}, \bar{v} \in C(E_0 \cup E, R^m)$ are solutions of (7.2), (7.3) then the function $\mu(x) = V(\bar{u} - \bar{v})(x)$, $x \in [-r, a]$, satisfies the integral inequality

$$\mu(x) \leq \int_0^c \sigma(s, \mu(x), \mu)\, ds, \ x \in [0, a],$$

and $\mu(x) = 0$ on $[-r_0, 0]$. Assumption H $[\sigma]$ and Lemma 7.3 imply $\mu(x) = 0$ on $[0, a]$, which completes the proof of Theorem 7.8.

Lemma 7.9 *If Assumption H $[\sigma]$ and conditions 1), 2) of Theorem 7.8 are satisfied then the operator F is quasi-equicontinuous on $C_{\phi.\varrho}(E_0 \cup E, R^m)$.*

PROOF. There is $\varepsilon > 0$ such that for $0 \leq \varepsilon \leq \varepsilon_0$ the right-hand maximum solution $\omega(\cdot, \varepsilon)$ of the problem

$$\omega'(x) = \sigma(x, \omega(x), \omega), \quad \omega(x) = \varepsilon \ \text{ on } \ [-r_0, 0],$$

is defined on $[-r_0, a]$ and $\lim_{\varepsilon \to 0} \omega(x, \varepsilon) = 0$ uniformly on $[0, a]$. Let $\varepsilon^{(k)}$ be the sequence given by (7.8) where $u^{(k)} \in C_{\phi.\varrho}(E_0 \cup E, R^m)$. For given $\varepsilon \in (0, \varepsilon_0)$ we choose $N(\varepsilon)$ such that

$$\| \varepsilon^{(k)}(x, y) - \varepsilon^{(n)}(x, y) \| < \varepsilon \ \text{ for } (x, y) \in E, \ k, n \geq N(\varepsilon).$$

Let $\mu^{(k,n)}(x) = V(u^{(k)} - u^{(n)})(x)$, $x \in [-r_0, a]$. It follows from (7.8) that

$$\mu^{(k,n)}(x) \leq \varepsilon + \int_0^x \sigma(s, \mu^{(k,n)}(s), \mu^{(k,n)})\, ds, \quad x \in [0, a].$$

Therefore $\mu^{(k,n)}(x) \leq \omega(x, \varepsilon)$ on $[0, a]$ for $k, n \geq N(\varepsilon)$. Since $\lim_{\varepsilon \to 0} \omega(x, \varepsilon) = 0$ uniformly on $[0, a]$, then the sequence $\{ u^{(k)} \}$ is equicontinuous on E. This proves the lemma.

Remark 7.10 *Suppose that all the assumption of Theorem 7.8 are satisfied. Then there exists exactly one local solution of problem (7.2), (7.3).*

7.4 FUNCTIONAL DIFFERENTIAL INEQUALITIES

In this Section we deal with functional differential inequalities generated by semilinear systems (7.2) and initial condition (7.3). The problem is considered on the sets E and E_0 defined by (7.1). We assume that E is the domain of determinacy for the system (7.2). If $\psi \in C(E_0 \cup E, R)$ and $(x, y) \in E$, $0 < x \leq a$, then we denote by $D_{(i,-)}\psi(x, y)$ the left-hand lower Dini derivative of ψ along i - th characteristic of (7.2) at the point (x, y) :

$$D_{(i,-)}\psi(x, y) = \liminf_{h \to 0-} \frac{1}{h} [\psi(x + h, g_i(x + h, x, y)) - \psi(x, y)],$$

(the values $+\infty$ and $-\infty$ being not excluded).
In a similar way we define the left-hand upper derivative $D^{(i,-)}\psi(x, y)$ and the right-hand derivatives

$$D_{(i,+)}\psi(x, y), \quad D^{(i,+)}\psi(x, y).$$

For a function $z = (z_1, \dots, z_m) \in C(E_0 \cup E, R^m)$ and for a point $(x, y) \in E$, $0 < x \le a$, we denote

$$D_- z(x,y) = \left(D_{(1,-)} z_1(x,y),\ D_{(2,-)} z_2(x,y), \dots, D_{(m,-)} z_m(x,y) \right).$$

Now we can formulate a theorem on functional differential inequalities.

Theorem 7.11 *Suppose that*

1) the sets E_0 and E are given by (7.1) and E is the domain of determinacy for the system (7.2),

2) the function $f : \Omega \to R^m$ satisfies the Volterra condition and is nondecreasing with respect to the functional variable,

3) f satisfies the following quasi-monotonicity condition: for each i, $1 \le i \le m$, if (x, y, p, z), $(x, y, \bar{p}, z) \in \Omega$ and $p \le \bar{p}$, $p_i = \bar{p}_i$ then $f_i(x,,y,p,z) \le f_i(x,y,\bar{p},z)$,

4) the function u, $v \in C(E_0 \cup E, R^m)$ satisfy the initial inequality

$$u(x,y) < v(x,y) \ \text{ on } J_0 \ \text{ and } u(x,y) \le v(x,y) \ \text{ on } E_0, \tag{7.17}$$

5) the functional differential inequality

$$D_- u(x,y) - f(x,y,u(x,y),u) < D_- v(x,y) - f(x,y,v(x,y),v) \tag{7.18}$$

is satisfied on $E \setminus J_0$.

Under these assumptions we have

$$u(x,y) < v(x,y) \ \text{ for } (x,y) \in E. \tag{7.19}$$

PROOF. It follows from (7.17) that there exists $\bar{x} > 0$ such that $u(x,y) < v(x,y)$ for $0 \le x \le \bar{x}$, $(x,y) \in E$. Suppose that the assertion (7.19) is not true, in this case there exists an integer i, $1 \le i \le m$, and a point $(\tilde{x}, \tilde{y}) \in E$, $\tilde{x} > 0$, such that

$$u(x,y) < v(x,y) \ \text{ for } 0 \le x < \tilde{x},\ (x,y) \in E, \tag{7.20}$$

$$u(\tilde{x},\tilde{y}) \le v(\tilde{x},\tilde{y}),\ \ u_i(\tilde{x},\tilde{y}) = v_i(\tilde{x},\tilde{y}). \tag{7.21}$$

It follows from (7.20) that $u_i(x, g_i(x,\tilde{x},\tilde{y})) < v_i(x, g_i(x,\tilde{x},\tilde{y}))$ for $x \in [0, \tilde{x})$, and then

$$D_{(i,-)} u_i(\tilde{x},\tilde{y}) \ge D_{(i,-)} v_i(\tilde{x},\tilde{y}). \tag{7.22}$$

The Volterra condition and the monotonicity property of f together with (7.18), (7.20), (7.21) yield

$$D_{(i,-)} u_i(\tilde{x},\tilde{y}) - D_{(i,-)} v_i(\tilde{x},\tilde{y}) < f_i\left(\tilde{x},\tilde{y},u(\tilde{x},\tilde{y}),u\right) - f_i\left(\tilde{x},\tilde{y},v(\tilde{x},\tilde{y}),v\right) \le 0.$$

This inequality contradicts (7.22). Thus Theorem 7.11 is proved.

Remark 7.12 *Let*

$$T^* = \{ (x,y) \in E \setminus J_0 : \ u(t,s) < v(t,s) \ \textit{for}\ 0 \le t < x,\ (t,s) \in E,$$

$$\textit{and there exists an index } i,\ 1 \le i \le m, \ \textit{such that } u_i(x,y) = v_i(x,y) \}.$$

In Theorem 7.11 we can assume instead of 5) that (7.18) holds true on T^*.

Now we deal with weak functional differential inequalities. We will have to make more restrictive assumptions on the right-hand sides of (7.2). Additional assumptions imply the uniqueness of solutions of (7.2), (7.3).

Assumption H $[\sigma, P]$. Suppose that

1) the function $\sigma : [0,a] \times R_+ \times C([-r_0,a], R_+) \to R_+$ is continuous and $\sigma(x,0,\theta) = 0$ for $x \in [0,a]$ where $\theta(t) = 0$ for $t \in [-r_0,a]$,

2) the function σ is nondecreasing with respect to the functional variable,

3) σ satisfies the Volterra condition and the function $\bar{\omega}(t) = 0$ for $t \in [-r_0,a]$ is the unique solution of the problem

$$\omega'(t) = \sigma(t, \omega(t), \omega), \quad \omega(t) = 0 \;\text{ for } t \in [-r_0, 0]. \tag{7.23}$$

Theorem 7.13 *Suppose that*

1) the conditions 1) - 3) of Theorem 7.11 and Assumption H $[\sigma, P]$ are satisfied,

2) if $p \in R^m$ and $\bar{p} = (p_1+\xi, \dots, p_m+\xi)$ where $\xi \in R_+$ and $z, \bar{z} \in C(E_0 \cup E, R^m)$ where $\bar{z}(t,s) \geq z(t,s)$ on E_x then

$$f_i(x,y,\bar{p},\bar{z}) - f_i(x,y,p,z) \leq \sigma(x,\xi,V(\bar{z}-z)),$$

3) the functions $u, v \in C(E_0 \cup E, R^m)$ satisfy the initial inequality

$$u(x,y) \leq v(x,y) \quad \text{on } E_0 \tag{7.24}$$

and the functional differential inequality

$$D_-u(x,y) - f(x,y,u(x,y),u) \leq D_-v(x,y) - f(x,y,v(x,y),v) \tag{7.25}$$

is satisfied on $E \setminus J_0$.

Under these assumptions we have

$$u(x,y) \leq v(x,y) \;\; \text{for} \;\; (x,y) \in E_0 \cup E. \tag{7.26}$$

PROOF. For $\varepsilon > 0$ denote by $\omega(\cdot, \varepsilon)$ the right-hand maximum solution of the problem

$$\omega'(t) = \sigma(t, \omega(t), \omega) + \varepsilon, \quad \omega(t) = \varepsilon \;\text{ for } t \in [-r_0, 0]. \tag{7.27}$$

There exists $\varepsilon_0 > 0$ such that for $0 < \varepsilon < \varepsilon_0$ the solution $\omega(\cdot, \varepsilon)$ is defined on $[-r_0, a]$ and

$$\lim_{\varepsilon \to 0} \omega(t, \varepsilon) = 0 \;\text{ uniformly on } [-r_0, a]. \tag{7.28}$$

Consider now the function $\tilde{u}(x,y) = u(x,y) - \Omega(x,\varepsilon)$, $(x,y) \in E_0 \cup E$, where $0 < \varepsilon < \varepsilon_0$ and $\Omega(\cdot,\varepsilon) = (\omega(\cdot,\varepsilon), \dots, \omega(\cdot,\varepsilon)) : [-r_0,a] \to R^m$. It follows from (7.24) that $\tilde{u}(x,y) < v(x,y)$ on E_0. We prove that

$$D_-\tilde{u}(x,y) - f(x,y,\tilde{u}(x,y),\tilde{u}) < D_-v(x,y) - f(x,y,v(x,y),v) \tag{7.29}$$

on $E \setminus J_0$. It follows from Assumption H $[\sigma, P]$ that for $1 \le i \le m$ we have

$$
\begin{aligned}
D_{(i,-)}\tilde{u}_i(x,y) &- f_i(x,y,\tilde{u}(x,y),\tilde{u}) \\
&= D_{(i,-)}u_i(x,y) - f_i(x,y,u(x,y),u) - \omega'(x,\varepsilon) \\
&\quad + [\, f_i(x,y,u(x,y),u) - f_i(x,y,\tilde{u}(x,y),\tilde{u})\,] \\
&\le D_{(i,-)}v_i(x,y) - f_i(x,y,v(x,y),v) - \omega'(x,\varepsilon) + \sigma(\,x,\omega(x,\varepsilon),\omega(\,\cdot\,,\varepsilon)\,) \\
&< D_{(i,-)}v_i(x,y) - f_i(x,y,v(x,y),v)
\end{aligned}
$$

which completes the proof of (7.29). It follows from Theorem 7.11 that

$$u(x,y) - \Omega(x\,,\varepsilon) < v(x,y) \quad \text{on} \quad E_0 \cup E.$$

From the above inequality and from (7.28) we obtain in the limit (letting ε tend to 0) inequality (7.26).

Remark 7.14 *Suppose that conditions 1) and 2) of Theorem 7.13 hold. Then the initial problem (7.2), (7.3) admits at most one solution.*

Remark 7.15 *Suppose that all the assumptions of Theorem (7.13) are satisfied with the exception that the functional differential inequality (7.25) is replaced by (7.18). Then the estimation $u(x,y) < v(x,y)$ for $(x,y) \in E \setminus J_0$ holds. This assertion can be proved by applying Theorem 7.13 and then by repeating the argument used in the proof of Theorem 7.11.*

7.5 Extremal solutions of semilinear systems

We prove a theorem on the local existence of extremal solutions.

Theorem 7.16 *Suppose that*

1) the assumptions 1), 2) of Theorem 7.13 are satisfied,

2) the matrix Λ is continuous on $\bar{\Xi}$,

3) $f \in C(\Omega, R^m)$ and the corresponding operator F is quasi-equicontinuous.

Under these assumptions there exist local extremal solutions of problem (7.2), (7.3).

Proof. We prove the existence of the maximum solution. Let us consider the system of equations

$$D_x z(x,y) + \Lambda(x,y)\, D_y z(x,y) = f(\,x,y,z,(x,y),z\,) + \theta_n \tag{7.30}$$

where $\theta_n = (n^{-1}, \ldots, n^{-1}\,) \in R^m$, with the initial condition

$$z(x,y) = \phi(x,y) + \theta_n \quad \text{on} \quad E_0 \tag{7.31}$$

It follows that there is a constant $L_0 > 0$ such that

$$\|\, f(\,x,y,\Phi(x,y),\Phi\,) + \theta_n \,\| \le L_0 \quad \text{on} \quad E. \tag{7.32}$$

For $z \in C_{\phi.\varrho}(E_0 \cup E, R^m)$ we have

$$\| f(x,y,z(x,y),z) + \theta_n \| \leq L \text{ on } E$$

with $L = L_0 + 1$. If we suppose that L, ϱ, a are such that $La \leq \varrho$, by Theorem 7.7 there exists a solution $z^{(n)}$ of problem (7.30), (7.31). The domain of $z^{(n)}$ is independent on n. Let us consider the sequence $\{ z^{(n)} \}$ of solutions of problems (7.30), (7.31). Now

$$z^{(n)}(x,y) = \Phi(x,y) + \theta_n + \int_0^x F[z^{(n)}](s, g(s,x,y))\, ds + x\,\theta_n.$$

So, by the quasi-equicontinuity hypothesis of the operator F, $\{ z^{(n)} \}$ is a sequence of equicontinuous functions. Moreover, by Theorem 7.13 we can prove that $\{ z^{(n)} \}$ is a decreasing sequence of functions. Thus, by Ascoli-Arzelá theorem, the sequence $\{ z^{(n)} \}$ is uniformly convergent to z^+. We prove that z^+ is the maximum solution of problem (7.2), (7.3). Let $\bar{z}$ be an arbitrary solution of this problem. Suppose that z^+ and $\bar{z}$ exist on E_c with $0 < c \leq a$. For every $n \in \mathbf{N}$ we have $\bar{z}(x,y) < z^{(n)}(x,y)$ on E_0 and

$$D_- z^{(n)}(x,y) > f(x,y,z^{(n)}(x,y),z^{(n)}),$$
$$D_- \bar{z}(x,y) \leq f(x,y,\bar{z}(x,y),\bar{z})$$

for $(x,y) \in E \cap ((0,c] \times R)$. So, all the hypothesis of Theorem 7.11 are satisfied, hence for every $n \in \mathbf{N}$ we have $\bar{z}(x,y) < z^{(n)}(x,y)$ on E_c. Therefore $\bar{z}(x,y) \leq z^+(x,y)$ on E_c. The proof of the existence of the minimum solution is analogous. Thus the Theorem 7.16 is proved.

The next theorem enable us to estimate a function satisfying the differential inequality by the maximum solution of an adequate initial problem.

Theorem 7.17 *Under the assumptions of Theorem 7.16, if $\tilde{z} \in C(E_0 \cup E, R^m)$ and*

$$D_- \tilde{z}(x,y) \leq f(x,y,\tilde{z}(x,y),\tilde{z}) \text{ on } E$$

and $\tilde{z}(x,y) \leq \phi(x,y)$ on E_0 then $\tilde{z}(x,y) \leq z^+(x,y)$ on E_c, $0 < c \leq a$, where z^+ is the maximum solution of (7.2) (7.3) defined on E_c.

PROOF. In the proof of Theorem 7.16 we have considered problems (7.30), (7.31) and the sequence $\{ z^{(n)} \}$ which is uniformly convergent on E_c to the maximum solution z^+ of (7.2), (7.3). The function $\tilde{z}$ satisfies the differential inequality

$$D_- \tilde{z}(x,y) < f(x,y,\tilde{z}(x,y),\tilde{z}) + \theta_n \text{ on } E_c \cap ((0,c] \times R),$$

and $\tilde{z}(x,y) < \phi(x,y) + \theta_n$ on E_0 for every $n \in \mathbf{N}$. Thus by Theorem 7.11 we have $\tilde{z}(x,y) < z^{(n)}(x,y)$ on E_c and consequently $\tilde{z}(x,y) \leq z^+(x,y)$ on E_c. This completes the proof of Theorem 7.17.

Remark 7.18 *It is easy to prove an analogous theorem for the minimum solution.*

Remark 7.19 *A theorem on continuous dependence of extremal solutions on the right-hand side of (7.2) is given in [42].*

7.6 Carathéodory solutions of functional differential inequalities

We formulate the problem. Let $a > 0$, $r_0 \in R_+$, $r = (r_1, \ldots, r_n) \in R^n_+$ and denote

$$E_0 = [-r_0, 0] \times R^n, \quad E = [0, a] \times R^n, \quad B = [-r_0, 0] \times [-r, r].$$

Let $\Omega = E \times R \times C(B, R) \times R^n$. Assume that $f : \Omega \to R$ and $\varphi : E_0 \to R$ are given functions. We consider the functional differential equation

$$D_x z(x, y) = f(x, y, z(x, y), z_{(x,y)}, D_y z(x, y)) \tag{7.33}$$

where $D_y z = (D_{y_1} z, \ldots, D_{y_n} z)$, with the initial condition

$$z(x, y) = \varphi(x, y) \text{ on } E_0. \tag{7.34}$$

Let $0 < \bar{a} \leq a$. A function $\bar{z} : E_0 \cup ([0, \bar{a}] \times R^n) \to R$ is a solution of problem (7.33), (7.34) in the sense "almost everywhere" or Carathéodory sense if
(i) the function $\bar{z}$ is continuous on $E_0 \cup ([0, \bar{a}] \times R^n)$,
(ii) the derivatives $D_x \bar{z}(x, y)$, $D_y \bar{z}(x, y)$ exist almost everywhere on $[0, \bar{a}] \times R^n$,
(iii) the function $\bar{z}$ satisfies equation (7.33) almost everywhere on $[0, \bar{a}] \times R^n$ and condition (7.34) holds.

Solutions of problem (7.33), (7.34) in the Carathéodory sense are not unique. Now, we give an example of the Cauchy problem for which there are at least two Carathéodory solutions. For $X \subset ([0, 1] \times R)$ we denote by κ_X the characteristic function of X. Put $n = 1$ and

$$X_1 = \{(x, y) : x \in [0, 1], \ 2 \leq y < x + 2\},$$
$$X_2 = \{(x, y) : x \in [0, 1], \ -x + 2 \leq y < 2\},$$
$$X_3 = \{(x, y) : x \in [0, 1], \ x - 2 \leq y < -x + 2\},$$
$$X_4 = \{(x, y) : x \in [0, 1], \ -2 \leq y < x - 2\},$$
$$X_5 = \{(x, y) : x \in [0, 1], \ -x - 2 \leq y < -2\}.$$

We define the function $g : [0, 1] \times R \to R$ by

$$g(x, y) = -0.5\,(x - y + 2)^2 \kappa_{X_1}(x, y) + [\,0.5\,(y - 2 + x)^2 - x^2\,] \kappa_{X_2}(x, y)$$
$$-x^2 \kappa_{X_3}(x, y) + [\,0.5\,(x - y - 2)^2 - x^2\,] \kappa_{X_4}(x, y) - 0.5\,(y + 2 + x)^2 \kappa_{X_5}(x, y).$$

Then $g \in C([0, 1] \times R, R)$. Consider the functional differential equation

$$D_x z(x, y) = \left[\int_{-2}^{2} z(x, y + s)\, ds - g(x, y) \right] \sin z(x, y) - (D_y z(x, y))^2$$

on $[0, 1] \times R$ with the initial condition $z(0, y) = 0$ for $y \in R$. Then functions $u(x, y) = 0$ for $(x, y) \in [0, 1] \times R$ and $v : [0, 1] \times R \to R$ given by

$$v(x, y) = 0 \text{ for } x \in [0, 1], \ y < -x \text{ or } y > x,$$

$$v(x,y) = y - x \text{ for } x \in [0,1],\ 0 \le y \le x,$$
$$v(x,y) = -y - x \text{ for } x \in [0,1],\ -x \le y < 0$$

are Carathéodory solutions of the above problem.

Let F be the Niemycki operator corresponding to (7.33), (7.34):

$$F[z](x,y) = f(x,y,z(x,y),z_{(x,y)},D_yz(x,y)).$$

We shall prove a theorem on functional differential inequalities generated by (7.33), (7.34). As a consequence of this theorem we get a uniqueness criterion for generalized entropy solutions to the Cauchy problem. Our Theorems 7.20 and 7.21 are some extensions of Kružkov results [136, 137] on the case of functional differential problems. Essentially, roots of our proof methods, similarly as in [136, 137] are related to Holmgren principle: *uniqueness of the solutions to some linear problems follows from the existence for the adjoint problem.* The proof of Theorem 7.20 is a nonlinear realization of this principle with the Cauchy problem (7.40), (7.41) as the *adjoint problem.*

If $G : [c,d] \times R^n \to R$, where c, $d \in R_+$ and $c \le d$, then supp G denote the support of function G. Given any linear normed spaces X and Y let $C^1(X,Y)$ for the set of all continuously differentiable functions defined on X and taking values in Y. For b, $\varepsilon \in R_+$ such that b, $b+\varepsilon \in [0,a]$ we shall denote

$$F^{1,0}_{b,\varepsilon} = \{ G \in C^1([b,b+\varepsilon] \times R^n, R_+) : \ G(b+\varepsilon,y) = 0 \text{ for } y \in R^n,$$
$$\text{supp}\, G \text{ is a compact set} \}$$

and

$$F_{b,\varepsilon} = \{\phi \in C([b,b+\varepsilon] \times R^n, R_+) : \ \text{supp}\,\phi \subset E \text{ is a compact set}\}.$$

Let $K(0,M_1) = \{y \in R^n : \ \|y\| \le M_1\}$ where $M_1 \in R_+$. The symbol $\|\cdot\|$ denotes here the Euclidean norm. If $M = (M_0,M_1) \in R^2_+$ then let

$$\Omega_M = E \times [-M_0,M_0] \times C(B,[-M_0,M_0]) \times K(0,M_1).$$

If X is a nonempty linear normed space and $U : C(X,R) \to R$ is a continuous linear functional, then

$$\|U\|_* = \sup \{ \|T\phi\| : \ \phi \in C(X,R),\ \|\phi\| \le 1 \}.$$

Assumption H $[f]$. Suppose the following:

1) $f : \Omega \to R$ and for all $P = (x,y,p,w,q) \in \Omega$ there exist the derivatives

$$D_yf(P),\ D_qf(P),\ D_wf(P),\ D_pf(P),$$
$$D_{yq}f(P),\ D_{qq}f(P),\ D_{qw}f(P),\ D_{qp}f(P),$$

2) there exist constants $(M_0,M_1) = M \in R^2_+$ such that the functions

$$D_yf,\ D_qf,\ D_wf,\ D_pf,\ D_{yq}f,\ D_{qq}f,\ D_{qw}f$$

are continuous on Ω_M,

3) there exist constants C_1, $C_2 \in R_+$ such that

$$\|D_y f(P)\|, \ \|D_q f(P)\|, \ \|D_w f(P)\|_*, \ D_p f(P) \le C_1$$

$$\|D_{qp} f(P)\|, \ \|D_{yq} f(P)\|, \ \|D_{qw} f(P)\|_* \le C_2$$

for every $P = (x, y, , p, w, q) \in \Omega_M$,

4) f is nondecreasing with respect to the functional variable on Ω_M,

5) there exist nonnegative constants C_3, C_4 with $C_3 > C_4$ such that we have the estimates

$$C_3\|\eta\|^2 \ge -\sum_{i,j=1}^{n} D_{q_i q_j} f(x, y, p, w, q)\, \eta_i \eta_j \ge C_4\|\eta\|^2 \text{ on } \Omega_M,$$

for arbitrary $\eta = (\eta_1, \ldots, \eta_n) \in R^n$.

Let $\lambda : [0, a] \to R_+$ be a bounded function on $[\delta, a]$ for every $\delta \in (0, a]$ and $M = (M_0, M_1)$. A function $z : E_0 \cup E \to R$ is said to be of class $C_{M,\lambda}(E_0 \cup E, R)$ if the following conditions are satisfied:

(i) $z \in C(E_0 \cup E, R)$ and $|z(x, y)| \le M_0$ on $E_0 \cup E$,

(ii) the derivatives $D_x z$, $D_y z$ exist a.e. on E and $D_x z \in L^2_{loc}(E, R)$, $D_y z \in L^2_{loc}(E, R^n)$,

(iii) $\|D_y z(x, y)\| \le M_1$ for almost all $(x, y) \in E$,

(iv) for every $\eta \in R^n$, $(x, y) \in E$ we have

$$z(x, y + \eta) - 2z(x, y) + z(x, y - \eta) \le \lambda(x)\, \|\eta\|^2.$$

Now we formulate a theorem on functional differential inequalities.

Theorem 7.20 *Suppose that the following conditions are true:*

1) Assumption H [f] is satisfied,

2) there are constants $(M_0, M_1) = M \in R_+^2$ *and a function* $\lambda : [0, a] \to R_+$, *bounded on* $[\delta, a]$ *for every* $\delta \in (0, a]$ *such that* u, $v \in C_{M,\lambda}(E_0 \cup E, R)$,

3) the functional differential inequality

$$D_x u(x, y) - F[u](x, y) \le D_x v(x, y) - F[v](x, y) \tag{7.35}$$

is satisfied almost everywhere on E and

$$u(x, y) \le v(x, y) \text{ on } E_0 \tag{7.36}$$

Under these assumptions we have the inequality

$$u(x, y) \le v(x, y) \tag{7.37}$$

on E provided either $D_w f(x, y, p, w, q) = 0$ *on* Ω_M, *or the function* $\lambda : [0, a] \to R_+$ *is integrable on the interval* $[0, a]$.

PROOF. Let $\{u^{(\nu)}\}_{\nu\geq 1}$ and $\{v^{(\nu)}\}_{\nu\geq 1}$ be two fixed sequences satisfying the following conditions:

(i) for every ν the functions $u^{(\nu)},\ v^{(\nu)} : E_0 \cup E \to R$ are of class C^2,

(ii) for arbitrary compact subset $X \subset E_0 \cup E$ we have

$$\lim_{\nu\to\infty} u^{(\nu)}|_X = u|_X, \quad \lim_{\nu\to\infty} v^{(\nu)}|_X = v|_X \quad \text{uniformly on } X,$$

$$\lim_{\nu\to\infty} \|D_y u^{(\nu)}|_X - D_y u|_X\|_{L^2(X,R^n)} = 0,$$

$$\lim_{\nu\to\infty} \|D_y v^{(\nu)}|_X - D_y v|_X\|_{L^2(X,R^n)} = 0,$$

(iii) for every $(x,y) \in E$ and $\eta \in R^n$ we have the estimates

$$D_{\eta\eta} u^{(\nu)}(x,y) \leq \lambda(x) \quad \text{and} \quad D_{\eta\eta} v^{(\nu)}(x,y) \leq \lambda(x)$$

where $D_{\eta\eta} z$ denotes the second order derivative of the function z in the direction of the vector η. Note that sequences $\{u^{(\nu)}\}$ and $\{v^{(\nu)}\}$ can be constructed by the regularization method, see [136]. Denote $\omega = v - u$ and $\omega^{(\nu)} = v^{(\nu)} - u^{(\nu)}$ for $\nu \geq 1$. Then from inequality (7.36) it follows that $\omega(x,y) \geq 0$ on E_0. Suppose that $u(x,y) \leq v(x,y)$ on $[-r_0, b] \times R^n$ for some $b \in [0,a)$. We shall show that $u(x,y) \leq v(x,y)$ on $[-r_0, \bar{a}] \times R^n$ for some $\bar{a} \in (b,a]$. Let $\varepsilon > 0$ and $b+\varepsilon \in [0,a]$. Suppose that $G \in F^{1,0}_{b,\varepsilon}$. Then from inequality (7.35) we get the estimate

$$\int_{R^n} \int_b^{b+\varepsilon} G(x,y) \Big\{ D_x \omega^{(\nu)}(x,y) - F[v^{(\nu)}](x,y) + F[u^{(\nu)}](x,y) \Big\}\, dxdy$$

$$\geq \int_{R^n} \int_b^{b+\varepsilon} G(x,y)\, \Gamma^{(\nu)}(x,y)\, dx\, dy \tag{7.38}$$

for $\nu \geq 1$, where the function $\Gamma^{(\nu)}$ is defined by

$$\Gamma^{(\nu)}(x,y) = D_x v^{(\nu)}(x,y) - F[v^{(\nu)}](x,y) - [D_x v(x,y) - F[v](x,y)]$$
$$- \Big[D_x u^{(\nu)}(x,y) - F[u^{(\nu)}](x,y) \Big] + D_x u(x,y) - F[u](x,y).$$

It follows from assumptions (i) - (iii) and from Assumption H $[f]$ that

$$\lim_{\nu\to\infty} \|\Gamma^{(\nu)}|_X\|_{L^2(X,R)} = 0$$

for arbitrary compact subset $X \subset E$. Define

$$P^{(\nu)}(x,y,\theta)$$

$$= \Big(x, y, (u^{(\nu)} + \theta\,\omega^{(\nu)})(x,y), (u^{(\nu)} + \theta\,\omega^{(\nu)})_{(x,y)}, D_y(u^{(\nu)} + \theta\,\omega^{(\nu)})(x,y) \Big).$$

Applying the Hadamard mean value theorem and integrating by parts in (7.38) we obtain the inequality

$$\int_{R^n} \int_b^{b+\varepsilon} \Big\{ -\omega^{(\nu)}(x,y) \Big[D_x G(x,y) + G(x,y) \int_0^1 D_p f(P^{(\nu)}(x,y,\theta))\, d\theta \tag{7.39}$$

$$-\sum_{j=1}^{n} \frac{\partial}{\partial y_j} \left(G(x,y) \int_0^1 D_{q_j} f(\, P^{(\nu)}(x,y,\theta)\,)\, d\theta \right) \Bigg]$$

$$-G(x,y) \int_0^1 D_w f(\, P^{(\nu)}(x,y,\theta)\,)\, d\theta \, (\omega^{(\nu)})_{(x,y)} \Bigg\} \, dxdy$$

$$\geq \int_{R^n} \int_b^{b+\varepsilon} G(x,y)\, \Gamma^{(\nu)}(x,y)\, dxdy + \int_{R^n} G(b,y)\, \omega(b,y)\, dy$$

for $\nu \geq 1$. Take arbitrary $\phi \in F_{b,\varepsilon}$ and consider the differential equation

$$D_x z(x,y) + z(x,y) \int_0^1 D_p f(\, P^{(\nu)}(x,y,\theta)\,)\, d\theta \tag{7.40}$$

$$-\sum_{j=1}^{n} \frac{\partial}{\partial y_j} \left\{ z(x,y) \int_0^1 D_{q_j} f(\, P^{(\nu)}(x,y,\theta)\,)\, d\theta \right\} = -\phi(x,y)$$

with the initial condition

$$z(b+\varepsilon, y) = 0. \tag{7.41}$$

Let

$$g^{(\nu)}(\,\cdot\,,x,y) : \; [b, b+\varepsilon] \to R^n \;\text{ for }\; (x,y) \in [b,b+\varepsilon] \times R^n$$

be the solution of the characteristic initial value problem corresponding to problem (7.40), (7.41):

$$\eta'(t) = -\int_0^1 D_q f(\, P^{(\nu)}(t, \eta(t), \theta)\,)\, d\theta, \quad \eta(x) = y. \tag{7.42}$$

For $(x,y) \in [b,b+\varepsilon] \times R^n$ we denote

$$H^{(\nu)}(x,y) = \int_0^1 D_p f(\, P^{(\nu)}(x,y,\theta)\,)\, d\theta$$

$$-\sum_{j=1}^{n} \int_0^1 A_{jj} \left(P^{(\nu)}(x,y,\theta); u^{(\nu)} + \theta\, \omega^{(\nu)} \right) d\theta$$

where the matrix

$$A(\,P; z\,) = [\, A_{ij}(\,P; z\,)\,]_{i,j=1,\ldots,n}$$

is defined by

$$A_{ij}(\,P; z\,) = D_{q_i q_j} f(\,P\,) + D_{q_i p} f(\,P\,) D_{y_j} z(x,y) \tag{7.43}$$

$$+ D_{q_i w} f(\,P\,)\, (D_y z)_{(x,y)} + \sum_{\mu=1}^{n} D_{q_\mu q_i} f(\,P\,)\, D_{y_\mu y_j} z(x,y)$$

where $z \in C_{M,\lambda}(E_0 \cup E, R)$ and $P = (x,y,p,w,q) \in \Omega_M$. Consider inequality (7.39) with $G^{(\nu)}$ satisfying problem (7.40), (7.41). Then we have the inequality

$$\int_{R^n} \int_b^{b+\varepsilon} \Big\{ \omega^{(\nu)}(x,y)\, \phi(x,y) \tag{7.44}$$

$$- G^{(\nu)}(x,y) \int_0^1 D_w f(\, P^{(\nu)}(x,y,\theta)\,)\, d\theta\, (\omega^{(\nu)})_{(x,y)} \Big\}\, dx\, dy$$

$$\geq \int_{R^n} \int_b^{b+\varepsilon} G^{(\nu)}(x,y)\, \Gamma^{(\nu)}(x,y)\, dxdy + \int_{R^n} G(b,y)\, \omega(b,y)\, dy$$

for $\nu \geq 1$, where

$$G^{(\nu)}(x,y) = \int_b^{b+\varepsilon} \phi\left(t, g^{(\nu)}(t,x,y)\right) \exp\left\{ \int_x^t H^{(\nu)}\left(\theta, g^{(\nu)}(\theta,x,y)\right) d\theta \right\} dt.$$

The above formula defines a transformation of the set $F_{b,\varepsilon}$ onto the set $F^{1,0}_{b,\varepsilon}$. For $(x,y) \in [b, b+\varepsilon] \times R^n$ the matrix

$$D_y g^{(\nu)}(\,\cdot\,,x,y) = \left[\, D_{y_i} g_j^{(\nu)}(\,\cdot\,,x,y) \,\right]_{i,j=1,\ldots,n}$$

satisfies the equation

$$\frac{d}{dt}\zeta(\,t,x,y\,) = -A\left(P^{(\nu)}(\,t, g^{(\nu)}(t,x,y),\theta); u^{(\nu)} + \theta\,\omega^{(\nu)}\right)\, \zeta(\,t,x,y\,) \tag{7.45}$$

with the initial condition

$$\zeta_{i,j}(x,x,y) = \delta_{ij}, \quad i,j = 1,\ldots,n, \tag{7.46}$$

where $[\,\zeta(\,t,x,y\,)\,] = [\,\zeta_{ij}(\,t,x,y\,)\,]_{i,j=1,\ldots,n}$, and δ_{ij} is the Kronecker symbol. Write

$$J^{(\nu)}(t,x,y) = \det\left[D_y g^{(\nu)}(t,x,y)\right] \text{ for } (x,y) \in [b,b+\varepsilon] \times R^n,\ t \in [b,b+\varepsilon].$$

Observe that $J^{(\nu)}(t,x,y) > 0$, moreover we have the relation

$$J^{(\nu)}(t,x,y) = \exp\left[\int_t^x \sum_{j=1}^n A_{jj}\left(P^{(\nu)}(\tau, g^{(\nu)}(\tau,x,y),\theta); u^{(\nu)} + \theta\,\omega^{(\nu)}\right) d\tau \right].$$

Now we use the change of variables

$$[b,b+\varepsilon] \times R^n \ni (x,y) \to (x,\eta) = \left(x, g^{(\nu)}(t,x,y)\right) \in [b,b+\varepsilon] \times R^n$$

and we obtain

$$\int_{R^n} \int_b^t \phi(\,t, g^{(\nu)}(t,x,y)\,)\, \exp\left[\int_x^t H^{(\nu)}\left(\tau, g^{(\nu)}(\tau,x,y)\right) d\tau \right]$$

$$\times \int_0^1 D_w(\, P^{(\nu)} x,y,\theta)\,)\, d\theta\, (\omega^{(\nu)})_{(x,y)}\, dx\, dy$$

$$= \int_{R^n} \int_b^t \phi(t,\eta) J^{(\nu)}(t,x,g^{(\nu)}(x,t,\eta)\,)$$

$$\times \exp\left[\int_x^t H^{(\nu)}\left(\tau, g^{(\nu)}(\tau, x, g^{(\nu)}(x,t,\eta)\,)\right) d\tau\right]$$

$$\times \int_0^1 D_w f(\,P^{(\nu)}(x, g^{(\nu)}(x,t,\eta), \theta)\,)\, d\theta\, (\,\omega^{(\nu)}\,)_{(x, g^{(\nu)}(x,t,\eta))}\, dx\, d\eta.$$

From the above relation and from (7.39), (7.40), (7.44) it follows that for every $\phi \in F_{b,\varepsilon}$ we have the inequality

$$\int_{R^n}\int_b^{b+\varepsilon} \phi(x,y)\left\{\,\omega^{(\nu)}(x,y) - \int_0^x J^{(\nu)}(x,t,g^{(\nu)}(t,x,y)\,)\right. \tag{7.47}$$

$$\times \exp\left[\int_t^x H^{(\nu)}\left(\tau, g^{(\nu)}(\tau, t, g^{(\nu)}(t,x,y)\,)\,\right) d\tau\right]$$

$$\left.\times \int_0^1 D_w f(\,P^{(\nu)}(t, g^{(\nu)}(t,x,y), \theta)\,)\,(\,\omega^{(\nu)}\,)_{(t, g^{(\nu)}(t,x,y))}\, d\theta\, dt\right\} dx\, dy$$

$$\geq \int_{R^n}\int_b^{b+\varepsilon} G(x,y)\,\Gamma^{(\nu)}(x,y)\, dx\, dy + \int_{R^n} G(b,y)\,\omega(b,y)\, dy$$

for $\nu \geq 1$. For $P^{(\nu)}(x,y,\theta)$ there are orthonormal vectors

$$\tilde{\eta}^{(j)} = \tilde{\eta}^{(\nu,j))}(x,y,\theta) \in R^n, \quad j = 1, \dots, n,$$

and eigenvalues $\lambda^{(\nu,j)}(x,y,\theta) \in R_+$ such that $C_3 \geq \lambda^{(\nu,j)}(x,y,\theta) \geq C_4 \geq 0$ and

$$-\sum_{i,j=1}^n D_{q_i q_j} f(\,P^{(\nu)}(x,y,\theta)\,) D_{y_i y_j} z^{(\nu)}(x,y)$$

$$= \sum_{j=1}^n \lambda^{(\nu,j)}(x,y,\theta) D_{\tilde{\eta}^{(j)}\tilde{\eta}^{(j)}} z^{(\nu)}(x,y).$$

Thus from Assumption H $[\,f\,]$ it follows that the function $H^{(\nu)}$ is upper-bounded by a measurable function independent of $\nu \geq 1$. The same property for the function $J^{(\nu)}$ can be stated. Let $\delta \in (0,\varepsilon)$ be fixed. Since for $\phi \in F_{b,\varepsilon}$, then the right-hand side of inequality (7.47) tends to 0 as $\nu \to \infty$, we have the inequality

$$\int_{R^n}\int_b^{b+\varepsilon} \phi(x,y)\left\{\,\omega(x,y) - \liminf_{\nu\to\infty}\int_0^x J^{(\nu)}\left(x,t,g^{(\nu)}(t,x,y)\right)\right. \tag{7.48}$$

$$\times \exp\left[\int_t^x H^{(\nu)}\left(\tau, g^{(\nu)}(\tau, t, g^{(\nu)}(t,x,y)\,)\,\right) d\tau\right]$$

$$\left.\times \int_0^1 D_w f(\,P^{(\nu)}(t, g^{(\nu)}(t,x,y), \theta)\,)\,)\,(\,\omega^{(\nu)}\,)_{(t, g^{(\nu)}(t,x,y))}\, d\theta\, dt\right\} dx\, dy \geq 0.$$

If $D_w f = 0$, then we get (7.37) what was to be proved. If $D_w f \neq 0$, then from inequality (7.48) we get the Volterra integral inequality

$$\omega(x,y) \;\geq\; \liminf_{\nu\to\infty}\int_0^x J^{(\nu)}(\,x,t,g^{(\nu)}(t,x,y)\,) \tag{7.49}$$

$$\times \exp\left[\int_t^x H^{(\nu)}(\tau, g^{(\nu)}(\tau, t, g^{(\nu)}(t,x,y)))\, d\tau\right]$$

$$\times \int_0^1 D_w f(P^{(\nu)}(t, g^{(\nu)}(t,x,y),\theta))\,(\omega^{(\nu)})_{(t,g^{(\nu)}(t,x,y))}\, d\theta\, dt$$

for $(x,y) \in [b+\delta, b+\varepsilon]$, where $\delta \in (0,\varepsilon)$. Denote by ω^- and $(\omega^{(\nu)})^-$ for $\nu \geq 1$ the functions that are given by

$$\omega^-(x,y) = \min\{0,\ \omega(x,y)\},\quad (\omega^{(\nu)})^-(x,y) = \min\{0,\ \omega^{(\nu)}(x,y)\},$$

where $(x,y) \in E_0 \cup E$. From inequality (7.49) it is easy to see that

$$\omega^-(x,y) \geq \liminf_{\nu\to\infty} \int_0^x J^{(\nu)}(x,t,g^{(\nu)}(t,x,y)) \tag{7.50}$$

$$\times \exp\left[\int_t^x H^{(\nu)}\left(\tau, g^{(\nu)}(\tau,t,g^{(\nu)}(t,x,y))\right) d\tau\right]$$

$$\times \int_0^1 D_w f(P^{(\nu)}(t,g^{(\nu)}(t,x,y),\theta))\,((\omega^{(\nu)})^-)_{(t,g^{(\nu)}(t,x,y))}\, d\theta\, dt$$

for $(x,y) \in [b+\delta, b+\varepsilon] \times R^n$, where $\delta \in (0,\varepsilon)$. In view of (7.43) and Assumption H $[f]$ we have the inequality

$$J^{(\nu)}\left(x, g^{(\nu)}(t,x,y)\right)\, \exp\left[\int_t^x H^{(\nu)}\left(t, g^{(\nu)}(\tau,t,g^{(\nu)}(t,x,y))\right)\, d\tau\right] \tag{7.51}$$

$$\leq \exp\left[\int_t^x (C_1 + 2nC_2(1+2M_1) + C_3\lambda(\tau))d\tau\right].$$

Let a function $W : [b, b+\varepsilon] \to R_-$ be defined by

$$W(x) = \inf\{\omega^-(t,y) :\ (t,y) \in [-r_0, x] \times R^n\},\quad x \in [a, a+\varepsilon].$$

Next, from (7.50), (7.51) and Assumption H $[f]$ we obtain the inequality

$$W(x) \geq \int_0^x \exp\left[\int_t^x (C_1 + 2nC_2(t+2M_1) + C_3\lambda(\tau))\, d\tau\right] W(t)\, dt$$

where $x \in [a, a+\varepsilon]$. From definition of ω^- and $(\omega^{(\nu)})^-$ we get the inequalities

$$\omega(x,y) \geq W(x) \geq 0,\quad (x,y) \in [-r_0, b+\varepsilon] \times R^n.$$

As $b \in [0,a)$ was taken to be arbitrary, we have obtained inequality (7.37). This proves the Theorem 7.20.

The following theorem on the uniqueness of generalized entropy solutions of problem (7.33), (7.34) is a simple consequence of Theorem 7.20.

Theorem 7.21 *Suppose that the following assumptions are true:*

1) Assumption H [f] is satisfied and $u,\ v \in C(E_0 \cup E, R)$,

2) there are constants $(M_0, M_1) = M \in R_+^2$ *and a function* $\lambda : [0,\ a] \to R_+$, *bounded on* $[\delta,\ a]$ *for every* $\delta \in (0,\ a]$, *such that* $u,\ v \in C_{M,\lambda}(E_0 \cup E, R)$,

3) the functions $u,\ v$ *are solutions of the Cauchy problem (7.33), (7.34) where* $\varphi \in C(E_0, R)$.

Then for every $(x, y) \in [0, a] \times R^n$ *we have* $u(x, y) = v(x, y)$ *provided either* $D_w f(x, y, p, w, q) = 0$ *on* Ω_M *or the function* $\lambda : [0,\ a] \to R_+$ *is integrable on* $[0,\ a]$.

PROOF. The functions $u,\ v$ satisfy the assumptions of Theorem 7.20, thus we have $u(x, y) \leq v(x, y)$ for $(x, y) \in [0,\ a] \times R^n$. The inverse inequality is obtained in the same way.

Remark 7.22 *If we omit condition (iv) in the definition of the class* $C_{M,\lambda}(E_0 \cup E, R)$ *then Theorem 7.21 is not true, because there may exist two different solutions of problem (7.33), (7.34). An adequate example of f without a functional variable is given in [136].*

Remark 7.23 *Our uniqueness result and the theorem on functional differential inequalities remain valid if we replace problem (7.33), (7.34) by the following Cauchy problem for the weakly coupled system*

$$D_x z_i(x, y) = f_i\left(x, y, z(x, y), z_{(x,y)}, D_y z_i(x, y)\right), \quad i = 1, \dots, k,$$

$$z(x, y) = \Phi(x, y) \ \text{ on } E_0.$$

We assume in this case that the function

$$f(x, y, p, w, q) = (f_1(x, y, p, w, q), \dots, f_k(x, y, p, w, q)),$$

satisfies the quasi-monotonicity condition with respect to p. While proving a theorem on functional differential inequalities we get a system of Volterra integral inequalities, which implies an appropriate inequality for functions.

7.7 Existence of Carathéodory solutions

We consider the Cauchy problem (7.33), (7.34) with $n = 1$. The existence result is based on a difference method. Write $C(B, R; t) = \{w \in C(B, R) : \ \|w\|_0 \leq t\}$, $t \in R_+$, where $\|\cdot\|_0$ is the supremum norm. We denote by $C_{0.L}(B, R)$ the class of all continuous functions $w : B \to R$ which satisfy the Lipschitz condition on B. Furthermore, for any $t \in R_+$ let

$$C_{0.L}(B, R; t) = \{w \in C_{0.L}(B, R) : \ \|w\|_{0.L} = \|w\|_0 + \|w\|_L \leq t\}$$

where

$$\|w\|_L = \sup\left\{ |w(t, s) - w(\bar{t}, \bar{s})|(|t - \bar{t}| + |s - \bar{s}|)^{-1} : \ (t, s), (\bar{t}, \bar{s}) \in B \right\}.$$

Assumption H $[\phi]$. Suppose that

1) $\phi \in C(E_0, R)$ and there are constants $M, L \in R_+$ such that

$$|\phi(x,y)| \leq M, \quad |\phi(x,y) - \phi(x,\bar{y})| \leq L|y - \bar{y}| \text{ on } E_0,$$

2) if $r_0 > 0$ then there is a constant $\tilde{K} \in R_+$ such that

$$\eta^{-2}\left[\phi(x, y+\eta) - 2\phi(x,y) + \phi(x, y-\eta)\right] \leq K, \ (x,y) \in E_0, \eta \in R.$$

Assumption $\tilde{H}$ $[f]$. Suppose that the function $f : [0,a] \times R \times R \times C(B,R) \times R \to R$ is of class C^2 and

1) there are a constant $N \geq M$ and a function $\gamma \in C([M,N], R_+)$ such that γ is nondecreasing and $|f(x,y,p,w,q)| \leq \gamma(t)$ on $[0,a] \times R \times [-t,t] \times C(B,R;t) \times R$ and

$$\int_M^N \frac{dt}{\gamma(t)} \geq a,$$

2) there are constants $\tilde{N} \geq L$, $A > 0$ and a nondecreasing function $\beta \in C([L, \tilde{N}], R_+)$ such that

$$|D_q(P)| \leq A, \ |D_y(P)|, \ |D_p(P)|, \ \|D_w(P)\|_* \leq \beta(t)$$

for $P = (x,y,p,w,q) \in [0,a] \times R \times [-N,N] \times C_{0.L}(B,R;t)$ and

$$\int_L^{\tilde{N}} \frac{dt}{(2t+1)\beta(N+3t)} \geq a,$$

3) the Fréchet derivative $D_w f$ satisfies the condition

$$D_w f(x,y,p,w,q)\bar{w} \geq 0 \text{ for } \bar{w} \in C(B, R_+),$$

4) the derivatives of the second order

$$D^2_{yy}f, \ D^2_{yp}f, \ D^2_{yw}f, \ D^2_{yq}f, \ D^2_{pp}f, \ D^2_{pw}f, \ D^2_{pq}f, \ D^2_{ww}f, \ D^2_{wq}f$$

are bounded and $D_{qq}f \leq 0$ on $[0,a] \times R \times [-N,N] \times C_{0.L}(B,R;N+3\tilde{N}) \times [-\tilde{N}, \tilde{N}]$,

5) if $r_0 = 0$ then there are $\varepsilon \in (0,a]$ and $\mu > 0$ such that $D_{qq}f(P) \leq -\mu$, on $[0,\varepsilon] \times R \times [-N,N] \times C_{0.L}(B,R;N+3\tilde{N}) \times [-\tilde{N}, \tilde{N}]$.

Now we formulate a difference method for problem (7.33), (7.34). For $h, k > 0$ we define $x_i = ih$, $i = 0,1,\ldots,n_0$, $n_0 h = a$, and $y_j = jk$, $j \in \mathbf{Z}$. If $r_0 > 0$ then there is an integer $n_1 > 0$ such that $-n_1 h \leq -r_0 < (-n_1+1)h$. We denote $x_i = ih$ for $i = -n_1+1, \ldots, -1$ and $x_{-n_1} = -r_0$. Let $U = \{(h,k) : A < \frac{k}{h}\}$. If $0 \leq i \leq n_0 - 1$ and j is an even number then we put

$$P_{ij} = [x_i, x_{i+1}] \times [y_{j-1}, y_{j+1}] \text{ and } Q_i = \bigcup_{j \in 2\mathbf{Z}} P_{ij},$$

where $2\mathbf{Z}$ is a set of even numbers. Let δ_0, δ be the difference operators

$$\delta_0 z^{(i,j)} = \frac{1}{h}\left(z^{(i+1,j)} - z^{(i,j)}\right), \quad \delta z^{(i,j)} = \frac{1}{2k}\left(z^{(i,j+1)} - z^{(i,j-1)}\right),$$

where $z^{(i,j)} = z(x_i, y_j)$. Let $\phi_{hk} : E_0 \to R$ be a function defined in the following way:

(i) If $r_0 > 0$ then for each $(x, y) \in E_0$ there are (i, j) such that $-n_1 \le i \le 0$, $j \in 2\mathbf{Z}$ and $(x, y) \in [x_i, x_{i+1}] \times [y_{j-1}, y_{j+1}]$. Then we write

$$\phi_{hk}(x, y) = \phi^{(i,j-1)} + (x - x_i)\delta_0\phi^{(i,j-1)}$$

$$+(y - y_{j-1})\delta\varphi^{(i,j)} + (x - x_i)(y - y_{j-1})\delta_0\,\delta\,\phi^{(i,j)}.$$

(ii) If $b_0 = 0$ then for each $y \in R$ there is an even number j such that $y \in [y_{j-1}, y_{j+1}]$. Then we write

$$\phi_{hk}(y) = \phi^{(j-1)} + (y - y_{j-1})\delta\phi^{(j)}.$$

For any $(h, k) \in U$ we define the function $u_{hk} : E_0 \cup E \to R$ in the following way.

I. Let $v^{(0,j)} = \phi^{(0,j)}$, $j \in \mathbf{Z}$, and $u_{hk}(x, y) = \phi_{hk}(x, y)$ on E_0.

II. If for some i, $0 \le i \le n_0 - 1$ we have defined $v^{(i,j)}$, $j \in \mathbf{Z}$, and u_{hk} is given on $E_0 \cup Q_0 \cup \ldots \cup Q_{i-1}$ then

$$v^{(i+1,j)} = 0.5(v^{(i,j+1)} + v^{(i,j-1)})$$

$$+h\,f(\,x_i, y_j, 0.5(v^{(i,j+1)} + v^{(i,j-1)}), (u_{hk})_{(x_i,y_j)}, \,\delta v^{(i,j)}\,),\;\; j \in \mathbf{Z},$$

and

$$u_{hk}(x, y) = v^{(i,j-1)} + (x - x_i)\delta_0 v^{(i,j-1)} + (y - y_{j-1})\delta v^{(i,j)} + (x - x_i)(y - y_{j-1})\,\delta_0\,\delta v^{(i,j)},$$

where $(x, y) \in P_{ij}$ and $j \in 2\mathbf{Z}$.

Now we formulate the main result in this Section.

Theorem 7.24 *Suppose that Assumptions $H\,[\phi]$ and $\tilde{H}\,[\,f\,]$ are satisfied. Then there is a sequence $\{\,(h_\nu, k_\nu)\,\}$, $(h_\nu, k_\nu) \in U$, and a function $u : E_0 \cup E \to R$ such that*

$$\lim_{\nu\to\infty} u_{h_\nu k_\nu}(x, y) = u(x, y) \;\; \textit{uniformly on}\; [0, a] \times [-b, b]$$

for every $b > 0$ and u is the Carathéodory solution of (7.33), (7.34) with $n = 1$ satisfying the entropy condition $\eta^{-2}\,[u(x, y + \eta) - 2u(x, y) + u(x, y - \eta)] \le \lambda(x)$ with $\lambda \in L([0, a], R_+)$.

The proof of the theorem can be found in [59].

Remark 7.25 *The question of the existence of Carathéodory solutions of (7.33), (7.34) with $n > 1$ is an open problem, see [110].*

7.8 Functional Differential Problems with Unbounded Delay

We start the investigations of first order partial functional differential equations with unbounded delay. We give sufficient conditions for the existence and uniqueness of Carathéodory solutions of initial problems for quasilinear equations with unbounded delay. We consider functional differential equations in a Banach space.

The theory of ordinary functional differential equations with unbounded delay is given in monographs [95, 141]. Our results are based on [112]. Initial boundary value problems for quasilinear equations with unbounded delay were considered in [66]. The construction of a phase space for the Darboux problem can be found in [62].

We formulate the problem. Let Y be a Banach space with the norm $\|\cdot\|$ and $D = (-\infty, 0\,] \times [-r, r] \subset R^{1+n}$, $r = (r_1, \ldots, r_n) \in R^n_+$. The norm in R^n will also be denoted by $\|\cdot\|$. For a function $z : (-\infty, b] \times R^n \to Y$, $b \geq 0$, and for a point $(x, y) \in [0, b] \times R^n$ we define a function $z_{(x,y)} : D \to Y$ by $z_{(x,y)}(\tau, s) = z(x + \tau, y + s)$, $(\tau, s) \in D$. The phase space X for equations with unbounded delay is a linear space, with a norm $\|\cdot\|_X$, consisting of functions mapping the set D into Y. Let $a > 0$ be fixed and suppose that

$$\varrho = (\,\varrho_1, \ldots, \varrho_n\,) : [0, a] \times R^n \times X \to R^n,$$

$$f : [0, a] \times R^n \times X \to Y, \quad \varphi : (-\infty, 0\,] \times R^n \to Y$$

are given functions. We consider the quasilinear equation

$$D_x z(x, y) + \sum_{i=1}^{n} \varrho_i(x, y, z_{(x,y)})\, D_{y_i} z(x, y) = f(x, y, z_{(x,y)}) \tag{7.52}$$

with the initial condition

$$z(x, y) = \varphi(x, y) \;\text{ on }\; (-\infty, 0\,] \times R^n. \tag{7.53}$$

We will deal with Carathéodory solutions of (7.52), (7.53). A function $\bar{u} : (-\infty, b] \times R^n \to Y$ where $0 < b \leq a$, is a solution of the above problem provides

(i) $\bar{u}$ is continuous on $[0, b] \times R^n$ and the derivatives $D_x\bar{u}(x, y)$, $D_y\bar{u}(x, y)$ exist for almost all $(x, y) \in [0, b] \times R^n$,

(ii) $\bar{u}$ satisfies (7.52) almost everywhere on $[0, b] \times R^n$ and condition (7.53) holds.

We adopt the following notations. If $z : (-\infty, b\,] \times R^n \to Y$, $0 < b \leq a$, is a function such that z is continuous on $[0, b] \times R^n$ then we put for $0 \leq x \leq b$, $y \in R^n$:

$$\|z\|_{[0,x;y]} = \max\{\|z(\tau, s)\| : \;\; (\tau, s) \in [0, x] \times [y - r, y + r]\,\}$$

$$\|z\|_{[0,x;R^n]} = \sup\{\|z(\tau, s)\| : \;\; (\tau, s) \in [0, x] \times R^n\,\}$$

and

$$\operatorname{Lip} z\;|_{[0,x;y]} =$$

$$\sup\left\{\frac{\|z(\tau, s) - z(\tau, \bar{s})\|}{\|s - \bar{s}\|} : \;\; (\tau, s), (\tau, \bar{s}) \in [0, x] \times [y - r, y + r]\right\}.$$

The fundamental axioms assumed on X are the following

Assumption H [X]. Suppose that

1) $(\,X, \|\cdot\|_X\,)$ is a Banach space,

2) if $z : (-\infty, b] \times R^n \to Y$, $0 < b \leq a$, is a function such that $z_{(0,y)} \in X$ for $y \in R^n$, z is continuous on $[0, b] \times R^n$ then $z_{(x,y)} \in X$ for $(x, y) \in (0, b] \times R^n$ and

(i) for $(x,y) \in [0,b] \times R^n$ we have

$$\|z_{(x,y)}\|_X \leq K\,\|z\|_{[0,x;y]} + M\,\|z_{(0,y)}\|_X,$$

where $K, L \in R_+$ are constant independent on z,

(ii) the function $(x,y) \to z_{(x,y)}$ is continuous on $[0,b] \times R^n$,

3) the linear subspace $X_L \subset X$ is such that

(i) X_L endowed with the norm $\|\ \|_{X_L}$ is a Banach space,

(ii) if $z : (-\infty, b\,] \times R^n \to Y$, $0 < b \leq a$, is a function such that $z_{(0,y)} \in X_L$ for $y \in R^n$, z is continuous on $[0,b] \times R^n$ and $z(x,\cdot\,) : R^n \to Y$ satisfies the Lipschitz condition with a constant independent on x, $x \in [0,b]$, then

(α) $z_{(x,y)} \in X_L$ for $(x,y) \in (0,b] \times R^n$,

(β) for $(x,y) \in [0,b] \times R^n$ we have

$$\|z_{(x,y)}\|_{X_L} \leq K_0\ \left(\,\|z\|_{[0,x;y]} + \operatorname{Lip} z\ |_{[0,x;y]}\,\right) + L_0\,\|z_{(0,y)}\|_{X_L},$$

where $K_0\, L_0 \in R_+$ are constants independent on z.

Let us denote by $L([\alpha,\beta], R)$ the class of all functions $\mu : [\alpha,\beta] \to R$, $[\alpha,\beta] \subset R$, which are integrable on $[\alpha,\beta]$. We will use the symbol Θ to denote the set of functions $\gamma : [0,a] \times R_+ \to R_+$ such that $\gamma(t,\cdot\,) : R_+ \to R_+$ is nondecreasing for almost all $t \in [0,a]$ and $\gamma(\,\cdot\,,\tau) \in L([0,a], R_+)$ for $\tau \in R_+$. Write

$$X[\kappa] = \{\, w \in X :\ \ \|w\|_X \leq \kappa \,\} \ \text{ and } \ X_L[\kappa] = \{\, w \in X_L :\ \ \|w\|_{X_L} \leq \kappa \,\}$$

where $\kappa \in R_+$. W start with assumptions on the initial function φ.

Assumption H$_0$. Suppose that $\varphi : (-\infty, 0\,] \times R^n \to Y$ and

(i) $\varphi_{(0,y)} \in X_L$ for $y \in R^n$,

(ii) there are $\tilde{L}$, $L \in R_+$ such that $\|\varphi(0,y)\| \leq \tilde{L}$ for $y \in R^n$ and

$$\|\varphi_{(0,y)} - \varphi_{(0,\bar{y})}\|_X \leq L\|y - \bar{y}\| \ \text{ for } \ y, \bar{y} \in R^n.$$

Assumption H $[\varrho]$**.** Suppose that

1) the function $\varrho(\,\cdot\,, y, w) : [0,a] \to R^n$ is measurable for $(y,w) \in R^n \times X$ and $\varrho(x,\cdot\,) : R^n \times X \to R^n$ is continuous for almost all $x \in [0,a]$,

2) there exist α, $\beta \in \Theta$ such that $\|\varrho(x,y,w)\| \leq \alpha(x,\kappa)$ for $(y,w) \in R^n \times X[\kappa]$ almost everywhere on $[0,a]$ and

$$\|\varrho(x,y,w) - \varrho(x,\bar{y},\bar{w})\| \leq \beta(x,\kappa)\,[\,\|y - \bar{y}\| + \|w - \bar{w}\|_X\,] \tag{7.54}$$

for $(y,w), (\bar{y},\bar{w}) \in R^n \times X_L[\kappa]$ almost everywhere on $[0,a]$.

Suppose that $\varphi : (-\infty, 0\,] \times R^n \to Y$ and $\varphi_{(0,y)} \in X_L$ for $y \in R^n$. Let $c \in [0,a]$ and $d = (d_0, d_1) \in R^2_+$, $\omega \in L([0,a], R_+)$. The symbol $Y_{c.\varphi}[\omega, d]$ denotes the class of all functions $z : (-\infty, c\,] \times R^n \to Y$ such that

(i) $z(x,y) = \varphi(x,y)$ on $(-\infty, 0\,] \times R^n$,

(ii) $\|z(x,y)\| \leq d_0$ on $[0,c] \times R^n$ and

$$\|z(x,y) - z(\bar{x},\bar{y})\| \leq \left|\int_x^{\bar{x}} \omega(\tau)\, d\tau\right| + d_1\|y - \bar{y}\|$$

for (x,y), $(\bar{x},\bar{y}) \in [0,c] \times R^n$.

For the above φ and for $z \in Y_{c.\varphi}[\omega, d]$ consider the Cauchy problem

$$\eta'(\tau) = \varrho(\tau, \eta(\tau), z_{(\tau,\eta(\tau))}), \quad \eta(x) = y, \tag{7.55}$$

where $(x,y) \in [0,c] \times R^n$. We consider Carathéodory solutions of (7.55). Denote by $g[z](\cdot, x, y)$ the solution of the above problem. The function $g[z]$ is the bicharacteristic of equation (7.52) corresponding to $z \in Y_{c.\varphi}[\omega, d]$. Let $\Delta_c = [0,c] \times [0,c] \times R^n$. For φ satisfying Assumption H_0 define

$$||\varphi||_{(X,\infty)} = \sup\{ ||\varphi_{(0,y)}||_X : \ y \in R^n \}.$$

Lemma 7.26 *Suppose that Assumptions H [X], H [ϱ] are satisfied and*

1) the functions $\varphi, \bar{\varphi} : (-\infty, 0] \times R^n \to Y$ satisfy Assumption H_0,

2) $c \in [0,a]$ and $z \in Y_{c.\varphi}[\omega, d]$, $\bar{z} \in Y_{c,\bar{\varphi}}[\omega, d]$.

Then the solutions $g[z](\cdot, x, y)$ and $g[\bar{z}](\cdot, x, y)$ are defined on $[0,c]$ and they are unique. Moreover we have the estimates on Δ_c

$$||g[z](\tau, x, y) - g[z](\tau, \bar{x}, \bar{y})|| \le \tag{7.56}$$

$$\left[||y - \bar{y}|| + \left| \int_x^{\bar{x}} \alpha(\xi, \bar{\kappa})\, d\xi \right| \right] \exp\left[\bar{d} \left| \int_x^{\tau} \beta(\xi, \kappa_0)\, d\xi \right| \right]$$

and

$$||g[z](\tau, x, y) - g[\bar{z}](\tau, x, y)|| \le \tag{7.57}$$

$$\left| \int_x^{\tau} \beta(\xi, \kappa_0) \left[K||z - \bar{z}||_{[0,\xi;R^n]} + M||\varphi - \bar{\varphi}||_{(X,\infty)} \right] d\xi \ \exp\left[\bar{d} \int_x^{\tau} \beta(\xi, \kappa_0)\, d\xi \right] \right|,$$

where $\bar{d} = 1 + Kd_0 + ML$ and

$$\bar{\kappa} = Kd_0 + M\, ||\varphi||_{(X,\infty)}, \quad \kappa_0 = K_0(d_0 + d_1) + L_0 \sup\{ ||\varphi_{(0,y)}||_{X_L} : \ y \in R^n \}.$$

PROOF. Suppose that (ξ, η), $(\xi, \bar{\eta}) \in [0,c] \times R^n$ and a function $\tilde{z} : (-\infty, c] \times R^n \to Y$ is defined by $\tilde{z}(\tau, s) = z(\tau, s + \bar{\eta} - \eta)$, $(\tau, s) \in (-\infty, 0] \times R^n$. Then $\tilde{z}_{(\xi,\eta)} = z_{(\xi,\bar{\eta})}$. It follows from Assumptions H $[X]$ and H_0 that

$$||z_{(\xi,\eta)} - z_{(\xi,\bar{\eta})}||_X = ||(z - \tilde{z})_{(\xi,\eta)}||_X \le (Kd_0 + ML)\, ||\eta - \bar{\eta}||.$$

The existence and uniqueness of the solutions of (7.55) follows from classical theorems. On this purpose, note that the right-hand side of the differential system satisfies the Carathéodory assumptions and the following Lipschitz condition

$$||\varrho(\tau, \eta, z_{(\tau,\eta)}) - \varrho(\tau, \bar{\eta}, z_{(\tau,\bar{\eta})})|| \le \bar{d}\, \beta(\tau, \kappa_0)\, ||\eta - \bar{\eta}||$$

holds on $[0,c] \times R^n$. The function $g[z](\cdot, x, y)$ satisfies the integral equation

$$g[z](\tau, x, y) = y + \int_x^{\tau} \varrho(\xi, g[z](\xi, x, y), z_{(\xi, g[z](\xi,x,y))})\, d\xi.$$

For (τ, x, y), $(\tau, \bar{x}, \bar{y}) \in \Delta_c$ we have

$$\|z_{(\tau, g[z](\tau,x,y))}\|_X \leq \bar{\kappa}, \quad \|z_{(\tau, g[z](\tau,x,y))}\|_{X_L} \leq \kappa_0 \tag{7.58}$$

and

$$\|z_{(\tau, g[z](\tau,x,y))} - z_{(\tau, g[z](\tau,\bar{x},\bar{y}))}\|_X \leq (Kd_1 + ML)\|g[z](\tau, x, y) - g[z](\tau, \bar{x}, \bar{y})\|.$$

Then it follows from Assumption H $[X]$ that the following integral inequality is satisfied

$$\|g[z](\tau, x, y) - g[z](\tau, \bar{x}, \bar{y})\| \leq \|y - \bar{y}\| + \left| \int_x^{\bar{x}} \alpha(\xi, \bar{\kappa})\, d\xi \right| +$$

$$\bar{d} \left| \int_x^{\tau} \beta(\xi, \kappa_0)\, \|g[z](\xi, x, y) - g[z](\xi, \bar{x}, \bar{y})\|\, d\xi \right|.$$

Now we obtain (7.56) by the Gronwall inequality. For $z \in Y_{c.\varphi}[\omega, d]$, $\bar{z} \in Y_{c.\bar{\varphi}}[\omega, d]$ we have

$$\|z_{(\xi, g[z](\xi,x,y))} - \bar{z}_{(\xi, g[\bar{z}](\xi,x,y))}\|_X \leq +K\|z - \bar{z}\|_{[0,\xi;R^n]} \tag{7.59}$$

$$+M\,\|\varphi - \bar{\varphi}\|_{(X,\infty)} + (Kd_1 + M\tilde{L})\,\|g[z](\xi, x, y) - g[\bar{z}](\xi, x, y)\|.$$

It follows from Assumption H $[X]$ and from (7.59) that the integral inequality

$$\|g[z](\tau, x, y) - g[\bar{z}](\tau, x, y)\| \leq$$

$$\left| \int_x^{\tau} \beta(\xi, \kappa_0)\, \left[\|z - \bar{z}\|_{[0,\xi;R^n]} + M\|\varphi - \bar{\varphi}\|_{(X,\infty)} \right] d\xi \right| +$$

$$+\bar{d} \left| \int_x^{\tau} \beta(\xi, \kappa_0) \|g[z](\xi, x, y) - g[\bar{z}](\xi, x, y)\|\, d\xi \right|$$

is satisfied. Now we obtain (7.57) by the Gronwall inequality. This completes the proof of the lemma.

Now we construct an integral operator corresponding to problem (7.52), 7.53). Suppose that the function φ satisfies Assumption H_0, $c \in (0, a]$, $z \in Y_{c.\varphi}[\omega, d]$ and $g[z](\,\cdot\,, x, y)$ is the bicharacteristic corresponding to z. Let us define the operator U_φ for all $z \in Y_{c.\varphi}[\omega, d]$ be the formulas

$$U_\varphi z(x, y) = \varphi(0, g[z](0, x, y)) + \int_0^x f(\,\tau, g[z](\tau, x, y), z_{(\tau, g[z](\tau,x,y))}\,)\, d\tau \tag{7.60}$$

where $(x, y) \in [0, c] \times R^n$ and

$$U_\varphi z(x, y) = \varphi(x, y) \quad \text{on} \quad (-\infty, 0\,] \times R^n. \tag{7.61}$$

The operator U_φ is obtained by integration of equation (7.52) along bicharacteristics. Now we give sufficient conditions for the solvability of the equation $z = U_\varphi z$ on $Y_{c.\varphi}[\omega, d]$.

Assumption H $[f]$. Suppose that

1) the function $f(\cdot, y, w) : [0,a] \to Y$ is measurable for $(y,w) \in R^n \times X$ and $f(x, \cdot) : R^n \times X \to Y$ is continuous for almost all $x \in [0,a]$,

2) for $(y,w) \in R^n \times X[\kappa]$ and for almost all $x \in [0,a]$ we have

$$\|f(x,y,w)\| \le \alpha(x,\kappa), \tag{7.62}$$

3) for (y,w), $(\bar{y},\bar{w}) \in R^n \times X_L[\kappa]$ and for almost all $x \in [0,a]$ we have

$$\|f(x,y,w) - f(x,\bar{y},\bar{w})\| \le \beta(x,\kappa)\, [\, \|y-\bar{y}\| + \|w-\bar{w}\|_X \,].$$

Remark 7.27 *We prove a theorem on the existence and uniqueness of solutions of problem (7.52), (7.53). For simplicity of notations, we have assumed the same estimation for ϱ and for f. We have assumed also the Lipschitz condition for these functions with the same coefficient.*

Lemma 7.28 *Suppose that Assumptions H [X], H_0, H [ϱ], H [f] are satisfied. Then there are $(d_0,d_1) = d \in R_+^2$, $c \in (0,a]$ and $\omega \in L([0,c],R_+)$ such that $U_\varphi : Y_{c.\varphi}[\omega,d] \to Y_{c.\varphi}[\omega,d]$.*

PROOF. Suppose that the constants $(d_0,d_1) = d$, $c \in (0,a]$ and the function $\omega \in L([0,c]\, R_+)$ satisfy the conditions:

$$d_0 \ge \tilde{L} + \int_0^c \alpha(\tau,\bar{\kappa})\, d\tau, \quad d_1 \ge \Gamma_c,$$

$$\omega(t) \ge (1+\Gamma_c)\, \alpha(t,\bar{\kappa}),$$

where

$$\Gamma_c = \left[L + \bar{d} \int_0^c \beta(\tau,\kappa_0)\, d\tau \right] \exp\left[\bar{d} \int_0^c \beta(\tau,\kappa_0)\, d\tau \right]. \tag{7.63}$$

Suppose that $z \in Y_{c.\varphi}[\omega,d]$. Then we have

$$\|U_\varphi z(x,y)\| \le \tilde{L} + \int_0^c \alpha(\tau,\bar{\kappa})\, d\tau \le d_0 \;\text{ on }\; [0,c] \times R^n. \tag{7.64}$$

If (x,y), $(\bar{x},\bar{y}) \in [0,c] \times R^n$ then using Lemma 7.26 and from (7.60) we obtain

$$\|\, U_\varphi z(x,y) - U_\varphi z(\bar{x},\bar{y}) \,\| \le$$

$$\|\varphi(0,g[z](0,x,y) - \varphi(0,g[z](0,\bar{x},\bar{y})\| + \left| \int_x^{\bar{x}} \alpha(\tau,\bar{\kappa})\, d\tau \right|$$

$$+ \int_0^x \beta(\tau,\kappa_0)\, [\, \|g[z](\tau,x,y) - g[z](\tau,\bar{x},\bar{y})\|$$

$$+ \|z_{(\tau,g[z](\tau,x,y))} - z_{\tau,g[z](\tau,\bar{x},\bar{y}))}\|_X \,]\, d\tau +$$

$$\le \Gamma_c \left[\|y-\bar{y}\| + \left| \left| \int_x^{\bar{x}} \alpha(\tau,\bar{\kappa})\, d\tau \right| \right| \right] + \left| \int_x^{\bar{x}} \alpha(\tau,\bar{\kappa})\, d\tau \right|.$$

Thus we see that

$$||U_\varphi z(x,y) - U_\varphi z(\bar{x},\bar{y})|| \le d_1||y-\bar{y}|| + \left|\int_x^{\bar{x}} \omega(\tau)\,d\tau\right|. \tag{7.65}$$

It follows from (7.64), (7.65) that $U_\varphi z \in Y_{c.\varphi}[\omega, d]$, which completes the proof of the lemma.

Next we will show that there exists exactly one solution of problem (7.52), (7.53). The solution is local with respect to x.

Theorem 7.29 *Suppose that Assumptions $H\,[X]$, H_0, $H\,[\varrho]$, $H\,[f]$ are satisfied.*

Then there are $(d_0, d_1) = d$, $c \in (0,a]$ and $\omega \in L([0,c], R_+)$ such that problem (7.52), (7.53) has exactly one solution $u \in Y_{c.\varphi}[\omega, d]$. If $\bar{\varphi} : (-\infty, 0] \times R^n \to Y$ satisfies Assumption H_0 and $\bar{u} \in Y_{c.\varphi}[\omega, d]$ is a solution of equation (7.52) with the initial condition $z = \bar{\varphi}$ on $(-\infty, 0] \times R^n$ then there is $\Lambda_c \in R_+$ such that for $x \in [0,c]$ we have

$$||u-\bar{u}||_{[0,x;R^n]} \le \Lambda_c\left[\,||\varphi-\bar{\varphi}||_{(X,\infty)}\right. \tag{7.66}$$

$$\left. + \sup\{\,||\varphi(0,s) - \bar{\varphi}(0,s)|| :\ \ y \in R^n\,\}\right].$$

PROOF. Lemma 7.28 shows that there are $(d_0, d_1) = d \in R_+^2$, $c \in (0,a]$ and $\omega \in L([0,c], R_+)$ such that $U_\varphi : Y_{c.\varphi}[\omega, d] \to Y_{c.\varphi}[\omega, d]$. Write

$$\lambda_c = K(1+\Gamma_c)\int_0^c \beta(\tau, \kappa_0)\,d\tau,$$

where Γ_c is given by (7.63). Let $c \in (0,a]$ be such a constant that $\lambda_c < 1$. Now we prove that U_φ is a contraction on $Y_{c.\varphi}[\omega, d]$. If z, $\tilde{z} \in Y_{c.\varphi}[\omega, d]$ then

$$||U_\varphi z(x,y) - U_\varphi \tilde{z}(x,y)|| \le L\,||g[z](0,x,y) - g[\tilde{z}](0,x,y)|| +$$

$$\int_0^x \beta(\tau,\kappa_0)\left[\,||g[z](\tau,x,y) - g[\tilde{z}](\tau,x,y)|| + ||z_{(\tau,g[z](\tau,x,y))} - \tilde{z}_{(\tau,g[z](\tau,x,y))}||_X\,\right]d\tau.$$

The estimate

$$||z_{(\tau,g[z](\tau,x,y))} - \tilde{z}_{(\tau,g[z](\tau,x,y))}||_X \le$$

$$(Kd_1 + M\tilde{L})||g[z](\tau,x,y) - g[\tilde{z}](\tau,x,y)|| + K||z-\tilde{z}||_{[0,\tau;R^n]}$$

and lemma 7.26 imply

$$||U_\varphi z(x,y) - U_\varphi \tilde{z}(x,y)|| \le$$

$$K(1+\Gamma_c)\int_0^x \beta(\tau,\kappa_0)||z-\tilde{z}||_{[0,\tau;R^n]}\,d\tau,\ \ (x,y) \in [0,c] \times R^n,$$

and consequently

$$||U_\varphi z - U_\varphi \tilde{z}||_{[0,c;R^n]} \le \lambda_c||z-\tilde{z}||_{[0,c;R^n]}.$$

By the Banach fixed point theorem there exists a unique solution $u \in Y_{c.\varphi}[\omega, d]$ of the equation $z = U_\varphi z$. Now we prove that u is a solution of (7.52). We have proved that

$$u(x,y) = \varphi(0, g[u](0,x,y)) + \int_0^x f(\tau, g[u](\tau,x,y), u_{(\tau, g[u](\tau,x,y))})\, d\tau \tag{7.67}$$

on $[0,c] \times R^n$. For given $y \in R^n$ let us put $\eta = g[u](0,x,y)$. It follows that $g[u](\tau,x,y) = g[u](\tau,0,\eta)$ for $\tau \in [0,c]$ and that $y = g[u](x,0,\eta)$. The relations $\eta = g[u](0,x,y)$ and $y = g[u](x,0,\eta)$ are equivalent for $y, \eta \in R^n$. It follows from (7.67) that

$$u(x, g[u](x,0,\eta)) = \varphi(0,\eta) + \int_0^x f(\tau, g[u](\tau,0,\eta), u_{(\tau, g[u](\tau,0,\eta))})\, d\tau, \tag{7.68}$$

where $(x,\eta) \in [0,c] \times R^n$. By differentiating (7.68) with respect to x and by using the transformation $\eta = g[u](0,x,y)$ which preserves the sets of measure zero, we obtain that u satisfies (7.52) for almost all $(x,y) \in [0,c] \times R^n$. It follows from (7.61) that u satisfies (7.53). Now we prove relation (7.66). If $u = U_\varphi u$ and $\bar{u} = U_{\bar{\varphi}} \bar{u}$ then

$$\|u(x,y) - \bar{u}(x,y)\| \leq$$

$$\sup\{\|\varphi(0,s) - \bar{\varphi}(0,s)\| : s \in R^n\} + L\,\|g[u](0,x,y) - g[\bar{u}](0,x,y)\| +$$

$$\int_0^x \beta(\tau,\kappa_0)\,\big[\bar{d}\,\|g[u](\tau,x,y) - g[\bar{u}](\tau,x,y)\| + K\|u - \bar{u}\|_{[0,\tau;R^n]}$$

$$+ M\|\varphi - \bar{\varphi}\|_{(X,\infty)}\big]\, d\tau,$$

where $(x,y) \in [0,c] \times R^n$. Put

$$A_c = (1 + \Gamma_c) M \int_0^c \beta(\tau,\kappa_0)\, d\tau \quad \text{and} \quad \gamma(t) = K(1+\Gamma_c)\beta(t,\kappa_0).$$

Then we get the integral inequality

$$\|u - \bar{u}\|_{[0,x;R^n]} \leq \sup\{\|\varphi(0,s) - \bar{\varphi}(0,s)\| : s \in R^n\} + A_c\|\varphi - \bar{\varphi}\|_{(X,\infty)} +$$

$$\int_0^x \gamma(\tau)\|u - \bar{u}\|_{[0,\tau;R^n]}\, d\tau, \quad x \in [0,c].$$

It follows from the Gronwall inequality that we have estimate (7.66) for

$$\Lambda_c = \exp\left[\int_0^c \gamma(\tau)\, d\tau\right].$$

This completes the proof of the theorem.

We give examples of spaces X satisfying Assumption H $[X]$.

Example 7.30 *Let X be the class of all function $w : (-\infty, 0] \times [-r, r] \to Y$ such that w are uniformly continuous and bounded on $(-\infty, 0] \times [-r, r]$. For $w \in X$ we write*

$$\|w\|_X = \sup \{\|w(\tau, s)\| : \ (\tau, s) \in (-\infty, 0] \times [-r, r]\}.$$

Let $X_L \subset X$ denote the set of all $w \in X$ such that

$$|w|_L = \tag{7.69}$$

$$= \sup \left\{ \frac{\|w(\tau, s) - w(\tau, \bar{s})\|}{\|s - \bar{s}\|} : \ (\tau, s), (\tau, \bar{s}) \in (-\infty, 0] \times [-r, r] \right\} < +\infty.$$

Write $\|w\|_{X_L} = \|w\|_X + |w|_L$, where $w \in X_L$. Then Assumption H [X] is satisfied.

Example 7.31 *Let X be the class of all functions $w : (-\infty, 0] \times [-r, r] \to Y$ such that*

(i) w is continuous and bounded on $(-\infty, 0] \times [-r, r]$,

(i) there exists the limit

$$\lim_{x \to -\infty} w(x, y) \quad \text{uniformly with respect to } y \in [-r, r].$$

Put

$$\|w\|_X = \sup \{ \|w(\tau, s)\| : \ (\tau, s) \in (-\infty, 0] \times [-r, r] \}.$$

Let $X_L \subset X$ denote the class of all $w \in X$ such that the Lipschitz condition (7.69) is satisfied. Write $\|w\|_{X_L} = \|w\|_X + |w|_L$, where $w \in X_L$. Then Assumption H [X] is satisfied.

Example 7.32 *Let $\gamma : (-\infty, 0] \to (0, +\infty)$ be a continuous function. Assume also that γ is nonincreasing on $(-\infty, 0]$. Let X be the space of continuous functions $w : (-\infty, 0] \times [-r, r] \to Y$ such that*

$$\lim_{\tau \to \infty} \frac{\|w(\tau, y)\|}{\gamma(\tau)} = 0, \ \ y \in [-r, r].$$

Put

$$\|w\|_X = \sup \left\{ \frac{\|w(\tau, s)\|}{\gamma(\tau)} : \ (\tau, s) \in (-\infty, 0] \times [-r, r] \right\}.$$

Denote by $X_L \subset X$ the set of all $w \in X$ such that

$$|w|_{\gamma.L} = \sup \left\{ \frac{\|w(\tau, s) - w(\tau, \bar{s})\|}{\gamma(\tau) \, \|s - \bar{s}\|} : \ (\tau, s), (\tau, \bar{s}) \in (-\infty, 0] \times [-r, r] \right\} < +\infty.$$

For $w \in X_L$ put $\|w\|_{X_L} = \|w\|_X + |w|_{\gamma.L}$. Then Assumption H [X] is satisfied.

Example 7.33 *Let $\delta \in R_+$ and $p \geq 1$ be fixed. Denote by X the class of all functions $w : (-\infty, 0] \times [-r, r] \to Y$ such that*

(i) w is continuous on $[-\delta, 0] \times [-r, r]$,

(ii) for $x \in [-r, r]$ we have

$$\int_{-\infty}^{-\delta} \|w(\tau, x\|^p \, d\tau < +\infty,$$

(iii) $w(t, \cdot) : [-r, r] \to Y$ is continuous for $t \in (-\infty, -\delta]$.
Write

$$\|w\|_X = \sup \{ \|z(\tau, s)\| : (\tau, s) \in [-\delta, 0] \times [-r, r] \} +$$

$$\sup \left[\left(\int_{-\infty}^{-\delta} \|w(\tau, x\|^p \, d\tau \right)^{\frac{1}{p}} : \quad x \in [-r, r] \right]$$

Let $X_L \subset X$ be the set of functions $w \in X$ such that the Lipschitz condition (7.69) is satisfied. Write $\|w\|_{X_L} = \|w\|_X + |w|_L$, where $w \in X_L$. Then Assumption H [X] is satisfied.

Remark 7.34 *Differential equations with a deviated argument and differential integral equations can be obtained from (7.52) by specializing operators ϱ and f.*

Remark 7.35 *It is important in our considerations that we have assumed the Lipschitz condition for given functions on some special function spaces. More precisely, we have assumed that the functions $\varrho(x, \cdot)$ and $f(x, \cdot)$ satisfy the Lipschitz condition on the space $R^n \times X_L$ for almost all $x \in [0, a]$ and the condition is local with respect to the functional variable. Let us consider simplest assumption on ϱ and f. Suppose that there is $P \in R_+$ such that for almost all $t \in [0, a]$ we have*

$$\|\varrho(x, y, w) - \varrho(x, \bar{y}, \bar{w})\| \le P [\|y - \bar{y}\| + \|w - \bar{w}\|_X], \tag{7.70}$$

$$\|f(x, y, w) - f(x, \bar{y}, \bar{w})\| \le P [\|y - \bar{y}\| + \|w - \bar{w}\|_X], \tag{7.71}$$

where $(y, w), (\bar{y}, \bar{w}) \in R^n \times X$. Of course, our results are true if we assume (7.70), (7.71) instead of (7.54), (7.62). Now we show that formulations (7.54), (7.62) are important. We show that there is a class of quasilinear equations satisfying (7.54), (7.62) and do not satisfying (7.70), (7.71). Let X and X_L be the spaces given in Example 7.30. Consider the equation with a deviated argument

$$D_x z(x, y) + \sum_{i=1}^{n} \tilde{\varrho}_i(x, y, z(\psi_0(x), \psi(x, y))) \, D_{y_i} z(x, y) \tag{7.72}$$

$$= \tilde{f}(x, y, z(\psi_0(x), \psi(x, y)))$$

where $\tilde{\varrho} = (\tilde{\varrho}_1, \dots, \tilde{\varrho}_n) : [0, a] \times R^n \times Y \to R^n$, $f : [0, a] \times R^n \times Y \to Y$ and $\psi_0 : [0, a] \to (-\infty, a]$, $\psi : [0, a] \times R^n \to R^n$. We assume that $\psi(x) \le x$ and $-r \le \psi(x, y) - y \le r$ for $(x, y) \in [0, a] \times R^n$. We get (7.72) by putting in (7.52)

$$\varrho(x, y, w) = \tilde{\varrho}(x, y, w(\psi_0(x) - x, \psi(x, y) - y))$$

$$f(x, y, w) = \tilde{f}(x, y, w(\psi_0(x) - x, \psi(x, y) - y)).$$

From now we consider the function ϱ *only. Suppose that there are* $\bar{C}, \tilde{C} \in R_+$ *such that*

$$\|\tilde{\varrho}(x,y,\zeta) - \tilde{\varrho}(x,\bar{y},\bar{\zeta})\| \leq \bar{C}\left[\, \|y-\bar{y}\| + \|\zeta - \bar{\zeta}\|\,\right]$$

and

$$\|\psi(x,y) - \psi(x,\bar{y})\| \leq \tilde{C}\, \|y-\bar{y}\|.$$

It is evident that for (y,w), $(\bar{y},\bar{w}) \in R^n \times X_L[\kappa]$ *and for almost all* $x \in [0,a]$ *we have*

$$\|\varrho(x,y,w) - \varrho(x,\bar{y},\bar{w})\| \leq \bar{C}\left[1 + \kappa(1+\tilde{C})\right] \|y-\bar{y}\| + \bar{C}\, \|w - \bar{w}\|_X.$$

Then condition (7.54) is satisfied. We see at once the the function $\varrho(\,x\,,\cdot\,)$ *does not satisfy the global Lipschitz condition (7.70) for* (y,w), $(\bar{y},\bar{w}) \in R^n \times X$. *Similar considerations apply to* f.

7.9 Viscosity solutions of functional differential inequalities

Let $\Gamma \subset R^n$ be an open and bounded domain. Suppose that $a > 0$, r_0, $r \in R_+$ are given constants. Define

$$\Xi = \{\, y \in R^n :\ \mathrm{dist}\,(\,y,\Gamma\,) \leq r\,\},\quad \partial_0\Gamma = \Xi \setminus \Gamma,\quad E = (0,a)\times\Gamma,$$

$$E_0 = [-r_0, 0]\times \Xi,\quad \partial_0 E = (0,a)\times\partial_0\Gamma,\quad E^* = E_0 \cup \partial_0 E \cup E_0.$$

Let $B = [-r_0,0] \times \{y \in R^n :\ \|y\| \leq r\}$ where $\|\cdot\|$ is the Euclidean norm. Put $\Omega = E \times C(B,R) \times R^n$ and suppose that $f : \Omega \to R$ and $\psi : E_0 \cup \partial_0 E \to R$ are given function. For a function $z : E^* \to R$ and a point $(x,y) \in [0,a) \times \bar{\Gamma}$, ($\bar{\Gamma}$ is the closure of Γ) we define the function $z_{(x,y)} : B \to R$ by $z_{(x,y)}(t,s) = z(x+t, y+s)$, $(t,s) \in B$. Let $\|\cdot\|_0$ be the supremum norm in the space $C(B,R)$. In this Section we consider functional differential inequalities generated by initial-boundary value problem

$$D_x z(x,y) = f(\,x, y, z_{(x,y)}, D_y z(x,y)\,) \text{ on } E, \tag{7.73}$$

$$z(x,y) = \psi(x,y) \text{ on } E_0 \cup \partial_0 E. \tag{7.74}$$

We investigate functional differential inequalities in a viscosity sense. The notion of viscosity solutions for Hamilton - Jacobi equations was first introduced by M. G. Crandal and P.L. Lions in [76, 155] and developed in [74, 75, 77]. A function $u \in C(E^*, R)$ is a viscosity subsolution (resp. supersolution) of (7.73) provided for all $\phi \in C^1(E,R)$ if $u - \phi$ attains a local maximum (resp. minimum) at $(x_0, y_0) \in E$ then

$$D_x\phi(x_0,y_0) \leq f(\,x_0, y_0, z_{(x_0,y_0)}, D_y\phi(x_0,y_0)\,)$$

$$\text{resp. } D_x\phi(x_0,y_0) \geq f(\,x_0, y_0, z_{(x_0,y_0)}, D_y\phi(x_0,y_0)\,).$$

A function $u \in C(E^*, R)$ is a viscosity solution of (7.73) if u is both a viscosity subsolution and supersolution of (7.73). We denote by $SUB\{\,f\,\}$ and $SUP\{\,f\,\}$ the sets of all viscosity subsolutions and supersolutions of (7.73) respectively.

Lemma 7.36 *Let $a > 0$ and $h, H \in C([0,a],R)$. Assume that h is a viscosity solution of $h'(x) \leq H(x)$ (i.e. h is a viscosity subsolution of $h'(x) = H(x)$ on $(0,a)$. Then*

$$h(x) \leq h(\xi) + \int_{\xi}^{x} H(\tau)\, d\tau \text{ for } 0 \leq \xi \leq x \leq a.$$

The proof can be found in [76]. For w, $\bar{w} \in C(B,R)$ we write $w \leq \bar{w}$ if $w(t,s) \leq \bar{w}(t,s)$ on B.

Assumption H $[C_\eta]$. Define $C_\eta(\varrho) = \sup\{\, f(x,y,w+h,q) - f(x,s,w,q) : 0 \leq h \leq \varrho,\ \|h\|_0 \leq \eta,\ \|y-s\|(1+\|q\|) \leq \varrho \,\}$. We assume that $\lim_{\varrho \to 0} C_\eta(\varrho) = 0$.

Theorem 7.37 *Suppose that*

1) the function f satisfies the following condition of monotonicity: if w, $\bar{w} \in C(B,R)$ and $w \leq \bar{w}$, $w(0,0) = \bar{w}(0,0)$ then $f(x,y,w,q) \leq f(x,y,\bar{w},q)$,

2) Assumption H $[C_\eta]$ is satisfied,

3) there exists $\varepsilon > 0$ such that $u \in SUB\{\, f - \varepsilon \,\}$, $v \in SUP\{\, f \,\}$ and $u(x,y) < v(x,y)$ on $E_0 \cup \partial_0 E$. Then $u(x,y) < v(x,y)$ on E.

Proof. Let $0 < \tilde{a} < a$. We will show that $u(x,y) < v(x,y)$ for $0 < x < \tilde{a}$, $y \in \Gamma$. Indeed, there exists $x_0 > 0$ such that $u(x,y) < v(x,y)$ for $0 < x < x_0$, $y \in \Gamma$. Let

$$\bar{x} = \sup\{\, \tau :\ u(x,y) < v(x,y)\ \ for - r_0 < x < \tau,\ y \in \Gamma \,\}.$$

Clearly $\bar{x} > 0$. We claim that $\bar{x} = \tilde{a}$. Indeed, suppose that $\bar{x} < \tilde{a}$. Put

$$\Phi(x,y,s) = u(x,y) - v(x,s) - \frac{\alpha}{2}\|y-s\|^2,\ \ \alpha > 0,$$

$$M_\alpha = \max\{\, \Phi(x,y,s) :\ (x,y,s) \in [-r_0,\bar{x}] \times \bar{\Gamma} \times \bar{\Gamma} \,\}.$$

For every α we choose $(x_\alpha, y_\alpha, s_\alpha) \in [-r_0,\bar{x}] \times \bar{\Gamma} \times \bar{\Gamma}$ such that $M_\alpha = \Phi(x_\alpha, y_\alpha, s_\alpha)$. Since Γ is bounded and $u(x,y) < v(x,y)$ on $E_0 \cup E$, $u(\bar{x},\bar{y}) - v(\bar{x},\bar{y}) = 0$ for some $\bar{y} \in \bar{\Gamma}$. This implies $M_\alpha \geq \Phi(\bar{x},\bar{y},\bar{y}) = 0$. On the other hand, we get

$$0 \leq M_\alpha \leq m_v(\|x_\alpha - y_\alpha\|) - \frac{\alpha}{2}\|x_\alpha - y_\alpha\|^2,$$

where m_v is the modulus of continuity of v. This implies

$$\alpha\|y_\alpha - s_\alpha\|^2 \leq 2\, m_v(\|y_\alpha - s_\alpha\|). \tag{7.75}$$

Since Γ is bounded and $\|y_\alpha - s_\alpha\| \to 0$ as $\alpha \to \infty$, without loss of generality we can assume that y_α, $s_\alpha \to \hat{y}$, $x_\alpha \to \hat{x}$. It is easy to show that $\hat{x} = \bar{x}$. Indeed,

$$0 \leq \Phi(x_\alpha, y_\alpha, s_\alpha) \to u(\hat{x},\hat{y}) - v(\hat{x},\hat{y}),$$

this yields that $u(\hat{x},\hat{y}) - v(\hat{x},\hat{y}) \geq 0$ and in view of the definition of $\bar{x}$, we have $\hat{x} = \bar{x}$. Since $u(\hat{x},y) - v(\hat{x},y) < 0$ for $y \in \partial_0 E$, it is obvious that $\hat{y} \in \Gamma$. Put

$$\tilde{\phi}(x,y) = v(x_\alpha, s_\alpha) + \frac{\alpha}{2}\|y - s_\alpha\|^2.$$

The function $u - \tilde{\phi}$ attains its local maximum at $(x_\alpha, y_\alpha) \in (0, \bar{x}] \times \Gamma$. Then we get

$$\varepsilon \le f(\, x_\alpha, y_\alpha, u_{(x_\alpha, y_\alpha)}, \alpha(y_\alpha - s_\alpha)\,).$$

Put

$$\bar{\phi}(x, s) = u(x_\alpha, y_\alpha) - \frac{\alpha}{2}\|y_\alpha - s\|^2.$$

Then the function $v(x, s) - \bar{\phi}(x, s)$ attains its maximum at $(x_\alpha, s_\alpha) \in (0, \bar{x}] \times \Gamma$ and we get

$$0 \ge f(\, x_\alpha, s_\alpha, v_{(x_\alpha, s_\alpha)}, \alpha(x_\alpha - s_\alpha)\,).$$

Then we have

$$\varepsilon \le f(\, x_\alpha, y_\alpha, u_{(x_\alpha, y_\alpha)}, \alpha(y_\alpha - s_\alpha)\,) - f(\, x_\alpha, s_\alpha, v_{(x_\alpha, s_\alpha)}, \alpha(x_\alpha - s_\alpha)\,).$$

It is easy to verify that

$$u_{(x_\alpha, y_\alpha)}(t, s) \le v_{(x_\alpha, s_\alpha)}(t, s) + P_\alpha \;\text{ for }\; (t, s) \in B,$$

and equality holds for $(t, s) = (0, 0)$ where

$$P_\alpha = M_\alpha + \frac{\alpha}{2}\|y_\alpha - s_\alpha\|^2.$$

and $P_\alpha \to 0$ as $\alpha \to \infty$. It follows from assumption 1) that

$$\varepsilon \le f(x_\alpha, y_\alpha, v_{(x_\alpha, s_\alpha)} + P_\alpha, \alpha(y_\alpha - s_\alpha)\,) - f(\, x_\alpha, s_\alpha, v_{(x_\alpha, s_\alpha)}, \alpha(y_\alpha - s_\alpha)\,).$$

Put $\eta = \|v\|_{E^*}$. Letting $\alpha \to \infty$ we get in view of Assumption H $[\,C_\eta\,]$ that $\varepsilon \le 0$, which is a contradiction. Thus $\bar{x} = \tilde{a}$. Since $0 < \tilde{a} < a$ is arbitrary the proof of Theorem 7.37 is complete.

Let $V : C(B, R) \to C([-r_0, 0], R_+$ be the the operator defined by

$$(\, Vw\,)(x) = \max\{\, |\, w(x, y)\,| : \; \|y\| \le r\,\}.$$

Assumption H $[\,\sigma\,]$**.** Suppose that the function $\sigma : [0, a) \times C([-r_0, 0], R_+) \to R_+$ is continuous and that the function $\bar{\omega}(x) = 0$, $x \in [0, a)$, is the unique solution of the problem

$$\omega'(x) = \sigma(\, x, \omega_{(x)}\,), \;\; \omega(x) = 0 \text{ for } x \in [-r_0, 0].$$

Theorem 7.38 *Suppose that*

1) the function f satisfies condition 1), 2) of Theorem 7.37,

2) there exists σ satisfying Assumption H $[\sigma\,]$ and such that

$$f(\, x, y, w, q\,) - f(\, x, y, \bar{w}, q\,) \le \sigma(\, x, V(w - \bar{w})\,) \quad \textit{on } \Omega$$

for $w \ge \bar{w}$,

4) u, $v \in C(E^*, R)$, $u \in SUB\{\, f\,\}$, $v \in SUP\{\, f\,\}$ and $u(x, y) \le v(x, y)$ on $E_0 \cup \partial_0 E$.

Under these assumptions $u(x, y) \le v(x, y)$ on E.

PROOF. Let $\tilde{a} < a$. We will show that $u(x,y) \le v(x,y)$ for $(x,y) \in (0,\tilde{a}) \times \Gamma$. Consider the Cauchy problem

$$\omega'(x) = \sigma(\,x, \omega_{(x)}\,) + \varepsilon, \;\; \omega(x) = \varepsilon \;\; \text{for } x \in [-r_0, 0]. \tag{7.76}$$

This is well-known fact that there exists $\varepsilon > 0$ such that for $0 < \varepsilon < \varepsilon_0$ problem (7.76) has the maximum solution $\omega(\,\cdot\,,\varepsilon)$ defined on $[-r_0,\tilde{a})$ and $\lim_{\varepsilon\to 0}\omega(x,\varepsilon) = 0$ uniformly on $[-r_0,\tilde{a})$. Put $\tilde{u}(x,y) = u(x,y) - \omega(x,\varepsilon)$, $(x,y) \in [-r_0,\tilde{a}) \times \Xi$. We will show that $\tilde{u}$ satisfies the functional differential inequality in the viscosity sense

$$D_x\tilde{u}(x,y) < f(\,x, y, \tilde{u}_{(x,y)}, D_y\tilde{u}(x,y)\,) - \varepsilon \;\; \text{on} \;\; (0,\tilde{a}) \times \Gamma. \tag{7.77}$$

Suppose that $\psi \in C^1(E,R)$ and that $\tilde{u} - \psi$ has a local maximum at the point $(\tilde{x},\tilde{y})$. Then

$$u(x,y) - [\,\psi(x,y) + \omega(x,\varepsilon)\,] \le u(\tilde{x},\tilde{y}) - [\,\psi(\tilde{x},\tilde{y}) + \omega(\tilde{x},\varepsilon)\,] \text{ near } (\tilde{x},\tilde{y}).$$

Thus since $u \in SUB\{\,f\,\}$, we get

$$D_x\psi(\tilde{x},\tilde{y}) + \omega'(\tilde{x},\varepsilon) \le f(\,\tilde{x},\tilde{y}, u_{(\tilde{x},\tilde{y})}, D_y\psi(\tilde{x},\tilde{y})\,).$$

This implies that

$$\begin{aligned} D_x\psi(\tilde{x},\tilde{y}) &\le f(\,\tilde{x},\tilde{y},\tilde{u}_{(\tilde{x},\tilde{y})}, D_y\psi(\tilde{x},\tilde{y})\,) \\ &+ f(\,\tilde{x},\tilde{y},u_{(\tilde{x},\tilde{y})}, D_y\psi(\tilde{x},\tilde{y})\,) - f(\,\tilde{x},\tilde{y},\tilde{u}_{(\tilde{x},\tilde{y})}, D_y\psi(\tilde{x},\tilde{y})\,) - \omega'(\tilde{x},\varepsilon) \\ &\le f(\,\tilde{x},\tilde{y},\tilde{u}_{(\tilde{x},\tilde{y})}, D_y\psi(\tilde{x},\tilde{y})\,) + \sigma(\tilde{x}, V(u_{(\tilde{x},\tilde{y})} - \tilde{u}_{(\tilde{x},\tilde{y})}\,) - \omega'(\tilde{x},\varepsilon). \end{aligned}$$

It follows that $V(u_{(\tilde{x},\tilde{y})} - \tilde{u}_{(\tilde{x},\tilde{y})}) = (\,\omega(\,\cdot\,,\varepsilon)\,)_{(\tilde{x})}$. Thus from (7.76) we get

$$D_x\psi(\tilde{x},\tilde{y}) \le f(\,\tilde{x},\tilde{y},u_{(\tilde{x},\tilde{y})}, D_y\psi(\tilde{x},\tilde{y})\,) - \varepsilon.$$

Hence $\tilde{u} \in SUB(f - \varepsilon)$. Since $\tilde{u}(x,y) < v(x,y)$ on $E_0 \cup (\,\partial_0 E \cap ((0,\tilde{a}) \times R^n)\,)$, it follows from Theorem 7.37 that $\tilde{u}(x,y) < v(x,y)$ on $(0,\tilde{a}) \times \Gamma$. Letting $\varepsilon \to 0$ we get $u(x,y) \le v(x,y)$ on $(0,\tilde{a}) \times \Gamma$. Since $0 < \tilde{a} < a$ is arbitrary, then Theorem 7.38 follows.

Remark 7.39 *Suppose that f satisfies all the assumptions of Theorem 7.38 and $f(\,\cdot\,,\cdot\,,0,0) \ge 0$. It easily seen that putting $v = 0$ we can draw the maximum principle from the above theorem.*

Remark 7.40 *Theorem 7.38 also yields the uniqueness of the viscosity solution of problem (7.73), (7.74).*

Now we investigate equation (7.73) for f nondecreasing with respect to the functional variable. This assumption which is stronger than condition 1) of Theorem 7.37 implies better result for strong inequalities than Theorem 7.37.

Theorem 7.41 *Suppose that*

1) the function f is nondecreasing with respect to w, and Assumption $H\,[C_\eta\,]$ is satisfied,

2) $u,\, v \in C(E^*,\, R$ and $u \in SUB\{\,f\,\}$, $v \in SUP\{\,f\,\}$,

3) for $(x,y) \in E_0 \cup E$ we have $u(x,y) < v(x,y)$.

Then $u(x,y) < v(x,y)$ on E.

PROOF. Let $0 < \tilde{a} < a$. We will show that $u(x,y) < v(x,y)$ for $0 < x < \tilde{a}$, $y \in \Gamma$. Since u and v are continuous functions, there exists $\varepsilon > 0$ such that

(i) $u(x,y) + \varepsilon < v(x,y)$ for $(x,y) \in E_0$,

(ii) for $(x,y) \in \partial_0 E \cap ([0,\tilde{a}) \times R^n)$ we have

$$u(x,y) + \frac{\varepsilon^2}{x+\varepsilon} < v(x,y).$$

Define

$$\tilde{u}(x,y) = u(x,y) + \varepsilon \ \text{ on } E_0,$$

$$\tilde{u}(x,y) = u(x,y) + \frac{\varepsilon^2}{x+\varepsilon} \ \text{ on } (E \cup \partial_0 E) \cap ([0,\tilde{a}) \times R^n)).$$

It is suffices to show that $\tilde{u}(x,y) < v(x,y)$ for $0 < x < \tilde{a}$, $y \in \Gamma$. Indeed, there exists $x_0 > 0$ such that $\tilde{u}(x,y) < v(x,y)$ for $0 < x < x_0$, $y \in \Gamma$. Let

$$\bar{x} = \sup\{\tau : \tilde{u}(x,y) < v(x,y) \text{ for } 0 < x < \tau,\ y \in \Gamma.\}.$$

We claim that $\bar{x} = \tilde{a}$. Suppose that $\bar{x} < \tilde{a}$. Put

$$\Phi(x,y,s) = \tilde{u}(x,y) - v(x,s) - \frac{\alpha}{2}\|y-s\|^2, \ \alpha > 0.$$

$$M_\alpha = \max\{\Phi(x,y,s) : \ 0 \le x \le \bar{x},\ x,y \in \Gamma \cup \partial_0\Gamma\}.$$

For every α we choose $(x_\alpha, y_\alpha, s_\alpha)$ such that $M_\alpha = \Phi(x_\alpha, y_\alpha, s_\alpha)$. We can assume that

$$0 < x_\alpha < \bar{x},\ y_\alpha,\ s_\alpha \in \Gamma,$$

$$\lim_{\alpha\to\infty} x_\alpha = \bar{x},\ \lim_{\alpha\to\infty} y_\alpha = \lim_{\alpha\to\infty} s_\alpha = \hat{y},$$

and

$$\lim_{\alpha\to\infty} \alpha\|y_\alpha - s_\alpha\|^2 = 0,$$

$$\lim_{\alpha\to\infty} M_\alpha = \tilde{u}(\bar{x},\hat{y}) - v(\bar{x},\hat{y})$$

$$= \max\{\tilde{u}(x,y) - v(x,y) : (x,y) \in E^* \cap ([-r_0,\bar{x}] \times R^n\} = 0.$$

Write

$$\tilde{\phi}(x,y) = v(x_\alpha, s_\alpha) + \frac{\alpha}{2}\|y - s_\alpha\|^2 - \frac{\varepsilon^2}{x+\varepsilon}.$$

The function $u - \tilde{\phi}$ attains its maximum at (x_α, y_α) and we get

$$\frac{\varepsilon^2}{(x+\varepsilon)^2} \le f(x_\alpha, y_\alpha, u_{(x_\alpha,y_\alpha)}, \alpha(y_\alpha - s_\alpha)).$$

Put

$$\bar{\phi}(x,s) = u(x_\alpha, y_\alpha) - \frac{\alpha}{2}\|y_\alpha - s\|^2 + \frac{\varepsilon^2}{x_\alpha + \varepsilon}.$$

Then the function $v(x,s) - \bar{\phi}(x,s)$ attains its local minimum at $(x_\alpha, s_\alpha$ and we get

$$0 \ge f(x_\alpha, s_\alpha, v_{(x_\alpha,s_\alpha)}, \alpha(y_\alpha - s_\alpha))$$

and consequently

$$\frac{\varepsilon^2}{(x_\alpha+\varepsilon)^2} \le f(x_\alpha, y_\alpha, u_{(x_\alpha,y_\alpha)}, \alpha(y_\alpha - s_\alpha)) - f(x_\alpha, s_\alpha, v_{(x_\alpha,s_\alpha)}, \alpha(y_\alpha - s_\alpha)).$$

Letting $\alpha \to \infty$ we obtain from Assumption H $[C_\eta]$ that $\frac{\varepsilon^2}{(\bar{x}+\varepsilon)^2} \le 0$. which is impossible. Since $a < \tilde{a} < a$ is arbitrary then the proof of Theorem 7.41 is completed.

Theorem 7.42 *Suppose that*

1) the function f is continuous and nondecreasing with respect to the functional variable,

2) Assumption H $[C_\eta]$ is satisfied and there exists a function σ satisfying Assumption H $[\sigma]$ and such that

$$f(x,y,w,q) - f(x,y,\bar{w},q) \le \sigma(x, V(w-\bar{w})) \text{ on } \Omega$$

for $w \ge \bar{w}$,

3) $u, v \in C(E^*, R)$ and $u \in SUB\{f\}$, $v \in SUP\{f\}$,

4) the initial inequality $u(x,y) \le v(x,y)$ is satisfied on $E_0 \cup \partial_0 E$.

Under these assumptions we have $u(x,y) \le v(x,y)$ on E.

The proof of the theorem is similar to the proof of Theorem 7.38. We omit details.

Remark 7.43 *Viscosity solutions of functional differential problems were first considered in [205 – 207]. Our results are based on [206].*

Chapter 8

Functional Integral Equations

8.1 Introduction

Let X be an arbitrary Banach space with the norm $||\cdot||$. We denote the Euclidean norm in R^n and the norm in the Banach space X by the same symbol. Elements of the space R^n will be denoted by $x = (x_1, \ldots, x_n)$, $s = (s_1, \ldots, s_n)$. Let $E \subset R^n_+$ be a compact set and $G(x) = \{\xi \in E : \xi \leq x\}$. Assume that functions

$$F \in C(E \times X^m \times X, X), \quad f = (f_1, \ldots, f_m) \in C(E \times E \times X, X^m),$$

$$\beta \in C(E, E), \quad \alpha = (\alpha_1, \ldots, \alpha_m) \in C(E, E^m)$$

are given and $\beta(x) \leq x$, $\alpha_i(x) \leq x$, $1 \leq i \leq m$, for $x \in E$. Suppose that the sets $H_j(x) \subset G(x)$ for $x \in E$, $1 \leq j \leq m$, are given. We assume further that $H_j(x)$ is contained in a p_j - dimensional hyperplane, $1 \leq p_j \leq n$, parallel to the coordinate axes, and it is Lebesgue - measurable, considered as a p_j - dimensional set. Let $L_{p_j}(H_j(x))$ denotes the p_j - dimensional Lebesgue measure of $H_j(x)$. We assume that p_j does not depend on x. If the p_j - dimensional hyperplane containing the set $H_j(x)$ and being parallel to the coordinate axes is defined by the equations

$$x_{t_1} = \bar{x}_{t_1}, \; x_{t_2} = \bar{x}_{t_2}, \ldots, x_{t_r} = \bar{x}_{t_r}, \quad r = n - p_j,$$

then

$$\int_{H_j(x)} g(x, s)\,(ds)_{p_j}, \quad \text{where } s = (s_1, \ldots, s_n),$$

denotes the p_j dimensional Lebesgue integral in the space

$$0\, x_{m_1}\, x_{m_2} \ldots x_{m_{p_j}}, \quad m_i \in \{\, \{1, \ldots, n\} \setminus \{t_1, \ldots t_r\}\, \},$$

and $s_{t_1} = \bar{x}_{t_1}, s_{t_2} = \bar{x}_{t_2}, \ldots, s_{t_r} = \bar{x}_{t_r}$. We adopt the following notation:

$$\int_{H(x)} f(x, s, z(\alpha(s)))\, ds =$$

$$\left(\int_{H_1(x)} f_1(\,x, s, z(\alpha_1(s))\,)\,(ds)_{p_1}, \ldots, \int_{H_m(x)} f_m(\,x, s, z(\alpha_m(s))\,)\,(ds)_{p_m} \right).$$

We will consider the nonlinear Volterra functional integral equation

$$u(x) = F(\,x, \int_{H(x)} f(\,x, s, u(\alpha(s))\,)\,ds,\; u(\,\beta(x)\,)\,). \tag{8.1}$$

There are various problems which lead to Volterra functional integral equations of type (8.1). Perhaps the simplest problem in the theory of differential equations which leads to such an equation is the initial value problem for the ordinary functional differential equation of the neutral type

$$y'(t) = F\,(\,t, y(\alpha_1(t)), \ldots, y(\alpha_m(t)), y'(\beta(t))\,)\,,\; t \in [0,a],\;\; y(0) = y_0.$$

The substitution $y'(t) = u(t)$, $t \in [0,a]$, transforms the above problem into a functional integral equation of type (8.1) with $n = 1$. Functional integral equations generated by ordinary functional differential equations of the neutral type were investigated in [113]. Various initial value problems for the hyperbolic functional differential equation of the neutral type

$$D_{xy}z(x,y) = F(x,y,z(\alpha^{(0)}(x,y)), D_x z(\alpha^{(1)}(x,y)), D_y z(\alpha^{(2)}(x,y)), D_{xy}z(\beta(x,y)))$$

where $\alpha^{(i)} = (\alpha_1^{(i)}, \alpha_2^{(i)})$, i = 0,1,2, and $\beta = (\beta_1, \beta_2)$, can be reformulated in terms of Volterra functional integral equations. Let us consider as an example the Darboux problem, where the domain is the rectangle $[0,a] \times [0,b]$ and where the boundary values $u(x,0) = \sigma(x)$ for $x \in [0,a]$, $u(0,y) = \tau(y)$ for $y \in [0,b]$ are prescribed. The Volterra functional integral equation corresponding to that problem is

$$u(x,y) = F(\,x, y, \gamma_0(x,y) + \int_{H_0(x,y)} u(s,t)\,dsdt,$$

$$\gamma_1(x,y) + \int_{H_1(x,y)} u(s,t)\,dt,\; \gamma_2(x,y) + \int_{H_2(x,y)} u(s,t)\,ds, u(\,\beta(x,y)\,)\,),$$

for $(x,y) \in [0,a] \times [0,b]$ where

$$\gamma_0(x,y) = \sigma(\alpha_1^{(0)}(x,y)) + \tau(\alpha_2^{(0)}(x,y)) - \sigma(0),$$

$$\gamma_1(x,y) = \sigma'(\alpha_1^{(1)}(x,y)),\;\; \gamma_2(x,y) = \tau'(\alpha_2^{(2)}(x,y))$$

and

$$H_0(x,y) = \left\{ (s,t) :\; s \in \left[0,\, \alpha_1^{(0)}(x,y)\right],\; t \in \left[0, \alpha_2^{(0)}(x,y)\right] \right\},$$

$$H_1(x,y) = \left\{ (s,t) :\; s = \alpha_1^{(1)}(x,y),\;\; t \in \left[0, \alpha_2^{(1)}(x,y)\right] \right\},$$

$$H_2(x,y) = \left\{ (s,t) :\; s \in \left[0, \alpha_1^{(2)}(x,y)\right],\; t = \alpha_2^{(2)}(x,y)] \right\}.$$

The Cauchy problem and the Goursat problem for hyperbolic functional differential equations lead to the Volterra functional integral equations of type (8.1), see

[221, 225]. Initial - value problems for equations in more than two variables and problems for equations of higher order can be reformulated in terms of Volterra functional integral equations.

As a particular case of equation (8.1) we can obtain the system of Volterra integral equations which was considered by W. Walter in the paper [221] and the monograph [225].

In this Chapter we give theorems concerning the existence and uniqueness of continuous solutions of (8.1) in a certain class of functions. We shall consider a method of successive approximations. The essential fact in our considerations is that we do not assume that the Lipschitz coefficient L_0 of the function F with respect to the last variable satisfies the condition $L_0 < 1$ (see Lemmas 8.11 - 8.16). If $L_0 < 1$, then we have a theorem on the existence and uniqueness of solutions of (8.1), which can be obtained by means of the Banach fixed point theorem. The second part of this Chapter concerns equation (8.1) considered in a finite dimensional Banach space. We prove a theorem on the existence of at least one solution of equation (8.1). In this case it is an important fact that we also do not assume that the Lipschitz coefficient L_0 of the function F with respect to the last variable satisfies the condition $L_0 < 1$, (see Lemma 8.25). This part of Chapter is an extension of the result contained in the paper [113], where an equation of type (8.1) with unknown function of one variable was considered. Our results are based on [114]. We consider also difference methods for functional integral equations which are generated by the Darboux problem.

Let

$$A = \{\, i: \ 1 \le i \le m, \ p_i = n \,\}, \ B = \{\, i: \ 1 \le i \le m, \ 1 \le p_i < n \,\}.$$

By changing of notation, if necessary, we may assume that $A = \{\, 1, \ldots, \kappa \,\}$, $B = \{\, \kappa+1, \ldots, m \,\}$ where $0 \le \kappa \le m$. We put $B = \emptyset$ if $\kappa = m$ and $A = \emptyset$ if $\kappa = 0$. We define the sets $\sigma_j \subset \{\, 1, \ldots, n \,\}$, $1 \le j \le m$, in the following way: if the axis $0\,x_i$ is parallel to the p_j - dimensional hyperplane in which the set $H_j(x)$ is contained, then $i \in \sigma_j$. Put $\bar{\sigma}_j = \{\, 1, \ldots, n \,\} \setminus \sigma_j$. For each $x \in E$ and $1 \le j \le m$ we introduce the set $E_j(x)$ by

$$E_j(x) = \left\{ s = (s_1, \ldots, s_n) : \ s_{t_i} = \bar{x}_{t_i} \text{ for } t_i \in \bar{\sigma}_j, \ \ 0 \le s_{t_i} \le \varphi^{(j)}_{t_i}(x) \text{ for } t_i \in \sigma_j \right\},$$

where

$$(\, \varphi^{(j)}_{t_1}, \ldots, \varphi^{(j)}_{t_{p_j}} \,) = \varphi_j \in C(E, R^{p_j}_+), \ t_i \in \sigma_j, \ \text{ and } \ H_j(x) \subset E_j(x) \subset E(x).$$

The p_j - dimensional Lebesgue measure of $E_j(x)$ satisfies

$$L_{p_j}(\, E_j(x) \,) = \prod_{r \in \sigma_j} \varphi^{(j)}_r(x).$$

We adopt the following notations:

$$L\,(\, E(x) \,) = (\, L_{p_1}(E_1(x)), \ldots, L_{p_m}(E_m(x)) \,)\,.$$

If $K = (K_1, \dots, K_m) \in C(E, R^m)$ then

$$K(x) \diamond \int_{H(x)} f(x, s, z(\alpha(s)))\, ds = \sum_{j=1}^{m} K_j(x) \int_{H_j(x)} f_j(x, s, z(\alpha_j(s)))\, (ds)_{p_j},$$

and

$$K(x) \diamond \int_{H(x)} z(\alpha(s))\, ds = \sum_{j=1}^{m} K_j(x) \int_{H_j(x)} z(\alpha_j(s))\, (ds)_{p_j}.$$

For $K \in C(E, R^m)$ we define

$$K(x) \diamond L(E(x)) = \sum_{j=1}^{m} K_j(x)\, L_{p_j}(E_j(x)).$$

We will consider linear integral functional comparison problems.

8.2 Properties of a comparison problem

We formulate assumptions of the sets $H_i(x)$, $1 \le i \le m$. Let $x \in E$, $h \in R^n, x + h \in E$ and $\kappa + 1 \le i \le m$. Suppose that the set $H_i(x)$ is contained in p_i - dimensional hyperplane, $1 \le p_i \le n$, parallel to the $n - p_i$ coordinate axes. We denote this hyperplane by $S_i(x)$. Let the set $H_i(x + h)$ be contained in a p_i - dimensional hyperplane $S_i(x + h)$ parallel to the hyperplane $S_i(x)$. There exists a vector $t_i(x, h) \in R^n$ such that the set $-t_i(x, h) + H_i(x + h)$ is contained in $S_i(x)$.

Assumption $\mathbf{H_1}$. Suppose that

1) for $1 \le i \le \kappa$ we have

$$\lim_{h \to 0} L_n[\, H_i(x) \div H_i(x + h)\,] = 0$$

uniformly with respect to $x \in E$,

2) for $\kappa + 1 \le i \le m$ we have

$$\lim_{h \to 0} t_i(x, h) = 0 \text{ and } \lim_{h \to 0} L_{p_i}\left[\, H_i(x) \div (\, -t_i(x, h) + H_i(x + h)\,)\,\right] = 0$$

uniformly with respect to $x \in E$.

Remark 8.1 *If $x \in H_i(x)$ for $x \in E$ and $\kappa + 1 \le i \le m$ then we can assume that $t_i(x, h) = h$. Condition $\lim_{h \to 0} t_i(x, h) = 0$ is satisfied in this case.*

Assumption $\mathbf{H_2}$. Suppose that

1) the functions $k, \tilde{h} \in C(E, R_+)$, $K = (K_1, \dots, K_m) \in C(E, R^m_+)$, $\beta \in C(E, E)$ are given and $\beta(x) \le x$ for $x \in E$,

2) for $x \in E$ we have

$$\tilde{m}(x) = \sum_{i=0}^{\infty} k^{(i)}(x)\, \tilde{h}(\beta^{(i)}(x)) < +\infty \tag{8.2}$$

where

$$k^{(0)}(x) = 1 \text{ for } x \in E, \quad k^{(i+1)}(x) = k(x)\, k^{(i)}(\beta(x)) \text{ for } x \in E,\ i \geq 0,$$

$$\beta^{(0)}(x) = x \text{ for } x \in E, \quad \beta^{(i+1)}(x) = \beta(\beta^{(i)}(x)) \text{ for } x \in E,\ i \geq 0,$$

3) for $x \in E$ we have

$$M(x) = \sum_{i=0}^{\infty} k^{(i)}(x) \left[K(\beta^{(i)}(x)) \diamond L\left(E(\beta^{(i)}(x)) \right) \right] < +\infty, \tag{8.3}$$

4) M, $\tilde{m} \in C(E, R_+)$ and the function

$$\tilde{M}(x) = \sum_{i=0}^{\infty} k^{(i)}(x) \left[\sum_{j=1}^{m} K_j(\beta^{(i)}(x))\, L_{p_j}\left(E_j(\beta^{(i)}(x)) \right) \left[\prod_{r \in \sigma_j} x_r \right]^{-1} \right]$$

where $x \in E$, is bounded on E,

5) $\alpha \in C(E, E^m)$ and $\alpha_i(x) \leq x$, $1 \leq i \leq m$, for $x \in E$.

We adopt the following notations:

$$\bar{m}(x) = \sum_{i=0}^{\infty} k^{(i)}(x)\, \bar{h}(\beta^{(i)}(x)) \text{ where } \bar{h} \in C(E, R_+),$$

$$(Vz)(x) = \sum_{i=0}^{\infty} k^{(i)}(x) \left[K(\beta^{(i)}(x)) \diamond \int_{H(\beta^{(i)}(x))} z(\alpha(s))\, ds \right].$$

Remark 8.2 *Suppose that condition 1) - 3) of Assumption H_2 are satisfied and*

1) $\bar{h} \in C(E, R_+)$, $\bar{h}(x) \leq \tilde{h}(x)$ for $x \in E$,

2) z is nonnegative and upper-semicontinuous on E.

Then $\bar{m}$ and Vz functions well defined on E.

The existence and uniqueness results for equation (8.1) are based on the following theorems.

Theorem 8.3 *If Assumptions H_1, H_2 are satisfied and $\bar{h} \in C(E, R_+)$, $\bar{h}(x) \leq \tilde{h}(x)$ for $x \in E$ then there exist solutions $\bar{z}$, $\tilde{z} \in C(E, R_+)$ of the equations*

$$z(x) = \bar{m}(x) + (Vz)(x), \quad x \in E, \tag{8.4}$$

and

$$z(x) = \tilde{m}(x) + (Vz)(x), \ x \in E, \tag{8.5}$$

respectively.

The solutions $\bar{z}$ and $\tilde{z}$ of (8.4) and (8.5) are unique in the set $M(E, R_+)$ of nonnegative and upper-semicontinuous functions.

PROOF. We shall show that equation (8.4) has exactly one solution in the set $M(E, R_+)$. Let U be the operator defined by the right-hand side of equation (8.4). We prove that $U : M(E, R_+) \to M(E, R_+)$. Let

$$z \in M(E, R_+) \text{ and } v_{ij} = \int_{H_j(\beta^{(i)}(x))} z(\alpha_j(s))\,(ds)_{p_j}.$$

Then there is a sequence $\{ z_r \}$ such that $z_r \in C(E, R_+)$ and

$$z_{r+1}(x) \le z_r(x),\ x \in E,\ r \ge 1,\quad z(x) = \lim_{r\to\infty} z_r(x),\ x \in E. \tag{8.6}$$

Write

$$v_{ij}^{(r)} = \int_{H_j(\beta^{(i)}(x))} z_r(\alpha_j(s))\,(ds)_{p_j},\ x \in E,\ i = 1,\dots,n,\ j = 1,\dots,m,$$

where $r \ge 1$.The functions $v_{ij}^{(r)}$ are continuous on E and $v_{ij}^{(r+1)}(x) \le v_{ij}^{(r)}(x)$ for $x \in E$. From (8.6) and by the Lebesgue theorem on integration of nonincreasing sequences we have

$$v_{ij}(x) = \lim_{r\to\infty} v_{ij}^{(r)}(x),\ x \in E,\ i = 1,\dots,n,\ j = 1,\dots,m.$$

Since v_{ij} is the limit of the nonincreasing sequence of continuous functions, we see that $v_{ij} \in M(E, R_+)$. It follows from Dini theorem and from Assumptions H_1 and H_2 that series (8.2) and (8.3) are uniformly convergent on E. From this fact and by the conditions

$$k^{(i)}(x)\,\bar h(\beta^{(i)}(x)) \le k^{(i)}(x)\,\tilde h(\beta^{(i)}(x)),\ x \in E,\ i \ge 0,$$

and

$$k^{(i)}(x)\left[K(\beta^{(i)}(x)) \diamond \int_{H(\beta^{(i)}(x))} z(\alpha(s))\,ds\right]$$
$$\le k^{(i)}(x)\left[K(\beta^{(i)}(x)) \diamond L\left(E(\beta^{(i)}(x))\right)\right][\sup\{ z(x) :\ x \in E \}]$$

where $x \in E$, $i \ge 0$, we infer the uniform convergence on E of the series:

$$\sum_{i=0}^{\infty} k^{(i)}(x)\,\bar h(\beta^{(i)}(x))$$

and

$$\sum_{i=0}^{\infty} k^{(i)}(x)\left[K(\beta^{(i)}(x)) \diamond \int_{H(\beta^{(i)}(x))} z(\alpha(s))\,ds\right].$$

Hence we get $\bar m \in C(E, R_+)$, $Vz \in M(E, R_+)$ and consequently, $U : M(E, R_+) \to M(E, R_+)$. Now we prove that the operator U is a contraction. Let

$$\|z\|_0 = \max\{ \exp[-\lambda(x_1 + \dots + x_n)]\,|z(x)| :\ x \in E \}.$$

where $z \in M(E, R_+)$ and

$$\lambda > \Lambda = \max \left\{ 1,\ \sup \{ \tilde{M}(x) : x \in E \} \right\}.$$

For $z, w \in M(E, R_+)$ we get

$$\|(Uz)(x) - (Uw)(x)\|$$

$$\leq \sum_{i=0}^{\infty} k^{(i)}(x) \left[K(\beta^{(i)}(x)) \diamond \int_{H(\beta^{(i)}(x))} | z(\alpha(s)) - w(\alpha(s)) | \, ds \right]$$

$$\leq \|z - w\|_0 \sum_{i=0}^{\infty} k^{(i)}(x) \left[K(\beta^{(i)}(x)) \diamond \int_{H(\beta^{(i)}(x))} \exp\{ \lambda(s_1 + \ldots + s_n) \} \, ds \right].$$

We have the following estimates

$$\int_{H_j(\beta^{(i)}(x))} \exp\{ \lambda(s_1 + \ldots + s_n) \} \, ds$$

$$\leq \exp \left[\lambda \sum_{r \in \bar{\sigma}_j} x_r \right] \int_{E_j(\beta^{(i)}(x))} \exp \left[\lambda \sum_{r \in \sigma_j} x_r \right] (ds)_{p_j}$$

$$= \exp \left[\lambda \sum_{r \in \bar{\sigma}_j} x_r \right] \prod_{r \in \sigma_j} \left[\frac{1}{\lambda} \left[\exp\{ \lambda \varphi_r^{(j)}(\beta^{(i)}(x)) \} - 1 \right] \right]$$

$$\leq \frac{1}{\lambda} \exp\{ \lambda(x_1 + \ldots + x_n) \} \, L_{p_j}(E_j(\beta^{(i)}(x))) \left[\prod_{r \in \sigma_j} x_r \right]^{-1}.$$

Finally, we obtain

$$\| (Uz)(x) - (Uw)(x) \|$$

$$\leq \frac{1}{\lambda} \|z - w\|_0 \exp \{ \lambda(x_1 + \ldots + x_n) \} \sum_{i=0}^{\infty} k^{(i)}(x)$$

$$\times \left[\sum_{j=1}^{m} K_j(\beta^{(i)}(x)) \, L_{p_j}(E_j(\beta^{(i)}(x))) \left[\prod_{r \in \sigma_j} x_r \right]^{-1} \right]$$

$$\leq \frac{1}{\lambda} \Lambda \|z - w\|_0 \exp \{ \lambda(x_1 + \ldots + x_n) \},$$

and consequently $\|Uz - Uw\|_0 \leq \frac{\Lambda}{\lambda} \|z - w\|_0$. Since $\Lambda < \lambda$ then by the Banach fixed point theorem we infer that equation (8.4) has a unique solution $\bar{z}$ being an upper-semicontinuous function. We prove that $\bar{z} \in C(E, R_+)$.The solution $\bar{z}$ of equation (8.4) is the limit of the sequence $\{ z_r \}$ which is defined in the following way: $z_0 \in M(E, R_+)$ is arbitrary and $z_{r+1}(x) = \bar{m}(x) + (Vz_r)(x)$, $x \in E$, $r \geq$

0. For $z \in M(E,R_+)$ we define $(V^0 z)(x) = z(x)$, $x \in E$, and $(V^{i+1}z)(x) = (V(V^i z))(x)$, $x \in E$, $i \geq 0$. It is easily seen that

$$z_{r+1}(x) = \sum_{i=0}^{r}(V^i\bar{m})(x) + (V^{r+1}z_0)(x), \; x \in E. \tag{8.7}$$

Then $\{V^r z_0\}$ is the sequence of successive approximations for the equation $z(x) = (Vz)(x)$. Since this equation has a solution $z(x) = 0$, $x \in E$, which is unique in the set $M(E,R_+)$ it follows that $\lim_{r\to\infty}(V^r z_0)(x) = 0$ uniformly with respect to $x \in E$. Since the functions $V^i\bar{m}$ are continuous on E and the sequence $\{z_r\}$ is uniformly convergent, it follows from (8.7) that

$$\bar{z}(x) = \sum_{i=0}^{\infty}(V^i\bar{m})(x)$$

is a continuous function on E. This completes the proof of Theorem 8.3.

Theorem 8.4 *If Assumptions H_1, H_2 are satisfied and $\bar{h} \in C(E,R_+)$, $\bar{h}(x) \leq \tilde{h}(x)$ for $x \in E$ then the functions $\bar{z}$ and $\tilde{z}$ given in Theorem 8.3 are solutions of equations*

$$z(x) = K(x) \diamond \int_{H(x)} z(\alpha(s))\,ds + k(x)\,z(\beta(x)) + \bar{h}(x), \quad x \in E, \tag{8.8}$$

and

$$z(x) = K(x) \diamond \int_{H(x)} z(\alpha(s))\,ds + k(x)\,z(\beta(x)) + \tilde{h}(x), \quad x \in E, \tag{8.9}$$

respectively. Moreover, these solutions are unique in the class $\tilde{M}(E,R_+,\tilde{z})$, where

$$\tilde{M}(E,R_+,\tilde{z}) = \{\, z : \; z \in M(E,R_+) \text{ and } \inf\{\, c : \; z(x) \leq c\,\tilde{z}(x)\,\} < +\infty \,\}.$$

The function $\tilde{z}$ satisfies the condition

$$\lim_{r\to\infty} k^{(r)}(x)\,\tilde{z}(\beta^{(r)}(x)) = 0 \;\; \textit{uniformly with respect to } \; x \in E. \tag{8.10}$$

PROOF. At first we prove that condition (8.10) is satisfied. It is easy to check that functions $k^{(r)}$ and $\beta^{(r)}$ satisfy the conditions

$$k^{(r)}(x)\,k^{(i)}(\beta^{(r)}(x)) = k^{(r+i)}(x), \quad \beta^{(i)}(\beta^{(r)}(x)) = \beta^{(r+i}(x), \tag{8.11}$$

where $x \in E$ and i, $r \geq 0$. Formulas (8.5) and (8.11) imply

$$k^{(r)}(x)\,\tilde{z}(\beta^{(r)}(x)) = \sum_{i=0}^{\infty} k^{(r+i)}(x)\,\tilde{h}(\beta^{(r+i)}(x))$$

$$+\sum_{i=0}^{\infty} k^{(r+i)}(x)\left[K(\beta^{(r+i)}(x)) \diamond \int_{H(\beta^{(r+i)}(x))} \tilde{z}(\alpha(s))\,ds\right].$$

The above relation and the uniform convergence of series (8.2) and (8.3) imply the uniform convergence on E of the sequence

$$\{\, k^{(r)}(x)\, \bar{z}(\beta^{(r)}(x))\, \}.$$

We observe that any solution of equation (8.4) is a solution of (8.8). Indeed, if $\bar{z}$ is a solution of equation (8.4) then we have

$$\bar{z}(x) - K(x) \diamond \int_{H(x)} \bar{z}(\alpha(s))\, ds - k(x)\, \bar{z}(\beta(x))$$

$$= \sum_{i=0}^{\infty} k^{(i)}(x)\, \bar{h}(\beta^{(i)}(x)) + \sum_{i=0}^{\infty} k^{i}(x) \left[K(\beta^{(i)}(x)) \diamond \int_{H(\beta^{(i)}(x))} \bar{z}(\alpha(s))\, ds \right]$$

$$-K(x) \diamond \int_{H(x)} \bar{z}(\alpha(s))\, ds - k(x) \left\{ \sum_{i=0}^{\infty} k^{(i)}(\beta(x))\, \bar{h}(\beta^{(i)}(\beta(x))) \right.$$

$$\left. + \sum_{i=0}^{\infty} k^{i}(\beta(x)) \left[K(\,\beta^{(i)}(\beta(x))\,) \diamond \int_{H(\beta^{(i)}(\beta(x))\,)} \bar{z}(\alpha(s))\, ds \right] \right\} = \bar{h}(x),\ x \in E,$$

which means that $\bar{z}$ is a solution of (8.8). Now we prove that $\bar{z}$ is a unique solution of (8.8) in the class $\tilde{M}(E, R_+, \tilde{z})$. In fact, if $\bar{w} \in C(E, R_+, \tilde{z})$ is a solution of (8.8) then for $r \geq 1$ and for $x \in E$ the equality

$$\bar{w}(x) = \sum_{i=0}^{r-1} k^{(i)}(x) \left[K(\beta^{(i)}(x)) \diamond \int_{H(\beta^{(i)}(x))} \bar{w}(\alpha(s))\, ds \right] \tag{8.12}$$

$$+ \sum_{i=0}^{r-1} k^{(i)}(x)\, \bar{h}(\beta^{(i)}(x)) + k^{(r)}(x)\, \bar{w}(\beta^{(r)}(x))$$

holds. Since $\bar{w} \in \tilde{M}(E, R_+, \tilde{z})$, we have for some $c \in R_+ : 0 \leq \bar{w}(x) \leq c\,\tilde{z}(x),\ x \in E$. Now, according to (8.10), we obtain

$$\lim_{r\to\infty} k^{(r)}(x)\, \bar{w}(\beta^{(r)}(x)) = 0 \text{ uniformly with respect to } x \in E. \tag{8.13}$$

If we let $r \to \infty$ in relation (8.12) then we obtain

$$\bar{w}(x) = \bar{m}(x) + (\,V\bar{w}\,)(x),\ \ x \in E.$$

Then the function $\bar{w}$ is the solution of equation (8.4). This equation has only the solution $\bar{z}$, hence it results that $\bar{z} = \bar{w}$. Thus the proof of Theorem 8.4 is completed.

Theorem 8.5 *If Assumptions H_1, H_2 are satisfied and $\bar{h} \in C(E, R_+)$, $\bar{h}(x) \leq \tilde{h}(x)$ for $x \in E$ then the function $z(x) = 0$ for $x \in E$ is the unique solution of the inequality*

$$z(x) \leq K(x) \diamond \int_{H(x)} z(\alpha(s))\, ds + k(x)\, z(\beta(x)),\ \ x \in E, \tag{8.14}$$

in the class $\tilde{M}(E, R_+, \tilde{z})$.

PROOF. Let us suppose that $\tilde{w} \in \tilde{M}(E, R_+, \tilde{z})$ and $\tilde{w}$ is a solution of inequality (8.14). We obtain easily for $x \in E$ and $r \geq 1$:

$$\tilde{w}(x) \leq \sum_{i=0}^{r-1} k^{(i)}(x) \left[K(\beta^{(i)}(x)) \diamond \int_{H(\beta^{(i)}(x))} \tilde{w}(\alpha(s))\, ds \right] \tag{8.15}$$

$$+k^{(r)}(x)\, \tilde{w}(\beta^{(r)}(x)).$$

Since $\tilde{w} \in \tilde{M}(E, R_+, \tilde{z})$, we have for some $c \in R_+$: $0 \leq \tilde{w}(x) \leq c\,\tilde{z}(x)$ for $x \in E$. By (8.10) the last inequality implies $\lim_{r\to\infty} k^{(r)}(x)\, \tilde{w}(\beta^{(r)}(x)) = 0$ uniformly with respect to $x \in E$. Letting r in (8.15) tend to ∞ we get

$$\tilde{w}(x) \leq (\, V\tilde{w}\,)(x), \; x \in E. \tag{8.16}$$

Let $\{\, z_r \,\}$ be the sequence defined in the following way:

$$z_0(x) = \tilde{z}(x), \; x \in E, \;\; z_{r+1}(x) = (\, V z_r \,)(x), \; x \in E, \; r \geq 0. \tag{8.17}$$

From Theorems 8.3 and 8.4 and from (8.17) it follows that

$$0 \leq z_{r+1}(x) \leq z_r(x), \; x \in E, \; r \geq 0, \tag{8.18}$$

and

$$\lim_{r\to\infty} z_r(x) = 0 \text{ uniformly in } E. \tag{8.19}$$

Further, by (8.16) we obtain $\tilde{w}(x) \leq c\, z_r(x)$ for $x \in E$, $r \geq 0$. The last formula together with (8.19) gives $\tilde{w}(x) = 0$ for $x \in E$, which completes the proof of Theorem 8.5.

Lemma 8.6 *Suppose that*

1) Assumptions H_1, H_2 are satisfied and $\bar{h} \in C(E, R_+)$, $\bar{h}(x) \leq \tilde{h}(x)$ for $x \in E$,

2) the functions k, $\tilde{h}$, $\bar{h}$, K, β are nondecreasing on E and $H_j(x) \subset H_j(\bar{x})$ for $x < \bar{x}$, $x, \bar{x} \in E$, $1 \leq j \leq m$.

Then the function $\bar{z}$ and $\tilde{z}$ given in Theorem 8.3 are nondecreasing on E.

We omit the simple prof of the above Lemma.

Lemma 8.7 *If Assumptions H_1 and H_2 are satisfied and the sequence $\{\, w_r \,\}$ is defined by the formulas: $w_0(x) = \tilde{z}(x), x \in E$, and*

$$w_{r+1}(x) = K(x) \diamond \int_{H(x)} w_r(\alpha(s))\, ds + k(x)\, w_r(\beta(x)), \tag{8.20}$$

where $r \geq 0$, $x \in E$, then

$$0 \leq w_{r+1}(x) \leq w_r(x) \leq w_0(x), \; x \in E, \; r \geq 0, \tag{8.21}$$

and

$$\lim_{r\to\infty} w_r(x) = 0 \;\; \textit{uniformly with respect to } x \in E. \tag{8.22}$$

PROOF. Relations (8.21) follow by induction. The convergence of the sequence $\{ w_r \}$ is implied by (8.21). Since $w_r \in C(E, R_+)$ it follows that the limit $\bar{w}$ of the sequence $\{ w_r \}$ is an upper-semicontinuous function. From (8.20) it follows that $\bar{w}$ satisfies inequality (8.14). According to Theorem 8.4 we have $\bar{w}(x) = 0$ for $x \in E$. The uniform convergence of the sequence $\{ w_r \}$ follows from Dini theorem. This completes the proof of Lemma 8.7.

8.3 THE EXISTENCE AND UNIQUENESS OF SOLUTIONS

We introduce the following assumptions on F.

Assumption H$_3$. Suppose that

1) there exist functions $l = (l_1, \ldots, l_m) \in C(E, R_+^m)$, $k \in C(E, R_+)$ such that

$$\| F(x, u, v) - F(x, \bar{u}, \bar{v}) \| \leq \sum_{i=1}^{m} l_i(x) \|u_i - \bar{u}_i\| + k(x) \|v - \bar{v}\|,$$

where $u = (u_1, \ldots, u_m), \bar{u} = (\bar{u}_1, \ldots, \bar{u}_m),\ u, \bar{u} \in X^m,\ v,\ \bar{v} \in X,\ x \in E,$

2) there exists a function $\bar{l} = (\bar{l}_1, \ldots, \bar{l}_m) \in C(E, R_+^m)$ such that

$$\|f_i(x, s, z) - f_i(x, s, \bar{z})\| \leq \bar{l}_i(x) \|z - \bar{z}\|, \quad x, s \in E, \quad z, \bar{z} \in X, \quad 1 \leq i \leq m,$$

3) there exists a function $u_0 \in C(E, X)$ such that Assumption H$_2$ is satisfied for $\tilde{h}$ and K defined by relations

$$\tilde{h}(x) = \| F(x, \int_{H(x)} f(x, s, u_0(\alpha(s)))\, ds,\ u_0(\beta(x))) - u_0(x) \|,$$

$$K(x) = (l_1(x) \bar{l}_1(x), \ldots, l_m(x) \bar{l}_m(x)),$$

and for k given in condition 1) of Assumption H$_3$.

Lemma 8.8 *If Assumptions H_1 and H_3 are satisfied and the sequence $\{ u_r \}$ is defined by the relations*

$$u_{r+1}(x) = F(x, \int_{H(x)} f(x, s, u_r(\alpha(s)))\, ds,\ u_r(\beta(x))), \quad x \in E, \tag{8.23}$$

where $r \geq 0$ and u_0 is given by condition 3) of Assumption H_3, then

$$\|u_r(x) - u_0(x)\| \leq \tilde{z}(x), \quad x \in E, \quad r \geq 0, \tag{8.24}$$

and

$$\|u_{r+p}(x) - u_r(x)\| \leq w_r(x), \quad x \in E, \quad r, p \geq 0, \tag{8.25}$$

where $\tilde{z}$ is defined in Theorem 8.3 and the sequence $\{w_r\}$ is given be relations (8.20).

PROOF. We prove that (8.24) is fulfilled. For $r = 0$ this inequality is evidently satisfied. If we assume that $\|u_r(x) - u_0(x)\| \leq \tilde{z}(x)$ for $x \in E$, then

$$\|u_{r+1}(x) - u_0(x)\| \leq \tilde{h}(x)$$

$$+\| F(x, \int_{H(x)} f(x,s,u_r(\alpha(s)))\,ds,\; u_r(\beta(x)))$$

$$-F(x, \int_{H(x)} f(x,s,u_0(\alpha(s)))\,ds,\; u_0(\beta(x)))\,\|$$

$$\leq K(x) \diamond \int_{H(x)} \tilde{z}(\alpha(s))\,ds + k(x)\,\tilde{z}(\beta(x)) + \tilde{h}(x) = \tilde{z}(x), \;\; x \in E.$$

Now we obtain (8.24) by induction. Next we prove (8.25). From (8.20) and (8.24) it follows that (8.25) is satisfied for $r = 0$, $p \geq 0$, $x \in E$. If we assume that (8.25) holds for an arbitrary fixed r and for any $p \geq 0$, $x \in E$, then

$$\| u_{r+1+p}(x) - u_{r+1}(x) \|$$

$$\leq \| F(x, \int_{H(x)} f(x,s,u_{r+p}(\alpha(s)))\,ds,\; u_{r+p}(\beta(x)))$$

$$-F(x, \int_{H(x)} f(x,s,u_r(\alpha(s)))\,ds,\; u_r(\beta(x)))\,\|$$

$$\leq K(x) \diamond \int_{H(x)} w_r(\alpha(s))\,ds + k(x)\,w_r(\beta(x)) = w_{r+1}(x), \;\; x \in E.$$

Now (8.25) follows by induction with respect to r. This proves Lemma 8.8.

Theorem 8.9 *If Assumptions H_1 and H_3 are satisfied, then there exists a solution $\bar{u} \in C(E,X)$ of equation (8.1) such that*

$$\|\bar{u}(x) - u_0(x)\| \leq \tilde{z}(x), \;\; x \in E, \tag{8.26}$$

and

$$\|\bar{u}(x) - u_r(x)\| \leq w_r(x), \;\; x \in E, \;\; r \geq 0, \tag{8.27}$$

where $\{u_r\}$ and $\{w_r\}$ are defined by (8.23) and (8.20) respectively. The solution $\bar{u}$ of (8.1) is unique in the class

$$\Xi(E,X) = \bigcup_{c \geq 0} \{ u : \; u \in C(E,X), \; \|u(x) - u_0(x)\| \leq c\,\tilde{z}(x), \; x \in E \}.$$

PROOF. It follows from (8.22) and (8.25) that the sequence $\{u_r\}$ is uniformly convergent on E to a certain function $\bar{u} \in C(E,X)$. Obviously $\bar{u}$ is a solution of (8.1). The estimates (8.26) and (8.27) are implied by (8.24) and (8.25) respectively. To prove that the solution $\bar{u}$ of (8.1) is unique in the class considered, let us suppose that there is another solution $\tilde{u}$ of equation (8.1) and $\tilde{u} \in \Xi(E,X)$. It is easy to check that the function $\bar{z}(x) = \|\tilde{u}(x) - \bar{u}(x)\|$ is an element of the set $\tilde{M}(E, R_+, \tilde{z})$ and

$$\bar{z}(x) \leq K(x) \diamond \int_{H(x)} \bar{z}(\alpha(s))\, ds + k(x)\, \bar{z}(\beta(x)), \ x \in E.$$

By Theorem 8.5 we get $\bar{z}(x) = 0$ for $x \in E$ and hence $\tilde{u}(x) = \bar{u}(x)$ for $x \in E$. Thus the proof of Theorem 8.9 is completed.

Now we give a theorem on the continuous dependence of solutions on the right-hand side of equation (8.1). Let us consider another equation

$$v(x) = \tilde{F}(\, x, \int_{\tilde{H}(x)} \tilde{f}(\, x, s, v(\tilde{\alpha}(s))\,)\, ds, \ v(\tilde{\beta}(x))\,), \ x \in E, \tag{8.28}$$

where the functions $\tilde{F}$, $\tilde{f} = (\,\tilde{f}_1, \ldots, \tilde{f}_m\,)$, $\tilde{\alpha} = (\,\tilde{\alpha}_1, \ldots, \tilde{\alpha}_m\,)$, $\tilde{\beta}$ and the sets $\tilde{H}(x) = (\,\tilde{H}_1(x), \ldots, \tilde{H}_m(x)\,)$ have the same properties as F, f, α, β, $H(x)$, which are formulated in Assumptions H_1, H_3. Suppose that $\bar{u}$ and $\bar{v}$ are solutions of equations (8.1) and (8.28) respectively. Let $\tilde{r} \in C(E, R_+)$ be a function such that $\|\bar{u}(x) - \bar{v}(x)\| \leq \tilde{r}(x)$ for $x \in E$. Let

$$q(x) = \|\, F(\, x, \int_{H(x)} f(\, x, s, \bar{v}(\alpha(s))\,)\, ds, \ \bar{v}(\beta(x))\,)$$

$$-\tilde{F}(\, x, \int_{\tilde{H}(x)} \tilde{f}(\, x, s, \bar{v}(\tilde{\alpha}(s))\,)\, ds\, \bar{v}(\beta(x))\,)\, \|,$$

$$\tilde{h}_1(x) = \max\{\, \tilde{r}(x), \, q(x), \, \tilde{h}(x)\,\}, \ x \in E.$$

Now we give an estimate of the difference between solutions of two problems.

Theorem 8.10 *Suppose that the functions F, $f\, \alpha$, β and $\tilde{F}$, $\tilde{f}$, $\tilde{\alpha}$, $\tilde{\beta}$, $\tilde{h}_1$ satisfy Assumptions H_3 and the sets $H(x)$, $\tilde{H}(x)$, $x \in E$, satisfy Assumption H_1. Then there exists a solution $\bar{w} \in C(E, R_+)$ of the equation*

$$z(x) = K(x) \diamond \int_{H(x)} z(\alpha(s))\, ds + k(x)\, z(\beta(x)) + q(x), \ x \in E, \tag{8.29}$$

such that

$$\|\bar{u}(x) - \bar{v}(x)\| \leq \bar{w}(x) \ for \ x \in E. \tag{8.30}$$

PROOF. Let $\bar{z}$ be a solution of the equation

$$z(x) = K(x) \diamond \int_{H(x)} z(\alpha(s))\, ds + k(x)\, z(\beta(x)) + \tilde{h}_1(x), \ x \in E.$$

Put $w_0(x) = \bar{z}(x)$, $x \in E$, and

$$w_{r+1}(x) = K(x) \diamond \int_{H(x)} w_r(\alpha(s))\, ds + k(x)\, w_r(\beta(x)) + q(x), x \in E,$$

for $r \geq 0$. By induction we get $0 \leq w_{r+1}(x) \leq w_r(x) \leq \bar{z}(x)$ for $x \in E$, $r \geq 0$. From these inequalities we see that the sequence $\{ w_r \}$ is convergent to the solution $\bar{w}$ of equation (8.29) and $0 \leq \bar{w}(x) \leq \bar{z}(x)$ for $x \in E$. However, in view of Theorem 8.4, there exists only one solution of this equation in the class of upper-semicontinuous functions satisfying the condition $0 \leq \bar{w}(x) \leq \bar{z}(x)$ for $x \in E$. Now we show that

$$\|\bar{u}(x) - \bar{v}(x)\| \leq w_r(x) \text{ for } x \in E,\ r \geq 0. \tag{8.31}$$

Since $\tilde{r}(x) \leq \tilde{h}_1(x) \leq \bar{z}(x) = w_0(x)$, $x \in E$, it follows that (8.31) is satisfied for $r = 0$ and $x \in E$. If we assume that $\|\bar{u}(x) - \bar{v}(x)\| \leq w_r(x)$ for $x \in E$ and for some $r \geq 0$, then

$$\begin{aligned}
\|\bar{u}(x) - \bar{v}(x)\| \leq \| & F(x, \int_{H(x)} f(x, s, \bar{u}(\alpha(s)))\, ds,\ \bar{u}(\beta(x))) \\
& -F(x, \int_{H(x)} f(x, s, \bar{v}(\alpha(s)))\, ds,\ \bar{v}(\beta(x))) \| \\
& +\| F(x, \int_{H(x)} f(x, s, \bar{v}(\alpha(s)))\, ds,\ \bar{v}(\beta(x))) \\
& -\tilde{F}(x, \int_{\tilde{H}(x)} \tilde{f}(x, s, \bar{v}(\tilde{\alpha}(s)))\, ds,\ \bar{v}(\tilde{\beta}(x))) \| \\
\leq K(x) \diamond \int_{H(x)} \| \bar{u}(\alpha(s)) - \bar{v}(\alpha(s)) \|\, ds & + k(x) \| \bar{u}(\beta(x)) - \bar{v}(\beta(x)) \| + q(x) \\
\leq K(x) \diamond \int_{H(x)} w_r(\alpha(s))\, ds & + k(x)\, w_r(\beta(x)) + q(x) = w_{r+1}(x),
\end{aligned}$$

where $x \in E$. Now (8.31) follows by induction. Letting $r \to \infty$ in (8.31) we get inequality (8.30). This proves Theorem 8.10.

8.4 Examples of comparison problems

Note that if there is $\tilde{k} \in (0,1)$ such that the Lipschitz coefficient k satisfies the inequality $k(x) \leq \tilde{k}$, $x \in E$, then conditions 2) - 4) of Assumption H_2 are fulfilled. Now we omit the assumption that $k(x) \leq \tilde{k} < 1$ on E.

Lemma 8.11 *Assume that*

1) there are $\bar{k} \in R_+$ *and* $(\bar{K}_1, \ldots, \bar{K}_m) \in R_+^m$ *such that* $k(x) \leq \bar{k}$ *and*

$$K(x) = (K_1(x), \ldots, K_m(x)) \leq (\bar{K}_1, \ldots, \bar{K}_m),$$

2) there are $(\bar{\alpha}_{t_1}^{(i)}, \ldots, \bar{\alpha}_{t_{p_i}}^{(i)}) \in R_+^{p_i}$, $\bar{\alpha}_{t_j}^{(i)} \leq 1$, *such that*

$$\varphi_i(x) = (\varphi_{t_1}^{(i)}(x), \ldots, \varphi_{t_{p_i}}^{(i)}(x)) \leq (\bar{\alpha}_{t_1}^{(i)} x_{t_1}, \ldots, \bar{\alpha}_{t_{p_i}}^{(i)} x_{t_{p_i}}),\ t_j \in \sigma_i,\ 1 \leq i \leq m,$$

3) there is $(\bar{\beta}_1, \ldots, \bar{\beta}_n) \in R^n_+$, $\bar{\beta}_i \le 1$, *such that*

$$\beta(x) = (\beta_1(x), \ldots, \beta_n(x)) \le (\bar{\beta}_1 x_1, \ldots, \bar{\beta}_n x_n),$$

4) for $x \in E$ *we have*

$$\sum_{i=0}^{\infty} \bar{k}^i \, \tilde{h}(\bar{\beta}_1^i x_1, \ldots, \bar{\beta}_n^i x_n) < +\infty,$$

5) for $1 \le j \le m$ *we have*

$$\bar{k} \prod_{r \in \sigma_j} \bar{\beta}_s < 1.$$

Under these assumptions condition 2) - 4) of Assumption H_2 *are satisfied.*

PROOF. By induction we easily obtain the estimates

$$k^{(i)}(x) \le \bar{k}^i \text{ and } \beta^{(i)}(x) \le (\bar{\beta}_1^i x_1, \ldots, \bar{\beta}_n^i x_n), \ x \in E, \ i \ge 0.$$

From these inequalities we get the following estimates for series (8.2):

$$\sum_{i=0}^{\infty} k^{(i)}(x) \, \tilde{h}(\beta^{(i)}(x)) \le \sum_{i=0}^{\infty} \bar{k}^i \, \tilde{h}(\bar{\beta}_1^i x_1, \ldots, \bar{\beta}_n^i x_n), \ x \in E.$$

Now 4) implies condition 2) of Assumption H_2. Since

$$\sum_{i=0}^{\infty} k^{(i)}(x) \left[K(\beta^{(i)}(x)) \diamond L(E(\beta^{(i)}(x))) \right]$$

$$= \sum_{j=1}^{m} \left[\sum_{i=0}^{\infty} k^{(i)}(x) \, K_j(\beta^{(i)}(x)) \, L_{p_j}(E(\beta^{(i)}(x))) \right],$$

then for a fixed index j we have

$$\sum_{i=0}^{\infty} k^{(i)}(x) \, K_j(\beta^{(i)}(x)) \, L_{p_j}(E(\beta^{(i)}(x))) \le \sum_{i=0}^{\infty} \bar{k}^i \bar{K}_j, \prod_{r \in \sigma_j} \varphi_r^{(j)}(\beta^{(i)}(x))$$

$$\le \bar{K}_j \sum_{i=0}^{\infty} \bar{k}^i \prod_{r \in \sigma_j} \bar{\alpha}_r^j \, \bar{\beta}_r^i \, x_r \le \bar{K}_j \left[\prod_{r \in \sigma_j} \bar{\alpha}_r^j \right] \left[\prod_{r \in \sigma_j} x_r \right] \sum_{i=0}^{\infty} \left[\bar{k} \prod_{r \in \sigma_j} \bar{\beta}_r \right]^i.$$

Hence and from assumption 5) of this Lemma it follows that condition 3) of Assumption H_2 is satisfied. From Dini theorem and from the last inequalities it follows that condition 4) of Assumption H_2 is satisfied, too. By a similar argument we can prove the following Lemmas.

Lemma 8.12 *Suppose that*

1) there are $\bar{k} \in R_+$ *and* $(\bar{K}_1, \ldots, \bar{K}_n) \in R^n_+$ *such that* $k(x) \le \bar{k}$, *and* $K_j(x) \le \bar{K}_1 x_1 + \ldots + \bar{K}_n x_n$, *where* $x \in E$, $j = 1, \ldots, m$,

2) assumptions 2) - 4) of Lemma 8.11 are satisfied.

Then conditions 2) - 4) of Assumption H_2 *are fulfilled.*

Lemma 8.13 *Suppose that*

1) $E = [0, a]$, $a = (a_1, \ldots, a_n)$, $a_i > 0$ *for* $1 \le i \le n$ *and there are* $(\bar{k}_1, \ldots, \bar{k}_n)$, $(\bar{K}_1, \ldots, \bar{K}_m) \in R_+^m$ *such that for* $x \in E$ *we have*

$$k(x) \le \bar{k}_1 x_1 + \ldots + \bar{k}_n x_n,\ K(x) = (K_1(x), \ldots K_m(x)) \le (\bar{K}_1, \ldots, \bar{K}_m),$$

2) assumptions 2), 3) of Lemma 8.11 are satisfied and $\bar{k}_1 \bar{\beta}_1 a_1 + \ldots + \bar{k}_n \bar{\beta}_n a_n < 1$.

Then conditions 2) - 4) of Assumption H_2 *are fulfilled.*

Lemma 8.14 *Suppose that*

1) there are $\bar{k} \in R_+$ *and* $(\bar{K}_1, \ldots, \bar{K}_n) \in R_+^n$ *such that* $k(x) \le \bar{k}$ *and* $K_j(x) \le \bar{K}_1 x_1 + \ldots + \bar{K}_n x_n)$ *for* $x \in E$, $1 \le j \le m$, *and for* $x \in E$ *we have* $\varphi_j(x) \le (\bar{\alpha}_{t_1}^{(j)} x_{t_1}^2, \ldots, \bar{\alpha}_{t_j}^{(j)} x_{t_{p_j}}^2)$ *where* $t_i \in \sigma_j$, $\bar{\alpha}_{t_r}^{(j)} \in R_+$,

2) assumptions 3), 4) of Lemma 8.11 are satisfied and the estimates

$$\bar{k}\bar{\beta}_i \left[\prod_{r \in \sigma_j} \bar{\beta}_r \right]^2 < 1$$

hold for $1 \le i \le n$, $1 \le j \le m$.

Then conditions 2) - 4) of Assumption H_2 *are fulfilled.*

Lemma 8.15 *Suppose that*

1) $E = [0, a]$, $a = (a_1, \ldots, a_n)$, $0 < a_i \le 1$ *for* $1 \le i \le n$ *and*

$$\prod_{r \in \sigma_j} a_r^2 < 1 \quad for\ \ 1 \le j \le m,$$

2) $\beta(x) \le (x_1^2, \ldots, x_n^2)$ *and assumptions 1), 2) of Lemma 8.11 are satisfied,*

3) for $x \in E$ *we have*

$$\sum_{i=0}^{\infty} \bar{k}^i\, \tilde{h}(x_1^{2^i}, \ldots, x_n^{2^i}) < +\infty.$$

Then conditions 2) - 4) of Assumption H_2 *are fulfilled.*

Lemma 8.16 *Suppose that*

1) assumptions 1) - 3) of Lemma 8.11 are satisfied and $\tilde{h}(x) \le h\, x^p$, *where* $h =$ const *and* $x^p = \prod_{i=1}^n x_i^p$,

2) for $\mu = \min[1, p]$ *we have*

$$\bar{k} \left[\prod_{r \in \sigma_j} \bar{\beta}_r \right]^{\mu} < 1.$$

Then conditions 2) - 4) of Assumption H_2 *are fulfilled.*

8.5 A CERTAIN FUNCTIONAL EQUATION

In this part of Chapter we give sufficient conditions for the existence of at least one continuous solution of equation (8.1) considered in a finite dimensional Banach space X. Now we do not assume the Lipschitz condition for the function F with respect to u_i for $1 \le i \le \kappa$ and for the functions f_i with respect to z for $1 \le i \le \kappa$.

Assumption $\mathbf{H_4}$. Suppose that

1) there exist functions

$$(h_0, h_1, \ldots, h_m) \in C(E, R_+^{m+1}), \quad g \in C(E, R_+),$$

and

$$(\bar{h}_1, \ldots, \bar{h}_m) \in C(E, R_+^m), \quad (\bar{g}_1, \ldots, \bar{g}_m) \in C(E, R_+^m)$$

such that

$$\| F(x, u, v) \| \le \sum_{i=1}^{m} h_i(x) \, \|u_i\| + h_0(x) \, \|v\| + g(x), \quad x \in E,$$

where $u = (u_1, \ldots, u_m) \in X^m$, $v \in X$, X is a finite dimensional Banach space and

$$\| f_i(x, s, z) \| \le \bar{h}_i(x) \, \|z\| + \bar{g}_i(x), \quad 1 \le i \le m, \quad x, s \in E, \quad z \in X,$$

2) there exist functions

$$(l_{\kappa+1}, \ldots, l_m) \in C(E, R_+^{m-\kappa}), \ (\bar{l}_{\kappa+1}, \ldots, \bar{l}_m) \in C(E, R_+^{m-\kappa}), \ k \in C(B, R_+),$$

such that

$$\|F(x, u_A, u_B, v) - F(x, u_A, \bar{v}_B, \bar{v}) \| \le \sum_{i=\kappa+1}^{m} l_i(x) \, \|u_i - \bar{u}_i\| + k(x) \|v - \bar{v}\|,$$

where $x \in E$, $u_A = (u_1, \ldots, u_\kappa)$, $u_B = (u_{\kappa+1}, \ldots, u_m)$, u_i, $\bar{u}_i$, v, $\bar{v} \in X$, and

$$\|f_i(x, s, z) - f_i(x, s, \bar{z})\| \le \bar{l}_i \, \|z - \bar{z}\| \ \text{ for } \kappa + 1 \le i \le m, \ x, s, \in B, \ z, \ \bar{z} \in X,$$

3) Assumption $\mathrm{H_2}$ is fulfilled for $\tilde{h}$, K defined by

$$K(x) = (h_1(x) \, \bar{h}_1(x), \ldots, h_m(x) \, \bar{h}_m(x)),$$

$$\tilde{h}(x) = g(x) + \sum_{i=1}^{m} \{ \, h_i(x) \, \bar{g}_i(x) \ [\sup\{ L_{p_i}(H_i(x)) : \ x \in E \}] \},$$

and for k defined in condition 2) of Assumption $\mathrm{H_4}$.

Let Θ be the class of all functions $\gamma \in C(R_+, R_+)$ which are subadditive, nondecreasing on R_+ and such that $\gamma(0) = 0$.

Assumption $\mathbf{H_5}$. Suppose that

1) there are functions $D_j \in \Theta$, $0 \le j \le m + 1$, such that

$$\| F(x, u, v) - F(\bar{x}, \bar{u}, \bar{v}) \| \le D_0(\|x - \bar{x}\|) + \sum_{i=1}^{m} D_i(\|u - \bar{u}_i\|) + D_{m+1}(\|v - \bar{v}\|)$$

for $x,\ \bar{x} \in E,\ \|u_i\|,\ \|\bar{u}_i\| \le \tilde{R},\ \|v\|,\ \|\bar{v}\| \le \tilde{r}$, where

$$\tilde{r} = \sup\{\tilde{z}(x) : x \in E\}, \quad \tilde{R} = \max_{1\le i\le m} \{ R_{1.i}\, R_{2.i}\, R_4 + R_{3.i} \}$$

and

$$R_{1.i} = \sup\{L_{p_i}(H_i(x)) : x \in E\}, \quad R_{2.i} = \sup\{\bar{h}_i(x) : x \in E\}$$

$$R_{3.i} = \sup\{\bar{g}_i(x) : x \in E\}, \quad R_4 = \sup\{\tilde{z}(x) : x \in E\},$$

and $\tilde{z}$ is defined in Theorem 8.3,

2) there are functions $d_i,\ \bar{d}_i,\ \hat{d}_i \in \Theta$, $1 \le i \le m$, such that the estimates

$$\|f_i(x,s,z) - f_i(\bar{x},\bar{s},\bar{z})\| \le d_i(\|x-\bar{x}\|) + \bar{d}_i(\|s-\bar{s}\|) + \hat{d}_i(\|z-\bar{z}\|),$$

are satisfied for $x,\ \bar{x},\ s,\bar{s}, \in E,\ \|z\|,\ \|\bar{z}\| \le \tilde{r},\ 1 \le i \le m,$

3) there are functions $\tilde{d}_i \in \Theta$, $1 \le i \le m$, such that

$$L_n[H_i(x+h) \div H_i(x)] \le \tilde{d}_i(\|h\|) \quad \text{for } 1 \le i \le \kappa,$$

and

$$L_{p_i}[H_i(x) \div (-t_i(x,h) + H_i(x+h))] \le \tilde{d}_i(\|h\|) \quad \text{for } \kappa+1 \le i \le m,$$

where $x,\ x+h \in E$,

4) there are $\bar{\omega}_i,\ \hat{\omega}_i \in \Theta$, $\kappa+1 \le i \le m$, such that $\|t_i(x,h)\| \le \bar{\omega}_i(\|h\|)$ for $\kappa+1 \le i \le m$, $\|h\| \in [0, r_0,]$ where r_0 is the diameter of the set E, and $\|\alpha_i(x+h) - \alpha_i(x)\| \le \hat{\omega}_i(\|h\|)$ for $\kappa+1 \le i \le m$, $x \in E$, $\|h\| \in [0,r_0]$.

Write $\tilde{\omega}_i(t) = \hat{\omega}_i(\bar{\omega}_i(t))$, $\kappa+1 \le i \le m$, $t \in [0,r_0]$, and $\tilde{\omega} = (\tilde{\omega}_{\kappa+1},\ldots,\tilde{\omega}_m)$. Now we prove a theorem on the existence and uniqueness of solutions of a certain functional equation.

Theorem 8.17 *Suppose that*

1) Assumption H_5 and condition 1), 3) of Assumption H_4 are satisfied,

2) the Lipschitz condition holds

$$\|F(x,u,v) - F(x,u,\bar{v})\| \le k(x)\|v-\bar{v}\|,\ x \in E,\ u \in X^m,\ v,\bar{v} \in X. \tag{8.32}$$

Then for any $y \in W$ where

$$W = \{ y \in C(E,X) : \|y(x)\| \le \tilde{z}(x) \ for\ x \in E \},$$

there exists the unique solution $u(\cdot\,,y) \in W$ of the equation

$$u(x) = F(x, \int_{H(x)} f(x,s,,y(\alpha(s)))\,ds,\ u(\beta(x))),\ x \in E. \tag{8.33}$$

PROOF. Put $u_0(x) = 0$ for $x \in E$ and

$$u_{r+1}(x) = F(x, \int_{H(x)} f(x,s,,y(\alpha(s)))\,ds,\ u_r(\beta(x))) \tag{8.34}$$

where $x \in E$, $r \geq 0$. We prove that

$$\| u_r(x) \| \leq \tilde{z}(x) \ \text{ for } \ x \in E, \ r \geq 0. \tag{8.35}$$

For $r = 0$ this inequality is evidently satisfied. If we assume that $\|u_r(x)\| \leq \tilde{z}(x)$ for $x \in E$, then

$$\| u_{r+1}(x) \| \leq \| F(x, \int_{H(x)} f(x,s,y(\alpha(s)))\,ds,\ 0) \|$$

$$+\| F(x, \int_{H(x)} f(x,s,y(\alpha(s)))\,ds,\ u_r(\beta(x))) - F(x, \int_{H(x)} f(x,s,y(\alpha(s)))\,ds,\ 0) \|$$

$$\leq k(x)\,\tilde{z}(\beta(x)) + K(x) \diamond \int_{H(x)} \tilde{z}(\alpha(s))\,ds + \tilde{h}(x) = \tilde{z}(x), \ \ x \in E.$$

Now we obtain (8.35) by induction. Next we prove that u_r are continuous functions on E. Since u_0 is continuous in E, it is sufficient to prove that the continuity of u_r implies the continuity of u_{r+1}. Let

$$\tilde{R}_1 = \max_{1 \leq j \leq m} \{ [\sup\{\bar{h}_j(x) : \ x \in E\}] \ [\sup\{ \|y(x)\| : \ x \in E\}]$$

$$+ \sup\{ \bar{g}_j(x) : \ x \in E\} \}.$$

By Assumption H_5 we have

$$\| u_{r+1}(x+h) - u_{r+1}(x) \| \leq D_0(\|h\|) + \sum_{j=1}^{m} D_j(\|\gamma_j(x,h)\|) \tag{8.36}$$

$$+D_{m+1}(\| u_r(\beta(x+h)) - u_r(\beta(x)) \|)$$

where

$$\gamma_j(x,h) = \int_{H_j(x+h)} f_j(x+h,s,y(\alpha_j(s)))\,(ds)_{p_j}$$

$$- \int_{H_j(x)} f_j(x,s,y(\alpha(s)))\,(ds)_{p_j}$$

and $1 \leq j \leq m$. Now for $1 \leq j \leq \kappa$ we get, writing ds instead of $(ds)_n$

$$\| \int_{H_j(x+h)} f_j(x+h,s,y(\alpha_j(s)))\,ds - \int_{H_j(x)} f_j(x,s,y(\alpha(s)))\,ds) \| \tag{8.37}$$

$$\leq L_n(E) d_j(\|h\|) + \tilde{R}_1\,\tilde{d}_j(\|h\|).$$

We define the sets $H_{1.j}(x,h)$, $H_{2.j}(x,h)$, $H_{3.j}(x,h)$, $\kappa+1 \le j \le m$, by

$$\begin{aligned} H_{1.j}(x,h) &= [\,-t_j(x,h) + H_j(x+h)\,] \cap H_j(x), \\ H_{2.j}(x,h) &= [\,-t_j(x,h) + H_j(x+h)\,] \setminus H_j(x), \\ H_{3.j}(x,h) &= H_j(x) \setminus [\,-t_j(x,h) + H_j(x+h)\,]. \end{aligned} \tag{8.38}$$

Then we have the estimates

$$\| \int_{H_j(x+h)} f_j(\,x+h, s, y(\alpha_j(s))\,)\,(ds)_{p_j} - \int_{H_j(x)} f_j(\,x, s, y(\alpha(s))\,)\,)\,(ds)_{p_j} \|$$

$$\le \int_{H_{1.j}(x,h)} \|f_j(\,x+h, s+t_j(x,h), y(\alpha_j(s+t_j(x,h)))\,)\,) - f_j(\,x,s,y(\alpha(s))\,)\,\|\,(ds)_{p_j}$$

$$+ \int_{H_{2.j}(x,h)} \|f_j(\,x+h, s+t_j(x,h), y(\alpha_j(s+t_j(x,h)))\,)\,)\,\|\,(ds)_{p_j}$$

$$+ \int_{H_{3.j}(x,h)} \|f_j(\,x,s,y(\alpha(s))\,)\,\|\,(ds)_{p_j}$$

$$\le L_{p_j}(\,H_j(x)\,)[\,d_j(\|h\|) + \bar{d}_j(\,\tilde{\omega}_j(\|h\|)\,) + \hat{d}_j(\,\hat{e}(\,\tilde{\omega}_j(\|h\|)\,)\,) + \tilde{R}_1 \tilde{d}_j(\|h\|)\,]$$

where $\hat{e}$ is the modulus of continuity for the function y. Then it follows from (8.36), (8.37) and from the continuity of y, u_r, α_j, β that u_{r+1} is continuous on E.

We put $z_0(x) = \tilde{z}(x)$ for $x \in E$ and $z_r(x) = k(x)\,z_{r-1}(\beta(x))$ for $x \in E$, $r \ge 1$. By induction we get

$$z_r(x) = k^{(r)}(x)\,\tilde{z}(\beta^{(r)}(x)), \; x \in E, \; r \ge 1. \tag{8.39}$$

In virtue of condition (8.10) of Theorem 8.4 it follows from (8.39) that

$$\lim_{r\to\infty} z_r(x) = 0 \text{ uniformly with respect to } x \in E. \tag{8.40}$$

Further, we get easily

$$\|\,u_{r+p}(x) - u_r(x)\,\| \le z_r(x), \; x \in E, \tag{8.41}$$

for $r, p \ge 0$. Indeed, from (8.34) and (8.35) it follows that (8.41) is satisfied for $r = 0$ and $p \ge 0$. If we assume that (8.41) holds for a fixed r and for $p \ge 0$, then we get

$$\|\,u_{r+1+p}(x) - u_{r+1}(x)\,\|$$

$$\le k(x)\,\|u_{r+p}(\beta(x)) - u_r(\beta(x))\,\| \le k(x)\,z_r(\beta(x)) = z_{r+1}(x).$$

Now we obtain (8.41) by induction. By (8.34), (8.40), (8.41) we infer that the sequence $\{u_r\}$ is uniformly convergent in E to the solution $\bar{u}$ of equation (8.33). Since $u_r \in C(E,X)$, we conclude by (8.35) that $\bar{u} \in W$. To prove that the solution $\bar{u}$ of (8.33) is unique in W, let us suppose that there exists another solution $\tilde{u} \ne \bar{u}$

and $\tilde{u} \in W$. Now, from (8.32) we have $\|\bar{u}(x) - \tilde{u}(x)\| \le k(x)\,\|\bar{u}(\beta(x)) - \tilde{u}(\beta(x))\|$, $x \in E$, and by induction we get

$$\|\bar{u}(x) - \tilde{u}(x)\| \le k^{(r)}(x)\,\|\bar{u}(\beta^{(r)}(x)) - \tilde{u}(\beta^{(r)}(x))\|,\ x \in E,\ r \ge 0. \qquad (8.42)$$

Since $\lim_{r\to\infty} k^{(r)}(x)\,\tilde{z}(\beta^{(r)}(x)) = 0$ uniformly with respect to $x \in E$, we infer by (8.42) that $\bar{u} = \tilde{u}$. This contradiction proves the uniqueness of $\bar{u}$ in W. The proof of Theorem 8.17 is completed.

Assumption $\mathbf{H}_6$. Suppose that
1) for $x \in E$, $\delta = (\delta_1, \ldots, \delta_m) \in R^m_+$, we have

$$m_0(x,\delta) = \sum_{i=0}^{\infty} k^{(i)}(x) \sum_{j=1}^{m} D_j(\,\delta_j L_{p_j}(\,E_j(\beta^{(i)}(x))\,)\,) < +\infty,$$

2) the function m_0 is continuous with respect to $(x,\delta) \in E \times R^m_+$.

Remark 8.18 *It is obvious that Assumption H_6 is fulfilled if, for instance,*

$$k(x) \le \bar{k} < 1 \ \text{ for } \ x \in E.$$

If condition 1) of H_6 is satisfied and the functions k, β, α_j are nondecreasing on E then condition 2) of H_6 is fulfilled.

Remark 8.19 *If the functions φ_i, $1 \le i \le m$, and β satisfy conditions 2) and 3) of Lemma 8.11 respectively, and*

$$k(x) \le \bar{k} = \text{const},\ D_j(t) \le D\,t,\ D = \text{const},\ \textit{and}\ \bar{k} \prod_{r\in\sigma_j} \bar{\beta}_r < 1 \ \textit{for}\ 1 \le j \le m,$$

then Assumption H_6 is fulfilled, (see the proof of Lemma 8.11).

We adopt the following notations:

$$\bar{K} = (\,K_{\kappa+1}, \ldots, K_m),\ \bar{L}(\,E(x)\,) = (L_{_{\kappa+1}}(\,E_{_{\kappa+1}}(x)\,), \ldots, L_{p_m}(\,E_m(x)\,)\,),$$

$$\bar{K}(x) \diamond \bar{L}(\,E(x)\,) = \sum_{j=\kappa+1}^{m} K_j(x)\, L_{p_j}(\,E_j(x)\,),$$

$$\int_{\bar{H}(x)} d(\,\alpha(s),\, \tilde{\omega}(t)\,)\, ds =$$

$$\left(\int_{H_{\kappa+1}(x)} d(\,\alpha_{\kappa+1}(s),\, \tilde{\omega}_{\kappa+1}(t))\,)\,(ds)_{p_{\kappa+1}}, \ldots, \int_{H_m(x)} d(\,\alpha_m(s),\, \tilde{\omega}_m(t)\,)\,(ds)_{p_m} \right)$$

and

$$\bar{K}(x) \diamond \int_{\bar{H}(x)} d(\,\alpha(s),\, \tilde{\omega}(t)\,)\, ds = \sum_{j=\kappa+1}^{m} K_j(x) \int_{H_j(x)} d(\,\alpha_j(s),\, \tilde{\omega}_j(t)\,)\,(ds)_{p_j},$$

where $\tilde{\omega} = (\tilde{\omega}_{\kappa+1}, \ldots, \tilde{\omega}_m)$.

Assumption H$_7$. Suppose that

1) there is $\omega \in \Theta$ such that $\|\beta(x+h) - \beta(x)\| \leq \omega(\|h\|)$ for $x, x+h \in E$,

2) there is a function $p : E \times [0, r_0] \to R_+$ such that

(i) $p \in C(E \times [0, r_0], R_+)$, where $r_0 \in R_+$ is defined in condition 4) of Assumption H$_5$,

(ii) $p(x, \cdot) \in \Theta$ for every $x \in E$,

3) for $(x,t) \in E \times [0, r_0]$ we have

$$\bar{m}(x,t) = \sum_{i=0}^{\infty} k^{(i)}(x)\, p(\, \beta^{(i)}(x), \omega^{(i)}(t)\,) < +\infty,$$

where $\omega^{(0)}(t) = t,\ \omega^{(i+1)}(t) = \omega(\omega^{(i)}(t)),\ i \geq 0,\ t \in [0, r_0]$,

4) for $x \in E$ we have

$$\bar{M}(x) = \sum_{i=0}^{\infty} k^{(i)}(x) \left[\bar{K}(\beta^{(i)}(x)) \diamond \bar{L}(\, E(\beta^{(i)}(x))\,) \right] < +\infty,$$

4) $\bar{m} \in C(E \times [0, r_0], R_+)$ and $\bar{M} \in C(E, R_+)$,

5) the function

$$M^*(x) = \sum_{i=0}^{\infty} k^{(i)}(x) \left[\sum_{j=\kappa+1}^{m} K_j(\beta^{(i)}(x))\, L_{p_j}(\, E(\beta^{(i)}(x))\,) \left[\prod_{r \in \sigma_j} x_r \right]^{-1} \right]$$

is bounded on E.

Theorem 8.20 *If Assumption H_7 is satisfied and condition 2) of Assumption H_1 holds then:*

1. There exists a solution $\tilde{d} \in C(E \times [0, r_0], R_+)$ of the equation

$$d(x,t) = \sum_{i=0}^{\infty} k^{(i)}(x)\, p(\, \beta^{(i)}(x), \omega^{(i)}(t)\,) \tag{8.43}$$

$$+ \sum_{i=0}^{\infty} k^{(i)}(x) \left[\bar{K}(\beta^{(i)}(x)) \diamond \int_{\bar{H}(\beta^{(i)}(x))} d(\, \alpha(s), \bar{\omega}(\omega^{(i)}(t))\,)\, ds \right]$$

where $(x,t) \in E \times [0, r_0]$. The solution $\tilde{d}$ of (8.43) is unique in the class $M(E \times [0, r_0], R_+)$ of nonnegative and upper-semicontinuous functions defined on $E \times [0, r_0]$. The function $\tilde{d}$ is nondecreasing and subadditive with respect to the last variable and $\tilde{d}(x, 0) = 0$ for $x \in E$.

2. The function $\tilde{d}$ is a solution of the equation

$$d(x,t) = \bar{K}(x) \int_{\bar{H}(x)} d(\, \alpha(s), \tilde{\omega}(t)\,)\, ds + k(x)\, d(\beta(x),\, \omega(t)) + p(x,t), \tag{8.44}$$

where $(x,t) \in E \times [0,r_0]$. *Moreover, this solution is unique in the class* $\tilde{M}(E \times [0,r_0], R_+, \tilde{d})$, *where*

$$\tilde{M}(E \times [0,r_0], R_+, \tilde{d})$$
$$= \Big\{ z \in M(E \times [0,r_0], R_+),\ \inf\{c:\ z(x,t) \le c\,\tilde{d}(x,t)\} < +\infty \Big\}.$$

The proof of this Theorem is similar to the proof of Theorems 8.3 and 8.4. We omit details.

8.6 Properties of the operator U.

Let

$$\tilde{W} = \{\, y \in C(E,X):\ \|y(x)\| \le \tilde{z}(x),\ \|y(x+h) - y(x)\| \le \tilde{d}(x, \|h\|) \,\}$$

where the function $\tilde{z}$ is given by Theorem 8.3 with $\tilde{h}$, K defined in Assumption H_4 and $\tilde{d}$ is given by Theorem 8.20. We consider the operator U defined by the formula: $Uy = u(\,\cdot\,, y)$, where $u(\,\cdot\,, y)$ is the solution of functional equation (8.33).

Theorem 8.21 *If Assumptions H_5, H_6, conditions 1) and 3) from Assumption H_4 and the Lipschitz condition (8.32) are satisfied, then the operator U is continuous on the set W.*

Proof. Let $y_1, y_2 \in W$, $u_1 = u(\,\cdot\,, y_1)$, $u_2 = u(\,\cdot\,, y_2)$ and $v(x) = \|u_1(x) - u_2(x)\|$, $\delta_j = \hat{d}_j(\sup\{\|y_1(x) - y_2(x)\| : x \in E\})$, $1 \le j \le m$. Then we have the inequality

$$v(x) \le \sum_{j=1}^{m} D_j(\,\delta_j(L_{p_j}(E_j(x))\,) + k(x)\,v(\beta(x)),\ x \in E,$$

and we get by induction

$$v(x) \le \sum_{i=0}^{r-1} k^{(i)}(x) \left[\sum_{j=1}^{m} D_j(\,\delta_j L_{p_j}(\,E_j(\beta^{(i)}(x))\,)\,) \right] \tag{8.45}$$
$$+k^{(r)}(x)\,v(\beta^{(r)}(x)),\ x \in E,\ r \ge 1.$$

Since

$$0 \le k^{(r)}(x)\,v(\beta^{(r)}(x)) \le 2\,k^{(r)}(x)\,\tilde{z}(\beta^{(r)}(x)),\ x \in E,\ r \ge 0,$$

and $\lim_{r\to\infty} k^{(r)}(x)\,\tilde{z}(\beta(x)) = 0$ uniformly with respect to $x \in E$, we get, making $r \to \infty$ in (8.45), that

$$v(x) \le \sum_{i=0}^{\infty} k^{(i)}(x) \left[\sum_{j=1}^{m} D_j\left(\delta_j L_{p_j}(\,E_j(\beta^{(i)}(x))\,)\right) \right] = m_0(x,\delta).$$

In view of the continuity of the function m_0 the assertion of Theorem 8.21 follows.

Lemma 8.22 *Suppose that*

1) Assumptions H_4, H_5, H_6, are satisfied and Assumption H_7 is fulfilled for p, $\bar{K}$ defined by the relations

$$p(x,t) = D_0(t) + \sum_{j=1}^{\kappa} D_j(L_n(E)d_j(t) + P_j\tilde{d}_j(t)) \tag{8.46}$$

$$+ \sum_{j=\kappa+1}^{m} l_j(x) \left[L_{p_j}(H_j(x))\,(d_j(t) + \bar{d}_j(\bar{\omega}_j(t))\,) + P_j\tilde{d}_j(t) \right]$$

where

$$P_j = \left[\sup\{\bar{h}_j(x) : \; x \in E\}\right] \left[\sup\{\tilde{z}(x) : \; x \in E\}\right] + \sup\{\bar{g}_j(x) : \; x \in E\},$$

and

$$\bar{K}(x) = (\,l_{\kappa+1}(x)\,\bar{l}_{\kappa+1}(x), \ldots, l_m(x)\,\bar{l}_m(x)\,), \; x \in E, \tag{8.47}$$

2) for x, $x+h \in E$ we have

$$\lim_{r\to\infty} k^{(r)}(x)\,\tilde{z}(\beta^{(r)}(x+h)) = 0 \; uniformly\, with\, respect\, to\; x,\; x+h \in E. \tag{8.48}$$

Under these assumptions the operator U maps $\tilde{W}$ into itself.

PROOF. In virtue of Theorem 8.17 it follows that for each $y \in \tilde{W}$ the function Uy satisfies the condition $\|(Uy)(x)\| \leq \tilde{z}(x)$ for $x \in E$. To prove that $Uy \in \tilde{W}$ for $y \in \tilde{W}$, it is sufficient to show that $\|(Uy)(x+h) - (Uy)(x)\| \leq \tilde{d}(x, \|h\|)$ for x, $x+h \in E$. Let us suppose that $y \in \tilde{W}$ and $u(x) = (Uy)(x)$. We show that

$$\|u(x+h) - u(x)\| \leq \tilde{d}(x, \|h\|) \;\text{ for } x,\; x+h \in E. \tag{8.49}$$

For $1 \leq j \leq \kappa$ we get, writing ds instead of $(ds)_n$ for simplicity,

$$\| \int_{H_j(x+h)} f_j(x+h, s, y(\alpha_j(s))\,)\,ds - \int_{H_j(x)} f_j(x, s, y(\alpha_j(s))\,)\,ds \,\|$$

$$\leq L_n(E)\,d_j(\|h\|) + P_j\tilde{d}(\|h\|).$$

In the case $\kappa+1 \leq j \leq m$ we consider the sets $H_{k.j}(x,h)$, $k = 1,2,3$, given by (8.38) an note that integration over the set $H_j(x+h)$ is equivalent to integration over $-t_j(x,h) + H_j(x+h)$ if we replace the variable s by $s + t_j(x,h)$. In this case we arrive at

$$\| \int_{H_j(x+h)} f_j(\,x+h, s, y(\alpha_j(s))\,)\,(ds)_{p_j} - \int_{H_j(x)} f_j(\,x, s, y(\alpha_j(s))\,)\,(ds)_{p_j} \,\|$$

$$\leq L_{p_j}(H_j(x))\left[\,d_j(\|h\|) + \bar{d}_j(\,\bar{\omega}_j(\|h\|)\,)\,\right]$$

$$+\bar{l}_j(x) \int_{H_j(x)} \|\,y(\alpha_j(\,s + t_j(x,h))\,) - y(\,\alpha_j(s)\,)\,\|\,(ds)_{p_j} + P_j\tilde{d}_j(\|h\|).$$

It follows from Assumption H_5 and from the above estimates that

$$\|u(x+h)-u(x)\| \leq D_0(\|h\|)+\sum_{j=1}^{\kappa} D_j\left(\|\gamma_j(x,h)\|\right)$$

$$+\sum_{j=\kappa+1}^{m} l_j(x)\,\|\,\gamma_j(x,h)\,\|+k(x)\,\|u(\beta(x+h))-u(\beta(x))\|$$

$$\leq p(x,\|h\|)+\sum_{j=\kappa+1}^{m} K_j(x)\int_{H_j(x)}\|y(\alpha_j(\,s+t_j(x,h))\,)-y(\,\alpha_j(s)\,)\,\|\,(ds)_{p_j}$$

$$+k(x)\,\|\,u(\beta(x+h))-u(\beta(x))\,\|.$$

Since $\|y(x+h)-y(x)\| \leq \tilde{d}(x,\|h\|)$, we have $\|\,y(\alpha_j(s+t_j(x,h))\,)-y(\alpha_j(s))\,\| \leq \tilde{d}(\alpha_j(s),\tilde{\omega}_j(\|h\|)\,)$ and consequently

$$\|u(x+h)-u(x)\| \leq p(x,\|h\|)+\bar{K}(x)\diamond\int_{\bar{H}(x)}\tilde{d}(\,\alpha(s),\tilde{\omega}(\|h\|)\,)\,ds$$

$$+k(x)\|u(\beta(x+h))-u(\beta(x))\,\|.$$

The last inequality implies the following:

$$\|u(x+h)-u(x)\| \leq \sum_{i=0}^{r-1} k^{(i)}(x)\,p(\,\beta^{(i)}(x),\|\beta^{(i)}(x+h)-\beta^{(i)}(x)\|\,) \tag{8.50}$$

$$+\sum_{i=0}^{r-1} k^{(i)}(x)\left[\bar{K}(\beta^{(i)}(x))\diamond\int_{\bar{H}(\beta^{(i)}(x))}\tilde{d}(\,\alpha(s),\tilde{\omega}(\,\|\beta^{(i)}(x+h)-\beta^{(i)}(x)\,\|)\,)\,ds\right]$$

$$+k^{(r)}(x)\,\|\,u(\beta^{(r)}(x+h))-u(\beta^{(r)}(x))$$

where $x, x+h\in E,\ r\geq 1$. It follows from the inequalities

$$k^{(r)}(x)\,\|\,u(\beta^{(r)}(x+h))-u(\beta^{(r)}(x))\,\|$$

$$\leq k^{(r)}(x)\,\tilde{z}(\beta^{(r)}(x))+k^{(r)}(x)\,\tilde{z}(\,\beta^{(r)}(x+h)\,),\ \ x,x+h\in E,\ r\geq 0,$$

and from conditions (8.10), (8.48) that

$$\lim_{r\to\infty} k^{(r)}(x)\,\|\,u(\beta^{(r)}(x+h))-u(\beta^{(r)}(x))\,\|=0 \text{ uniformly on } E. \tag{8.51}$$

By induction we easily obtain

$$\|\beta^{(i)}(x+h)-\beta^{(i)}(x)\| \leq \omega^{(i)}(\|h\|),\ \ x,x+h\in E,\ i\geq 0. \tag{8.52}$$

Now, from (8.51), (8.52) and by the definition of $\tilde{d}$, we have, letting $r\to\infty$ in (8.50)

$$\|u(x+h)-u(x)\| \leq \sum_{i=0}^{\infty} k^{(i)}(x)\,p(\,\beta^{(i)}(x),\omega^{(i)}(\|h\|)\,)$$

$$+\sum_{i=0}^{\infty} k^{(i)}(x)\left[\bar{K}(\beta^{(i)}(x))\diamond\int_{\bar{H}(\beta^{(i)}(x))}\tilde{d}(\,\alpha(s),\tilde{\omega}(\omega^{(i)}(\|h\|))\,)\,ds\right]=\tilde{d}(x,\|h\|),$$

which proves (8.49). Then the proof of Lemma 8.22 is completed.

Remark 8.23 *If the function k, $\tilde{h}$, K, β are nondecreasing on E and $H_j(x) \subset H_j(\bar{x})$ for $x < \bar{x}$, $x, \bar{x} \in E$, $1 \le j \le m$, then assumption 2) of Lemma 8.22 is satisfied. This fact follows from assertion 4) of Theorem 8.3 and from (8.10).*

Now we formulate the main result in this Section.

Theorem 8.24 *Suppose that Assumptions H_4, H_5, H_6 are satisfied and*

1) Assumption H_7 is fulfilled for p, $\bar{K}$ defined by relations (8.46), (8.47),

2) condition (8.48) of Lemma 8.22 holds.

Under these assumptions equation (8.1) has at least one solution $\tilde{u} \in \tilde{W}$.

PROOF. It follows from Theorems 8.17 and 8.21 and from Lemma 8.22 that the continuous operator U maps the compact and convex set $\tilde{W} \subset C(E, X)$ into itself. By the Schauder fixed point theorem there exists at least one solution $\tilde{u} \in \tilde{W}$ of equation (8.1).

Lemma 8.25 *If*

1) $k(x) \le \bar{k} =$ const, and

$$\bar{K}(x) = (\, K_{\kappa+1}(x), \ldots, K_m(x)\,) \le (\, \bar{K}_{\kappa+1}, \ldots, \bar{K}_m \,) = \text{const},$$

2) the functions φ_i, $1 \le i \le m$, and β satisfy conditions 2) and 3), respectively, of Lemma 8.11,

3) there exist constants ω_0 and D such that

$$D_i(t),\; d_j(t),\; \tilde{d}_j(t),\; \bar{d}_r(t),\; \bar{\omega}_r(t) \le D\, t,$$

where $1 \le i \le \kappa$, $1 \le j \le m$, $\kappa + 1 \le r \le m$, and $\omega(t) \le \omega_0\, t$,

4) $\bar{k}\, \omega_0 < 1$ and for $\kappa + 1 \le j \le m$ we have

$$\bar{k} \prod_{r \in \sigma_j} \bar{\beta}_r < 1,$$

then conditions 3) - 5) of Assumption H_7 are fulfilled.

The proof of this Lemma is similar to the proof of Lemma 8.11. Using this Lemma we can easily formulate a theorem which is more effective than Theorem 8.24.

Remark 8.26 *Theorems 8.9 and 8.24 can be extended on equations of the type*

$$u(x) = F(\,x, \int_{H(x)} f(\,x, s, u(\alpha(s))\,)\, ds,\; u(\,\beta(x))\,)$$

with $u(\,\beta(x)\,) = u(\,\beta_1(x)\,), \ldots, u(\,\beta_k(x)\,)$ and $\beta = (\beta_1, \ldots, \beta_k) : E \to E^k$, $k \ge 1$.

8.7 Nonlinear functional integral equations

Let $a = (a_1, \dots, a_n)$, $a_i > 0$ for $1 \le i \le n$ and $r = (r_1, \dots, r_n) \in R^n_+$ be fixed. Let X be the Banach space with the norm $\| \cdot \|$. Write $E = [0, a]$, $B = [-r, 0]$, $E_0 = [-r, a] \setminus (0, a]$. For a function $u : [-r, a] \to X$ and $x \in [0, a]$ we define the function $z_{(x)} : B \to X$ by $u_{(x)}(s) = u(x + s)$, $s \in B$. Let $\| \cdot \|_B$ be the supremum norm in the space $C(B, X)$. Assume that functions

$$F : E \times X^m \times X \to X, \quad f = (f_1, \dots, f_m) : E \times E \times C(B, X) \to X^m$$

and $\phi : E_0 \to X$, $\beta : E \to E$ are given. Suppose that the sets $(H_1(x), \dots, H_m(x))$, $H_j(x) \subset E$, satisfy Assumption H_1. Write

$$\int_{H(x)} f(x, s, u_{(s)})\, ds$$

$$= \left(\int_{H_1(x)} f(x, s, u_{(s)})\, (ds)_{p_1}, \dots, \int_{H(x)} f(x, s, u_{(s)})\, (ds)_{p_m} \right).$$

Consider the nonlinear integral functional equation

$$u(x) = F(x, \int_{H(x)} f(x, s, u_{(s)})\, ds, u_{(\beta(x))}), \ x \in E, \tag{8.53}$$

with the boundary condition

$$u(x) = \phi(x) \ \text{ for } \ x \in E_0. \tag{8.54}$$

Existence and uniqueness results given in Theorem 8.9 can be easily extended for problem (8.53), (8.54).

Assumption H$_8$. Suppose that the functions F, f, α, β, ϕ are continuous and

1) the sets $(H_1(x), \dots, H_m(x))$ satisfy Assumption H_1 with $E = [0, a]$ and $\alpha_i(x) = x$, $1 \le i \le m$, $\beta(x) \le x$ for $x \in E$,

2) there exist functions $l = (l_1, \dots, l_m) \in C(E, R^m_+)$ and $k \in C(E, R)$ such that

$$\| F(x, u, v) - F(x, \bar{u}, \bar{v}) \| \le \sum_{i=1}^{m} l_i(x) \|u_i - \bar{u}_i\| + k(x) \|v - \bar{v}\|$$

where $u = (u_1, \dots, u_m)$, $\bar{u} = (\bar{u}_1, \dots, \bar{u}_m)$, $u, \bar{u} \in X^m$, $v, \bar{v} \in X$, $x \in E$,

3) there exists a function $\bar{l} = (\bar{l}_1, \dots, \bar{l}_m) \in C(E, R^m_+)$ such that

$$\|f_i(x, s, w) - f_i(x, s, \bar{w})\| \le \bar{l}_i \|w - \bar{w}\|_B, \ \ x, s \in E, \ w, \bar{w} \in C(B, X),$$

4) the consistency condition holds true

$$\phi(x, y) = F(x, \int_{H(x)} f(x, s, \phi_{(s)})\, ds, \ \phi_{(\beta(x))}), \ x \in E \cap E_0,$$

5) let $u_0 \in C([-r,a],X)$ be such a function that $u_0(x) = \phi(x)$ on E_0. Write

$$\tilde{h}(x) = \| F(x, \int_{H(x)} f(x,s,(u_0)_{(s)})\,ds, (u_0)_{(\beta(x))}) - u_0(x) \|,\ x \in E.$$

Suppose that Assumption H_2 is satisfied for the above $\tilde{h}$, $\alpha_i(x) = x$, $1 \le i \le m$, $x \in E$, and for K given by $K(x) = (\, l_1(x)\,\bar{l}_1(x), \dots, l_m(x)\,\bar{l}_m(x)\,)$, $x \in E$.

Theorem 8.27 *If Assumption H_8 is satisfied then there exists a solution $\tilde{u} \in C([-r,a],X)$ of problem (8.53), (8.54) such that*

$$\|\tilde{u}(x) - u_0(x)\| \le \tilde{z}(x),\ x \in E,$$

$$\|\tilde{u}(x) - u_r(x)\| \le w_r(x),\ x \in E,\ r \ge 0,$$

where $\{\, w_r \,\}$ is defined by (8.20) and $\{\, u_r \,\}$ is given by

$$u_{r+1}(x) = F(x, \int_{H(x)} f(x,s,(u_r)_{(s)})\,ds, (u_r)_{(\beta(x))}),\ x \in E,$$

$$u_r(x) = \phi(x) \text{ for } x \in E_0$$

with u_0 satisfying condition 5) of Assumption H_8.

The proof of the above theorem is similar to the proof of Theorem 8.9. Details can be found in [61].

Remark 8.28 *The existence Theorem 8.24 can be extended on problem (8.53), (8.54). Details can be found in [65, 66].*

Now we give an example of a functional differential problem which can be transformed into (8.53), (8.54). Consider the Darboux problem for the hyperbolic functional differential equation of the neutral type with two independent variables

$$D_{xy}z(x,y) = F(x,y,z_{(x,y)},(D_xz)_{(x,y)},(D_yz)_{(x,y)},(D_{xy}z)_{(\beta(x,y))}) \tag{8.55}$$

where $(x,y) \in [0,a]$ $a = (a_1,a_2)$, with the boundary condition

$$z(x,y) = \phi(x,y),\quad (x,y) \in E_0, \tag{8.56}$$

where $E_0 = [-r,a] \setminus (0,a]$, $r = (r_1,r_2)$ and $\beta = (\beta_1,\beta_2)$. Put $D_{xy}z(x,y) = u(x,y)$. Then u is a solution of

$$u(x,y) = F(x,y,\psi_{(0.x,y)} + \int_{H_0(x,y)} u_{(s,t)}\,dsdt, \tag{8.57}$$

$$\psi_{(1.x,y)} + \int_{H_1(x,y)} u_{(s,t)}\,dt,\ \psi_{(2.x,y)} + \int_{H_2(x,y)} u_{(s,t)}\,dt,\ u_{(\beta(x,y))}),$$

$$u(x,y) = D_{xy}\psi(x,y) \text{ on } E_0, \tag{8.58}$$

where

$$\psi_{(0.x,y)} = -\phi_{(0,0)} + \phi_{(x,0)} + \phi_{(0,y)},\ \ \psi_{(1.x,y)} = (D_x\phi)_{(x,0)},\ \ \psi_{(2.x,y)} = (D_y\phi)_{0,y)}$$

and

$$H_0(x,y) = [0,x] \times [0,y],\ \ H_1(x,y) = \{\,x\,\} \times [0,y],\ \ H_2(x,y) = [0,x] \times \{\,y\,\}.$$

If given functions in (8.57), (8.58) are continuous the problems (8.55) (8.56) and (8.57), 8.58) are equivalent.

8.8 Discretization of the Darboux problem

Take a, $b > 0$ and a_0, $b_0 \in R_+$. Define $E = (0,a]\times(0,b]$, $E_0 = ([-a_0,a] \times [-b_0,b])\setminus E$ and $B = [-a_0,0] \times [-b_0,0]$. Given a function $z : E_0 \cup E \to R$ and a point $(x,y) \in E$, we define the function $z_{(x,y)} : B \to R$ by

$$z_{(x,y)}(t,s) = z(x+t, y+s),\ \ (t,s) \in B. \tag{8.59}$$

Suppose that $f : \bar{E} \times C(B,R) \to R$ and $\phi : E_0 \to R$ are given functions. Consider the Darboux problem

$$D_{xy}z(x,y) = f(x,y,z_{(x,y)}),\ \ (x,y) \in E, \tag{8.60}$$

$$z(x,y) = \phi(x,y)\ \text{ for }\ (x,y) \in E_0 \tag{8.61}$$

We consider classical solutions of problem (8.60), (8.61). A function $v \in C(E_0 \cup E, R)$ is regarded as a solution of (8.60), (8.61) if $D_{xy}v$ exists on E, $D_{xy}v \in C(E,R)$, and v satisfies (8.60), (8.61).

For a few recent years, a certain number of papers concerning numerical methods for functional partial differential equations have been published. Difference methods for nonlinear parabolic functional differential problems were considered in [145, 161, 162]. The main problem in these investigations is to find a difference approximation which is stable and satisfies consistency conditions with respect to the original problem. A method of difference inequalities or simple theorems on linear recurrent inequalities are used in the investigation of the stability. The semidiscretizations in space variables of linear parabolic Volterra integral differential equations (the method of lines, Galerkin or collocation techniques) lead to large systems of stiff ordinary integral functional equations. The analysis of spatial and temporal discretizations of linear integral functional parabolic problems has received considerable attention during a few last years [43, 86, 130, 191, 234, 235]. Most of these contributions seem to focus on the convergence theory. There are very few numerical studies. Difference methods for first order functional differential equations with initial or initial-boundary conditions were studied in [28, 106, 111]. The proofs of the convergence were based either on functional difference inequalities or on a general theorem on error estimates for approximate solutions of functional difference equations of the Volterra type with initial or initial-boundary conditions and with unknown function of several variables.

Convergence of difference methods for functional hyperbolic systems in the Schauder canonic form was studied in [148]. For further bibliographical information concerning numerical methods for functional partial differential equations we suggest to see the survey paper [33] and [111, 117, 151]. Our presentation is organized as follows. In this Section we construct a general class of difference schemes for (8.60), (8.61). This leads to implicit functional difference problems. The existence and uniqueness of solutions of such problems are considered in Section 8.9. The comparison method of the investigation of functional difference equations is used. The next Section deals with a theorem on the convergence of explicit difference schemes with nonlinear estimates for given functions. We assume that increment functions satisfy the nonlinear estimates of the Perron type with respect to the functional variable. In Section 8.11, we establish an error estimate for implicit difference methods. A numerical example can be found in [117]. Differential equations with a deviated argument and integral differential problems can be obtained from (8.60), (8.61) by a specification of given operators.

We construct a mesh in $E_0 \cup E$ in the following way. Suppose that $(h,k) \in (0,a] \times (0,b]$ stand for steps of the mesh. Write $x_i = ih$, $y_j = jk$, $x_{i+\frac{1}{2}} = ih + \frac{h}{2}$, $y_{j+\frac{1}{2}} = jk + \frac{k}{2}$, $i, j \in \mathbf{Z}$. Denote by Δ the set of all $(h,k) \in (0,a] \times (0,b]$ such that there exist $M_0, N_0 \in \mathbf{N}$ such that $M_0 h = a_0$, $N_0 k = b_0$. We assume that $\Delta \neq \emptyset$ and that there is a sequence $\{(h_n, k_n)\}$, $(h_n, k_n) \in \Delta$, such that $\lim_{n\to\infty}(h_n, k_n) = (0,0)$. For $(h,k) \in \Delta$ we put $Z_{hk} = \{(x_i, y_j) : \ i, j \in \mathbf{Z}\}$, and $E^0_{hk} = Z_{hk} \cap E_0$, $E_{hk} = Z_{hk} \cap E$. There are $M, N \in \mathbf{N}$ such that $Mh \le a < (M+1)h$, $Nk \le b < (N+1)k$. Let $A_{hk} = \{(x_i, y_j) : 0 \le i \le M-1,\ 0 \le j \le N-1\}$. Using the above definitions of M and N, we have $E_{hk} = \{(x_i, y_j) : 1 \le i \le M,\ 1 \le j \le N\}$. Let $K, L \in \mathbf{Z}$ be fixed and assume that $-M_0 \le K \le 1$, $-N_0 \le L \le 1$. Write

$$D_{hk} = \{(x_i, y_j) : \ -M_0 \le i \le K,\ -N_0 \le j \le L\}.$$

For $z \in \mathbf{F}(E^0_{hk} \cup E_{hk}, R)$ we write $z^{(i,j)} = z(x_i, y_j)$, $(x_i, y_j) \in E^0_{hk} \cup E_{hk}$. In the same way we define $w^{(i,j)}$ for $w \in \mathbf{F}(D_{hk}, R)$. We will need a discrete version of the restriction operator given by (8.59). If $z : E^0_{hk} \cup E_{hk} \to R$ and $0 \le i \le M-1$, $0 \le j \le N-1$, then the function $z_{[i,j]} : D_{hk} \to R$ is defined as follows: $z_{[i,j]}(t,s) = z(x_i + t, y_j + s)$, $(t,s) \in D_{hk}$. We consider the difference operator δ given by

$$\delta z^{(i,j)} = \frac{1}{hk}\left[z^{(i+1,j+1)} - z^{(i+1,j)} - z^{(i,j+1)} + z^{(i,j)} \right].$$

Suppose that $F_{hk} : A_{hk} \times \mathbf{F}(D_{hk}, R) \to R$, $\phi_{hk} : E^0_{hk} \to R$ are given functions. Consider the problem

$$\delta z^{(i,j)} = F_{hk}(x_i, y_j, z_{[i,j]}), \quad (x_i, y_j) \in A_{hk}, \tag{8.62}$$

$$z^{(i,j)} = \phi^{(i,j)}_{hk} \text{ for } (x_i, y_j) \in E^0_{hk}. \tag{8.63}$$

Remark 8.29 *If $K = 1$ and $L = 1$ then problem (8.62), (8.63) turns out to be an implicit difference method. If $K \le 0$ or $L \le 0$, then problem (8.62), (8.63) represents a simple functional difference equation of the Volterra type. It is obvious that in this case there exists exactly one solution $u_{hk} : E^0_{hk} \cup E_{hk} \to R$ of (8.62), (8.63).*

Example 8.30 *Suppose that* $F : \bar{E} \times R \to R$, $\varphi : [0,a] \to R$, $\psi : [0,b] \to R$ *are given functions. We assume that* $\varphi(0) = \psi(0)$. *Put* $a_0 = 0$, $b_0 = 0$ *and* $f(x,y,w) = F(x,y,w(0,0))$, $(x,y,w) \in \bar{E} \times C(B,R)$. *Then (8.60), (8.61) reduces to the classical Darboux problem*

$$D_{xy}z(x,y) = F(x,y,z(x,y)), \quad (x,y) \in E, \tag{8.64}$$

$$z(x,0) = \varphi(x) \text{ for } x \in [0,a], \quad z(0,y) = \psi(y) \text{ for } y \in [0,b]. \tag{8.65}$$

One of the implicit difference schemes for (8.64), (8.65) takes the form

$$\delta z^{(i,j)} = F(x_{i+\frac{1}{2}}, y_{j+\frac{1}{2}}, \frac{1}{4}(z^{(i,j)} + z^{(i+1,j)} + z^{(i,j+1)} + z^{(i+1,j+1)}))$$

where $(x_i,y_j) \in A_{hk}$, *and* $z^{(i,0)} = \varphi(x_i)$ *for* $0 \le i \le M$, $z^{(0,j)} = \psi(y_j)$ *for* $0 \le j \le N$. *The most natural explicit difference method for (8.64), (8.65) takes the form*

$$\delta z^{(i,j)} = F(x_i, y_j, z^{(i,j)}), \quad (x_i,y_j) \in A_{hk},$$

with the above boundary condition.

Example 8.31 *Suppose that* $a_0 < 0$, $b_0 < 0$. *For the same* F *we put*

$$f(x,y,w) = F(x,y,\int_B w(t,s)\,dtds), \quad (x,y,w) \in \bar{E} \times C(B,R).$$

Then problem (8.60), (8.61) is equivalent to the integral differential equation

$$D_{xy}z(x,y) = F(x,y,\int_B z(x+t,y+s)\,dt\,ds), \quad (x,y) \in E, \tag{8.66}$$

with boundary condition (8.61). Now, we construct an explicit difference method for (8.66), (8.61). Let

$$B_{hk} = \{(x_i,y_j) : -M_0 \le i \le 0, \ -N_0 \le j \le 0\}.$$

We define the operator $T_{hk} : \mathbf{F}(B_{hk},R) \to \mathbf{F}(B,R)$ *in the following way. Suppose that* $w \in \mathbf{F}(B_{hk},R)$ *and* $(t,s) \in B$. *Then there is* $(x_i,y_j) \in B_{hk}$ *such that* $i < 0$, $j < 0$ *and* $x_i \le t \le x_{i+1}$, $y_j \le s \le y_{j+1}$. *We put*

$$T_{hk}w(t,s) = w^{(i,j)}\,[1 - \frac{t-x_i}{h}]\,[1 - \frac{s-y_j}{k}] + w^{(i+1,j+1)}\,\frac{t-x_i}{h}\,\frac{s-y_j}{k} \tag{8.67}$$

$$+w^{(i+1,j)}\,\frac{t-x_i}{h}\,[1 - \frac{s-y_j}{k}] + w^{(i,j+1)}\,[1 - \frac{t-x_i}{h}]\,\frac{s-y_j}{k}.$$

It is easy to see that $T_{hk} : \mathbf{F}(B_{hk},R) \to C(B,R)$. *We assume that* $D_{hk} = B_{hk}$ *i.e.* $K = 0$, $L = 0$. *We will approximate solutions of problem (8.66), (8.61) by means of solutions of the equation*

$$\delta z^{(i,j)} = F(x_i, y_j, \int_B T_{hk}z_{[i,j]}(t,s)\,dt\,ds), \quad (x_i,y_j) \in A_{hk}, \tag{8.68}$$

with boundary condition (8.63).

Example 8.32 *Suppose that $F : \bar{E} \times R \to R$, $\varphi : \bar{E} \to R$, $\psi : \bar{E} \to R$, $\phi : E_0 \to R$ are given functions, and $-a_0 \le \varphi(x,y) - x \le 0$, $-b_0 \le \psi(x,y) - y \le 0$ for $(x,y) \in \bar{E}$. Let $f(x,y,w) = F(x,y,w(\varphi(x,y)-x, \psi(x,y)-y))$ on $\bar{E} \times C(B,R)$. Then problem (8.60), (8.61) reduces to the differential equation with a deviated argument*

$$D_{xy}z(x,y) = F\left(x,y,z(\varphi(x,y),\psi(x,y))\right), \quad (x,y) \in E,$$

with boundary condition (8.61). It is easy to construct difference methods for the above equation using ideas from Example 8.31.

8.9 Solvability of difference problems

The existence and uniqueness of a solution of problem (8.62), (8.63) is investigated by the comparison method. This method is based on the association of the operator F_{hk} with another operator σ_{hk}, which is followed by the thorough analysis of a comparison equation. If the last equation possesses adequate properties, then problem (8.62), (8.63) has exactly one solution which is the limit of a sequence of successive approximations. We obtain the simplest case of the operator σ_{hk} corresponding to equation (8.62) if the function F_{hk} satisfies the Lipschitz condition. The comparison problem is linear in this case.

The following property of the operator δ is important:

Lemma 8.33 *Problem (8.62), (8.63) is equivalent to*

$$z^{(i,j)} = hk \sum_{\mu=0}^{i-1} \sum_{\nu=0}^{j-1} F_{hk}(x_\mu, y_\nu, z_{[\mu,\nu]}) + \phi_{hk}^{(i,0)} + \phi_{hk}^{(0,j)} - \phi_{hk}^{(0,0)} \tag{8.69}$$

where $1 \le i \le M$, $1 \le j \le N$, and

$$z^{(i,j)} = \phi_{hk}^{(i,j)} \text{ for } (x_i, y_j) \in E_{hk}^0. \tag{8.70}$$

We omit the simple proof of the above Lemma. Let $\eta_{hk} : E_{hk}^0 \cup E_{hk} \to R$ be a function given by

$$\eta_{hk}(x,y) = \phi_{hk}(x,y) \text{ for } (x,y) \in E_{hk}^0 \text{ and } \eta_{hk}(x,y) = 0 \text{ for } (x,y) \in E_{hk}.$$

We define a sequence $\{z_n\}$, $z_n : E_{hk}^0 \cup E_{hk} \to R$, in the following way:

$$z_0^{(i,j)} = \eta_{hk}^{(i,j)} \text{ for } (x_i, y_j) \in E_{hk}^0 \cup E_{hk} \tag{8.71}$$

and

$$z_{n+1}^{(i,j)} = hk \sum_{\mu=0}^{i-1} \sum_{\nu=0}^{j-1} F_{hk}(x_\mu, y_\nu, (z_n)_{[\mu,\nu]}) + \phi_{hk}^{(i,0)} + \phi_{hk}^{(0,j)} - \phi_{hk}^{(0,0)} \tag{8.72}$$

where $1 \le i \le M$, $1 \le j \le N$, and

$$z_{n+1}^{(i,j)} = \phi_{hk}^{(i,j)} \text{ for } (x_i, y_j) \in E_{hk}^0, \quad n \ge 0. \tag{8.73}$$

We prove that, under suitable assumptions on the function F_{hk}, the sequence $\{z_n\}$ converges to the unique solution of problem (8.62), (8.63). For $w \in \mathbf{F}(D_{hk}, R)$ we put

$$\|w\|_{hk} = \max\{\, |w^{(i,j)}| : (x_i, y_j) \in D_{hk}\}.$$

For the above w we define a function $|w|_{hk} : D_{hk} \to R_+$ by $|w|_{hk}(x,y) = |w(x,y)|$, $(x,y) \in D_{hk}$. We will consider a comparison function $\sigma_{hk} : A_{hk} \times \mathbf{F}(D_{hk}, R_+) \to R_+$ corresponding to the function F_{hk}.

Assumption H $[\sigma_{hk}]$. Suppose that

1) for each $(x,y) \in A_{hk}$ the function $\sigma_{hk}(x,y,\cdot) : \mathbf{F}(D_{hk}, R_+) \to R_+$ is nondecreasing and $\sigma_{hk}(x,y,\theta_{hk}) = 0$ for $(x,y) \in A_{hk}$, where $\theta_{hk}(x,y) = 0$ for $(x,y) \in D_{hk}$,

2) for $(x,y,w) \in A_{hk} \times \mathbf{F}(D_{hk}, R)$, $\bar{w} \in \mathbf{F}(D_{hk}, R)$ we have

$$|\, F_{hk}(x,y,w) - F_{hk}(x,y,\bar{w})\,| \le \sigma_{hk}(x,y,|w-\bar{w}|_{hk}), \tag{8.74}$$

3) there exists a function $\bar{g}_{hk} : E^0_{hk} \cup E_{hk} \to R_+$ which is a solution of the problem

$$\omega^{(i,j)} \ge hk \sum_{\mu=0}^{i-1} \sum_{\nu=0}^{j-1} \sigma_{hk}(\, x_\mu, y_\nu, \omega_{[\mu,\nu]}\,) + \bar{\eta}^{(i,j)}_{hk}, \quad 1 \le i \le M,\ 1 \le j \le N, \tag{8.75}$$

$$\omega^{(i,j)} = |\phi^{(i,j)}_{hk}| \ \text{ for } (x_i, y_j) \in E^0_{hk} \tag{8.76}$$

with the function $\bar{\eta}_{hk}$ satisfying the condition

$$\bar{\eta}^{(i,j)}_{hk} \ge hk \sum_{\mu=0}^{i-1} \sum_{\nu=0}^{j-1} |\, F_{hk}(\, x_\mu, y_\nu, (\bar{\eta}_{hk})_{[\mu,\nu]}\,)\,| + |\phi^{(i,0)}_{hk}| + |\phi^{(0,j)}_{hk}| + |\phi^{(0,0)}_{hk}|, \tag{8.77}$$

where $1 \le i \le M$, $1 \le j \le N$,

4) the function $\omega(x,y) = 0$, $(x,y) \in E^0_{hk} \cup E_{hk}$, is the unique solution of the problem

$$\omega^{(i,j)} = hk \sum_{\mu=0}^{i-1} \sum_{\nu=0}^{j-1} \sigma_{hk}(\, x_\mu, y_\nu, \omega_{[\mu,\nu]}\,), \quad 1 \le i \le M,\ 1 \le j \le N, \tag{8.78}$$

$$\omega^{(i,j)} = 0 \ \text{ for } (x_i, y_j) \in E^0_{hk}, \tag{8.79}$$

in the class of functions satisfying the condition $0 \le \omega(x,y) \le \bar{g}_{hk}(x,y)$ on E_{hk}.

Theorem 8.34 *If Assumption H $[\sigma_{hk}]$ is satisfied then there exists a solution $\bar{z} : E^0_{hk} \cup E_{hk} \to R$ of problem (8.62), (8.63). The solution is unique in the class of functions $z : E^0_{hk} \cup E_{hk} \to R$ satisfying the condition $|z(x,y)| \le \bar{g}_{hk}(x,y)$ for $(x,y) \in E_{hk}$.*

PROOF. Consider the sequence $\{g_n\}$, $g_n : E^0_{hk} \cup E_{hk} \to R_+$, where $g_0 = \bar{g}_{hk}$ and

$$g^{(i,j)}_{n+1} = hk \sum_{\mu=0}^{i-1} \sum_{\nu=0}^{j-1} \sigma_{hk}(\, x_\mu, y_\nu, (g_n)_{[\mu,\nu]}\,), \;\; 1 \le i \le M, \; 1 \le j \le N, \tag{8.80}$$

$$g^{(i,j)}_{n+1} = 0 \;\text{ for } (x_i, y_j) \in E^0_{hk}, \tag{8.81}$$

for $n \ge 0$. We prove that

$$g_{n+1}(x,y) \le g_n(x,y) \;\text{ for } (x,y) \in E_{hk},\; n \ge 0, \tag{8.82}$$

$$\lim_{n\to\infty} g_n(x,y) = 0, \;\; (x,y) \in E_{hk}, \tag{8.83}$$

$$|z_n(x,y)| \le \bar{g}_{hk}(x,y), \;\; (x,y) \in E_{hk},\; n \ge 0, \tag{8.84}$$

$$|z_{n+r}(x,y) - z_n(x,y)| \le g_n(x,y), \;\; (x,y) \in E_{hk}, \;\; n, r \ge 0. \tag{8.85}$$

It follows from condition 3) of Assumption H $[\sigma_{hk}]$ that $g_1(x,y) \le g_0(x,y)$ for $(x,y) \in E_{hk}$. Assume that, for fixed $n \in \mathbf{N}$, we have $g_n(x,y) \le g_{n-1}(x,y)$, $(x,y) \in E_{hk}$. It follows from the monotonicity of σ_{hk} with respect to the functional variable and from (8.80) that $g_{n+1}(x,y) \le g_n(x,y)$ for $(x,y) \in E_{hk}$. Then we have (8.82) by induction on $n \in \mathbf{N}$. Since $0 \le g_n(x,y)$ for $(x,y) \in E_{hk}$, $n \in \mathbf{N}$, then there exists $\bar{g}(x,y) = \lim_{n\to\infty} g_n(x,y)$, $(x,y) \in E_{hk}$. It follows from (8.80), (8.81) that $\bar{g}$ is a solution of problem (8.78), (8.79). Condition 4) of Assumption H $[\sigma_{hk}]$ implies $\bar{g}(x,y) = 0$ for $(x,y) \in E_{hk}$. Now we prove (8.84). This relation is obvious for $n = 0$. If we assume that $|z_n(x,y)| \le \bar{g}_{hk}(x,y)$, $(x,y) \in E_{hk}$, then we deduce from (8.74), (8.77) and from the monotonicity of σ_{hk} that

$$|z^{(i,j)}_{n+1}| \le hk \sum_{\mu=0}^{i-1} \sum_{\nu=0}^{j-1} |\, F_{hk}(\, x_\mu, y_\nu, (z_n)_{[\mu,\nu]}\,) - F_{hk}(\, x_\mu, y_\nu, (\eta_{hk})_{[\mu,\nu]}\,)\,|$$

$$+ hk \sum_{\mu=0}^{i-1} \sum_{\nu=0}^{j-1} |\, F_{hk}(\, x_\mu, y_\nu, (\eta_{hk})_{[\mu,\nu]}\,)\,| + |\phi^{(i,0)}_{hk}| + |\phi^{(0,j)}_{hk}| + |\phi^{(0,0)}_{hk}|$$

$$\le hk \sum_{\mu=0}^{i-1} \sum_{\nu=0}^{j-1} \sigma_{hk}(\, x_\mu, y_\nu, (\bar{g}_{hk})_{[\mu,\nu]}\,) + |\phi^{(i,0)}_{hk}| + |\phi^{(0,j)}_{hk}| + |\phi^{(0,0)}_{hk}| + \bar{\eta}^{(i,j)}_{hk} \le \bar{g}^{(i,j)}_{hk},$$

where $1 \le i \le M$, $1 \le j \le N$; and it is seen that inequality (8.84) is obtained by induction on $n \in \mathbf{N}$. We prove (8.85) by induction on $n \in \mathbf{N}$. Estimate (8.85) for $n = 0$ follows from (8.84). If we assume that, for fixed $n \in \mathbf{N}$, we have

$$|z^{(i,j)}_{n+r} - z^{(i,j)}_n| \le g^{(i,j)}_n, \;\; (x_i, y_j) \in E_{hk}, \;\; r \ge 0,$$

then, applying (8.72), (8.74), (8.80) and the monotonicity of σ_h, we get

$$|z^{(i,j)}_{n+r+1} - z^{(i,j)}_{n+1}| \le hk \sum_{\mu=0}^{i-1} \sum_{\nu=0}^{j-1} |\, F_{hk}(\, x_\mu, y_\nu, (z_{n+r})_{[\mu,\nu]}\,) - F_{hk}(\, x_\mu, y_\nu, (z_n)_{[\mu,\nu]}\,)\,|$$

$$\leq hk \sum_{\mu=0}^{i-1} \sum_{\nu=0}^{j-1} \sigma_{hk}(\, x_\mu, y_\nu, (g_n)_{[\mu,\nu]}\,) = g_{n+1}^{(i,j)},$$

which completes the proof of assertion (8.85). It follows from (8.83), (8.85) that there is $\bar{z} : E^0_{hk} \cup E_{hk} \to R$ such that $\bar{z}(x,y) = \lim_{n\to\infty} z_n(x,y)$, $(x,y) \in E^0_{hk} \cup E_{hk}$. Relations (8.72), (8.73) imply the function $\bar{z}$ is a solution of problem (8.69), (8.70). Suppose that $\bar{u} : E^0_{hk} \cup E_{hk} \to R$ is another solution of problem (8.69), (8.70), and that $|\bar{u}(x,y)| \leq \bar{g}_{hk}(x,y)$ for $(x,y) \in E_{hk}$. Then we obtain by induction on $n \in \mathbf{N}$ the relation $|\bar{u}(x,y) - z_n(x,y)| \leq g_n(x,y)$ for $(x,y) \in E_{hk}$, $n \geq 0$. It follows from (8.83) that $\bar{u} = \bar{z}$, which completes the proof of the Theorem 1.2

Now, we prove a result on the global uniqueness of a solution of (8.69), (8.70).

Lemma 8.35 *Suppose that Assumption H $[\sigma_{hk}]$ is satisfied and the function $\bar{\omega}(x,y) = 0$ for $(x,y) \in E^0_{hk} \cup E_{hk}$ is the only solution of the problem*

$$\omega^{(i,j)} \leq hk \sum_{\mu=0}^{i-1} \sum_{\nu=0}^{j-1} \sigma_{hk}(\, x_\mu, y_\nu, \omega_{[\mu,\nu]}\,), \;\; 1 \leq i \leq M,\; 1 \leq j \leq N, \tag{8.86}$$

$$\omega^{(i,j)} = 0 \;\text{ for } (x_i, y_j) \in E^0_{hk}. \tag{8.87}$$

Then the solution $\bar{z} : E^0_{hk} \cup E_{hk} \to R$ of problem (8.69), (8.70) is unique.

PROOF. If $\bar{z}$, $\bar{u} : E^0_{hk} \cup E_{hk} \to R$ are solutions of (8.69), (8.70), then $\tilde{\omega} = \bar{z} - \bar{u}$ satisfies (8.86), (8.87), and the assertion follows.

Now, we give sufficient conditions for the uniqueness of the solution $\bar{\omega} = 0$ of problem (8.86), (8.87).

Lemma 8.36 *Suppose that the function σ_{hk} satisfies the conditions:*

1) for each function $\lambda_{hk} : E_{hk} \to R_+$ there exists a solution of the problem

$$\omega^{(i,j)} \geq hk \sum_{\mu=0}^{i-1} \sum_{\nu=0}^{j-1} \sigma_{hk}(x_\mu, y_\nu, \omega_{[\mu,\nu]}) + \lambda_{hk}^{(i,j)}, \;\; 1 \leq i \leq M,\; 1 \leq j \leq N, \tag{8.88}$$

$$\omega^{(i,j)} = 0 \;\text{ for } (x_i, y_j) \in E^0_{hk}, \tag{8.89}$$

2) the function $\bar{\omega} = 0$ is a unique solution of problem (8.78), (8.79).

Under these assumptions, the function $\bar{\omega}(x,y) = 0$, $(x,y) \in E^0_{hk} \cup E_{hk}$, is the only solution of problem (8.86), (8.87).

PROOF. Suppose that $\tilde{\omega} : E^0_{hk} \cup E_{hk} \to R_+$ is a solution of problem (8.86), (8.87). Consider the sequence $\{\omega_n\}$, $\omega_n : E^0_{hk} \cup E_{hk} \to R_+$, given by:

(i) ω_0 is a solution of (8.88), (8.89) for $\lambda_{hk} = \tilde{\omega}$,

(ii) if ω_n is a given function then

$$\omega_{n+1}^{(i,j)} = hk \sum_{\mu=0}^{i-1} \sum_{\nu=0}^{j-1} \sigma_{hk}(\, x_\mu, y_\nu, (\omega_n)_{[\mu,\nu]}\,), \;\; 1 \leq i \leq M,\; 1 \leq j \leq N, \tag{8.90}$$

$$\omega_{n+1}^{(i,j)} = 0 \text{ for } (x_i, y_j) \in E_{hk}^0. \tag{8.91}$$

Then we have $\tilde{\omega}(x,y) \le \omega_n(x,y)$ for $(x,y) \in E_{hk}$, $n \ge 0$, and $0 \le \omega_{n+1}(x,y) \le \omega_n(x,y)$ for $(x,y) \in E_{hk}$, $n \ge 0$. The above relations can be proved by induction on $n \in \mathbf{N}$. Let $\bar{\omega} : E_{hk}^0 \cup E_{hk} \to R_+$ be defined by $\bar{\omega}(x,y) = \lim_{n\to\infty} \omega_n(x,y)$. It follows from (8.90), (8.91) that $\bar{\omega} = 0$. Since $\tilde{\omega} \le \bar{\omega}$ then the assertion follows.

8.10 Nonlinear estimates

In this Section we consider a particular case of the set D_{hk}. We assume that $K = 0$, $L = 0$. We will use the following comparison lemma

Lemma 8.37 *Suppose that $K = 0$ and $L = 0$ in the definition of D_{hk} and*

1) the function $G_{hk} : A_{hk} \times \mathbf{F}(B_{hk}, R) \to R$ is nondecreasing with respect to the functional variable,

2) the functions $u, v : E_{hk}^0 \cup E_{hk} \to R$ satisfy the relations:

$$u^{(i,j)} - hk \sum_{\mu=0}^{i-1} \sum_{\nu=0}^{j-1} G_{hk}(x_\mu, y_\nu, u_{[\mu,\nu]}) \le v^{(i,j)} - hk \sum_{\mu=0}^{i-1} \sum_{\nu=0}^{j-1} G_{hk}(x_\mu, y_\nu, v_{[\mu,\nu]}),$$

whenever $1 \le i \le M$, $1 \le j \le N$; and $u^{(i,j)} \le v^{(i,j)}$ on E_{hk}^0.

Then $u^{(i,j)} \le v^{(i,j)}$ for $(x_i, y_j) \in E_{hk}$.

We omit the simple proof of the lemma. Denote by Ξ the class of all functions $\alpha : \Delta \to R_+$ such that $\lim_{(h,k)\to(0,0)} \alpha(h,k) = 0$.

Assumption H $[\theta]$. Suppose that $K = 0, L = 0$ and

1) conditions 1), 2), 4) of Assumption H $[\sigma_{hk}]$ are satisfied,

2) the solution $\bar{\omega}(x_i, y_j) = 0$, $(x_i, y_j) \in E_{hk}^0 \cup E_{hk}$, of the problem

$$\delta\omega^{(i,j)} = \sigma_{hk}(x_i, y_j, \omega_{[i,j]}), \ (x_i, y_j) \in A_{hk}, \tag{8.92}$$

$$\omega^{(i,j)} = 0 \text{ for } (x_i, y_j) \in E_{hk}^0, \tag{8.93}$$

is stable in the following sense: if $\omega_{hk} : E_{hk}^0 \cup E_{hk} \to R_+$ is the solution of the problem

$$\delta\omega^{(i,j)} = \sigma_{hk}(x_i, y_j, \omega_{[i,j]}) + \alpha(h,k), \ (x_i, y_j) \in A_{hk}, \tag{8.94}$$

$$\omega^{(i,j)} = \alpha_0(h,k) \text{ for } (x_i, y_j) \in E_{hk}^0, \tag{8.95}$$

where α, $\alpha_0 \in \Xi$ then there is $\beta \in \Xi$ such that $\omega_{hk}^{(i,j)} \le \beta(h,k)$ for $(x_i, y_j) \in E_{hk}$.

Theorem 8.38 *Suppose that Assumption H $[\theta]$ is satisfied, and*

1) $u_{hk} : E_{hk}^0 \cup E_{hk} \to R$ is a solution of problem (8.62), (8.63) and there is $\alpha_0 \in \Xi$ such that

$$|\phi^{(i,j)} - \phi_{hk}^{(i,j)}| \le \alpha_0(h,k) \ for \ (x_i, y_j) \in E_{hk}^0, \tag{8.96}$$

2) $v : E_0 \cup E \to R$ is a solution of problem (8.60), (8.61) and v is of class C^3, on $\bar{E}$,

3) the following compatibility condition is satisfied: there is $\tilde{\alpha} \in \Xi$ such that

$$| F_{hk}(x_i, y_j, (v_{hk})_{[i,j]}) - f(x_i, y_j, v_{(x_i,y_j)}) | \leq \tilde{\alpha}(h,k) \tag{8.97}$$

where $(x_i, y_j) \in A_{hk}$ and the function v_{hk} is the restriction of the function v to the set $E^0_{hk} \cup E_{hk}$.

Under these assumptions there exists $\beta \in \Xi$ such that

$$|u^{(i,j)}_{hk} - v^{(i,j)}_{hk}| \leq \beta(h,k), \quad (x_i, y_j) \in E_{hk}. \tag{8.98}$$

PROOF. Let $\Gamma_{hk} : A_{hk} \to R$ be defined by

$$\delta v^{(i,j)}_{hk} = F_{hk}(x_i, y_j, (v_{hk})_{[i,j]}) + \Gamma^{(i,j)}_{hk}, \quad (x_i, y_j) \in A_{hk}. \tag{8.99}$$

It follows from assumption 2) that there is $\alpha_1 \in \Xi$ such that

$$|\delta v^{(i,j)}_{hk} - D_{xy} v^{(i,j)}| \leq \alpha_1(h,k), \quad (x_i, y_j) \in A_{hk}. \tag{8.100}$$

From the above inequality and from the compatibility condition (8.97) we deduce that there is $\alpha \in \Xi$ such that $|\Gamma^{(i,j)}_{hk}| \leq \alpha(h,k)$ for $(x_i, y_j) \in A_{hk}$. Let $\omega_{hk} = |u_{hk} - v_{hk}|$. Then the function ω_{hk} satisfies the relations

$$\omega^{(i,j)}_{hk} \leq hk \sum_{\mu=0}^{i-1} \sum_{\nu=0}^{j-1} \sigma_{hk}(x_\mu, y_\nu, (\omega_{hk})_{[\mu,\nu]}) + ih\, jk\, \alpha(h,k), \tag{8.101}$$

where $1 \leq i \leq M,\ 1 \leq j \leq N$, and

$$\omega^{(i,j)}_{hk} \leq \alpha_0(h,k) \ \text{ for } (x_i, y_j) \in E^0_{hk}. \tag{8.102}$$

Let $\tilde{\omega} : E^0_{hk} \cup E_{hk} \to R_+$ be a solution of the problem

$$\omega^{(i,j)} = hk \sum_{\mu=0}^{i-1} \sum_{\nu=0}^{j-1} \sigma_{hk}(x_\mu, y_\nu, (\omega)_{[\mu,\nu]}) + ab\alpha(h,k), \tag{8.103}$$

where $1 \leq i \leq M,\ 1 \leq j \leq N$, and

$$\omega^{(i,j)} = \alpha_0(h,k) \ \text{ for } (x_i, y_j) \in E^0_{hk}. \tag{8.104}$$

Relations (8.101), (8.102) and Lemma 8.37 imply $\omega^{(i,j)}_{hk} \leq \tilde{\omega}^{(i,j)}_{hk}$ for $(x_i, y_j) \in E_{hk}$. Now we obtain the assertion of our theorem from the stability of the functional difference problem (8.92), (8.93).

Remark 8.39 *If the assumptions of Theorem 8.38 are satisfied then we have the following estimate for the error of approximate solutions of problem (8.60), (8.61):*

$$| u^{(i,j)}_{hk} - v^{(i,j)}_{hk} | \leq \tilde{\omega}^{(i,j)}_{hk}, \quad (x_i, y_j) \in E_{hk},$$

where the function $\tilde{\omega}_h$ is the only solution of problem (8.103), (8.104) with α_0 given by (8.96), $\alpha = \tilde{\alpha} + \alpha_1$, and $\tilde{\alpha}$, α_1 are defined by (8.97), (8.100).

Now, we consider problem (8.60), (8.61) and the difference method

$$\delta z^{(i,j)} = f(x_i, y_j, T_{hk} z_{[i,j]}), \quad (x_i, y_j) \in A_{hk}, \tag{8.105}$$

$$z^{(i,j)} = \phi_{hk}^{(i,j)} \text{ for } (x_i, y_j) \in E_{hk}^0, \tag{8.106}$$

where the operator T_{hk} is defined in Example 8.31. It is obvious that there exists exactly one solution of problem (8.105), (8.106).

Assumption H $[\sigma]$. Suppose that the function $f : \bar{E} \times C(B, R) \to R$ is continuous, and there is a function $\sigma : \bar{E} \times R_+ \to R_+$ such that

1) σ is continuous, and $\sigma(x, y, 0) = 0$ for $(x, y) \in \bar{E}$,

2) σ is nondecreasing with respect to all variables and the function $\bar{\omega}(x, y) = 0$, $(x, y) \in \bar{E}$, is the unique solution of the problem

$$D_{xy} z(x, y) = \sigma(x, y, z(x, y)), \quad (x, y) \in \bar{E},$$

$$z(x, 0) = 0 \text{ for } x \in [0, a], \quad z(0, y) = 0 \text{ for } y \in [0, b],$$

3) the estimate

$$|f(x, y, w) - f(x, y, \bar{w})| \le \sigma(x, y, \|w - \bar{w}\|_B)$$

is satisfied on $\bar{E} \times C(B, R)$.

Theorem 8.40 *Suppose that Assumption H $[\sigma]$ is satisfied, and*

1) $u_{hk} : E_{hk}^0 \cup E_{hk} \to R$ is a solution of problem (8.105), (8.106) and there is $\alpha_0 \in \Xi$ such that estimate (8.96) holds,

2) $v : E_0 \cup E \to R$ is a solution of problem (8.60), (8.61), and v is of class C^3 on $\bar{E}$.

Then there is $\beta \in \Xi$ such that

$$|u_{hk}^{(i,j)} - v_{hk}^{(i,j)}| \le \beta(h, k), \quad (x_i, y_j) \in E_{hk}, \tag{8.107}$$

where v_{hk} is the restriction of the function v to the set $E_{hk}^0 \cup E_{hk}$.

PROOF. We apply Theorem 8.38 in the proof of assertion (8.107). Put $D_{hk} = B_{hk}$ and

$$F_{hk}(x, y, w) = f(x, y, T_{hk} w), \quad (x, y, w) \in A_{hk} \times \mathbf{F}(D_{hk}, R). \tag{8.108}$$

Then we have

$$|F_{hk}(x, y, w) - F_{hk}(x, y, \bar{w})| \le \sigma(x, y, \|T_{hk}(w - \bar{w})\|_B)$$
$$= \sigma(x, y, \|w - \bar{w}\|_{hk}) \text{ on } A_{hk} \times \mathbf{F}(D_{hk}, R).$$

Consider the problem

$$\delta \omega^{(i,j)} = \sigma(x_i, y_j, \omega^{(i,j)}), \quad 0 \le i \le M - 1, \ 0 \le j \le N - 1, \tag{8.109}$$

$$\omega^{(i,j)} = 0 \text{ for } (x_i, y_j) \in E_{hk}^0. \tag{8.110}$$

We prove that the solution $\bar{\omega}(x,y)=0$, $(x,y)\in E^0_{hk}\cup E_{hk}$, of the problem

$$\delta\omega^{(i,j)}=\sigma(x_i,y_j,\omega^{(i,j)}),\ \ (x_i,y_j)\in A_{hk},\quad \omega^{(i,j)}=0\ \text{ for }\ (x_i,y_j)\in E^0_{hk},$$

is stable in the sense of condition 2) of Assumption H $[\theta]$. Let $\bar{\omega}_{hk}:E^0_{hk}\cup E_{hk}\to R_+$ be a solution of the problem

$$\delta\omega^{(i,j)}=\sigma(x_i,y_j,\omega^{(i,j)})+\alpha(h,k),\ \ (x_i,y_j)\in A_{hk}, \tag{8.111}$$

$$\omega^{(i,j)}=\alpha_0(h,k)\ \text{ for }\ (x_i,y_j)\in E^0_{hk}, \tag{8.112}$$

where α, $\alpha_0\in\Xi$. Consider the Darboux problem

$$D_{xy}z(x,y)=\sigma(x,y,z(x,y))+\alpha(h,k),\ \ (x,y)\in\bar{E},$$

$$z(x,0)=\alpha_0(h,k)\ \text{ for } x\in[0,a],\quad z(0,y)=\alpha_0(h,k)\ \text{ for } y\in[0,b].$$

There is $\varepsilon_0>0$ such that if $h+k\le\varepsilon_0$, then there is also a solution $z_{hk}:\bar{E}\to R$ of the above problem, and

$$\lim_{(h,k)\to(0,0)} z_{hk}(x,y)=0\ \text{ uniformly on } \bar{E}. \tag{8.113}$$

It follows from the monotonicity of σ that for $h+k\le\varepsilon_0$, $(x_i,y_j)\in A_{hk}$, we have the relations

$$z^{(i,j)}_{hk}=\int_0^{x_i}\int_0^{y_j}[\sigma(t,s,z_{hk}(t,s))+\alpha(h,k)]\,dt\,ds+\alpha_0(h,k)$$

$$\ge hk\sum_{\mu=0}^{i-1}\sum_{\nu=0}^{j-1}\Big[\sigma(x_\mu,y_\nu,z^{(\mu,\nu)}_{hk})+\alpha(h,k)\Big]+\alpha_0(h,k).$$

Then the function z_{hk} satisfies the difference inequality

$$z^{(i,j)}_{hk}\ge hk\sum_{\mu=0}^{i-1}\sum_{\nu=0}^{j-1}\Big[\sigma(x_\mu,y_\nu,z^{(\mu,\nu)}_{hk})+\alpha(h,k)\Big]+\alpha_0(h,k)$$

where $1\le i\le M$, $1\le j\le N$. The function $\bar{\omega}_{hk}$ satisfies the equation

$$\bar{\omega}^{(i,j)}_{hk}=hk\sum_{\mu=0}^{i-1}\sum_{\nu=0}^{j-1}\Big[\sigma(x_\mu,y_\nu,\bar{\omega}^{(\mu,\nu)}_{hk})+\alpha(h,k)\Big]+\alpha_0(h,k)$$

where $1\le i\le M$, $1\le j\le N$. It follows from Lemma 8.37 that $\bar{\omega}^{(i,j)}_{hk}\le z^{(i,j)}_{hk}$ for $(x_i,y_j)\in E_{hk}$. Thus the stability of problem (8.109), (8.110) follows from condition (8.113). Now we prove the consistency condition for equation (8.108). We will use the following property of the operator T_{hk} : if $w\in\mathbf{F}(B,R)$, w is of class C^3 and w_{hk} is the restriction of w to the set B_{hk}, then there is $C>0$ such that

$$\|T_{hk}w_{hk}-w\|_B\le C(h^2+k^2).$$

It follows from the above property of T_{hk} and from assumption 2) that the operator F_{hk} given by (8.108) satisfies condition (8.97) with $\tilde{\alpha}\in\Xi$. This completes the proof of the Theorem.

8.11 Implicit difference methods

In this Section we consider a general class of difference problems consistent with (8.60), (8.61) which satisfy Assumption H $[\sigma_{hk}]$ and they are convergent. We formulate the functional difference equation. Let $K = 1,\ L = 1$ in the definition of D_{hk}. We define the operator

$$\tilde{T}_{hk} : \mathbf{F}(D_{hk}, R) \to \mathbf{F}([-a_0, h] \times [-b_0, k], R)$$

in the following way. Let $w \in \mathbf{F}(D_{hk}, R)$ and $(t, s) \in [-a_0, h] \times [-b_0, k]$. Then there is (x_i, y_j) such that $-M_0 \le i < 1,\ -N_0 \le j < 1$ and $x_i \le t \le x_{i+1},\ y_j \le s \le y_{j+1}$. We define $(\tilde{T}_{hk}w)(t, s)$ as the right-hand side of formula (8.67). Denote by $S_{hk} : \mathbf{F}(D_{hk}, R) \to \mathbf{F}(B, R)$ the operator given by

$$(S_{hk}w)(t, s) = (\tilde{T}_{hk}w)(t + \frac{h}{2}, s + \frac{k}{2}),\quad (t, s) \in B.$$

The function $S_{hk}w$ is the restriction of the function $\tilde{T}_{hk}w$ to the set $[-a_0 + \frac{h}{2}, \frac{h}{2}] \times [-b_0 + \frac{k}{2}, \frac{k}{2}]$ which is shifted to the set B. Consider problem (8.60), (8.61) and the difference equation

$$\delta z^{(i,j)} = f(x_{i+\frac{1}{2}}, y_{j+\frac{1}{2}}, S_{hk}z_{[i,j]}),\quad 0 \le i \le M - 1,\ 0 \le j \le N - 1, \tag{8.114}$$

with boundary condition (8.63).

Assumption H $[f]$. Suppose that the function $f : \bar{E} \times C(B, R) \to R$ is continuous and there is $\tilde{L} \in R_+$ such that

$$|\, f(x, y, w) - f(x, y, \bar{w})\, | \le \tilde{L}\, \|w - \bar{w}\|_B \ \text{ on } \bar{E} \times C(B, R).$$

Theorem 8.41 *Suppose that Assumption H [f] is satisfied and*

1) $v : E_0 \cup E \to R$ is a solution of problem (8.60), (8.61), and v is of class C^4 on $\bar{E}$,

2) $ab\tilde{L} < 1$, and there is $\alpha_0 \in \Xi$ such that inequality (8.96) holds true.
Then there exists exactly one solution $u_{hk} : E^0_{hk} \cup E_{hk} \to R$ of problem (8.114), (8.63), and there is $C \in R_+$ such that the following error estimate holds

$$|u^{(i,j)}_{hk} - v^{(i,j)}| \le \frac{Cx_iy_j(h^2 + k^2) + 2\alpha_0(h, k)}{1 - x_iy_j\tilde{L}},\quad (x_i, y_j) \in E_{hk}. \tag{8.115}$$

Proof. We put

$$F_{hk}(x_i, y_j, w) = f(x_{i+\frac{1}{2}}, y_{j+\frac{1}{2}}, S_{hk}w),\quad (x_i, y_j, w) \in A_{hk} \times \mathbf{F}(D_{hk}, R).$$

Then F_{hk} satisfies the Lipschitz condition with respect to the functional variable with the constant $\tilde{L}$. Put

$$\sigma(x, y, w) = \tilde{L}\, \|w\|_{hk},\quad (x, y, w) \in A_{hk} \times \mathbf{F}[D_{hk}, R_+). \tag{8.116}$$

Then equation (8.78) is equivalent to

$$\omega^{(i,j)} = hk \sum_{\mu=1}^{i} \sum_{\nu=1}^{j} \tilde{L} \| \omega_{[\mu-1,\nu-1]} \|_{hk}, \quad 1 \le i \le M, \ 1 \le j \le N.$$

The above equation with boundary condition (8.79) is equivalent to the problem

$$\omega^{(i,j)} = hk \sum_{\mu=1}^{i} \sum_{\nu=1}^{j} \tilde{L} \omega^{(\mu,\nu)}, \ 1 \le i \le M, \ 1 \le j \le N, \tag{8.117}$$

$$\omega^{(i,j)} = 0 \ \text{ for } \ (x_i, y_j) \in E_{hk}. \tag{8.118}$$

It follows from assumption 2) that problem (8.117), (8.118) satisfies conditions 3), 4) of Assumption H $[\sigma_{hk}]$ and that the unique solution of problem (8.86) (8.87) with σ given by (8.116) is $\omega(x, y) = 0$. Then there exists exactly one solution $u_{hk} : E^0_{hk} \cup E_{hk} \to R$ of problem (8.114), (8.63). Let $\Gamma_{hk} : A_{hk} \to R$ be defined by

$$\delta v^{(i,j)}_{hk} = f\left(x_{i+\frac{1}{2}}, y_{j+\frac{1}{2}}, S_{hk}(v_{hk})_{[i,j]}\right) + \Gamma^{(i,j)}_{hk}, \quad (x_i, y_j) \in A_{hk}.$$

There is $C \in R_+$ such that $|\Gamma^{(i,j)}_{hk}| \le C(h^2 + k^2)$ for $(x_i, y_j) \in A_{hk}$. Let $\omega_{hk} = u_{hk} - v_{hk}$. Obviously, the function ω_{hk} satisfies the inequalities

$$|\omega^{(i,j)}_{hk}| \le hk \sum_{\mu=0}^{i-1} \sum_{\nu=0}^{j-1} \left[\tilde{L} \| (\omega_{hk})_{[\mu,\nu]} \|_{hk} + C(h^2 + k^2) \right] + 2\alpha_0(h, k)$$

where $1 \le i \le M, 1 \le j \le N$, and

$$\omega^{(i,j)}_{hk} \le \alpha_0(h, k) \ \text{ for } \ (x_i, y_j) \in E^0_{hk}.$$

The function

$$\tilde{\omega}^{(i,j)}_{hk} = \frac{C x_i y_j (h^2 + k^2) + 2\alpha_0(h, k)}{1 - x_i y_j \tilde{L}}, \quad (x_i, y_j) \in E_{hk},$$

$$\tilde{\omega}^{(i,j)}_{hk} = 2\alpha_0(h, k) \ \text{for} \ (x_i, y_j) \in E^0_{hk},$$

satisfies the inequalities

$$\omega^{(i,j)}_{hk} \ge hk \sum_{\mu=0}^{i-1} \sum_{\nu=0}^{j-1} \left[\tilde{L} \| (\omega)_{[\mu,\nu]} \|_{hk} + C(h^2 + k^2) \right]$$

$$+ 2\alpha_0(h, k), \ \ 1 \le i \le M, \ 1 \le j \le N,$$

$$\omega^{(i,j)} \ge \alpha_0(h, k) \ \text{ for } \ (x_i, y_j) \in E^0_{hk}.$$

Consequently, we obtain assertion (8.115) from Lemma 8.37.

Remark 8.42 *All the results on difference methods for problem (8.60), (8.61) can be extended on the equation*

$$D_{xy}z(x,y) = F(x, y, z_{(x,y)}, (D_x z)_{(x,y)}, (D_y z)_{(x,y)})$$

with the boundary condition (8.61), see [66, 68, 154].

Many real processes and phenomena studied in mechanics, theoretical physics, population dynamics and economy are characterized by the fact that at certain moments of their development the system of parameters undergo rapid changes by jumps. In the mathematical simulation of such processes and phenomena the duration of these changes is usually neglected and the process is assumed to change its state impulsively. A natural tool for the mathematical simulation of such processes are impulsive differential equations. The paper V. Milman and A. Myshkis [165] initiated the theory of impulsive ordinary differential equations. Up to now numerous papers were published concerning various problems for classes of equations and also dealing with special problems appearing in this theory. It is not our aim to show a full review of papers concerning ordinary impulsive differential equations. We shall mention only monographs which contain reviews. They are [11, 139].

Partial differential equations with impulses were first treated in [85]. The authors therein have shown that parabolic equations with impulses provide natural framework for many evolutionary processes in the population dynamics. Hyperbolic differential equations with impulses were considered in [5, 6, 126]. Estimates of solutions, estimates of the difference between of two problems, uniqueness theory and continuous dependence on given functions were considered. The monotone iterative methods for impulsive nonlinear hyperbolic equations were investigated in [9]. Difference methods for first order partial functional differential equations with impulses were investigated in [7, 8]. The authors proved that there are natural classes of difference methods for such problems. Theorems on difference inequalities or recurrent inequalities were used in the investigation of the stability of difference schemes. Detailed bibliographical information can be found in [10, 129]. Numerical method of lines is studied in [240].

Bibliography

[1] Augustynowicz A., Kamont Z., On Kamke's functions in uniqueness theorems for first order partial differential-functional equations, Journ. Nonl. Anal., TMA, 14 (1990), 837 - 850.

[2] Augustynowicz A., Leszczyński H., On the existence of analytic solutions of the Cauchy problem for first order partial differential equations with retarded variables, Ann. Soc. Math. Polon., Comm. Math. 36 (1996), 11 - 25.

[3] Augustynowicz A., Leszczyński H., On x - analytic solutions to the Cauchy problem for partial differential equations with retarded variables, Zeit. Anal. Anwend. 15 (1996), 345 - 356.

[4] Augustynowicz A, Leszczyński H., Walter W., Cauchy-Kovalevskaya theory for equations with deviating arguments, Aeqat. Math., to appear.

[5] Bainov D., Kamont Z., Minchev E., On first order impulsive partial differential inequalities, Appl. Math. and Comp. 61 (1994), 207 - 230.

[6] Bainov D., Kamont Z., Minchev E., On the impulsive partial differential-functional inequalities of first order, Utilitas Math. 48 (1995), 107 - 128.

[7] Bainov D., Kamont Z., Minchev E., Difference methods for impulsive differential functional equations, Appl. Numer. Math. 16 (1995), 401 - 416.

[8] Bainov D., Kamont Z., Minchev E., The finite difference method for first order impulsive partial differential functional equations, Computing 55 (1995), 237 - 253.

[9] Bainov D., Kamont Z., Minchev E., Monotone iterative methods for impulsive hyperbolic differential functional equations, Journ. Comp. and Appl. Math. 70 (1996), 329 - 347.

[10] Bainov D., Kamont Z., Minchev E., Approximate Solutions of Impulsive Hyperbolic Equations, Academic Publishers, Calcuta, 1996.

[11] Bainov D., Simeonov P., Impulsive Differential Equations: Asymptotic Properties of Solutions, World Scientific, Singapore, 1995.

[12] Bardi M., Crandall M. G, Evans L. C., H. M. Soner, P. E. Souganidis, Viscosity Solutions and Applications, Lect. Notes Math., Springer, 1997.

[13] Bassanini P., On a boundary value problem for a class of quasilinear hyperbolic systems in two independent variables, Atti Sem. Mat. Fis. Univ. Modena 24 (1975), 343 - 372.

[14] Bassanini P., On a recent proof concerning a boundary value problem for quasilinear hyperbolic systems in the Schauder canonic form, Boll. Un. Mat. Ital. (5) 14 - A (1977), 325 - 332.

[15] Bassanini P., Iterative methods for quasilinear hyperbolic systems, Boll. Un. Mat. Ital. (6) 1 - B (1982), 225 - 250.

[16] Bassanini P., The problem of Graffi-Cesari, Proc. Internat. Conf. on Nonlinear Phenomena in Math. Sci., Arlington, USA, Acad. Press 1982, 87 - 101.

[17] Bassanini P., Cesari L., La duplicazione di frequenza nella radiazione laser, Atti Acad. Naz. Lincei 69 (1980), 166 - 173.

[18] Bassanini P., Galaverni M., Contrazioni multiple, sistemi iperbolici é problema del laser, Atti Sem. Mat. Fis. Univ. Modena 31 (1982), 32 - 50.

[19] Bassanini P., Salvatori M. C., Un problema ai limiti per sistemi integrodifferenziali non lineari di tipo iperbolico, Boll. Un. Mat. Ital. (5) 18 - B (1981), 785 - 798.

[20] Bassanini P., Turo J., Generalized solutions to free boundary problems for hyperbolic systems of functional partial differential equations, Ann. Mat. pura ed appl. 156 (1990), 211 - 230.

[21] Besala P., On solutions of first order partial differential equations defined in an unbounded zone, Bull. Acad. Polon. Sci, Ser. sci. math. astr. phys. 12 (1964), 95 - 99.

[22] Besala P., Finite difference approximation to the Cauchy problem for non-linear parabolic differential equations, Ann. Polon. Math. 46 (1985), 19 - 26.

[23] Besala P., On uniqueness of initial-value problems for partial differential equations of the first order, Ann. Polon. Math. 40 (1983), 105 - 108.

[24] Besala P, Observations on quasi-linear partial differential equations, Ann. Polon. Math. 53 (1991), 267 - 283.

[25] Brandi P., Sul problema di Cauchy per una equazione differenziale non lineare alle derivate parziali del primo ordine con argomento funzionale, Atti Sem. Mat. Fis. Univ. Modena 28 (1979), 86 - 101.

[26] Brandi P., Ceppitelli R., On the existence of solutions of a nonlinear functional partial differential equations of the first order, Atti Sem. Mat. Fis. Univ. Modena 29 (1980), 166 - 186.

[27] Brandi P., Ceppitelli R., Existence, uniqueness and continuous dependence for a first order nonlinear partial differential equations in a hereditary structure, Ann. Polon. Math. 47 (1986), 121 - 136.

[28] Brandi P., Kamont Z., Salvadori A., Differential and differential-difference inequalities related to mixed problems for first order partial differential functional equations, Atti Sem. Mat. Fis. Univ. Modena 39 (1991), 255 - 276.

[29] Brandi P., Kamont Z., Salvadori A., Approximate solutions of mixed problems for first order partial differential functional equations, Atti Sem. Mat. Fis. Univ. Modena 39 (1991), 277 - 302.

[30] Brandi P., Kamont Z., Salvadori A., Existence of weak solutions for partial differential functional equations. Preprint.

[31] Brandi P., Marcelli C., Extremal solutions and Gronwall inequality in functional setting, Boll. Un. Mat. Ital.(7) X - B (1996), 233 - 254.

[32] Brandi P., Marcelli C., Haar inequality in hereditary setting and applications, Rend. Sem. Mat. Univ. Padova 96 (1996), 177 - 194.

[33] Bruner H., The numerical treatment of ordinary and partial Volterra integro differential equations, Proceed. First Internat. Colloquium on Numerical Anal., Plovdiv, Eds: Bainov D., V. Covachev, 13 - 26, VSP Utrecht, Tokyo, 1993.

[34] Brzychczy S., Approximate iterative method and the existence of solutions of nonlinear parabolic differential functional equations, Ann. Polon. Math. 42 (1983), 37 - 43.

[35] Brzychczy S.. Chaplygin method for a system of nonlinear parabolic differential functional equations (Russian), Diff. Urav. 22 (1986), 705 - 708.

[36] Brzychczy S., Existence of solutions for nonlinear systems of differential functional equations of parabolic type in arbitrary domain, Ann. Polon. Math. 47 (1987), 309 - 317.

[37] Brzychczy S., On estimation of the speed of the convergence of Chaplygin successive approximations for a parabolic system of differential functional equations, (Russian), Diff. Urav. 25 (1989), 1050 - 1052.

[38] Brzychczy S., Existence of solutions and monotone iterative method for infinite systems of parabolic differential functional equations, Ann. Polon. Math. to appear.

[39] Burys S., On partial differential functional inequalities of the first order, Univ. Iagel. Acta Math. 16 (1974), 107 - 112.

[40] Byszewski L., Finite systems of strong nonlinear differential functional degenerate implicit inequalities with first order partial derivatives, Univ. Iagell. Acta Math. 29 (1992), 77 - 84.

[41] Carvalho A. N., Oliveira L. A., Delay partial differential equations with some large diffusion Nonl. Anal., TMA, 22 (1994), 1057 - 1095.

[42] Ceppitelli R., Kamont Z., Extremal solutions for semilinear differential functional systems in two independent variables, Atti Sem. Mat. Fis. Univ. Modena 42 (1994), 329 - 341.

[43] Chen C., Thomee T., Wahlbin L. B., Finite element approximation of a parabolic integro-differential equation with a weakly singular kernel, Math. Comp. 58 (1992), 587 - 602.

[44] Cazzani Nieri M. G., Nuovi teoremi di esistenza per un problema di Cauchy nella classe delle funzioni assolutamente continue nell singole variabili, Ann. Mat. pura ed appl. 167 (1994), 351 - 387.

[45] Cesari L., A boundary value problem for quasilinear hyperbolic systems in the Schauder canonic form, Ann. Scuola. Norm. Sup. Pisa IV, 1 (1974), 311 - 358.

[46] Cesari L., A boundary value problem for quasilinear hyperbolic systems, Riv. Mat. Univ. Parma 3 (1974), 107 - 131.

[47] Cesari L., Un problema ai limiti per sistemi di equazioni iperboliche quasi lineari nell forma canonica di Schauder, Rend. Accad. Naz. Lincei 57 (1975), 303 - 307.

[48] Chae D., Dubovskii P., Existence and uniqueness for spatially inhomogeneous coagulation-condensation equation with unbounded kernels, Journ. of Integral Equat. and Appl. 9, 1997, 219 - 236.

[49] Chaplygin S. A., Collected Papers on Mechanics and Mathematics, Moscow, 1954.

[50] Chyan C. J., Webb G. F., A model of proliferating cell populations with correlation of mother-daughter mitotic times. Ann. Mat. pura ed appl. 157 (1991), 1 - 11.

[51] Cinquini S., Sopra i sistemi iperbolici di equazioni a derivate parziali (nonlineari) in piú variabili indipendenti, Ann. Mat. pura ed appl. 120 (1979), 201 - 214.

[52] Cinquini Cibrario M., Teoremi di esitenza per sistemi di equazioni quasilineari a derivate parziali in piú variabili indipendenti, Ann. Mat. pura ed appl. 75 (1967), 1 - 46.

[53] Cinquini Cibrario M., Una classe di sistemi di equazioni a derivate parziali in piú variabili indipendenti, Rend. Mat. VII (2) (1982), 499 - 522.

[54] Cinquini Cibrario M., Nuove ricerche sui sistemi di equazioni non lineari a derivate parziali in piú variabili indipendenti, Rend. Sem. Mat. Fis. Univ. Milano, 52 (1982), 531 - 550.

[55] Cinquini Cibrario M., Sopra una classe di sistemi di equazioni non lineari a derivate parziali in piú variabili indipendenti, Ann. Mat. pura ed appl. 140 (1985), 223 - 253.

[56] Człapiński T., A boundary value problem for quasilinear hyperbolic differential functional systems, Atti Sem. Mat. Fis. Univ. Modena 38 (1990), 39 - 59.

[57] Człapiński T, A boundary value problem for quasilinear hyperbolic systems of partial differential functional equations of the first order, Boll. Un. Mat. Ital. (7) 5 - B (1991), 619 - 637.

[58] Człapiński T., On the Cauchy problem for quasilinear hyperbolic systems of partial differential functional equations of the first order, Zeit. Anal. Anwend. 10 (1991), 169 - 182.

[59] Człapiński T., On the existence of generalized solutions of nonlinear first order partial differential functional equations in two independent variables, Czechosl. Mat. Journ. 41 (1991), 490 - 506.

[60] Człapiński T., Generalized solutions to boundary value problems for quasilinear hyperbolic systems of partial differential functional equations, Ann. Polon. Math. 57 (1992), 177 - 191.

[61] Człapiński T., On some non-linear Volterra integral functional equations in several variables, Anal. Math. 20 (1994), 241 - 253.

[62] Człapiński T., Existence of solutions of the Darboux problem for partial differential functional equations with infinite delay in a Banach space. Ann. Soc. Math. Polon., Comm. Math. 35 (1995), 111 - 122.

[63] Człapiński T., On the Chaplygin method for partial differential functional equations of the first order, Univ. Iagell. Acta Math. 35 (1997), 137 - 149.

[64] Człapiński T., On the mixed problem for quasilinear partial differential functional equations of the first order, Zeit. Anal. Anwend. 16, (1997), 463 - 478.

[65] Człapiński T., On the existence of solutions of a certain nonlinear Volterra integral functional equation in several variables, Ann. Soc. Math. Polon., Comm. Math. 37 (1997), 61 - 79.

[66] Człapiński T., Hyperbolic Functional Differential Equations, Dissertation, University of Gdańsk, 1999.

[67] Człapiński T., On the mixed problem for hyperbolic partial differential functional equations of the first order, Czechosl. Math. Journ, to appear.

[68] Człapiński T., Difference methods for the Darboux problem to functional partial differential equations, Ann. Polon. Math., to appear

[69] Człapiński T., On the mixed problem for quasilinear partial functional differential equations with unbounded delay, Ann. Polon. Math., to appear.

[70] Człapiński T., Iterative methods for the Darboux problem for partial functional differential equations, Journ. Ineq. and Appl., to appear.

[71] Człapiński T., Kamont Z., Generalized solutions for quasilinear hyperbolic systems of partial differential functional equations, Journ. Math. Anal. Appl. 172 (1993), 353 - 370.

[72] Człapiński T., Kamont Z., On the local Cauchy problem for quasilinear hyperbolic functional differential systems, Appl. Anal. 64 (1997), 329 - 342.

[73] Człapiński T., Kamont Z., Generalized solutions of local initial problems for quasilinear hyperbolic functional differential systems, Studia Sci. Math. Hungarica, to appear.

[74] Crandal M. G, Evans L. C., Lions P. L., Some properties of viscosity solutions of Hamilton-Jacobi equations, Trans. Amer. Math. Soc. 282 (1984), 487 - 502.

[75] Crandal M. G., Ishii H., Lions P. L., User's guide to viscosity solutions of second order partial differential equations, Bull. Amer. Math. Soc. 27 (1992), 1 - 67.

[76] Crandal M. G., Lions P. L., Viscosity solutions of Hamilton-Jacobi equations, Trans. Amer. Math. Soc. 277 (1983), 1 - 42.

[77] Crandal M. G., Lions P. L., On the existence and uniqueness of solutions of Hamilton-Jacobi equations, Nonl. Anal., TMA, 10 (1986), 353 - 370.

[78] Czyżewska-Ważewska M, Lasota A., Mackey M. C., Maximizing chances of survival, Journ. Math. Biol. 13 (1981), 149 - 158.

[79] Dafermos C. M., Hyperbolic conservation laws with memory. Differential Equations (Xanti 1987), Lect. Not. in Pure and Appl. Math., 118 Dekker, New York, 1989., 157 - 166.

[80] Denkowski Z., Pelczar A., On the existence and uniqueness of solutions of some partial differential functional equations, Ann. Polon. Math. 35 (1978), 261 - 304.

[81] Diekman O., van Gils S. A., Verduyn Lunel S. M., H. O. Walther, Delay Equations. Functional, Complex and Nonlinear Analysis, Springer-Verlag, 1995.

[82] Diekman O., Heijmans H. J., Thieme H. R., On the stability of the cell size distribution, Journ. Math. Biol. 19 (1984), 227 - 248.

[83] Eichhorn W., Gleissner W., On a functional differential equation arising in the theory of the distribution of wealth, Aequat. Math. 28 (1985), 190 - 198.

[84] El-Doma M., Analysis of nonlinear integro differential equations arising in age dependent epidemic models, Nonl. Anal., TMA, 11 (1978), 913 - 937.

[85] Erbe L. II., Freedman H. I., Liu X. Z., Wu J. H., Comparison principles for impulsive parabolic equations with applications to models of single species growth, Journ. Austral. Math. Soc., Ser. B, 32 (1991), 382 - 400.

[86] Fairweather G., Galerkin and collocation methods for partial integro differential equations, Integral equations and Inverse Problems, Eds: V. Petkov, R. Lazarov, Pitman Research Notes in Math., vol. 235, 76 - 85, Lingman, Harlow, 1991.

[87] Fitzke A., On a system of difference inequalities of parabolic type, Ann. Polon. Math. 22 (1970), 299 - 302.

[88] Fitzke A., Method of difference inequalities for parabolic equations with mixed derivatives, Ann. Polon. Math. 31 (1975), 121 - 128.

[89] Franzone Colli P., Paganuzzi A., Stefanelli M., A mathematical model of iron metabolizm, Journ. Math. Biol. 15 (1982), 173 - 201.

[90] Gastaldi L., Tomarelli F., A nonlinear and nonlocal evolution equation describing the muscle contraction, Nonl. Anal., TMA 11 (1987), 163 - 182.

[91] Győry I., The method of lines for the solutions of some nonlinear partial differential equations, Comput. Math. Appl. 15 (1988), 635 - 648.

[92] Gurtin M. E., Murphy L. F., On the optimal harvesting of persistent age-structured population, Journ. Math. Biol. 13 (1981), 131 - 148.

[93] Hale J., Cruz M. A., Existence, uniqueness and continuous dependence for hereditary systems, Ann. Mat. pura ed appl. 4 (84) (1970), 63 - 81

[94] Hale J., Verdyun Lunel S. M., Introduction to Functional Differential Equations, Springer, 1993.

[95] Hino Y., Murakami S., Naito T., Functional Differential Equations with Infinite Delay, Lect. Notes Math. 1473, Springer, 1991.

[96] Hőrmander L., Lectures on Nonlinear Hyperbolic Differential Equations, Springer, 1997.

[97] Ivanauskas F. F., On solutions of the Cauchy problem for a system of differential integral equations, (Russian), Zh. Vychisl. Mat. i Mat. Fis. 18 (1978), 1025 - 1028.

[98] Jaruszewska-Walczak D., Existence of solutions of first order partial differential functional equations, Boll. Un. Mat. Ital. (7) 4 - B (1990), 57 - 82.

[99] Jaruszewska-Walczak D., The initial-value problem for first order partial differential functional equations, Acta Math. Hungar. 67 (1995), 301 - 313.

[100] Jaruszewska-Walczak D., Initial value problems for quasilinear hyperbolic systems of partial differential functional equations, Atti Sem. Mat. Fis. Univ. Modena 40 (1992), 229 - 246.

[101] Jaruszewska-Walczak D., Kamont Z., Existence of solutions of first order partial differential functional equations via the method of lines, Serdica Bul. Math. Publ. 16 (1990), 104 - 114.

[102] Kamenskii G. A., Myshkis A. D., Mixed functional-differential equations, Nonl. Anal., TMA, 30 (1997), 2577 - 2584.

[103] Kamont Z., On the Cauchy problem for non-linear partial differential functional equations of the first order, Serdica Bulg. Math. Publ. 5 (1979), 327 - 339.

[104] Kamont Z., On the Chaplygin method for partial differential functional equations of the first order, Ann. Polon. Math. 37 (1980), 27 - 46.

[105] Kamont Z., A difference method for the nonlinear partial differential equations of the first order with a retarded argument, Math. Nachr. 107 (1982), 87 - 93.

[106] Kamont Z., Existence of solutions of first order partial differential functional equations, Ann. Soc. Math. Polon., Comm. Math. 25 (1985), 249 - 263.

[107] Kamont Z., Semilinear differential functional systems in two independent variables, Math. Nachr. 127 (1986), 246 - 263.

[108] Kamont Z., On the line method approximations to the Cauchy problem for parabolic differential functional equations, Studia Math. Hung. 27 (1992), 313 - 330.

[109] Kamont Z., Iterative methods for hyperbolic differential functional problems, Discuss. Math. 13 (1993), 93 - 117.

[110] Kamont Z., Initial value problems for hyperbolic differential functional systems, Boll. Un. Mat. Ital. (7) 8 - B (1994), 965 - 984.

[111] Kamont Z., Finite difference approximations for first order partial differential functional equations, Ukr. Math. Journ. 46 (1994) 985 - 996.

[112] Kamont Z., Hyperbolic functional differential equations with unbounded delay, Zeit. Anal. Anwend. 18 (1999), 97 - 109.

[113] Kamont Z., Kwapisz M., On the Cauchy problem for differential-delay equations in a Banach space, Math. Nachr. 74 (1976), 173 - 190.

[114] Kamont Z., Kwapisz M., On nonlinear Volterra integral-functional equations, Ann. Polon. Math. 40 (1981), 1 - 29.

[115] Kamont Z., Kwapisz M., Zacharek S., On difference-functional inequalities related to some classes of partial differential functional equations, Math. Nachr. 146 (1990), 335 - 360.

[116] Kamont Z., Leszczyński H., Uniqueness result for the generalized entropy solutions to the Cauchy problem for first-order partial differential functional equations, Zeit. Anal. Anwend. 13 (1994), 477 - 491.

[117] Kamont Z., Leszczyński H., Numerical solutions of the Darboux problem with the functional dependence, Georgian Math. Journ. 5 (1998), 71 - 90.

[118] Kamont Z., Pawelski W., On the characteristics for a system of partial differential equations of the first order in a special case, Ann. Polon. Math. 29 (1974), 215 - 228.

[119] Kamont Z., Prządka K., On solutions of first order partial differential functional equations in an unbounded domain, Zeit. Anal. Anwend. 6 (1987), 121 - 132.

[120] Kamont Z., Prządka K., Difference methods for non-linear partial equations of the first order, Ann. Polon. Math. 48 (1988), 227 - 246.

[121] Kamont Z., Prządka K., Difference methods for first order partial differential functional equations with initial-boundary conditions, Journ. Vycisl. Mat. i Mat. Fis. 31 (1991), 1476 - 1488.

[122] Kamont Z., Salvadori A., Uniqueness of solutions to hyperbolic functional differential problems, Nonl. Anal., TMA, 30 (1997), 4584 - 4594.

[123] Kamont Z., Topolski K., Mixed problems for quasilinear hyperbolic differential functional systems, Math. Balk. 6 (1992), 315 - 324.

[124] Kamont Z., Turo J., On the Cauchy problem for quasi-linear hyperbolic systems of partial differential equations with a retarded argument, Boll. Un. Mat. Ital. (6), 4 - B (1985), 901 - 916.

[125] Kamont Z., Turo J., On the Cauchy problem for quasilinear hyperbolic systems with a retarded argument, Ann. Mat. pura ed appl. 143 (1986), 235 - 246.

[126] Kamont Z., Turo J., Zubik-Kowal B., Differential and difference inequalities generated by mixed problems for hyperbolic functional differential equations with impulses, Appl. Math. and Comp. 80 (1996), 127 - 154.

[127] Kamont Z., Zacharek S., On the existence of weak solutions of nonlinear first order partial differential equations in two independent variables, Boll. Un. Mat. Ital. (6) 5 - B (1986), 851 - 879.

[128] Kamont Z., Zacharek S., The line method for parabolic differential functional equations with initial boundary conditions of the Dirichlet type, Atti Sem. Mat. Fis. Univ. Modena 35 (1987), 249 - 262.

[129] Kamont Z., Zubik-Kowal B., Numerical methods for impulsive partial differential equations, Dynamic Syst. and Appl. 7 (1998), 29 - 52.

[130] Kauthen J. P., The method of lines for parabolic partial integrodifferential equations, Journ. Integral Equat. 4 (1992), 69 - 81.

[131] Kisyński J., Pelczar A., Comparison of solutions and successive approximations in the theory of the equation $\frac{\partial^2 z}{\partial x \partial y} = f(x, y, z, \frac{\partial z}{\partial x}, \frac{\partial z}{\partial y})$, Dissert. Math. 76 (1970), 3 - 74.

[132] Kolmanovski V. B., Myshkis A. D. Introduction to the Theory and Applications of Functional Differential Equations, Kluwer Academic Publishers, 1999.

[133] Kowalski Z., A difference method for the nonlinear partial differential equation, Ann. Polon. Math. 18 (1966), 235 - 242.

[134] Kowalski Z., A difference method for certain hyperbolic systems of nonlinear partial differential equations of first order, Ann. Polon. Math. 19 (1967), 313 - 322.

[135] Kowalski Z., A difference method for a non-linear parabolic differential equation without mixed derivatives, Ann. Polon. Math. 20 (1968), 167 - 177.

[136] Kružkov S. N., Generalized solutions of nonlinear first order partial differential equations, (Russian), Mat. Sb. 70 (1966), 394 - 415.

[137] Kružkov S. N., Generalized solutions in an unbounded domain of the Cauchy problem for nonlinear first order partial differential equations, (Russian), Dokl. Akad. Nauk SSSR 181 (1969), 29 - 32.

[138] Ladde G. S., Lakshmikantham V., Vatsala A. S., Monotone Iterative Techniques for Nonlinear Differential Equations, Pitman Advances Publishing Program, Boston, London, Melbourne, 1985.

[139] Lakshmikantham V., Bainov D., Simeonov P., Theory of Impulsive Differential Equations, World Scientific, Singapore, 1989.

[140] Lakshmikantham V., Leela S., Differential and Integral Inequalities, Academic Press, New York and London, 1969.

[141] Lakshmikantham V., Wen L. Z., Zhang B. G., Theory of Differential Equations with Unbounded Delay, Dordrecht, Kluwer Acad. Publ., 1994.

[142] Langan T. J., Differential inequalities for semilinear hyperbolic operators with two independent variables, Journ. Diff. Equat. 12 (1972), 148 - 161.

[143] Langan T. J., On the existence of not necessarily unique solutions of semilinear hyperbolic systems in two independent variables, Journ. Diff. Equat. 12 (1972), 262 - 290.

[144] Langan T. J., Existence and uniqueness of solutions to initial-boundary value problems for semilinear hyperbolic systems in two independent variables, Journ. Diff. Equat. 22 (1976), 74 - 88.

[145] Leszczyński H., Convergence of one-step difference methods for nonlinear parabolic differential functional systems with initial-boundary conditions of the Dirichlet type, Ann. Soc. Math. Polon., Comm. Math. 30 (1990), 357 - 375.

[146] H. Leszczyński, On the existence and uniqueness of weak solutions of the Cauchy problem for weakly coupled systems of non-linear partial differential functional equations of the first order, Boll. Un. Mat. Ital. (7) 7 - B (1993), 323 - 354.

[147] Leszczyński H., Uniqueness results for unbounded solutions of first order non-linear differential functional equations, Acta Math. Hungar. 64 (1994), 75 - 92.

[148] Leszczyński H., Discrete approximations to the Cauchy problem for hyperbolic differential functional systems in the Schauder canonic form, Comp. Math. and Math. Phys. 34 (1994) 151 - 164.

[149] Leszczyński H., On CC-solutions to the initial-boundary value problem for first order partial differential functional equations, Rend. Math. 15 (1995), 173 - 209.

[150] Leszczyński H., Fundamental solutions to linear first-order equations with a delay at derivatives, Boll. Un. Mat. Ital. (7) 10 - A (1996), 363 - 375.

[151] Leszczyński H., Convergence results for unbounded solutions of first order nonlinear differential functional equations, Ann. Polon. Math. 54 (1996), 1 - 16.

[152] Leszczyński H., Parabolic Equations with Functional Dependence at the Derivatives, Dissertation, University of Gdańsk, 1998.

[153] Leszczyński H., On the method of lines for a nonlinear heat equation with functional dependence, Ann. Polon. Math. 69 (1998), 61 - 74.

[154] Leszczyński H., Convergence of difference analogues to the Darboux problem with functional dependence, Bull. Belg. Math. Soc. 5 (1998), 39 - 57.

[155] Lions P. L., Generalized Solutions of Hamilton-Jacobi equations, Pitman, London, 1982.

[156] Lorenzi A., An identification problem related to a nonlinear hyperbolic integro differential equation, Nonl. Anal.,TMA 22 (1994), 21 - 44.

[157] Łojasiewicz S., Sur probléme de Cauchy pour les systémes d' équations aux dérivées partielles du premier ordre dans le cas hyperbolique de deux variables indépendantes, Ann. Polon. Math. 3 (1956), 87 - 117.

[158] Lusin N. N., On the Chaplygin method of integration, Collected Papers, vol. 3, Moscow, 1953, 146 - 167.

[159] Malec M., Systémes d' inégalites aux différences finies du type parabolique et application, Coll. Math. 35 (1976), 305 - 312.

[160] Malec M., Sur une méthode des différences finies pour une équation non linéaire intégro différentielle á argument retardé, Bull. Polon. Acad. Sci., Ser. sci. math. phys. astr., 26 (1978), 501 - 517.

[161] Malec M., Mączka C, Voigt W., Weak difference inequalities and their applications to the difference analogue of non-linear parabolic differential functional equations, Beit. Num. Math. 11 (1983), 63 - 79.

[162] Malec M., Rosati M., Weak monotonicity for non linear systems of functional-finite difference inequalities of parabolic type, Rend. Math. 3 (1983), 157 - 170.

[163] Malec M., Schafiano A., Méthode aux différences finies pour une équation non linéaire différentielle fonctionelle du type parabolic avec une condition initiale de Cauchy, Boll. Un. Mat. Ital. (7) I - B (1987), 99 - 109.

[164] Mattioli N. M., Salvatori M. C., A theorem of existence and uniqueness in nonlinear dispersive optics, Atti Sem. Mat. Fis. Univ. Modena 28 (1979), 405 - 424.

[165] Milman V., Myshkis A. D., On the stability of motion in the presence of impulses, Sib. Math. Journ. 1 (1960), 233 - 237.

[166] Mlak W., Schechter E., On the Chaplygin method for partial differential equations of the first order, Ann. Polon. Math. 22 (1969), 1 - 18.

[167] Myshkis A. D., Slopak A. S., A mixed problem for systems of differential functional equations with partial derivatives and with operators of the Volterra type, (Russian), Mat. Sb. (N. S.) 41 (1957), 239 - 256.

[168] Nagel R., Sinestrari E., Nonlinear hyperbolic Volterra integrodifferential equations, Nonl. Anal., TMA, 27 (1996), 167 - 186.

[169] Nagumo M., Über das Anfangswertproblem partieller Differentialgleichungen, Jap. Journ. Math. 18 (1942), 41 - 47.

[170] Nowotarska M., Chaplygin method for an infinite system of first order partial differential functional equations, Univ. Iagell. Acta Math. 22 (1981), 125 - 142.

[171] Pao C. V., Coupled nonlinear parabolic systems with time delays, Journ. Math. Anal. Appl. 196 (1995), 237 - 265.

[172] Pao C. V., Systems of parabolic equations with continuous and discrete delays, Journ. Math. Anal. Appl. 205 (1997), 157 - 185.

[173] Pawelski W., Remarques sur des inégalités mixtes entre les intégrales des équations aux dérivées partielles du premier ordre, Ann. Polon. Math. 13 (1963), 309 - 326.

[174] Pawelski W. Sur les ineégalites mixtex entre les intégrales de l'équation aux dérivées partielles $z_x = f(x, y, z, z_y)$, Ann. Polon. Math. 19 (1967), 235 - 247.

[175] Pawelski W. On a case of mixed inequalities between solutions of first order partial differential equations, Ann. Polon. Math. 20 (1968), 95 - 102.

[176] Pelczar A., Some functional differential equations. Dissert. Math. 100 (1973), 3 - 110.

[177] Pliś A., On the estimation of the existence domain for solutions of a non-linear partial differential equation of the first order, Bull. Acad. Polon. Sci., Ser. sci, math. astr. phys. 4 (3) (1966), 125 - 129.

[178] Pliś A., On difference inequalities corresponding to partial differential inequalities of first order, Ann. Polon. Math. 20 (1968), 179 - 181.

[179] Prządka K., Convergence of one step difference methods for first order partial differential functional equations, Atti Sem. Mat. Fis. Univ. Modena 35 (1987), 263 - 288.

[180] Prządka K., Difference methods for non-linear partial differential functional equations of the first order, Math. Nachr. 138 (1988), 105 - 123.

[181] Prządka K., The finite difference method in partial differential functional equations, Univ. Iagell. Acta Math. 29 (1992), 41 - 60.

[182] Redheffer R., Walter W., Stability of the null solution of parabolic functional inequalities, Trans. Amer. Math. Soc. 262 (1980), 285 - 302.

[183] Redlinger R., Existence theorems for semilinear parabolic systems with functionals, Nonl. Anal., TMA, 8 (1984), 667 - 682.

[184] Rozhdestvenskij B. L., Yanenko N. N., Systems of Quasilinear Equations and Their Applications to Gas Dynamics, Providence, Rhode Island, 1983.

[185] Salvadori A., Cauchy problem per una struttura ereditaria di tipo iperbolico. Esistenza, unicita e dipendenza continua, Atti Sem. Mat. Fis. Univ. Modena 32 (1983), 329 - 356.

[186] Sanz-Serna J. M., Verwer J. G., Convergence analysis of one step schemes in the method of lines, Appl. Math. Comput. 31 (1989), 183 - 196.

[187] Sayah A., Equations d' Hamilton Jacobi du premier ordre avec terms integro différentielles, I. Unicite des solutions de viscosité. Comm. Partial Diff. Equat. 16 (1991), 1057 - 1074.

[188] Sayah A, Equations d' Hamilton Jacobi du premier ordre avec terms integro différentielles, II. Existence de solutions de viscosité. Comm. Partial Diff. Equat. 16 (1991), 1075 - 1093.

[189] Schiesser W. E., The Numerical Method of Lines, Acad. Press, San Diego, 1991.

[190] Sinestrari E., Webb G. F., Nonlinear hyperbolic systems with nonlocal boundary conditions, Journ. Math. Anal. Appl. 121 (1987), 449 - 464.

[191] Sloar I. H., Thomee V., Time discretization of an integro-differential equation of parabolic type, SIAM Journ. Numer. Anal. 25 (1986), 1052 - 1061.

[192] Subbotin A., Generalized Solutions of First-Order PDEs, Birkhauser, Boston, 1995.

[193] Szarski J., Characteristics and Cauchy Problem for Nonlinear Partial Differential Equations of First Order, University of Kansas, Lawrence, Kansas, 1959.

[194] Szarski J., On an integro-differential equation, Ann. Polon. Math. 14 (1964), 321 - 333.

[195] Szarski J. Differential Inequalities, Polish Sci. Publ., Warszawa, 1967.

[196] Szarski J., Uniqueness of solutions of a mixed problem for parabolic differential functional equations, Ann. Polon. Math. 28 (1973), 57 - 65.

[197] Szarski J., Strong maximum principle for nonlinear parabolic differential functional inequalities, Ann. Polon. Math. 29 (1974), 207 - 214.

[198] Szarski J., Strong maximum principle for nonlinear parabolic differential functional inequalities in arbitrary domains, Ann. Polon. Math. 31 (1975), 197 - 203.

[199] Szarski J., Uniqueness of solutions to a mixed problem for parabolic functional differential equations in arbitrary domains, Bull. Acad. Polon. Sci., Ser. sci. math. astr. phys., 24 (1976), 841 - 849.

[200] Szarski J., Generalized Cauchy problem for differential functional equations with first order partial derivatives, Bull. Acad. Polon. Sci., Ser. sci. math. astr. phys., 24 (1976), 575 - 580.

[201] Szarski J., Cauchy problem for an infinite system of differential functional equations with first order partial derivatives, Ann. Soc. Math. Polon., Comm. Math. tom spec. I, (1978), 293 - 300.

[202] Szarski J., Comparison theorems for infinite systems of differential functional equations and strongly coupled infinite systems of first order partial differential equations, Rocky Mountain Journ. Math. 10 (1980), 237 - 246.

[203] Thieme H. R., A differential-integral equation modelling the dynamics of population with a rank structure, Lect. Notes Biomath. 68 (1986), 496 - 511.

[204] Topolski K., Generalized solutions of first order partial differential functional equations in two independent variables, Atti Sem. Mat. Fis. Univ. Modena 39 (1991), 669 - 684.

[205] Topolski K., On the uniqueness of viscosity solution for first order partial differential functional equations, Ann. Polon. Math. 59 (1994), 65 - 75.

[206] Topolski K., Classical methods for viscosity solutions of differential functional inequalities, Nonlinear World 4 (1997), 1 - 18.

[207] Topolski K., Parabolic differential functional inequalities in a viscosity sense, Ann. Polon. Math. 68, (1998), 17 - 25.

[208] Topolski K., On the existence of classical solutions for differential functional IBVP, Abstract and Appl. Anal, to appear.

[209] Torrejón R., Yong J., On a quasilinear wave equation with memory, Nonl. Anal. TMA 16 (1991), 61 - 78.

[210] Turo J., On some class of quasilinear hyperbolic systems of partial differential functional equations of the first order, Czech. Math. Journ. 36 (1986), 185 - 197.

[211] Turo J., Existence and uniqueness of solutions of quasilinear hyperbolic systems of partial differential functional equations, Math. Slov. 37 (1987), 375 - 387.

[212] Turo J., Generalized solutions of the Cauchy problem for nonlinear functional partial differential equations, Zeit. Anal. Anwend. 7 (1988), 127 - 133.

[213] Turo J., A boundary value problem for hyperbolic systems of differential functional equations, Nonl. Anal., TMA, 13 (1989), 7 - 18.

[214] Turo J., Local generalized solutions of mixed problems for quasilinear hyperbolic systems of functional partial differential equations in two independent variables, Ann. Polon. Math. 49 (1989), 259 - 278.

[215] Turo J., Global solvability of the mixed problem for first order functional partial differential equations, Ann. Polon. Math. 52 (1991), 231 - 238.

[216] Turo J., Nonlocal problems for quasilinear functional partial differential equations of the first order, Publ. Math. 41 (1997), 507 - 517.

[217] Turo J., Mixed problems for quasilinear hyperbolic systems, Nonl. Anal., TMA, 30 (1997), 2329 - 2340.

[218] Umanaliev M. I., Vied' J. A., On integral differential equations with first order partial derivatives, (Russian), Differen. Urav. 25 (1989), 465 - 477.

[219] Voigt A., Lines method approximations to the Cauchy problem for nonlinear parabolic differential equations, Numer. Math. 23 (1974), 23 - 36.

[220] Voigt A., The method of lines for nonlinear parabolic differential equations with mixed derivatives, Numer. Math. 32 (1979), 197 - 207.

[221] Walter W., On nonlinear Volterra integral equations in several variables, Journ. Math. Mech. 16 (9) (1967), 967 - 985.

[222] Walter W., Die Linienmethode bei nichtlinearen parabolischen Differentialgleichungen, Numer. Math. 12 (1968), 307 - 321.

[223] Walter W., Ein Existenzbeweis für nichtlineare parabolische Differentialgleichungen aufrung der Linienmethode, Math. Zeit. 107 (1968), 173 - 188.

[224] Walter W., Existenzsätze in Grossen für das Cauchy Problem bei nichtlinearen parabolischen Differentialgleichungen mit Linienmethode, Math. Ann. 183 (1969), 254 - 274.

[225] Walter W., Differential and Integral Inequalities, Springer-Verlag, 1970.

[226] Walter W., Functional differential equations of the Cauchy-Kovalevsky type, Aeqat. Math. 28 (1985), 102 - 113.

[227] Walter W., A. Acker, On a global existence of solutions of parabolic differential equations with a singular nonlinear term, Nonl. Anal., TMA 2 (1978), 499 - 504.

[228] Walter W., Schmit K., Thompson R. C., Existence of solutions of a nonlinear boundary value problem via method of lines, Nonl. Anal., TMA, 2 (1978), 519 - 535.

[229] Webb G. F., Theory of Nonlinear Age-Dependent Population Dynamics, Monographs and Textbooks in Pure and Applied Mathematics Series, Vol. 89, Dekker, New York, Basel, 1985.

[230] Wu J., Theory and Applications of Partial Functional Differential Equations, Springer, 1996.

[231] Ważewski T., Sur le probléme de Cauchy relatif á un systéme d' équations aux dérivеés partielles, Ann. Soc. Polon. Math. 15 (1936), 101 - 127.

[232] Ważewski T., Sur une procédé de prouver la convergence des approximations successives sans utilisation des séries de comparaison, Bull. Acad. Polon. Sci., Ser.sci. math. astr. phys., 8 (1960), 45 - 52.

[233] Yamanaka T., Tamaki H., Cauchy-Kovalevskaja theorem for functional partial differential equations, Comm. Math. Univ. St. Paul 29 (1980), 55 - 64.

[234] Yan Y., Fairweather G., Ortogonal spline collocation methods for some partial integrodifferential equations, SIAM Journ. Numer. Anal. (1992), 755- 768.

[235] Yanik E. G., Fairweather G., Finite element methods for parabolic partial integro-differential equations, Nonl. Anal., TMA, 12 (1988), 785 - 809.

[236] Zennaro M., Delay differential equations: theory and numerics, in: Theory and Numerics of Ordinary and Partial Differential Equations, Editors: Ainsworth M., Levesley J., Light W. A., Marletta M., Clarendon Press, Oxford, 1995.

[237] Zubik-Kowal B., Convergence of the lines method for first-order partial differential functional equations, Numer. Meth. Partial Equat. 10 (1994), 395 - 409.

[238] Zubik-Kowal B., The method of lines for parabolic differential functional equations, IMA Journ. Numer. Anal. 17 (1997), 103 - 123.

[239] Zubik-Kowal B., The method of lines for first order partial differential functional equations, Studia Scient. Math. Hung. 34 (1998), 413 - 428.

[240] Zubik-Kowal B., The method of lines for impulsive functional partial differential equations of the first order, Comm. Appl. Anal. 2 (1998), 111 - 128.

Index

Other *Mathematics and Its Applications* titles of interest:

S. Hu and N.S. Papageorgiou: *Handbook of Multivalued Analysis*. Volume I: Theory. 1997, 980 pp. ISBN 0-7923-4682-3 (Set of 2 volumes: 0-7923-4683-1)

L.A. Sakhnovich: *Interpolation Theory and Its Applications*. 1997, 216 pp. ISBN 0-7923-4830-0

G.V. Milovanović: *Recent Progress in Inequalities*. 1998, 531 pp. ISBN 0-7923-4845-1

V.V. Filippov: *Basic Topological Structures of Ordinary Differential Equations*. 1998, 530 pp. ISBN 0-7293-4951-2

S. Gong: *Convex and Starlike Mappings in Several Complex Variables*. 1998, 208 pp. ISBN 0-7923-4964-4

A.B. Kharazishvili: *Applications of Point Set Theory in Real Analysis*. 1998, 244 pp. ISBN 0-7923-4979-2

R.P. Agarwal: *Focal Boundary Value Problems for Differential and Difference Equations*. 1998, 300 pp. ISBN 0-7923-4978-4

D. Przeworska-Rolewicz: *Logarithms and Antilogarithms*. An Algebraic Analysis Approach. 1998, 358 pp. ISBN 0-7923-4974-1

Yu. M. Berezansky and A.A. Kalyuzhnyi: *Harmonic Analysis in Hypercomplex Systems*. 1998, 493 pp. ISBN 0-7923-5029-4

V. Lakshmikantham and A.S. Vatsala: *Generalized Quasilinearization for Nonlinear Problems*. 1998, 286 pp. ISBN 0-7923-5038-3

V. Barbu: *Partial Differential Equations and Boundary Value Problems*. 1998, 292 pp. ISBN 0-7923-5056-1

J. P. Boyd: *Weakly Nonlocal Solitary Waves and Beyond-All-Orders Asymptotics*. Generalized Solitons and Hyperasymptotic Perturbation Theory. 1998, 610 pp. ISBN 0-7923-5072-3

D. O'Regan and M. Meehan: *Existence Theory for Nonlinear Integral and Integrodifferential Equations*. 1998, 228 pp. ISBN 0-7923-5089-8

A.J. Jerri: *The Gibbs Phenomenon in Fourier Analysis, Splines and Wavelet Approximations*. 1998, 364 pp. ISBN 0-7923-5109-6

C. Constantinescu, W. Filter and K. Weber, in collaboration with A. Sontag: *Advanced Integration Theory*. 1998, 872 pp. ISBN 0-7923-5234-3

V. Bykov, A. Kytmanov and M. Lazman, with M. Passare (ed.): *Elimination Methods in Polynomial Computer Algebra*. 1998, 252 pp. ISBN 0-7923-5240-8

W.-H. Steeb: *Hilbert Spaces, Wavelets, Generalised Functions and Modern Quantum Mechanics*. 1998, 234 pp. ISBN 0-7923-5231-9

Other *Mathematics and Its Applications* titles of interest:

X. Xu: *Introduction to Vertex Operator Superalgebras and Their Modules*. 1998, 356 pp.
ISBN 0-7923-5242-4

E.E. Rosinger: *Parametric Lie Group Actions on Global Generalised Solutions of Nonlinear PDFs* including a Solution to Hilbert's Fifth Problem. 1998, 234 pp. ISBN 0-7923-5232-7

A. Khrennikov: *Superanalysis*. 1999, 370 pp. ISBN 0-7923-5607-1

V. Koshmanenko: *Singular Quadratic Forms in Perturbation Theory*. 1999, 316 pp.
ISBN 0-7923-5625-X

Z. Kamont: *Hyperbolic Functional Differential Inequalities and Applications*. 1999, 318 pp.
ISBN 0-7923-5791-4

SPRINGER NATURE

GPSR Compliance

The European Union's (EU) General Product Safety Regulation (GPSR) is a set of rules that requires consumer products to be safe and our obligations to ensure this.

If you have any concerns about our products, you can contact us on ProductSafety@springernature.com

In case Publisher is established outside the EU, the EU authorized representative is:

Springer Nature Customer Service Center GmbH
Europaplatz 3
69115 Heidelberg, Germany

FSC
www.fsc.org
MIX
Papier | Fördert
gute Waldnutzung
FSC® C083411

Kolibri 360 GmbH
Ferdinand-Jühlke-Straße 7
99095 Erfurt, Deutschland
produktsicherheit@kolibri360.de